Jiuping Xu and Xiaoyang Zhou

Fuzzy-Like Multiple Objective Decision Making

Studies in Fuzziness and Soft Computing, Volume 263

Editor-in-Chief

Prof. Janusz Kacprzyk
Systems Research Institute
Polish Academy of Sciences
ul. Newelska 6
01-447 Warsaw
Poland
E-mail: kacprzyk@ibspan.waw.pl

Further volumes of this series can be found on our homepage: springer.com

Jiuping Xu and Xiaoyang Zhou

Fuzzy-Like Multiple Objective Decision Making

 Springer

Authors

Prof. Dr. Dr. Jiuping Xu
Uncertainty Decision-Making Laboratory
Sichuan University
610064 Chengdu
P. R. China

Assis. Prof. Xiaoyang Zhou
Uncertainty Decision-Making Laboratory
Sichuan University
610064 Chengdu
P. R. China

ISBN 978-3-642-43702-1   ISBN 978-3-642-16895-6 (eBook)

DOI 10.1007/978-3-642-16895-6

Studies in Fuzziness and Soft Computing   ISSN 1434-9922

Typeset & Cover Design: Scientific Publishing Services Pvt. Ltd., Chennai, India.

Printed on acid-free paper

9 8 7 6 5 4 3 2 1

springer.com

Preface

Everyday people need to make decisions, and the decision makers usually face multiple, conflicting objectives and uncertain environments. The research about uncertain multi-objective decision making problems has been profolic. It mainly provides decision makers with the help to find optimal solutions for many objectives under uncertain environments, and has been a permanent focus for many years.

To trace the origins of the multi-objective decision making with certain parameters, we have to go back over the eighteenth century. B. Franklin introduced how to coordinate multiple objectives in 1772. A. A. Cournot proposed the multi-objective decision making model from the standpoint of the economics in 1836. V. Pareto firstly presented the optimal solution to the multi-objective decision making model from the standpoint of the mathematics in 1896 and then K. J. Arrow et al. proposed the concept of efficient points. Traditional multi-objective decision making is only aimed at problems with certain parameters, but as we know, it is usual that many decision making problems have uncertain factors. As people know more and more about the uncertain event, the research about random multi-objective decision making, fuzzy multi-objective decision making, rough multi-objective decision making and two-fold uncertain multi-objective decision making problems were gradually developed.

In the last 25 years, fuzzy set theory has been applied to many disciplines such as operations research, management science, control theory, artificial intelligent/expert systems, human behavior, etc. Growth of the applications of fuzzy set theory have been accumulating. In 1978, H. Kwakernaak combined randomness with fuzziness and initialized the concept of the fuzzy random variable, then introduced its basic definition and properties. This viewpoint which combined two different uncertain variables to describe complicated events received approval from many scholars and move forward a further step to uncertain events. Then many papers and books about the two-fold uncertain theory presented more and more, and therefore promoted the development of two-fold uncertain theory. This monograph presents systematically state-of-art of fuzzy-like multiple objective mathematical programming in both techniques and applications.

In real life, input data is usually imprecise or uncertain because of incomplete or non-obtainable information. So if we want to describe the uncertainty, especially subjective uncertainty, fuzzy variables will be used. For instance, If we ask people about the unit transport cost between two places, they say that the cost will be around a value, and they cannot give an absolute certain value. Since the information is described in linguistic terms, and not by the chance concept, conventional probability may not be a correct way to model these imprecise nature, so it can be derived that the unit transport cost could be a fuzzy variable rather than a constant. It is appropriate for us to use a LR fuzzy variables $(a, \theta, b)_{LR}$ to describe, where a, b are the left and right widths of the fuzzy variable, and θ is the middle value. Further more, we consider about the middle value θ, due to the conditions of road, traffic and weather, etc., this value may not be the constant either, but here may be a random variable or fuzzy variable or rough variable, thus is regarded as a fuzzy-like two-fold uncertain variable. This book mainly concentrates on one type of two-fold uncertain theory, that is, the fuzzy-like two-fold uncertain variable including fuzzy random variables, bifuzzy variables and fuzzy rough variables, and then deduces their properties and the application to the real world.

In the classical multi-objective decision making model, all data and information are assumed to be absolutely accurate, and the objectives and constraints are all assumed to be well expressed by mathematical formations. However, it is difficult to clearly describe the objective functions and constraints by mathematical equations for many realistic problems, thus, the multi-objective decision making model with certain parameters cannot deal with all real-life problems. So uncertain multi-objective decision making models are proposed. In 1965, after L.A. Zadeh proposed the fuzzy set theory, it was rapidly and widely applied in the filed of operations, management science, control theory and so on. In 1970, R. E. Bellman and L. A. Zadeh collaborated to propose the fuzzy decision making model based on multi-objective programming. That's the original work which is the basis of this monograph.

The multi-objective decision making model with fuzzy-like variables can be summarized as follows:

$$
\begin{cases}
\max \ f(x, \xi) = [f_1(x, \xi), f_2(x, \xi), \cdots, f_m(x, \xi)] \\
\text{s.t.} \begin{cases} g_r(x, \eta) \leq 0, r = 1, 2, \cdots, p \\ x \in X, \end{cases}
\end{cases}
$$

where x is an n-dimensional decision variable, f is the objective function, ξ and η are both fuzzy-like variables.

In this book, we take real-life problems as background and guidance, and present the application of the fuzzy-like multi-objective decision making (FLMODM) model to the real world, and finally construct a relative model. After that, basic theories, models and algorithms are proposed and applied to solve realistic problems. This book consists of 6 chapters. Chapter 1 reviews some relative preliminary knowledge such as fuzzy set, fuzzy numbers, fuzzy arithmetic and membership functions. Chapter 2 mainly introduces multi-objective decision making with fuzzy parameters and its application to farm structure optimization problem. Three sections introduce the fuzzy expected value model (Fuzzy EVM), the fuzzy chance-constrained model (Fuzzy CCM) and the fuzzy dependent-chance model (Fuzzy DCM). In each section, the equivalent model of those problems which have some special fuzzy variables are deduced and the traditional solution methods for the crisp equivalent models are introduced the technique of fuzzy simulations-based hybrid algorithms to deal with the general decision making problems under fuzzy environments. Finally, in the next section, the proposed models and algorithms have been applied to solve the realistic problems introduced in the first section. Chapters 3, 4 and 5 have the same structure as Chapter 2. Chapter 3 proposes a multi-objective decision making model with fuzzy random parameters and introduces its application to portfolio selection problems under a fuzzy random environment. The equivalent models of the fuzzy random expected value model (Fu-Ra EVM), fuzzy random chance-constrained model (Fu-Ra CCM) and fuzzy random dependent-chance model (Fu-Ra DCM) are deduced. Chapter 4 proposes the multi-objective decision making model with a bifuzzy parameters and introduces its application to material purchasing problems under a bifuzzy environment. The equivalent models of the bifuzzy expected value model (Fu-Fu EVM), bifuzzy chance-constrained model (Fu-Fu CCM) and bifuzzy dependent-chance model (Fu-Fu DCM) are deduced. Chapter 5 proposes a multi-objective decision making model with fuzzy rough parameters and introduces its application to logistics problems under a fuzzy rough environment. The equivalence models of the fuzzy rough expected value model (Fu-Ro EVM), fuzzy rough chance-constrained model (Fu-Ro CCM) and fuzzy rough dependent-chance model (Fu-Ro DCM) are deduced. Finally in Chapter 6, the main problems are concluded, models, methods and algorithms to obtain the organic systems are discussed, and a methodological system for the FLMODM is proposed.

The authors wish to thank the support from the National Science Foundation for Distinguished Young Scholars, P. R. China (Grant No. 70425005) and the National Natural Science Foundation of P. R. China (Grant No. 79760060, and No. 70171021). The authors also wish to acknowledge the assistance of Prof. Janusz Kacprzyk for his helpful comments and recommendations to publish this book in this series. This book benefited many literatures, and the authors also wish to take this opportunity to thank these researchers here. For discussion and advice, the authors are grateful to researchers from the Uncertainty Decision-Making Laboratory of Sichuan University, J. Li, Y. Liu, Z. Tao, L. Yao and others, who have done a lot of work in this field and have made a number of corrections. Finally the authors express their deepest gratitude to the staff of Springer, especially, Dr. Thomas Ditzinger and Renuka Devi for the wonderful cooperation.

Sichuan University,   Jiuping Xu
June 2009   Xiaoyang Zhou

Contents

List of Figures

List of Tables

Acronyms

MODM   Multiple objective decision making
FLMODM   Fuzzy-like multiple objective decision making
EVM   Expected value model
CCM   Chance-constrained model
DCM   Dependent-chance model
ECM   Expectation model with chance constraints
CEM   Chance-constrained model with expectation constraints
DEM   Dependent-chance model with expectation constraints
Fu-Ra EVM   Fuzzy random expected value model
Fu-Ra CCM   Fuzzy random chance-constrained model
Fu-Ra DCM   Fuzzy random dependent-chance model
Fu-Fu EVM   Bifuzzy expected value model
Fu-Fu CCM   Bifuzzy chance-constrained model
Fu-Fu DCM   Bifuzzy dependent-chance model
Fu-Ro EVM   Fuzzy rough expected value model
Fu-Ro CCM   Fuzzy rough chance-constrained model
Fu-Ro DCM   Fuzzy rough dependent-chance model
PSO   Particle swarm optimization algorithm
GA   Genetic algorithm
SA   Simulated annealing algorithm
TS   Tabu search algorithm

Chapter 1
Fuzzy Set Theory

Fuzzy set theory has been developed to solve problems where the descriptions of activities and observations are imprecise, vague, or uncertain. The term "fuzzy" refers to a situation in where there are no well defined boundaries of the set of activities or observations to which the descriptions apply. For example, one can easily assign a person 180cm tall to the "class of tall men". But it would be difficult to justify the inclusion or exclusion of a 173cm tall person to that class, because the term "tall" does not constitute a well defined boundary. This notion of fuzziness exists almost everywhere in our daily life, such as a "class of red flowers," a "class of good shooters," a "class of comfortable speeds for traveling, a "numbers close to 10," etc. These classes of objects cannot be well represented by classical set theory. In classical set theory, an object is either in a set or not in a set. An object cannot partially belong to a set.

To cope with this difficulty, Zadeh [9] proposed the fuzzy set theory in 1965. A fuzzy set is a class of objects with a continuum of membership grades. A membership function, which assigns to each object a grade of membership, is associated with each fuzzy set. Usually, the membership grades are in the interval [0, 1]. When the grade of membership for an object in a set is one, this object is absolutely in that set; when the grade of membership is zero, the object is absolutely not in that set. Borderline cases are assigned numbers between zero and one. Precise membership grades do not convey any absolute significance as they are context-dependent can be subjectively assessed.

In the following sections, we will present some basic definitions of fuzzy set and operations on fuzzy set from mathematical aspects. Subsequently, the decomposition theorem, the extension principle and fuzzy number operations, which are important to subsequent discussions, will be introduced. Special fuzzy numbers such as LR fuzzy numbers, triangular numbers, and trapezoidal numbers and their arithmetic operations are also presented. Numerical and graphical examples are used to make the contents more understandable.

J. Xu and X. Zhou: Fuzzy-Like Multiple Objective Decision Making, STUDFUZZ 263, pp. 1–55.
springerlink.com   © Springer-Verlag Berlin Heidelberg 2011

1.1   Fuzzy Sets

Precise mathematics are not sufficient to model a complex system because of the incomplete knowledge and information.

Traditionally, the probability theory is the prevailing approach to handle this incompleteness or uncertainty. One of the fundamentals in the probability theory is the law of the exclude middle ($p(A \cup A^c) = 1$) and contradiction ($p(A \cap A^c) = 0$). For instance, a fruit is either a banana or not a banana, an animal(normally speaking) is either male or female. For this case, the probability is certainly a good approach to represent some knowledge or information whose boundaries can be clearly defined. Throwing coins into air, you can guess that either heads or tails will be up. Rotating a dice, the result will be 1, 2, 3, 4, 5, or 6, but never be 3.5 or 2.8. So of course, there are a lot of problems satisfying the laws of the excluded middle and contradiction.

However, intuitively and common sensibly, this is not true in other problems. An evidence favoring a particular hypothesis to some degree does not simultaneously disconfirm it to any degree, because it may not give any support to the contrary. For example, a man may be smart, not smart, or a little smart. A color can be red, no red, or reddish. Therefore, it is difficult to define the sets of smart men and red colors with sharp/crisp boundaries. Similarly, how can we define the classify "good shooter", "beautiful lady", "good personality" and so on? Obviously, the probability theory can not model all the possible problems of incompleteness. The fuzzy set theory is developed to define and solve these problems without sharp boundaries. That is: fuzzy set theory considers the partial relationship or membership.

Examples of fuzzy sets

To clearly distinguish the fuzzy sets from the classically crisp sets, let us first consider the following examples.

Example 1.1. In the classical set theory, an element may either "belong to" set A or "not belong to" set A in the given universe. Suppose there is a rather large target and shooters always hit inside the target U (see Figure 1.1). A circle is located in the center of the target. If a shooter hits inside the circle, region A, he is given the title "good shooter." Otherwise, he is called a "poor shooter."

It is clear that shooter s^1 shoots inside region A, so he is a good shooter. On the other hand, shooter s^2, who shoots far away from region A, is a poor shooter. The classical set theory with binary relationship shows some problems. For instance, if there are three shooters, s^3, s^4 and s^5, who hit the target within close range of one another (see Figure 1.1), yet only s^3 is within the target, then shooter s^3 is a good shooter, and shooter s^4 and s^5 are poor shooters. This is obviously unreasonable.

As a matter of fact, these three people should obtain some similar designation, at least up to a certain degree. Therefore, a measure of degree to which the shooter belongs to the set "good shooters" should be developed in order to discern how good the shooter is. The set composed of good shooters is actually fuzzy because there is no crisp boundary. It is rational to consider the distance from the boundary of the

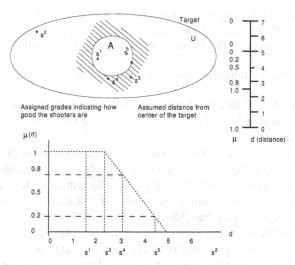

Fig. 1.1 Graphic explanation for the fuzzy set of good shooter

region A as a measure for indicating the degree to which shooter s^i belongs to the set of "good shooters". In Figure 1.1, s^1 and s^3 are absolutely good shooters. On the other hands, s^4 and s^5 are not absolutely good shooters or absolutely poor shooters. They are, to some degree, good shooters. By giving a numerical measure which is assumed linearly proportional to the distance, d, of each shooter from region A, one can say that shooter s^4 has 0.8 degree of membership in the set of good shooters versus 0.2 for shooter s^5. $\mu(s^4) = 0.8$ and $\mu(s^5) = 0.2$. Of course, the numerical measure, μ, can be any number. A normalized measure which is [0, 1] is always adapted. In the above example, the preference concept is implied while assigning the numerical measures for each shooter to represent the degree of " a good shooter belonging to the set of good shooters." The shorter the hit spot is from the center of the target, the larger the grade assigned (small d is preferred to large d). The grade values are actually preference values. The function μ (called membership function in fuzzy set theory), constituted by these grades as shown in Figure 1.1, is then a preference function.

Similarly, in practice, the decision maker may feel that: "Around $20,000 profit is acceptable;" "The budget should be around $9,000;" "Dividends should be higher than 6%;" "Overcome should be less than 5% of the regular man-hours;" and so on. Obviously, these linguistic statements cannot be described by probability. The fuzzy set theory, on the other hand, gives us a way to handle such linguistic situation. Meanwhile, the preference concept is often assumed in building the membership functions of the above linguistic statements.

In order to be able to more clearly understand the fuzzy sets concept, let us consider two more examples.

Example 1.2. Consider a universe composed of 4 female students with the same height of 165cm, Zhang, Wang, Li and Zhao. That is, $U = \{$Smith, Johnson, Carson, Williams$\}$. The weights for the 4 female students are given as follows:

$$\text{Smith} : 55\text{kg}, \quad \text{Johnson} : 75\text{kg}, \quad \text{Carson} : 65\text{kg}, \quad \text{Williams} : 50\text{kg}.$$

Now, let us consider the linguistic proposition "fat female students." The students who belong to "fat female students" then constitute a fuzzy set, A. Is Smith $\in A$, Johnson $\in A$, Carson $\in A$ or Williams $\in A$?

One plots the weights on a real line (see Figure 1.2) in order to present the relative differences. According to common sense, a female student weighting more than 75kg at 165cm in height is absolutely considered a fat female student (actually "fat" depends on culture, race, etc., which are beyond or research). On the other hand, a female student weighting less than 50kg is not fat at all. Therefore, it is obvious that Johnson is absolutely fat and Williams is absolutely not fat. How about Smith and Carson? Actually Smith approaches Williams with respect to weight and Carson is close to Wang. It is true that heavier the female students is, the greater the degree to which he belongs to fuzzy set A. Thus, a degree-scaled line (see Figure 1.2) can be drawn corresponding to previous weight-scaled line in order to represent the degree of membership indicating that a female student belongs to A. The scale on the degree-scaled line is linearly proportional to the weight-scaled line when the weight belongs to the interval [50, 75]. As a result, the following grades of membership are available:

$$\text{degree(Smith} \in A) = \mu(x = Smith) = 0.2,$$
$$\text{degree(Johnson} \in A) = \mu(x = Johnson) = 1,$$
$$\text{degree(Carson} \in A) = \mu(x = Carson) = 0.8,$$
$$\text{degree(Williams} \in A) = \mu(x = Williams) = 0$$

where $\mu(\cdot)$ (detailed definition is given in the next section) is the membership function of the fuzzy subset A of the set U. Like the previous example, the preference concept is used to elicit the membership function $\mu(\cdot)$.

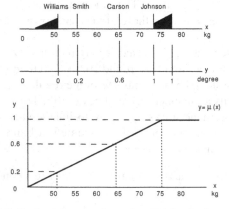

Fig. 1.2 Derived degree of membership for the fuzzy set A

Example 1.3. Consider the statement of "the interest rate will be around 8% to 8.5%." The most possible interest rate is between 6% (optimistic value) and 10% (pessimistic value). It is not like that the interest rate will be less than 6% or larger than 10% in the opinion of the decision maker. Thus, we give 8% - 8.5% interest rate as 1 possibility, and 6% or less than and 10% or more a 0 possibility. Between them, let us assume the grades are as shown in Figure 1.3. Here, we did not use the preference concept to grade various interest rates, but used the possibility concept: most possible, least possible, or in between. Therefore, we call this membership function a possibility function or possibility distribution, denoted by $\pi(\cdot)$.

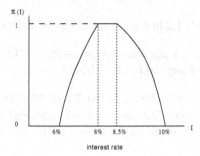

Fig. 1.3 The membership(possibility) function of the interest rate

Definition of a fuzzy set

Let U be the universe which is a classical set of object, and the generic elements are denoted by x. The membership in a crisp subset of A is often viewed as characteristic function π_A from U to $\{0,1\}$ such that:

$$\pi_A(x) = \begin{cases} 1 \text{ iff } x \in A \\ 0 \text{ otherwise} \end{cases} \tag{1.1}$$

where $\{0,1\}$ is called a valuation set.

If the valuation set is allowed to be the real interval $[0,1]$, A is called a fuzzy set. $\mu_A(x)$ is the degree of membership of x in fuzzy set A. The closer the value of $\mu_A(x)$ is to 1, the more x belongs to A. Therefore, A is characterized by the set of ordered pairs:

$$A = \{(x, \mu_A(x)) | x \in U\}. \tag{1.2}$$

It is worth noting that the characteristic function can be either a possibility distribution function as shown in Example 1.1, or a membership function as shown in Examples 1.2. If the possibility distribution is preferred, the characteristic function will be specified as $\pi(x)$. On the other hand, if the membership function is preferred, then the characteristic function will be denoted as $\mu(x)$.

Along with the expression of (1.2), the following notations are also proposed. When U is a finite set $\{x_1, x_2, \cdots, x_n\}$, a fuzzy set A is then expressed as:

$$A = \mu_A(x_1)/x_1 + \cdots + \mu_A(x_n)/x_n = \sum_i \mu_A(x_i)/x_i. \tag{1.3}$$

When U is not a finite set, A then can be written as:

$$A = \int_U \mu_A(x)/x. \tag{1.4}$$

Example 1.4. For Example 1.2 in section 1.1, A can be expressed as:

$$A = \mu_A(\text{Zhang})/\text{Zhang} + \mu_A(\text{Wang})/\text{Wang} + \mu_A(\text{Li})/\text{Li} + \mu_A(\text{Zhao})/\text{Zhao}$$
$$= 0.2/\text{Zhang} + 1/\text{Wang} + 0.6/\text{Li} + 0/\text{Zhao}.$$

Example 1.5. Let $U = \{10, 20, 30, 40, 50, 60, 70, 80, 90, 100\}$, the possible speed (mph) at which makes people feel comfortable in traveling a long distance. Then the fuzzy set "comfortable speed for long distance travel" may be defined by an individual as:

$$A = \{(0.7, 30), (0.75, 40), (0.8, 50), (0.8, 60), (1, 70), (0.8, 80), (0.3, 90)\}.$$

Example 1.6. Let $U = \{\text{Positive real numbers}\}$, which is an infinite set. Then, the fuzzy set $A = $ "real numbers close to 10" (see Figure 1.4) may be defined as:

$$A = \{(x, \mu_A(x))\}, \text{where } \mu_A(x) = \frac{1}{1 + [1/5(x - 10)]^2}.$$

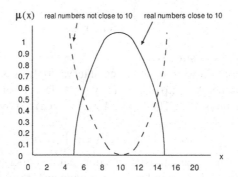

Fig. 1.4 The fuzzy set "real numbers close to 10"

1.2 Basic Concepts of Fuzzy Sets

The basic concepts presented here include complement, intersection, union, algebraic product, algebraic sum, difference, support, α-cut, convexity, normality, cardinality, and the mth power of a fuzzy set A.

Complement

The complement of fuzzy set A, denoted by $\bar{A}$, is defined as:

$$\mu_{\bar{A}}(x) = 1 - \mu_A(x), \ \forall x \in U. \tag{1.5}$$

Example 1.7. Consider Example 1.5 in section 1.2. The complement of fuzzy set A, "comfortable speed for long distance travel," $\bar{A}$, "uncomfortable speed for long distance travel" is computed as:

$$\bar{A} = \{(1,10),(1,20),(0.3,30),(0.25,40),(0.2,50),(0.2,60),(0.2,80),(0.7,90)\}.$$

For a case in this example, the membership grade $\mu_{\bar{A}}(x)$ at $x = 30$ is computed as:

$$\mu_{\bar{A}}(30) = 1 - \mu_A(30) = 0.3.$$

Since one has the most comfortable feeling at 70 mph, i.e., $\mu_A(70) = 1$, it is not possible to include $x = 70$ in $\bar{A}$.

Example 1.8. Consider the fuzzy set "real numbers close to 10" (see Figure 1.4). Its complement set is represented by the dashed curve in Figure 1.4. This complement set may be interpreted as "real numbers not close to 10."

Intersection

The intersection of A and B denoted by $A \cap B$ which is the largest fuzzy subset contained in both fuzzy subsets A and B. When the min operator is used to express the logic "and", its corresponding membership is then characterized by:

$$\begin{aligned} \mu_{A \cap B} &= \min(\mu_A(x), \mu_B(x)) \quad \forall x \in X \\ &= \mu_A(x) \wedge \mu_B(x), \end{aligned} \tag{1.6}$$

where $\wedge$ is conjunction here.

Example 1.9. Consider $U = \{\text{Smith,Johnson, Carson,Williams}\}$. Suppose A is the fuzzy subset of "good looking students" and B is the fuzzy subset of "intelligent students". The following data are also assumed:

$$\begin{aligned} A &= 0.2/\text{Smith} + 0.3/\text{Johnson} + 0.6/\text{Carson} + 0.8/\text{Williams}, \\ B &= 0.7/\text{Smith} + 0.4/\text{Johnson} + 0.1/\text{Carson} + 0.5/\text{Williams}. \end{aligned} \tag{1.7}$$

Then, the intersection of A and B (if the min-operator is adopted) $A \cap B = 0.2/\text{Smith} +$ $0.3/\text{Johnson} + 0.1/\text{Carson} + 0.5/\text{Williams}$ which is the fuzzy subset of "good look-ing and intelligent student". Besides, $\wedge$ is considered as a hard "and".

Union

The union of A and B which denoted by $A \cup B$ is dual to the notation of intersection. Thus, the union of A and B is defined as the smallest fuzzy set containing both A and B. The membership function of $A \cup B$ is given by:

$$
\begin{aligned}
\mu_{A \cup B} &= \max(\mu_A(x), \mu_B(x)) \quad \forall x \in X \\
&= \mu_A(x) \vee \mu_B(x),
\end{aligned}
\tag{1.8}
$$

where $\vee$ is disjunction.

For Example 1.9, the union of A and B (if the max-operator is adopted) $A \cup B = 0.7/\text{Smith} + 0.4/\text{Johnson} + 0.6/\text{Carson} + 0.8/\text{Williams}$ is the fuzzy subset of "good looking or intelligent student". Besides, $\vee$ is considered as a hard "or".

Algebraic product

The algebraic product AB of A and B is characterized by the following membership function:

$$
\mu_{AB} = \mu_A(x)\mu_B(x) \quad \forall x \in X.
\tag{1.9}
$$

This algebraic product is considered as a soft "and" .

For Example 1.9, $AB = 0.14/\text{Smith} + 0.12/\text{Johnson} + 0.06/\text{Carson} + 0.4/\text{Williams}$ is the fuzzy subset of "good looking and intelligent student".

Algebraic sum

The algebraic sum $A \oplus B$ of A and B is defined by the following membership func-tion:

$$
\mu_{A \oplus B} = \mu_A(x) + \mu_B(x) - \mu_A(x)\mu_B(x) \quad \forall x \in X.
\tag{1.10}
$$

This algebraic sum is considered as a soft "or".

For Example 1.9, $A \oplus B = 0.76/\text{Smith} + 0.58/\text{Johnson} + 0.64/\text{Carson} + 0.9/\text{Williams}$ is the fuzzy subset of "good looking or intelligent student".

Difference

The difference $A - B$ of A and B is defined by:

$$
\mu_{A \cap B^c}(x) = \min(\mu_A(x), \mu_{B^c}(x)) \quad \forall x \in X,
\tag{1.11}
$$

where B^c is the complement of B.

For Example 1.9, we have $A - B$ as shown in the following Table 1.1:

Table 1.1 Solutions of example 1.9

x	Smith	Johnson	Carson	Williams
$\mu_A(x)$	0.2	0.3	0.6	0.8
$\mu_B(x)$	0.7	0.4	0.1	0.5
$\mu_{A \cap B}(x)$	0.2	0.3	0.1	0.5
$\mu_{A \cup B}(x)$	0.7	0.4	0.6	0.8
$\mu_{AB}(x)$	0.14	0.12	0.06	0.4
$\mu_{A \oplus B}(x)$	0.76	0.58	0.64	0.9
$\mu_{B^c}(x)$	0.3	0.6	0.9	0.5
$\mu_{A \cap B^c}(x)$	0.2	0.3	0.6	0.5

Support and α-cut

Sometimes, we might only need objects of a fuzzy set instead of its characteristic function, that is, to transfer a fuzzy set into a crisp set. In order to do so, we need two concepts, support and α-cut.

It is often necessary to consider those elements in a fuzzy set which have non-zero membership grades. These element are the support of that fuzzy set.

Definition 1.1. (Zadeh [12]) Given a fuzzy set A, its support $S(A)$ is an ordinary crisp subset on U defined as

$$S(A) = \{x | \mu_A(x) > 0 \text{ and } x \in U\}. \tag{1.12}$$

Example 1.10. Consider Example 1.2 in section 1.1, supp $A = \{\text{Zhang, Wang, Li}\}$, where Zhao does not belong to supp A because of $\mu(\text{Zhao}) \not> 0$.

Definition 1.2. (Zadeh [12]) Given a fuzzy set A, its α-cut A_α defined as

$$A_\alpha = \{x | \mu_A(x) \geq \alpha \text{ and } x \in U\}, \tag{1.13}$$

where α is the confidence level.

It is obviously that the α-cut of a fuzzy set A is an ordinary crisp subset whose elements belongs to fuzzy set A - at least to the degree of α. That is, for fuzzy set A its α-cut is defining as (1.13) and is denoted by (see Figure 1.5):

The α-cut is a more general case of the support of a fuzzy set, when $\alpha = 0, A_\alpha = \text{supp}(A)$.

Example 1.11. Consider the Example 1.2 in section 1.1,

$$A_{0.2} = \{\text{Smith, Johnson, Carson}\}$$

and

$$A_{0.5} = \{\text{Johnson, Carson}\}.$$

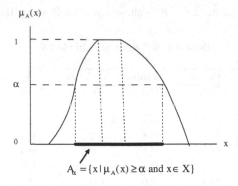

$$A_\alpha = \{x \mid \mu_A(x) \geq \alpha \text{ and } x \in X\}$$

Fig. 1.5 An α-set

Example 1.12. Consider the example 1.5 in section 1.2, the support of fuzzy set A, "comfortable speed for long distance", is given as:

$$S(A) = \{30, 40, 50, 60, 70, 80, 90\},$$

where all these x values have their corresponding $\mu_A(x) > 0$.

By setting $\alpha = 0.5$, we obtain

$$A_{0.5} = \{30, 40, 50, 60, 70, 80\},$$

where $x = 90$ is discarded since $\mu_A(90) < 0.5$. If we set $\alpha = 0.9$, $A_{0.9} = \{70\}$.

Example 1.13. Consider the fuzzy set "real numbers close to 10" used in Example 1.8. Its support is any real number between [5,15]. And its α-cut at degree of 0.5 is any real number between [6,14]. That is, the set of real numbers that have at least 0.5 membership value is between 6 and 14 (inclusive).

By definition 1.3, the α-cut A_α of the fuzzy number A is actually a close interval of the real number field, that is,

$$A_\alpha = \{x \in R \mid \mu_A(x) \geq \alpha\} = [A_\alpha^L, A_\alpha^R], \quad \alpha \in [0,1],$$

where A_α^L and A_α^R are the left and the right extreme points of the close interval.

Example 1.14. Given fuzzy number A with membership function

$$\mu_{\tilde{A}}(x) = \begin{cases} L(\frac{a-x}{l}), & \text{if } a-l \leq x < a, l > 0 \\ 1 & \text{if } x = a \\ R(\frac{x-a}{r}), & \text{if } a < x \leq a+r, r > 0, \end{cases}$$

and the basis functions $L(x)$, $R(x)$ are continuous un-increasing functions, and $L, R : [0,1] \rightarrow [0,1], L(0) = R(0) = 1, L(1) = R(1) = 0$, then $\tilde{A}$ is LR fuzzy number, denoted by $\tilde{A} = (a, l, r)_{LR}$, where a is the central value of $\widetilde{A}$, $l, r > 0$ is the left and the right spread. α-cut A_α of the LR fuzzy number $\widetilde{A}$ is

$$A_\alpha = [A_\alpha^L, A_\alpha^R] = [a - L^{-1}(\alpha)l, a + R^{-1}(\alpha)r], \quad \alpha \in [0,1].$$

We use the Figure. 1.6 to explain.

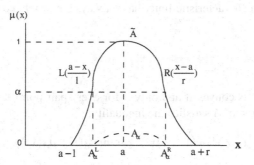

Fig. 1.6 α-cut of LR fuzzy number

Example 1.15. Especially, when $L(x) = R(x) = 1 - x$, this kind of LR-type fuzzy number is called triangular fuzzy number, denoted as $\widetilde{A} = (a - l, a, a + r)$.

The α-cut A_α of $\widetilde{A}$ is

$$A_\alpha = [A_\alpha^L, A_\alpha^R] = [a - (1 - \alpha)l, a + (1 - \alpha)r], \quad \alpha \in [0,1]. \tag{1.14}$$

And we use the Figure. 1.7 to explain.

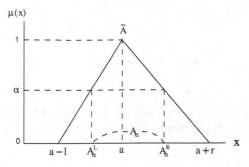

Fig. 1.7 α-cut of triangular fuzzy number

Although characteristic function can be assigned by any number, a normalized value between 0 and 1 is always preferred. Thus let us introduce the normality as follows.

Normality

A fuzzy set A is normal if and only if $\sup_x \mu_A(x) = 1$, that is, the supreme is unity (see Figure 1.5). A fuzzy set is subnormal if it is not normal. A non-empty subnormal fuzzy set can be normalized by dividing each $\mu_A(x)$ by the factor $\sup_x \mu_A(x)$. (A fuzzy set is empty if and only if $\mu_A(x) - 0, \forall x \subset U$).

It is noted that a characteristic function is always a normalized function through this study.

Convexity

A fuzzy set A in U is convex if and only if for every pair point x^1 and x^2 in U, the membership function of A satisfies the inequality:

$$\mu_A(\lambda x^1 + (1 - \lambda)x^2) \geq \min(\mu_A(x^1), \mu_A(x^2)), \tag{1.15}$$

where $\lambda \in [0, 1]$ (see Figure 1.8). Alternatively, a fuzzy set is convex if all α-level sets are convex.

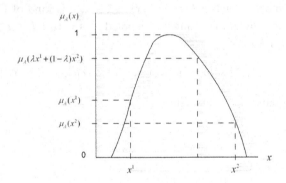

Fig. 1.8 A convex fuzzy set

Dually, A is concave if its complement A^c is convex. It is easy to show that if A and B are convex, so is $A \cap B$. Dually, if A and B are concave, so is $A \cup B$.

Cardinality of a fuzzy set

The cardinality of fuzzy set A evaluates the proportion of elements of U having the property A. When U is finite, it is defined as:

$$|A| = \sum \mu_A(x), \ x \in U. \tag{1.16}$$

For infinite U the cardinality is defined as:

$$|A| = \int_x \mu_A(x)dx. \tag{1.17}$$

The relative cardinality of A is defined as:

$$||A|| = \frac{|A|}{|U|}. \tag{1.18}$$

The relative cardinality can be interpreted as the proportion of elements of U being in A weighted by their degree of membership in A.

Example 1.16. For the fuzzy set A, "comfortable cruising speed for long distance travel" in Example 1.5 in section 1.1, its cardinality $|A|$ and relative cardinality $||A||$ are computed as:

$$|A| = 0.7 + 0.75 + 0.8 + 0.8 + 1 + 0.8 + 0.3 = 5.15,$$
$$||A|| = \frac{5.15}{10} = 0.515.$$

The mth power of a fuzzy set

The mth power of fuzzy set A is defined as:

$$\mu_A m = [\mu_A]^m. \tag{1.19}$$

It is very useful in modelling linguistic modifiers into fuzzy sets. For example, the second power of a fuzzy set, "good", is interpreted as "very good" where "very" is the linguistic modifier used to modify fuzzy set "good".

Example 1.17. Let fuzzy set A be in Table 1.2.

Table 1.2 Fuzzy set A in Example 1.17

x	3	4	5	6	7	8	9	10
μ_A	0	0	0.2	0.4	0.6	0.8	1	0

The second power of A is computed as in Table 1.3.

For instance, $\mu_{A^2} = [\mu_A(7)]^2 = 0.36$.

Table 1.3 Second power of A

x	3	4	5	6	7	8	9	10
μ_{A^2}	0	0	0.04	0.16	0.36	0.64	1	0

1.3   Fuzzy Number

Fuzzy arithmetics is a direct application of the Extension principle, and is used on fuzzy numbers. Some works related to fuzzy number operations are from Jain[300], Mizumoto and Tanaka[301, 302], Baas and Kwakernaak[303], Dubois and Prade[71, 305], Dijkman, Haeringen, and Delange[307], Gupta[308], Kaufmann and Gupta[309] among others have been.

Definition of fuzzy numbers

The term fuzzy number is used to handle imprecise numerical quantities, such as "close to 10," "about 60," "several," etc. A general definition of a fuzzy number is given by Dubois and Prade[71, 305]: any fuzzy subset $M = \{(x, \mu(x))\}$ where x takes its number on the real line R and $\mu_M(x) \in [0, 1]$.

Definition 1.3. (Dubois [71]) Let A be a fuzzy set, its membership function is $\mu_A :$ $R \rightarrow [0, 1]$, if
(i) A is upper semi-continuous, i.e., α-cut A_α is close set, for $0 < \alpha \leq 1$.
(ii) A is normal, i.e., $A_1 \neq \emptyset$.
(iii) A is convex, i.e., A_α is a convex subset of R, for $0 < \alpha \leq 1$.
(iv) The closed convex hull of A $A_0 = cl[co\{x \in R, \mu_A(x) > 0\}]$ is cored.
then A is a fuzzy number.

A fuzzy number may be represented in discrete or continuous form. For example, Let M be the fuzzy number "about 60" which may be given as either one of the following:
   (1) Discrete membership function: Given the universe

$$U = \{10, 20, 30, 40, 50, 60, 70, 80, 90, 100\},$$

the fuzzy number M may be represented as shown in Figure 1.9.
   (2) Continuous membership function: Given the universe $U = \{$real numbers$\}$, the continuous membership function for M may be represented as (see Figure 1.10):

$$\mu_M(x) = (1 + (\frac{x - 60}{10})^2)^{-1}.$$

Special fuzzy numbers

Since special fuzzy numbers are proposed to reduce the amount of computational effort, so let's focus on the following cases. So far, triangular numbers (Laarhoven and Pedrycz[310]), trapezoidal numbers (Buckley[311, 312]), LR fuzzy numbers (Dubois and Prade[305], Bonissone [313, 314]) have been applied to various decision models. Figure 1.11 and Figure 1.12 present some special fuzzy numbers. We can interpret the fuzzy number M with a unique peak as a fuzzy quantity "approximately m", and a trapezaoidal number may be seen as a fuzzy quantity "approximately in the interval of $[m_1, m_2]$."

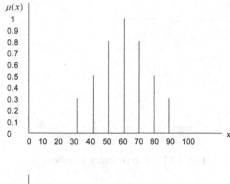

Fig. 1.9 A discrete fuzzy number M

x	30	40	50	60	70	80	90
$\mu_M(x)$	0.3	0.5	0.8	1	0.8	0.5	0.3

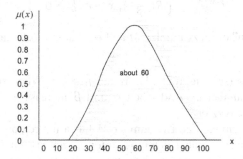

Fig. 1.10 A continuous fuzzy number M

LR fuzzy number

Definition 1.4. (Dubois [305]) A function denoted by L or R is a reference function of fuzzy numbers iff
(i) $L(x) = L(-x)$;
(ii) $L(0) = 1$;
(iii) $L(x)$ is nonincreasing on $[0, +\infty)$.

The following functions can be the reference function of LR fuzzy number:
(1) $L(x) = \max\{0, 1 - |x|^p\}$, $p \geq 0$.
(2) $L(x) = \begin{cases} 1, x \in [-1,1] \\ 0, x \notin [-1,1]. \end{cases}$
(3) $L(x) = \frac{1}{1+|x|^p}$, $p \geq 0$.
(4) $L(x) = exp\{-|x|^p\}$, $p \geq 0$.

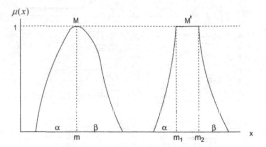

Fig. 1.11 LR type fuzzy numbers.

Definition 1.5. (Dubois [305]) Let $L(x), R(x)$ be the reference function of fuzzy number M, A fuzzy number M is said to be LR type iff

$$\mu_M(x) = \begin{cases} L(\frac{m-x}{\alpha}), & x \le m, \alpha > 0 \\ R(\frac{x-m}{\beta}), & x > m, \beta > 0, \end{cases} \tag{1.20}$$

where m is the "mean" of fuzzy number M and α, β are the left and right "spreads", respectively.

It is often written as (see Figure 1.11): $M = (m, \alpha, \beta)$. When $\alpha = \beta = 0$, M is considered a crisp number m. And when α and β increase gradually, the fuzzy number M turns to be fuzzier.

If the peak is not unique, the LR number M has a flat region. That is, the kernel $Ker(M)$ of LR fuzzy number is a close interval $[m_1, m_2]$, and the membership function has not only a peak value point (the point which make the value of the membership function the biggest), but a curve with flat. It can be written as (see Figure 1.11):

$$M' = (m_1, m_2, \alpha, \beta).$$

Triangular and trapezoidal fuzzy number

Let $x, l, m, n \in R$. A triangular fuzzy number M is defined as (see Figure 1.12):

$$\mu_M(x) = \begin{cases} 0, & x \le l \\ (x-l)/(m-l), & l < x \le m \\ (n-x)/(n-m), & m < x \le n \\ 0, & x \ge n. \end{cases} \tag{1.21}$$

In Figure 1.12, $M = (l, m, n)$ with l and n being the lower and upper bounds of fuzzy number M.

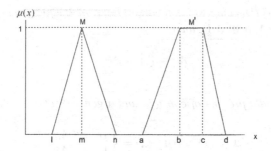

Fig. 1.12 Triangular and trapezoidal fuzzy numbers.

When there are multiple peaks, fuzzy number M is represented by

$$M' = (a, b, c, d),$$

with the $[b, c]$ interval being the most likely values for M' and any value below a and above d being totally impossible. The membership value decrease gradually (or linearly) from b to a and from c to d.

It is clear that the triangular (or trapezoidal) number is a more restricted from than the LR fuzzy number, in that all "legs" must be linear. Furthermore, we find that

$$M = (l, m, u) = (m, \alpha, \beta),$$

where $\alpha = m - l$ and $\beta = u - m$. Similarly, we find that

$$M' = (a, b, c, d) = (b, c, \alpha, \beta),$$

where $\alpha = b - a$ and $\beta = d - c$. The characteristics of M and M' in Figure 1.11 and Figure 1.12 remain the same.

We have four different formulas. Each has its own algebraic operation formula.

1.4 Fuzzy Arithmetic

After defining a fuzzy number, we now discuss the basic operations of two fuzzy numbers, which are based on Zadeh's Extension principle[22] as defined below.

Decomposition theorem and extension principle

Decomposition theorem and extension principle is the bridge to link the general sets and the fuzzy sets. By α-cut and Decomposition theorem, we can transform the fuzzy sets to general sets to tackle, its the basis of fuzzy arithmetics.

Theorem 1.1. *[19] (Decomposition theorem) Let A be the fuzzy set of the universe, A_α is the α-cut of A, $\alpha \in [0,1]$, then*

$$A = \bigcup_{\alpha \in [0,1]} \alpha A_\alpha,$$

where αA_α is the dot product of constant and a general set, it can be defines as a special set in U.

$$\mu_{\alpha A_\alpha}(x) = \alpha \mu_{A_\alpha}(x) = \begin{cases} \alpha, & x \in A_\alpha \\ 0, & x \notin A_\alpha. \end{cases}$$

We can get from the Decomposition theorem that any of the fuzzy sets can be seemed as a group of general sets.

In 1965, Zadeh introduced the Extension principle to extend the analytical methods of general sets to the theories of fuzzy sets. Same as decomposition theorem, extension principle is a basic theorem in fuzzy mathematics. As noted by Dubois and Prade[23], the Extension principle introduce by Zadeh[12, 13, 19] and others is one of the most basic ideas of fuzzy set theory. It is used to generalize non-fuzzy (crisp) mathematical concepts into fuzzy quantities. An important field of applications for the extension principle is given by algebraic operations such as addition and multiplication. We shall give the definition of the Extension principle first and extend from it to fuzzy algebraic operations.

Before introducing the Extension principle, we have to define the concept of cartesian product first.

Definition 1.6. (Zadeh [22]) Let U be a cartesian product of universe, $U = U_1 \times \cdots \times U_n$ and $A_1, \cdots, A_n$ be n fuzzy sets in $U_1, \cdots, U_n$, respectively. The cartesian product of $A_1, \cdots, A_n$ is defined as:

$$C = A_1 \times \cdots \times A_n = \int_{U_{A_1} \times \cdots \times U_{A_n}} (x_1, \cdots, x_n) / \min(\mu_{A_1}(x_1), \cdots, \mu_{A_n}(x_n)).$$

That is, the membership function of the cartesian product of $A_1, \cdots, A_n$ is

$$\mu_{A_1 \times \cdots \times A_n}(x_1, \cdots, x_n) = \min(\mu_{A_1}(x_1), \cdots, \mu_{A_n}(x_n)).$$

Example 1.18. Let fuzzy sets A and B be in Table 1.4

Table 1.4 Fuzzy set A and B in Example 1.18

x	3	4	5	6	7	8	9	10
μ_A	0	0	0.2	0.4	0.6	0.8	1	0
μ_B	0	0.5	0.7	1	0.7	0.5	0	0

The cartesian product of A and B is:

$$A \times B = \{[(5;4),0.2],[(5;6),0.2],[(5;8),0.2],[(6;4),0.4],\cdots$$
$$[(8;8),0.8],[(9;4),0.5],[(9;6),0.7],[(9;8),1]\}.$$

There is a total of 15 elements which are pairs of each element in A and each element in B. As a demonstration, we derive

$$\mu_{A \times B}(6;4) = \min[\mu_A(6),\mu_B(4)] = \min[0.4,0.5] = 0.4.$$

It follows that the Extension principle can be defined as follows.

Theorem 1.2. *[22](Extension principle) Let X be a Cartesian product of universes $X_i, i = 1,2,\cdots,r$ with $X = X_1 \times X_2 \times \cdots \times X_r$. $A_i, i = 1,2,\cdots,r$ are the corresponding fuzzy subsets in X_i. f is a mapping from X to a universe Y defined by $(x_1,\cdots,x_r) \rightarrow y = f(x_1,\cdots,x_r)$. Then*

$$\widetilde{B} = \{(y,\mu_{\widetilde{B}}(y)) | y = f(x_1,\cdots,x_r), and\ (x_1,\cdots,x_r) \in X\},$$

with

$$\mu_{\widetilde{B}} = \begin{cases} \sup_{(x_1,\cdots,x_r) \in f^{-1}(y)} \min\{\mu_{\widetilde{A_1}}(x_1),\cdots,\mu_{\widetilde{A_r}}(x_r)\}, & if\ f^{-1}(y) \neq \emptyset \\ 0, & otherwise \end{cases} \quad (1.22)$$

is a fuzzy set in Y.

Note that Equation (1.22) is true only when the inverse of f is not zero, i.e., $f^{-1}(y) \neq 0$. When $f^{-1}(y) = 0$, $\mu_{\widetilde{B}}(y) = 0$. $\mu_{\widetilde{B}}(y)$ is the greatest among the membership values $\mu_{A_1 \times \cdots \times A_n}(x_1, \cdots, x_n)$ of the realization of y using n-tuples $(x_1,\cdots,x_n)$.

The special case for $n = 1$ gives:

$$\mu_{\widetilde{B}}(y) = \begin{cases} \mu_A(f^{-1}(y)), & if\ f^{-1}(y) \neq 0 \\ 0, & otherwise. \end{cases} \quad (1.23)$$

Example 1.19. This example shows we can use the Extension principle to extend a crisp algebraic operation into a fuzzy one. Let fuzzy set A_1 and A_2 be in Table 1.5:

Table 1.5 Fuzzy set A_1 and A_2 in Example 1.19

x_1,x_2	2	3	4	5	6	7
$\mu_{A_1}(x_1)$	0	0.4	1	0.7	0	0
$\mu_{A_2}(x_2)$	0	0.1	0.8	1	0.3	0

Based on the crisp algebraic function, $f(x) = 2x_1 + x_2$, the composition of A_1 and A_2 is completed using the Extension principle as in Table 1.6:

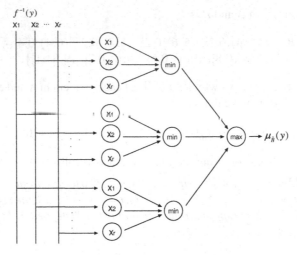

Fig. 1.13 Extension principle

Table 1.6 Composition of A_1 and A_2 in Example 1.19

$y = 2x_1 + x_2$	6	7	8	9	10	11	12	13	14	15	16	17
$\mu_B(y)$	0	0	0.1	0.1	0.4	0.4	0.8	1		0.7	0.7	0.3	0

For instance, to get $\mu_B(12)$ we know the possible (x_1, x_2) pairs such that $12 = 2x_1 + x_2$ are in Table 1.7:

Table 1.7 Possible (x_1, x_2) pairs in Example 1.19

x_1	3	4	5
x_2	6	4	2

The corresponding $\mu_{A_1}(x_1)$ and $\mu_{A_2}(x_2)$ and their minimums are in Table 1.8:

Table 1.8 $\mu_{A_1}(x_1)$ and $\mu_{A_2}(x_2)$ in Example 1.19

$\mu_{A_1}(x_1)$	0.4	1	0.7
$\mu_{A_2}(x_2)$	0.3	0.8	0
$\mu_{A_1}(x_1) \wedge \mu_{A_2}(x_2)$	0.3	0.8	0

Thus, $\mu_B(y = 12) = \max[0.3, 0.8, 0] = 0.8$.

In general,
$$\widetilde{B} = f(\widetilde{A}_1 \times \cdots \times \widetilde{A}_r).$$

Because
$$f^{-1}(y) = \bigcup_{f(x_1,\cdots,x_r)=y} \left(\bigcap_{k=1}^{r} (X_k = x_k) \right),$$

we have
$$\mu_{\widetilde{B}}(y) = \max \min \left\{ \mu_{\widetilde{A}_1(x_1)}, \cdots, \mu_{\widetilde{A}_r(x_r)} \right\}.$$

Therefore, Extension Principle extends set operation from crisp set to fuzzy set when algebra sum is replaced by sup(max) and algebra product is replaced by min.

When $r = 1$ in Theorem 1.2, then
$$\widetilde{B} = \left\{ (y, \mu_{\widetilde{B}}(y)) | y = f(x), x \in X \right\} = f(\widetilde{A}),$$

where
$$\mu_{\widetilde{B}} = \begin{cases} \sup_{f(x)=y} \mu_{\widetilde{A}}(x), & \text{if } f^{-1}(y) \neq \emptyset \\ 0, & \text{otherwise.} \end{cases}$$

Example 1.20. Consider a fuzzy set $\widetilde{A} = \{(-1,0.5),(0,0.8),(1,1),(2,0.4)\}$ and a function $f(x) = x^2 + 1$ as shown in Figure 1.14. Then, $\widetilde{B} = f(\widetilde{A}) = \{(1,0.8),(2,1),(5,0.4)\}$.

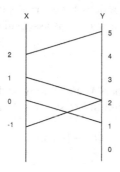

Fig. 1.14 Extension principle with $r = 1$

Example 1.21. Consider the following fuzzy sets:

$$F(R) = \{\text{fuzzy numbers}\},$$
$$F(R^+) = \{\text{positive fuzzy numbers}\},$$
$$F(R^-) = \{\text{negative fuzzy numbers}\}.$$

For a unary operation F defined by $F : R \to R$ and $\widetilde{M} \in F(R)$, Extension principle gives
$$\mu_{f(\widetilde{M})}(Z) = \sup_{f(x)=z} \mu_{\widetilde{M}}(x).$$

Property 1.1. The Extension principle has the following properties:
(i) $f(x) = -x$, then $\mu_{-\widetilde{M}}(x) = \mu_{\widetilde{M}}(-x)$, $\forall \widetilde{M} \in F(R)$;
(ii) $f(x) = \frac{1}{x}$, then $\mu_{\widetilde{M}^{-1}}(x) = \mu_{\widetilde{M}}\left(\frac{1}{x}\right)$, $\forall \widetilde{M} \in F(R^+)$ or $F(R^-)$;
(iii) $f(x) = \lambda \cdot x(\lambda \neq 0)$, then $\mu_{\lambda\widetilde{M}}(x) = \mu_{\widetilde{M}}\left(\frac{f(x)}{\lambda}\right)$, $\forall \widetilde{M} \in F(R)$.

Define a binary operation by $* \cdot R \times R \to R$, then for $\widetilde{M}, \widetilde{N} \in F(R)$, $\widetilde{M}(*)\widetilde{N}$ is a fuzzy number defined by

$$\mu_{\widetilde{M}(*)\widetilde{N}} = \sup_{z=x*y} \min\{\mu_{\widetilde{M}}(x), \mu_{\widetilde{N}}(y)\}.$$

Because, by Decomposition theory,

$$\mu_{\widetilde{M}}(x) = \bigvee_{\alpha} \alpha \cdot M_\alpha = \bigvee_{\alpha} \alpha \cdot [a_\alpha^{(L)}, a_\alpha^{(U)}],$$
$$\mu_{\widetilde{N}}(y) = \bigvee_{\alpha} \alpha \cdot [b_\alpha^{(L)}, b_\alpha^{(U)}].$$

Therefore,

$$\mu_{\widetilde{M}(*)\widetilde{N}}(Z) = \bigvee_{\alpha} \alpha \cdot (M_\alpha * N_\alpha)$$
$$= (\bigvee_{\alpha} \alpha \cdot M_\alpha) * (\bigvee_{\alpha} \alpha \cdot N_\alpha)$$
$$= \mu_{\widetilde{M}(Z)}(*)\mu_{\widetilde{N}(Z)}.$$

Lemma 1.1. *[19]* $(\widetilde{M}(*)\widetilde{N})_\alpha = M_\alpha * N_\alpha$.

This implies fuzzy number operations by interval-valued operations. Thus, if $*$ is commutative, $(*)$ is commutative, and if $*$ is associative, $(*)$ is associative too.

Addition of fuzzy numbers

The addition of two fuzzy numbers M and N may be done in two different ways.
(1) Use of α-cut: Let's define the α sets for M and N using the intervals of confidence as:

$$M_\alpha = [m_1, m_2] \tag{1.24}$$

and

$$N_\alpha = [n_1, n_2]. \tag{1.25}$$

The addition of M and N may be rewritten as:

$$M_\alpha(+)N_\alpha = [m_1 + n_1, n_1 + n_2]. \tag{1.26}$$

This is equivalent to adding two intervals of confidence level by Kaufmann and Gupta[309].
(2) Use of max-min convolution: Let $\forall x, y, z, \in R$. Then the addition of M and N equals

$$\mu_{M(+)N}(z) = \max_{z=x+y} (\mu_M(x) + \mu_N(y)). \tag{1.27}$$

One can see that equation 1.27 is an example of the Extension principle.

For fuzzy numbers similar to the one in Figure 1.9, we would use max-min convolution to get the sum of their addition. For fuzzy numbers similar to the one in Figure 1.10, we would use α-cut to get the sum of their addition. Note, however, that different addition operations may be used interchangeably.

It has been proved by Kaufmann and Gupta[309] that equation (1.26) and (1.27) describe the same operation. Let $x, y, z \in R$, then the addition of M and N can be computed using

$$\mu_{M_\alpha(+)N_\alpha}(z) = \max_{z=x+y} (\mu_{M_\alpha}(x) + \mu_{N_\alpha}(y)). \tag{1.28}$$

Assume that $\mu_M(x) = 1$, if $x \in [m_1, m_2]$. Otherwise, $\mu_M(x) = 0$. Similarly, $\mu_N(y) = 1$, if $y \in [n_1, n_2]$. Otherwise $\mu_N(y) = 0$. Thus, for all x and y such that $\mu_M(x) = 1$ and $\mu_N(y) = 1$, the right side of equation (1.28) gives 1. If not, equation (1.28) gives 0. And since $z = x + y$. We write

$$z = [m_1 + n_1, m_2 + n_2], \tag{1.29}$$

equation (1.29) may be regarded as another form of equation (1.28).

Property 1.2. The properties of fuzzy addition can be summarized as follows:
(i) Commutative: $M(+)N = N + M$;
(ii) Associative: $(M(+)N)(+)K = M(+)(N(+)K)$;
(iii) If a neutral exists at the left and the right, if is the real number 0. Thus $M(+)0 = 0(+)M = M$;
(iv) Nonsymmetric: $M(+)(-N) = (-N)(+)M \neq 0$ where $-N$ is the image of N with membership function $\mu_{-N}(x) = \mu_N(-x)$.

We shall use the following examples to show the computational procedure of each fuzzy addition operation.

Example 1.22. (Discrete case) Let M represent "integers close to 3" and N represent "integers close to 2", as shown in Table 1.11.

Table 1.9 Fuzzy numbers M and N

x, y	0	1	2	3	4	5	6
$\mu_M(x)$	0	0.3	0.8	1	0.5	0.1	0
$\mu_N(y)$	0	0.6	1	0.9	0.4	0	–

Their addition is summarized below.

Table 1.10 Addition of fuzzy numbers M and N

$z = x + y$	1	2	3	4	5	6	7	8	9	10
$\mu_{M(+)N}(z)$	0	0.3	0.6	0.8	1	0.9	0.5	0.4	0.1	0

Table 1.11 Membership values, and $\mu_M(x) \wedge \mu_N(y)$

x	0	1	2	3	4	5	6
y	6	5	4	3	2	1	0
$\mu_M(x)$	0	0.3	0.8	1	0.5	0.1	0
$\mu_N(y)$	–	0	0.4	0.9	1	0.6	0
$\mu_M(x) \wedge \mu_N(y)$	0	0	0.4	0.9	0.5	0.1	0

For instance, to get $\mu_{M(+)N}(z = 6)$, the possible (x,y) pairs, their corresponding membership values, and $\mu_M(x) \wedge \mu_N(y)$ are:

Thus, $\mu_{M(+)N}(z = 6) = \max[0, 0, 0.4, 0.9, 0.5, 0.1, 0] = 0.9$.

Example 1.23. (Continuous case) Let M represent "real numbers close to 2" and N represent "real numbers close to 8" (see Figure 1.15), where

$$\mu_M(x) = \begin{cases} 0, & x \leq 0 \\ x/2, & 0 < x \leq 2 \\ (4-x)/2, & 2 < x \leq 4 \\ 0, & x > 4, \end{cases} \quad \mu_N(y) = \begin{cases} 0, & y \leq 3 \\ (y-3)/5, & 3 < y \leq 8 \\ (11-y)/2, & 8 < y \leq 11 \\ 0, & y > 11. \end{cases}$$

The addition of M and N is illustrated as follows. The α-cut of M and N are:

$$M_\alpha = [m_1, m_2], N_\alpha = [n_1, n_2].$$

That is, at some α level, the x can be either m_1 or m_2, and y can take either n_1 or n_2. Thus, if we set $\alpha = 1/2$ for $\mu_M(x)$, we have $\alpha = m_1/2$, i.e., $m_1 = 2\alpha$. Similarly, we can obtain other α-cut values as:

The addition of M and N at α level is computed based on equation (1.27) as:

$$M + N = [2\alpha + (5\alpha + 3), (-2\alpha + 4) + (-3\alpha + 11)] = [7\alpha + 3, -5\alpha + 15].$$

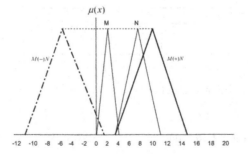

Fig. 1.15 The fuzzy numbers M, N, $M(+)N$, and $M(-)N$

Table 1.12 α-cut values

m_1	m_2	m_3	m_4
2α	$-2\alpha+4$	$5\alpha+3$	$-3\alpha+11$

Let $Z_\alpha = [z_1, z_2] = [7\alpha+3, -5\alpha+15]$, then

$$\alpha = (z_1 - 3)/7, \text{ and } \alpha = (-z_2 + 15)/5.$$

Consequently, we have $Z = \{(z, \mu_Z(z))\}$ (see Figure 1.15) where

$$\mu_Z(z) = \begin{cases} 0, & z \leq 3 \\ (z-3)/7, & 3 < z \leq 10 \\ (-z+15)/5, & 10 < z \leq 15 \\ 0, & z > 15. \end{cases}$$

From the example shown above, monotonicity, convexity, and normality are well preserved by addition. The proof can be seen in Kaufmann and Gupta[309].

Theorem 1.3. *[305] Let $M = (m, \alpha, \beta)_{LR}$, $N = (n, \gamma, \delta)_{LR}$, then we have $M(+)N = (m+n, \alpha+\gamma, \beta+\delta)_{LR}$.*

Example 1.24. Let $\widetilde{M} = (m, \alpha, \beta)_{LR} = (4, 2, 3)_{LR}$, $\widetilde{N} = (n, \gamma, \delta)_{LR} = (8, 3, 5)_{LR}$, and the left and right reference functions are

$$L(x) = \begin{cases} 0, & \text{if } x < -1 \\ (1+x)^{\frac{1}{2}}, & \text{if } -1 \leq x \leq 0, \end{cases} \quad R(x) = \begin{cases} 1-x^2, & \text{if } 0 \leq x \leq 1 \\ 0, & \text{if } x > 1. \end{cases}$$

The membership function of $\widetilde{M}$ and $\widetilde{N}$ are as follows respectively.

$$\mu_{\widetilde{M}}(x) = \begin{cases} 0, & \text{if } x < 2 \\ (1+\frac{x-4}{2})^{\frac{1}{2}}, & \text{if } 2 \leq x \leq 4 \\ 1, & \text{if } x = 4 \\ 1-(\frac{x-4}{3})^2, & \text{if } 4 < x \leq 7 \\ 0, & \text{if } x > 7, \end{cases} \quad \mu_{\widetilde{N}}(x) = \begin{cases} 0, & \text{if } x < 5 \\ (1+\frac{x-8}{3})^{\frac{1}{2}}, & \text{if } 5 \leq x \leq 8 \\ 1, & \text{if } x = 8 \\ 1-(\frac{x-8}{5})^2, & \text{if } 8 < x \leq 13 \\ 0, & \text{if } x > 13. \end{cases}$$

By Theorem 1.3, we have

$$\widetilde{M}(+)\widetilde{N} = (m+n, \alpha+\gamma, \beta+\delta)_{LR} = (12, 5, 8)_{LR},$$

and the membership function is

$$\mu_{\widetilde{M}(+)\widetilde{N}}(x) = \begin{cases} 0, & \text{if } x < 7 \\ (1+\frac{x-12}{5})^{\frac{1}{2}}, & \text{if } 7 \leq x \leq 12 \\ 1, & \text{if } x = 12 \\ 1-(\frac{x-12}{8})^2, & \text{if } 12 < x \leq 20 \\ 0, & \text{if } x > 20. \end{cases}$$

Substraction of fuzzy numbers

The definition of substraction can also be defined by either α-cut or max-min convolution.
(1) Use of α-cut:

$$M_\alpha(-)N_\alpha = [m_1 - n_2, m_2 - n_1]. \tag{1.30}$$

(2) Use of max-min convolution:

$$\mu_{M(-)N}(z) = \max_{z=x-y} (\mu_M(x) \wedge \mu_N(y)). \tag{1.31}$$

Since the image of fuzzy number N is given by

$$\mu_{-N}(x) = \mu_N(-x), \forall x. \tag{1.32}$$

equation (1.31) may be rewritten as:

$$\mu_{M(-)N}(z) = \max_{z=x+(-y)} (\mu_M(x) \wedge \mu_N(-y)) = \max_{z=x+y} (\mu_M(x) \wedge \mu_{-N}(y)). \tag{1.33}$$

The substraction, $M(-)N$, is equivalent to the addition of the image of N to M, $M(+)(-N)$.

Because a negative number may appear as a result of substraction, the commutative and associative properties cannot be preserved. However, since M and N are fuzzy numbers, $M(-)N$ must be a fuzzy number (Dubois and Prada[71, 305]).

Example 1.25. Let M and N be fuzzy numbers presented in Table. 1.11. The result of $M(-)N$, Z, is computed as:

Table 1.13 Result of $M(-)N$

z	-5	-4	-3	-2	-1	0	1	2	3	4	5
$\mu_Z(z)$	0	0	0.3	0.4	0.8	0.9	1	0.6	0.5	0.1	0

Note that $(-5,0)$ may be dropped from the fuzzy set since, by the definition of a fuzzy number, any number smaller than -4 must have a membership value of 0. The computational procedure for substraction is the same as for addition. For example, to get $\mu_{M(-)N}(z = -1)$, the possible (x,y) pairs, their corresponding membership values, and $(\mu_M(x) \wedge \mu_N(y))$ are:

Thus, $\mu_{M(-)N}(-1) = \max[0, 0.3, 0.8, 0.4, 0, 0] = 0.8$.

Example 1.26. Let M and N be fuzzy numbers presented in Figure 1.15. They are the same fuzzy numbers we used for addition. The α-cut of M and N are:

Table 1.14 Result of $(\mu_M(x) \wedge \mu_N(y))$

x	0	1	2	3	4	5
y	1	2	3	4	5	6
$\mu_M(x)$	0	0.3	0.8	1	0.5	0.1
$\mu_N(y)$	0.6	1	0.9	0.4	0	-
$\mu_M(x) \wedge \mu_N(y)$	0	0.3	0.8	0.1	0	-

Table 1.15 α-cut values

m_1	m_2	m_3	m_4
2α	$-2\alpha+4$	$5\alpha+3$	$-3\alpha+11$

Based on equation (1.30), we have

$$M_\alpha(-)N_\alpha = [m_1 - n_2, m_2 - n_1] = [5\alpha - 11, -7\alpha + 1].$$

Consequently, the membership function $\mu_{M(-)N}(z)$ is (see Figure 1.15):

$$\mu_{M(-)N}(z) = \begin{cases} 0, & z \leq -11 \\ (z+11)/5, & -11 < z \leq -6 \\ (1-z)/7, & -6 < z \leq 1 \\ 0, & z > 15. \end{cases}$$

Theorem 1.4. *[305] Let $M = (m, \alpha, \beta)_{LR}$, $N = (n, \gamma, \delta)_{LR}$, then we have $M(-)N = (m - n, \alpha + \delta, \beta + \gamma)_{LR}$.*

Example 1.27. Let $\widetilde{M} = (m, \alpha, \beta)_{LR} = (4, 2, 3)_{LR}$, $\widetilde{N} = (n, \gamma, \delta)_{LR} = (8, 3, 5)_{LR}$, and the left and right reference functions are

$$L(x) = \begin{cases} 0, & \text{if } x < -1 \\ (1+x)^{\frac{1}{2}}, & \text{if } -1 \leq x \leq 0, \end{cases} \quad R(x) = \begin{cases} 1-x^2, & \text{if } 0 \leq x \leq 1 \\ 0, & \text{if } x > 1. \end{cases}$$

By Theorem 1.4, we have

$$\widetilde{M}(-)\widetilde{N} = (m - n, \alpha + \delta, \beta + \gamma)_{LR} = (-4, 7, 6)_{LR},$$

and the membership function is

$$\mu_{\widetilde{M}(-)\widetilde{N}}(x) = \begin{cases} 0, & \text{if } x < -11 \\ (1 + \frac{x+4}{7})^{\frac{1}{2}}, & \text{if } -11 \leq x \leq -1 \\ 1, & \text{if } x = -4 \\ 1 - (\frac{x+4}{6})^2, & \text{if } -4 < x \leq 2 \\ 0, & \text{if } x > 2. \end{cases}$$

Multiplication of fuzzy numbers

The multiplication of fuzzy numbers is a little complicated because the signs of fuzzy numbers must be considered. We shall consider case in which both M and N are positive fuzzy numbers, i.e.,

$$\mu_M(x) = 0, \ \forall x < 0 \text{ and } \mu_N(y) = 0, \ \forall y < 0.$$

Let Z be the product of the multiplication of M and N. Since $\mu_Z(z)$ increase monotonically to the left of the peak $(\mu_Z(z) = 1)$ and decreases monotonically to the right of the peak, the multiplication is done in the following manner (Kaufmann and Gupta[309]):

   (1) At the left, we take into account all pairs (x,y) such that $xy \leq z$. That is ,the left leg of $\mu_Z(z)$ is defined as:

$$\mu_{M(\cdot)N}(z) = \max_{xy \leq z}(\mu_M(x) \wedge \mu_N(y)). \tag{1.34}$$

(2) At the right, we take into account all airs (x,y) such that $xy \geq z$. That is, the right leg of $\mu_Z(z)$ is defined as:

$$\mu_{M(\cdot)N}(z) = \max_{xy \geq z}(\mu_M(x) \wedge \mu_N(y)). \tag{1.35}$$

(3) To simplify the process, omit from consideration any (x,y) pair where either $\mu_M(x)$ or $\mu_N(y)$ is zero. Conversely, we compute z for which $\mu_{M(\cdot)N}(z) = 1$. This will show what value of z occurs when we pass from the left to the right of the peak.

   When both M and N are continuous membership functions, their multiplication is defined as:

$$M_\alpha(\cdot)N_\alpha = [m_1 n_1, m_2 n_2]. \tag{1.36}$$

equations (1.34),(1.35), and (1.36) are equivalent. This can be easily proved (as in the case of addition).

Property 1.3. The properties of fuzzy multiplication can be summarized as below:

(i) When both M and N have the same sign, $M(\cdot)N$ can also be a positive fuzzy number.
(ii) Since $(-M)(\cdot)N = -(M(\cdot)N)$, we know M and N can take different signs (Dubois and Prada[71, 305]).
(iii) The multiplication of fuzzy numbers M and N is commutative and associative, i.e.,

$$M(\cdot)N = N(\cdot)M,$$

and

$$(M(\cdot)N)(\cdot)K = M(\cdot)(N(\cdot)K)).$$

(iv) If a neutral exists at the left and at the right, it is the real number 1 (Kaufmann and Gupta[309]), i.e., $M(\cdot)1 = 1(\cdot)M = M$.

(v) The inverse of M is M^{-1} and $M(\cdot)M^{-1} \neq 1$ where

$$M_\alpha^{-1} = [\frac{1}{m_2}, \frac{1}{m_1}].$$

Example 1.28. (Discrete case) Let M and N be the fuzzy numbers shown in Table. 1.11. By applying equations (1.34) and (1.35), we can obtain $Z = M(\cdot)N$:

Table 1.16 Membership value of z

z	0	1	2	3	4	5	6	7	8	9	10	11	12	13	14	15	16	17
$\mu_Z(z)$	0	0.3	0.6	0.6	0.8	0.8	1	0.9	0.9	0.9	0.5	0.5	0.5	0.4	0.4	0.4	0.4	0

For instance, $\mu_{M(\cdot)N}(6)$ is calculated as:

$$\mu_{M(\cdot)N}(6) = \max[(\mu_M(3) \wedge \mu_N(2)), (\mu_M(2) \wedge \mu_N(3))] = 1.$$

There are no other z values where $\mu_{M(\cdot)N}(z) = 1$. Thus, for a z value less than 6, say 4, the membership function $\mu_{M(\cdot)N}(4)$ can be derived as follows. For the (x,y) pairs where $xy \leq 4$, we have $(\mu_M(x) \wedge \mu_N(y))$ as:

$$\mu_M(x) \wedge \mu_N(y) = \begin{array}{c c} & \begin{array}{c c c c} x \,|\, y & 1 & 2 & 3 & 4 \end{array} \\ \begin{array}{c} 1 \\ 2 \\ 3 \\ 4 \end{array} & \left[\begin{array}{c c c c} 0.3 & 0.3 & 0.3 & 0.3 \\ 0.6 & 0.8 & - & - \\ 0.6 & - & - & - \\ 0.5 & - & - & - \end{array} \right] \end{array}$$

Thus, $\mu_{M(\cdot)N}(4) = \max[0.3, 03, 0.3, 0.3, 0.6, 0.8, 0.6, 0.5] = 0.8$. Note that $(\mu_M(2), \mu_N(2))$, $(\mu_M(1), \mu_N(4))$, and $(\mu_M(4), \mu_N(1))$ are not the only pairs being evaluated.

For a z greater than 6, say 14, the membership value $\mu_{M(\cdot)N}(14)$ is derived as follows. For the (x,y) pairs where $xy \geq 14$ we have (see Table 1.17):

Table 1.17 Membership value $\mu_{M(\cdot)N}(14)$

x	4	5	5
y	4	4	3
$\mu_M(x) \wedge \mu_N(y)$	0.4	0.1	0.1

Thus, $\mu_{M(\cdot)N}(14) = \max\limits_{14 \leq xy}[0.4, 0.1, 0.1] = 0.4$. Note that the pairs such as $(\mu_M(3) \wedge \mu_N(5))$ are dropped from evaluation because $\mu_N(5) = 0$.

Example 1.29. (Continuous case) Let M and N be presented as in Figure 1.15. The α level sets for M and N are the same as in the addition case:

$$M_\alpha = [2\alpha, -2\alpha + 4], \text{and } N_\alpha = [5\alpha + 3, -3\alpha + 11].$$

According to equation (1.36), we can obtain

$$\begin{aligned} Z_\alpha = M_\alpha(\cdot)N_\alpha &= [(2\alpha)(5\alpha + 3), (-2\alpha + 4)(-3\alpha + 11)] \\ &= [10\alpha^2 + 6\alpha, 6\alpha^2 - 34\alpha + 44]. \end{aligned}$$

We now solve the following two equations,

$$10\alpha^2 + 6\alpha - z = 0, \tag{1.37}$$

and

$$6\alpha^2 - 34\alpha + 44 - z = 0. \tag{1.38}$$

The roots for equations (1.37) and (1.38) are:

$$\alpha = (-6 + (36 + 40z)^{0.5})/20, \text{and } \alpha = (34 - (100 + 24z)^{0.5})/12.$$

Thus, we have (see Figure 1.16),

$$\mu_{M(\cdot)N}(z) = \begin{cases} 0, & z \leq 0 \\ (-6 + (36 + 40z)^{0.5})/20, & 0 < z \leq 16 \\ (34 - (100 + 24z)^{0.5})/12, & 16 < z \leq 44 \\ 0, & z > 44. \end{cases}$$

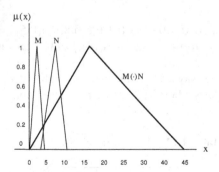

Fig. 1.16 Fuzzy number $M(\cdot)N$

Clearly, $\mu_{M(\cdot)N}(z)$ is still a fuzzy number even though its left and right "legs" are no longer linear.

Division of fuzzy numbers

Division of two positive fuzzy numbers M and N can be defined as follows:
(1) For the left leg of $M(:)N$, we have

$$\mu_{M(:)N}(z) = \max_{z \geq x/y} (\mu_M(x) \wedge \mu_N(y)), \ \forall x, y, z. \quad (1.39)$$

(2) For the right leg of $M(:)N$, we have

$$\mu_{M(:)N}(z) = \max_{z \leq x/y} (\mu_M(x) \wedge \mu_N(y)), \ \forall x, y, z. \quad (1.40)$$

If M and N are continuous membership functions, we define

$$M_\alpha(:)N_\alpha = [m_1/n_2, m_2/n_1], \ n_2 > 0 \quad (1.41)$$

The division operation is an extension of multiplication, i.e., $M(:)N = M(\cdot)N^{-1}$ where N^{-1} is the inverse of N. Recall that N^{-1} can be written as (Dubois and Prada [71, 305], Kaufmann and Gupta [309]):

$$\mu_{N^{-1}}(y) = \mu_N(1/y) \quad (1.42)$$

or

$$N_\alpha^{-1} = [\frac{1}{n_2}, \frac{1}{n_1}]. \quad (1.43)$$

Thus, equations (1.39) and (1.40) can be easily revised to

$$\mu_{M(\cdot)N^{-1}}(z) = \max_{z \geq x(1/y)} (\mu_M(x) \wedge \mu_N(1/y)) = \max_{z \geq xy} (\mu_M(x) \wedge \mu_{N^{-1}}(y)) \quad (1.44)$$

and

$$\mu_{M(\cdot)N^{-1}}(z) = \max_{z \leq x(1/y)} (\mu_M(x) \wedge \mu_N(1/y)) = \max_{z \leq xy} (\mu_M(x) \wedge \mu_{N^{-1}}(y)). \quad (1.45)$$

And equation (1.42) can easily be written as equation (1.41):

$$M_\alpha(\cdot)N_\alpha^{-1} = [m_1(\frac{1}{n_2}), m_2(\frac{1}{n_1})].$$

The computation of division is identical to that of multiplication. Because of this, we shall omit numerical examples of division altogether.

Fuzzy max and fuzzy min

Dubois and Prada [71, 305] pointed out that the fuzzy max is the dual operation with respect to union, while the fuzzy min is the dual operation with respect to intersection. It is easy to derive from foregoing statement the fuzzy max and the fuzzy min as:

$$M_\alpha(\vee)N_\alpha = [m_1 \vee n_1, m_2 \vee n_2], \tag{1.46}$$

$$M_\alpha(\wedge)N_\alpha = [m_1 \wedge n_1, m_2 \wedge n_2], \tag{1.47}$$

respectively, or

$$\mu_{M(\vee)N}(z) = \max_{z=x\vee y}(\mu_M(x) \wedge \mu_N(y)), \tag{1.48}$$

$$\mu_{M(\wedge)N}(z) = \max_{z=x\wedge y}(\mu_M(x) \wedge \mu_N(y)), \tag{1.49}$$

respectively. Graphically, the fuzzy max and the fuzzy min are presented in Figs. 1.17 and 1.18.

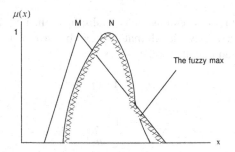

Fig. 1.17 Example of the fuzzy max

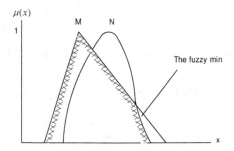

Fig. 1.18 Example of the fuzzy min

The properties of the fuzzy max and the fuzzy min are summarized as follows:
(1) The fuzzy max and min are commutative and associative operations.
(2) Distributive: Let M, N and K be fuzzy numbers, then

$$\min[M, \max(N, K)] = \max[\min(M, N), \min(M, K)]$$

and
$$\max[M, \min(N,K)] = \min[\max(M,N), \max(M,K)].$$

(3) Absorption: Given fuzzy numbers M and N,
$$\max[M, \max(M,N)] = M$$

and
$$\min[M, \max(M,N)] = M.$$

(4) De Morgan's law: Given fuzzy numbers M and N, then
$$1 - \min(M,N) = \max[1(-)M, 1(-)N]$$

and
$$1 - \max(M,N) = \min[1(-)M, 1(-)N].$$

(5) Idempotence:
$$\max(M,M) = M = \min(M,M).$$

(6) Given fuzzy numbers $M, N,$ and K,
$$M(+)\max(N,K) = \max[M(+)N, M(+)K].$$

The same property holds true for the fuzzy min.
(7) $\max(M,N)(+)\min(M,N) = M(+)N.$

Example 1.30. Let M and N be fuzzy numbers presented in Table 1.18.

Table 1.18 Fuzzy numbers M and N

x,y	1	2	3	4	5	6
$\mu_M(x)$	0	0.7	1	0.4	0.2	0
$\mu_N(y)$	0	0.3	1	0.6	0	–

The fuzzy max of M and N is computed as Table 1.19:
For instance, to obtain $\mu_{\max}(z=3)$, the (x,y) pairs that satisfy $3 = x \vee y$ are:

Table 1.19 Fuzzy max of M and N

$z = x \vee y$	1	2	3	4	5	6
$\mu_{\max}$	0	0.3	1	0.7	0.2	0

The corresponding μ_M and μ_N, and their minimum are in Table 1.20:
Thus, $\mu_{\max}(3) = \max[0, 0.7, 1, 0.3, 0] = 1.$
The fuzzy min of M and N is computed as in Table 1.21:
The computational procedure for $\mu_{\min}(z)$ is the same as that of $\mu_{\max}(z)$ expect that the fuzzy min uses $z = x \vee y$ but the fuzzy max uses $z = x \wedge y$.

Table 1.20 (x, y) pairs that satisfy $3 = x \vee y$

x	1	2	3	3	3
y	3	3	3	2	1

Table 1.21 Minimum of M and N

$\mu_M(x)$	0	0.7	1	1	1
$\mu_N(y)$	1	1	1	0.3	0
$\mu_M(x) \wedge \mu_N(y)$	0	0.7	1	0.3	0

Table 1.22 Fuzzy min of M and N

$z = x \wedge y$	1	2	3	4	5	6
$\mu_{\min}$	0	0.7	1	0.4	0	–

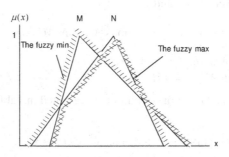

Fig. 1.19 Example of the fuzzy min

Example 1.31. Let M and N be fuzzy numbers presenting in Figure 1.19. By taking the α-cut, we have

$$M_\alpha = [4\alpha + 1, 8 - 3\alpha], \text{and } N_\alpha = [\alpha + 2, 9 - 6\alpha].$$

Based on equation (1.47), their maximum is defined as:

$$M_\alpha(\vee)N_\alpha = [(4\alpha + 1) \vee (\alpha + 2), (8 - 3\alpha) \vee (9 - 6\alpha)].$$

By changing the α value, equation (1.47) may yield different results. That is, when $0 \leq \alpha \leq 0.33$, we get $M_\alpha(\vee)N_\alpha = [\alpha + 2, 9 - 6\alpha]$.

similarly, when $0.33 \leq \alpha \leq 1$, we get $M_\alpha(\vee)N_\alpha = [4\alpha + 1, 8 - 3\alpha]$.

Thus, the computed membership function is

$$
\mu_{M(\vee)N}(z) = \begin{cases}
0, & z \leq 2 \\
(z-2)/1, & 2 < z \leq 2.33 \\
(z-1)/4, & 2.33 < z \leq 5 \\
(8-z)/3, & 5 < z \leq 7 \\
(9-z)/6, & 7 < z \leq 9 \\
0, & z > 9.
\end{cases}
$$

This membership function is illustrated in Figure 1.19 by the dashed line. Similarly, the fuzzy min can be obtained and illustrated as the dotted line in 1.19.

Table 1.23 Algebraic operation for $M = (m, \alpha, \beta)$, $N = (n, \gamma, \delta)$

Image of N: $-N = (-n, \delta, \gamma)$
Inverse of N: $N^{-1} = (\frac{1}{n}, \frac{\delta}{n^2}, \frac{\gamma}{n^2})$
Addition: $M(+)N = (m+n, \alpha+\gamma, \beta+\delta)$
Substraction: $M(-)N = (m-n, \alpha+\delta, \beta+\gamma)$
Scalar Multiplication:
  $k > 0, k \in R$: $k(\cdot)M = (km, k\alpha, k\beta)$
  $k < 0, k \in R$: $k(\cdot)M = (km, -k\beta, -k\alpha)$
Multiplication:
  $M > 0, N > 0$: $M(\cdot)N = (mn, m\gamma+n\alpha, m\delta+n\beta)$
  $M < 0, N > 0$: $M(\cdot)N = (mn, n\alpha-m\gamma, n\beta-m\delta)$
  $M < 0, N < 0$: $M(\cdot)N = (mn, -n\beta-m\delta, -n\alpha-n\gamma)$
Division:
  $M > 0, N > 0$: $M(:)N = (\frac{m}{n}, \frac{m\delta+n\alpha}{n^2}, \frac{m\gamma+n\beta}{n^2})$
  $M < 0, N > 0$: $M(:)N = (\frac{m}{n}, \frac{n\alpha-m\gamma}{n^2}, \frac{n\beta-m\delta}{n^2})$
  $M < 0, N < 0$: $M(:)N = (\frac{m}{n}, \frac{-n\beta-m\gamma}{n^2}, \frac{-n\alpha-m\delta}{n^2})$

Table 1.24 Algebraic operation for $M = (a, b, \alpha, \beta)$, $N = (c, d, \gamma, \delta)$

Image of N: $-N = (-d, -c, \delta, \gamma)$
Inverse of N: $N^{-1} = (\frac{1}{d}, \frac{1}{c}, \frac{\delta}{d(d+\delta)}, \frac{\gamma}{c(c-\gamma)})$
Addition: $M(+)N = (a+c, b+d, \alpha+\gamma, \beta+\delta)$
Substraction: $M(-)N = (a-d, b-c, \alpha+\delta, \beta+\gamma)$
Multiplication:
  $M > 0, N > 0$: $M(\cdot)N = (ac, bd, a\gamma+c\alpha-\alpha\gamma, b\delta+d\beta+\beta\delta)$
  $M < 0, N > 0$: $M(\cdot)N = (ad, bc, d\alpha-a\delta+\alpha\delta, -b\gamma+c\beta-\beta\gamma)$
  $M < 0, N < 0$: $M(\cdot)N = (bd, ac, -b\delta-d\beta-\beta\delta, -a\gamma-c\alpha+\alpha\gamma)$
Division:
  $M > 0, N > 0$: $M(:)N = (\frac{a}{d}, \frac{b}{c}, \frac{a\delta+d\alpha}{d(d+\delta)}, \frac{b\gamma+c\beta}{c(c-\gamma)})$
  $M < 0, N > 0$: $M(:)N = (\frac{a}{c}, \frac{b}{d}, \frac{c\alpha-a\gamma}{c(c-\gamma)}, \frac{d\beta-b\delta+}{d(d+\delta)})$
  $M < 0, N < 0$: $M(:)N = (\frac{b}{c}, \frac{a}{d}, \frac{-b\gamma-c\beta}{c(c-\gamma)}, \frac{-a\delta-d\alpha}{d(d+\delta)})$

Table 1.25 Algebraic operation for $M = (l,m,n), N = (a,b,c)$

Image of N: $-N = (-c,-b,-a)$
Inverse of N: $N^{-1} = (\frac{1}{c},\frac{1}{b},\frac{1}{a})$
Addition: $M(+)N = (l+a,m+b,n+c)$
Substraction: $M(-)N = (l-a,m-b,n-c)$
Scalar Multiplication:
$\quad k>0, k \in R$: $k(\cdot)M = (kl,km,kn)$
$\quad k<0, k \in R$: $k(\cdot)M = (kn,-km,-kl)$
Multiplication:
$\quad M>0, N>0$: $M(\cdot)N = (la,mb,nc)$
$\quad M<0, N>0$: $M(\cdot)N = (lc,mb,na)$
$\quad M<0, N<0$: $M(\cdot)N = (nc,mb,la)$
Division:
$\quad M>0, N>0$: $M(:)N = (\frac{l}{c},\frac{m}{b},\frac{n}{q})$
$\quad M<0, N>0$: $M(:)N = (\frac{n}{c},\frac{m}{b},\frac{l}{q})$
$\quad M<0, N<0$: $M(:)N = (\frac{n}{a},\frac{m}{b},\frac{l}{c})$

Algebraic operation formulas for special fuzzy numbers

Table 1.23 summarizes the algebraic operations for LR triangular numbers. Table 1.24 summarizes the algebraic operations for LR trapezoidal numbers. Table 1.25 and 1.26 summarizes the algebraic operations for triangular and trapezoidal numbers, respectively.

Table 1.26 Algebraic operation for $M = (a_1,b_1,c_1,d_1), N = (a_2,b_2,c_2,d_2)$

Image of N: $-N = (-d_2,-c_2,-b_2,-a_2)$
Inverse of N: $N^{-1} = (\frac{1}{d_2},\frac{1}{c_2},\frac{1}{b_2},\frac{1}{a_2})$
Addition: $M(+)N = (a_1+a_2,b_1+b_2,c_1+c_2,d_1+d_2)$
Substraction: $M(-)N = (a_1-d_2,b_1-c_2,c_1-b_2,d_1-a_2)$
Scalar Multiplication:
$\quad k>0, k \in R$: $k(\cdot)M = (ka_1,kb_1,kc_1,kd_1)$
$\quad k<0, k \in R$: $k(\cdot)M = (kd_1,kc_1,kb_1,ka_1)$
Multiplication:
$\quad M>0, N>0$: $M(\cdot)N = (a_1a_2,b_1b_2,c_1c_2,d_1d_2)$
$\quad M<0, N>0$: $M(\cdot)N = (d_1a_2,c_1b_2,b_1c_2,a_1d_2)$
$\quad M<0, N<0$: $M(\cdot)N = (d_1d_2,c_1c_2,b_1b_2,a_1a_2)$
Division:
$\quad M>0, N>0$: $M(:)N = (\frac{a_1}{d_2},\frac{b_1}{c_2},\frac{c_1}{b_2},\frac{d_1}{a_2})$
$\quad M<0, N>0$: $M(:)N = (\frac{d_1}{d_2},\frac{c_1}{c_2},\frac{b_1}{b_2},\frac{a_1}{a_2})$
$\quad M<0, N<0$: $M(:)N = (\frac{d_1}{a_2},\frac{c_1}{b_2},\frac{b_1}{c_2},\frac{a_1}{d_2})$

1.5 Membership Function

While classical (crisp) mathematics are dichotomous in character, fuzzy set theory considers the situations involving the human factor with all its vagueness of perception, subjectivity, attitudes, goals and conceptions. By introducing vagueness and linguistics into the crisp set theory, fuzzy set theory becomes more robust and flexible than the original dichotomous classical set theory.

There are two essential features of the fuzzy set theory[45]:

(i) Membership function of fuzzy set and operators play a crucial role in fuzzy set theory.

(ii) Fuzzy set theory is essentially a very general, flexible, formal theory. If it is to be applied to a real problem, it can and has to be adapted carefully. Neither the concept of membership nor the operator has a unique semantic interpretation. The context-dependent semantic interpretation will lead to different mathematical definitions and appropriate operators.

Thus membership functions and operators are actually the cornerstones of the fuzzy set theory. In this section , let us discuss some approaches to generate membership functions.

Membership function forms

In the following context, we will present a state-of-art survey of function forms of membership. Based on Dombi [7] , we classify all existing membership functions into the following four classed:

(1) Membership functions based on heuristic determination.

(a) Zadeh's unimodal functions:

$$\mu_{young} = \begin{cases} 1/\{1 + [(x-25)/5]^2\}, & \text{if } x > 25 \\ 1, & \text{if } x \leq 25, \end{cases}$$

$$\mu_{old} = \begin{cases} 1/\{1 + [(x-50)/5]^{-2}\}, & \text{if } x \geq 25 \\ 0, & \text{if } x < 50. \end{cases}$$

(b) Dimitru and Luban's power functions:

$$\mu(x) = x^2/a^2 + 1, \quad x \in [0,a],$$

$$\mu(x) = -x^2/a^2 - 2x/a + 1, \quad x \in [0,a].$$

(c) Svarowski's sin function:

$$\mu(x) = 1/2 + (1/2)[sin\{\pi/(b-a)\}[x - (a+b)/2]\}, \quad x \in [a,b].$$

(2) Membership functions based on reliability concerns with respect to the particular problem.

(a) Zimmermannn's linear function:

$$\mu(x) = 1 - x/a, \quad x \in [0, a].$$

(b) Tanaka, Uejima and Asai's symmetric triangular function:

$$\mu(x) = \begin{cases} 1 - \frac{|h-r|}{a}, & \text{if } b - u \leq x \leq b + u \\ 0, & \text{otherwise.} \end{cases}$$

(c) Hannan's piecewise linear function:

$$\mu(x) = \sum_j \alpha_j |x - a_j| + \beta x + r, \quad j = 1, 2, \cdots, N,$$
$$\alpha_j = (t_{j+1} - t_j)/2,$$
$$\beta = (t_{N+1} + t_1)/2,$$
$$r = (s_{N+1} + s_1)/2,$$

where $\mu(x) = t_i x + s_i$ for each segment i, $a_{i-1} \leq x \leq a_i$, $\forall i$. Here, t_i is the slope and s_i is the y-intercept for the section of the curve initiated at a_{i-1} and terminated at a_i.

(d) Dubois and Prada's LR fuzzy number:

$$\mu(x) = \begin{cases} L(\frac{a-x}{\alpha}), & \text{if } x < a \\ R(\frac{x-b}{\beta}), & \text{if } x > b \\ 1, & \text{if } a \leq x \leq b, \end{cases}$$

where $L(\cdot)$ and $R(\cdot)$ are reference functions.

(e) Leberling's hyperbolic function:

$$\mu(x) = 1/2 + (1/2)tanh(a(x-b)), \quad -\infty \leq x \leq \infty,$$

where a is a parameter.

(f) Sakawa and Yumine's exponential and hyperbolic inverse functions, respectively:

$$\mu(x) = c(1 - e^{(b-x)/(b-a)}), \quad x \in [a, b],$$
$$\mu(x) = 1/2 + ctanh^{-1}(d(x-b)),$$

where c and d are parameters.

(g) Dimitru and Luban's function:

$$\mu(x) = 1/(1 + x/a),$$

where a is a parameter.

(3) Membership functions based on more theoretical demand.
(a) Civanlar and Trussel's function:

$$\mu(x) = \begin{cases} ap(x), & \text{if } ap(x) \leq 1 \\ 0, & \text{otherwise,} \end{cases}$$

where $a \in [0,1]$ is a parameter and $p(x)$ is the probability density function.
(b) Svarovski's function:

$$\mu(x) = \begin{cases} 0, & \text{if } x < a \\ K(x-a)^2, & \text{if } a \leq x \leq b \\ K_2 x^2 + K_1 x + K_0, & \text{if } b < x \leq c \\ 1, & \text{if } x > c, \end{cases}$$

where K, k_0, k_1 and K_2 are parameters.
(4) Membership functions as a model for human concepts.
(a) Hersh and Caramazze's function (which was experimentally formed in order to determine the nature of context effects upon the interpretation of a set of nature language terms, such as short, very short, sort of short, etc.):

$$\mu(x) = 1/2 + d(r/10),$$

where $d(x) = 1$ for "yes" responses and $d(x) = -1$ for "no" responses and r is a confidence value.
(b) Zimmermann and Zysno's function:

$$\mu(x) = 1/2 + (1/d)[1/(1 + e^{-a(x-b)}) - c].$$

(c) Dombi's function:

$$\mu(x) = (1-s)x^2 / [(1-s)x^2 + s(1-x)^2\},$$

where s (the characteristic value of the shape) is the intersection value of $y = \mu(x)$ and $y = x$.

Membership functions can be distinguished into two classes of preference-based membership functions and possibility distributions. The preference-based membership function is constructed by eliciting the preference information from the decision makers. On the other hand, the possibility distribution, which is an analogous of probability distribution, is constructed by considering the possible occurrence of the events.

Approaches to generate membership functions

There are two approaches to generate membership functions: axiomatic and semantic approaches[14]. The axiomatic approach, which is similar to the approaches used in the utility theory[45], is centered ont the mathematical consideration. On the other hand, the sematic approach is concentrated on the practical interpretation of the terms but not the mathematical structure which is emphasized in the axiomatic

concept. The semantic approach actually follows the perceptions of pragmatism in
its insistence that all conclusions should be firmly based on the practical meaning of
the concepts involved. The practical meaning and interpretation of membership and
interpretation are considered more important that mathematical terms.

(1) Distance approach

When eliciting membership functions, the determination of the lowest necessary
scale level of membership for a specific application is very important. Generally, the
required scale level should be a low as possible in order to facilitate data acquisition
which usually affords the participation of human beings. Besides, a suitable numerical handling is desirable in order to insure mathematically appropriate operations.
Among the five classical scale levels (nominal, ordinal, interval, ratio, and absolute
scale), Zimmermann and Zysno proposed that the interval scale level seems to be
most adequate, since the intended mathematical operations require at least interval
scale quality.

The concept of using a distance $d(x)$ from a reference as an evaluation criterion
has been popular in many fields, such as multiple attribute decision-making, multiple objective decision-making, goal decision-making, et al. Zimmermann and Zysno
used this concept to derive membership functions. The rationality of this concept is
that if the object has all the ideal features, the distance should be zero. By contrast,
if no similarity between the object and the ideal exists, the distance then should be
∞. If this evaluation concept is formally represented by a fuzzy set A in X, then a
certain degree of membership $\mu_A(x)$ will be assigned to each element x. Membership can then be defined as a function of the distance between a given object x and a
standard (ideal). Thus Zimmermann and Zysno proposed the following relation for
μ and $d(x)$:

$$\mu = \frac{1}{1+d(x)},$$ (1.50)

where $d(x) = 0 \Rightarrow \mu = 1$, and $d(x) = \infty \Rightarrow \mu = 0$. Obviously, Equation (1.50) is just
a transformation rule (or a normalization process) which maps real number R into
the interval [0,1].

However, experience shows that ideals are very rarely ever fully realized and distance function is quite context dependent. The context-dependent parameters a and
b are then created to represent the evaluation unit and the reference/stansard, respectively. For instance, a may be the unit of length such as feet, meter, yards, etc.,
and b may be a fast and rough pre-evaluation such as "rather positive," "rather negative," etc. Since the relationship between physical units and perceptions if generally
exponential, Zimmermann and Zysno proposed the following distance function as:

$$d(x) = \frac{1}{e^{a(x-b)}},$$ (1.51)

and then:

$$\mu = \frac{1}{1+e^{a(x-b)}},$$ (1.52)

where a, b can be considered as sematic parameters from a linguistic point of view.

Since concepts or categories, which are formally represented by sets, are normally linguistically described, the membership function is the formed representation of meaning. The vagueness of the concept os operated by the slope a and the identification threshold by b. For managerial terms such as "appropriate dividend" or "good utilization of capacities", the parameter a models the slope of the membership function in the tolerance interval and b represents the point at which the tendency of the subject's attitude changes from rather positive into rather negative.

Equation (1.52), however, is still too general to fit subjective models of different persons. Frequently, only a certain part of the logistic function is needed to represent a perceived situation. In order to allow for such a calibration, it is assumed that only a certain interval of the physical scale of mapped into the open interval (0,1). And, the membership grades of the lower and upper bounds are assigned to be 0 and 1, respectively.

Since an interval scale is requested, the interval of the degrees of membership may be transformed linearly. On this scale level the ratios of two distances are in variant. Let μ^U and μ^L be the upper and lower bounds of the normalized membership scale, respectively, and μ_i is a degree of membership between these bounds, $\mu^L < \mu < \mu^U$, and let $\mu^{L'}$ and $\mu^{U'}$ be the corresponding values on the transformed scale. Then we have:

$$\frac{\mu_i - \mu^L}{\mu^U - \mu^L} = \frac{\mu_i' - \mu^{L'}}{\mu^{U'} - \mu^{L'}}.$$

For the normalized membership function, one may have $\mu^L = 0$ and $\mu^U = 1$. Hence,

$$\mu_i' = \mu_i(\mu^{U'} - \mu^{L'}) + \mu^{L'}. \tag{1.53}$$

Generally, it is preferable to define the range of validity by specifying the interval d with the center c (see Figure) as follows:

$$\mu^{U'} - \mu^{L'} = d \text{ and } (\mu^{U'} + \mu^{L'})/2 = c. \tag{1.54}$$

The Equation (1.53) will become:

$$\begin{aligned}
\mu_i' &= \mu_i(d) + \mu^{L'} \\
&= d\mu_i - d/2 + d/2 + \mu^{L'} \\
&= d\mu_i - d/2 + [(\mu^{U'} - \mu^{L'})/2 + \mu^{L'}] \\
&= d(\mu_i - 1/2) + c,
\end{aligned} \tag{1.55}$$

which can be specified by two parameters, a and b, if μ_i is replaced by μ_i'.

Solving Equation (1.55) for μ_i will lead to the following complete membership model:

$$\mu_i = \left| (1/d)\left(\frac{1}{1 + e^{-a(x-b)}} - c\right) + \frac{1}{2} \right|, \tag{1.56}$$

where $\mu_i \in [0, 1]$, $\mu_i(x) = 1$ for $x > x^U$, and $\mu_i(x) = 0$ for $x < x^L$.

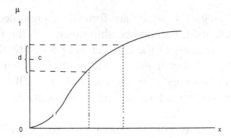

Fig. 1.20 Calibration of the interval for measurement

The determination of the parameters from an empirical data base does not pose any difficulties in the general model of Equation (1.52). That is:

$$\mu = 1/[1 + e^{a(x-b)}] \Rightarrow ln[\mu/(1-\mu)] = a(x-b).$$

Now, suppose $y = ln[\mu/(1-\mu)]$. A linear relationship between x and y is obvious. The straight line of the model is then defined by the least squares of deviations.

The estimation of the parameter c and d in the extended model still poses some problems. There is no direct way for a numerically optimal estimation. Thus Zimmermann and Zyson supposed the following interactive procedure. First, they assumed that a set of stimuli which is equally spread over the physical continuum was chosen such that the distance between any two of the neighboring stimuli is constant:

$$x_{i+1} - x_i = s,$$

where s is a constant distance. This condition serves as a criterion for precision. If c and d are correctly estimated, then those scale values x_i' are reproducible and are invariant with respect to x_i with the exception of the additive and multiplicative constant. It is obvious that Equation (1.56) can be written as:

$$d(\mu_i - 1/2) = 1/[1 + e^{-a(x-b)}]$$

or

$$ln\left\{ \frac{d(\mu_i - 1/2) + c}{1 - [d(\mu_i - 1/2) + c]} \right\} = x_i' = a(x_i - b). \tag{1.57}$$

Let s' be the distance between the pair x_{i+1}' and x_i', and M' be their mean value. If the estimated values d^e and c^e are equal to their true values, then the estimated distance $s^{e'}$ and the mean $M^{e'}$ are equal to their respective true values and vice versa:

$$(d^e = d) \text{ and } (c^e = c) \iff (s^{e'} = s) \text{ and } (M^{e'} = M).$$

Our aim, therefore, is to research the equivalent of $s^{e'}$ and s as well as $M^{e'} = M$.

The algorithm is as following:

Step 0. Determine the initial appropriate values c_1 and d_1 by the following formula:

$$c_1 = (\sum_i \mu_i)/n, \tag{1.58}$$

$$d_1 = \min(1 - 1/k, 2(1-c), 2c), \tag{1.59}$$

where n is the member of stimuli and k is the number of different degree of membership. If only the values 0 and 1 occur, $d_1 = 1/2$. Only the linear interval in the middle of the logistic function is used. With increasing k, d converges to 1; the entire range of the function is then used. In any case d must not exceed the minor part of $2c$ or $2(1-c)$.

Step 1. Determine the x_i' which corresponds to the empirically determined μ_i:

$$M^{e'} = (\sum_i x_i')/n \tag{1.60}$$

$$\begin{aligned} S^{e'} &= [\sum_i (x_{i+1} x_i')]/(n-1) \\ &= (x_n' - x_1')/(n-1). \end{aligned} \tag{1.61}$$

Step 2. Calculate the absolute difference between two estimates. That is:

$$|M^{e''} - M^{e'}| \leq \delta_M, \tag{1.62}$$

$$|S^{e''} - S^{e'}| \leq \delta_S, \tag{1.63}$$

where δ_M and δ_S are predetermined tolerances. If Equations (1.64) and (1.65) are satisfied, the last estimate is accepted as sufficiently exact and then stop. Otherwise, go to the next stop.

Step 3. Determine the interval of the base variable x_i' which corresponds to the $(0,1)$ interval of the membership value. That is to compute:

$$x^{U'} = M^{e'} + (n/2)S^{e'}, \tag{1.64}$$

$$x^{L'} = M^{e'} - (n/2)S^{e'}, \tag{1.65}$$

which is the upper and lower bounds, respectively.

Now, by Equation (1.57), the corresponding $x^{U'}$ and $x^{L'}$ can be computed, and then by Equation (1.54), new parameters c^e and d^e are estimated. Then go to Step 1.

Example 1.32. From the empirical evidence that 64 subjects (16 for each set) from 21 t 25 years of age individually rated 52 different statements of age concerning one of the four fuzzy sets: "very young man," "young man," "old man" and "very old man," Zimmermann and Zyson obtained monotonic membership functions as shown in Figure. 1.21, and unimodal membership functions as shown in Figure. 1.22.

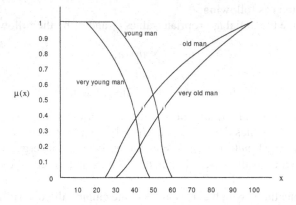

Fig. 1.21 Monotonic membership functions of 'very young man," "young man," "old man" and "very old man"

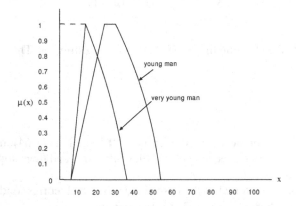

Fig. 1.22 Unimodal membership functions of "very young man" and "young man'

The monotonic membership functions of "old man" and "very old man" (see Figure. 1.21) are rather similar. They differ only with respect to their inflection points, indicating a difference of about five years between "old man" and "very old man". The same holds for the monotonic membership functions of "very young man" and "young man"; their inflection points differ by nearly 15 years. It is interesting to note that the modifier "very" has a greater effect on "young" than "old", but in both cases it can be formally represented by a constant.

Finally, the meaning of "young" is less vague than that of "old" if the slope is an indicator for vagueness as shown in Figure. 1.21. On the other hand, the variability of membership functions may be regarded as an indicator of ambiguity. Thus, though being less vague, "young" seems to be more ambiguous.

(2) True-valued approach

Smets and Magrez [8] provided a definition of what is meant by a proposition with a truth value of 0.35 or any other intermediate value between True (or 1) and False (or 0). A canonical scale for the truth values is defined, and sets of fuzzy propositions are constructed for which the truth value had a unique numerical value on the canonical scale. Such a set of propositions with well-defined intermediate truth values is necessary to give meaning to the assertion "the truth P is 0.35".

Fuzzy logic has two characteristics: the truth domain is the whole [0,1] interval, and the truth value can be a fuzzy subset of [0,1]. Smets and Magrez considered the first characteristic, that is to reduce fuzzy logic to its multi-valued logic components.

According to Zadeh[9], fuzzy sets are those for which one can define, for each element, a grade of membership in [0,1]. Smets and Magrez then postulated the relation between degree of truth and grade of membership so that the degree of membership $\mu_A(x)$ of an element x to a fuzzy set A is numerically equal to the degree of truth $v(x$ is $A')$ that the fuzzy predicate A' describing the fuzzy set A applies to the element x:

$$\mu_A(x) = a \Leftrightarrow v(x \text{ is } A) = a. \tag{1.66}$$

For instance, if A is the fuzzy set of "tall man", the degree of membership $\mu_A(\text{John})$ of John to the set A is numerically equal to the degree of truth of the proposition "John is tall". Therefore, the presentation could have been based on both approaches when fuzzy logic is used. Whatever is deduced of truth can be applied to degree of membership.

The semantical interpretation of "$A \rightarrow B$ is true" is that the consequent B is at least as true as the antecedent A. The degree of truth of $A \Rightarrow B$ quantifies the degree by which B is as least as true A. It will be Truth whenever B is truer that A. Otherwise, it will be False whenever A is False and B is Truth. When A is somehow truer than B, the degree by which B is at least as true as A might be intermediate between True and False.

Now let $v(A)$ be the truth value of proposition A with $A \in \Omega$, where Ω is a Boolean algebra of propositions with T its tautology and F its contradiction. Then $v : \Omega \rightarrow \sigma$ is a mapping from Ω to a truth domain σ, where σ is a bounded set with its greatest element Truth $= v(T)$, its least element False $= v(F)$ and the order relation "at least as true as". Thus Smets and Magrez defined a strictly increasing transformation w from σ to [0,1] such that $w \cdot v : \Omega \rightarrow [0,1]$ with $w \cdot v(T) = 1, w \cdot v(F) = 0$ and 'A us as true as B' is equivalent to "$w \cdot v(A) \geq w \cdot v(B)$". $v = w \cdot v$, then v will measure the degree of truth of propositions in Ω on the interval [0,1]. For simplicity, $v(A)$ is the truth value of A (in fact the truth value is defined on σ and $v(A)$ is defined on [0,1]), because they are uniquely related.

In order to construct a logic with multi-value truth domain for generating the degree of truth, the following axioms for the implication connective $\rightarrow$ should be first be considered:

(1) Truth-Functionality: $v(A \to B)$ depends only on $v(A)$ and $v(B)$.

(2) Contrapositive Symmetry: $v(A \to B) = v(\neg B \to \neg A)$, where the Trillas's strong negation is defined: there exists a strictly decreasing continuous function $n: [0,1] \to [0,1]$ such that $(\neg A) = n(v(A))$, $n(0) = 1$ and $n(n(a)) = a$.

(3) Exchange Principle: $v(A \to (B \to C)) = v(B \to (A \to C))$.

(4) Monotony: $v(A \to B) \geq v(C \to D)$ if $v(A) \leq v(C)$ and $v(B) \leq v(D)$.

(5) Boundary Condition: $v(A \to B) = 1$ iff $v(A) \leq v(B)$.

(6) Neutrality Principle: $v(T \to A) = v(A)$.

(7) Continuity: $v(A \to B)$ is a continuous function of $v(A)$.

With these axioms. Smets and Magrez have proved that the implication and the negation operators are necessary such that:

$$v(A \to B) = f^{-1}(\min\{f(1) - f(a) + f(b), f(1)\}), \qquad (1.67)$$

$$v(\neg A) = f^{-1}(f(1) - f(a)), \qquad (1.68)$$

where $a = v(A), b = v(B)$, and the generator f is any bounded, continuous, monotonically increasing function from $[0,1]$ to $[0,\infty)$ with $f(0) = 0$ and $f(1) < \infty$.

The f generator is defined up to any strictly monotonic transformation. As the truth scale $v(A)$ is also defined up to any strictly monotonic transformation, we can then define the canonical scale for the truth value $v(A)$ of a proposition by selecting the f generator such that $f(v(T) = f(1) = 1)$ and $f(v(A)) = f(a) = a$. In that case, we obtain the Lukasiewicz operator for the material implication connective:

$$v(A \to B) = \min\{1 - v(A) + v(B), 1\}, \qquad (1.69)$$

$$v(\neg A) = 1 - v(A). \qquad (1.70)$$

This conclusion is quite natural as f and v can always be adapted in order to obtain such a scale, and it is obviously the simplest we can construct.

However, this canonical scale dose not explain what is meant by a truth value of 0.35. It must mean something more than just that it is between 0.34 and 0.36. Thus we need a set of propositions for which the truth value is uniquely defined. These propositions could be used later as a reference scale to measure the truth of other propositions.

According to Gaines, Smets and Magrez constructed a reference scale which defines the meaning of any truth value in [0,1] as follows:

(a) Let $A = $ "Yao is tall" and $\neg A = $ "Yao is not tall." The truth of A and of $\neg A$ depends on the height h of Yao. If $h = 110cm$, A is false and $\neg A$ is true. If $h = 190cm$, A is true and $\neg A$ is false (see Table 1.27). Thus, when h increase from 110 to 190, $a = v(A)$ increases continuously from $0 = v(F)$ to $1 = v(T)$ (where T and F are the tautology and contradiction) and $na = v(\neg A)$ decreases continuously from 1 to 0.

There exists a value h' such that $a = na$. In such a case, $\neg A \to A$ is true, therefore $(\neg A \to A_{=} 1$ becomes $f^{-1}(f(1) - f(na) + f(a)) = 1$. As $f(ns) = f^{-1}(f(1) - f(a)) = f(1) - f(a)$, we have $f^{-1}(2f(a)) = 1$, thus $2f(a) = f(1)$.

Table 1.27 Heights and true values of $A = $ "Yao is tall" and $B = $ "Yi is tall"

height	110 cm	k'	h'	190 cm
$v(A)$	0		0.5	1
$v(\neg A)$	1		0.5	0
$v(B)$	0	0.25	0.5	1
$v(\neg B)$	1	0.75	0.5	0
$v(\neg B \to B)$	0		0.5	

As $f(1) < \infty$, we can use an f scale such that $f(1 = 1)$, in which case $f(a) = f(na) = 0.5$. It is noted that the result would have been directly obtained by considering the strong negation and the canonical scale as $a = na$ is identical to $a = 1 - a$.

(b) Let $B = $ "Yi is tall" and $\neg B = $ "Yi is not tall". The truth of B and $\neg B$ depend on the height k of Yi. Let us further postulate that $k = h'$, thus $b = v(B) \leq 0.5$.

Consider the proposition $\neg B \to B$, i.e., 'B is at least as true as $\neg B$'. Its truth value $c = v(\neg B \to B)$ depends on k. If $k = h'$, $c - 1$ as $v(B) = v(\neg B)$. If $k = 110cm$, B is false and $\neg B$ is true, therefore $c = 0$. By increasing k from 110 cm to h', c is increasing from 0 to 1. Thus, there exists a k' such that $v(\neg B \to B) = v(A) = 0.5$ implies $f(1) - f(nb) = f(b) = 2f(b) = f(a) = 0.5 \Rightarrow f(b) = 0.25$ and $f(nb) = 0.75$.

(c) Let $P_0 = A$, $P_1 = B$, $h_0 = h'$ and $h_1 = k'$. The procedure is iterated with the propositions $P_i \equiv $ "X_i is tall" such that $v(\neg P_{i-1} \to P_i) = v(P_{i-1}), i = 1, 2, \cdots$, and $v(P_i) < v(P_{i-1})$. The values of the heights h_i of X_i, all i, are derived as above. For such h_i and with $p_i = v(P_i)$, we obtain $f(p_i) = 2^{-i-1}$ and $f(np_i) = 1 - 2^{-i-1}$.

(d) Other values of f can be obtained from expression based on the truth of $\neg P_i \to P_j$, as it corresponds to $f(p_i) + f(p_j) = 2^{-i-1} + 2^{-j-1}$. Further values are obtained from expressions like $((\neg P_i \to P_j) \to P_k) \to P_s$ whose true value is $f(p_i) + f(p_j) + f(p_k) + f(p_s)$, etc.

Appropriate sequences of implications based on propositions p_i can be constructed on order to obtain any f value. For instance, let us construct the sequence to obtain $f = 0.65625$. As $0.65625 = 0.5 + 0.125 + 0.03125$, $0.65625 = f(p_0) + f(p_2) + f(p_4)$, it corresponds to the true value of the implication $((\neg P_0 \to P_2) \to P_4)$ with $P_i = $ "X_i is tall" and the height of X_i is h_i; thus, 0.65625 is the degree of truth of the proposition "P_4 is at least as true as not 'P_2 is at least as true as $\neg P_0$'" or "P_4 is at least as true as 'P_2' is less true than P_0".

Finally, it is noted that f function is defined up to any strictly monotonic transformation such that $f(0) = 0$ and $f(1) = 1$. Thus we can state without loss of generality that $f(x) = x$, for computational efficiency. By use of the previous canonical scale, $v(P_i) = 2^{-i-1}$, and all other truth values can be equated to some sequence of implication based on propositions P_i. We have derived a reference scale for the truth value of any proposition Q by use of the relation "Q is as true as" where P is a sequence of implication based on propositions P_j. Thus P and Q share the same numerical truth value.

(3) Payoff function

Giles[14] pointed out that the concepts of a set (fuzzy or not) are closely related to that of a property: any set determines a property which belongs to the set, and any property determines a set which contains all objects with the property. For example, to the property "tall", there corresponds the set of all tall people (objects). Then any statement about sets can be translated into a corresponding statement about properties, and vice versa. The concepts of the set and property thus appear to be equivalent. Nevertheless, the concept of a property is actually less abstract and closer to common usage that that of a set. Thus the analogue of degree of membership in a fuzzy set can be equivalent to that of the "degree of possession" of the corresponding property. For instance, the degree of membership of Yao in the set of all tall people may be identified with the degree of which Yao has the property, "tall".

There is a third notion related to these two concepts. This is "degree of truth"-for the previous example, the degree to which Yao is tall with the "degree of truth" of the statement, "Yao is tall". Every assignment of a degree of membership in a set of of a degree of truth to a corresponding statement. Indeed, we can hardly contemplate intermediate (other than 0 and 1) grades of membership in a set unless we are also prepared to consider intermediate (other than true and false) degrees of truth for a statement. Conversely, should we succeed in the task of attaching a well-defined meaning to intermediate grades of membership and intermediate "degree of possession" of a property.

Once we have attached a clear meaning to statement with other that classical truth values, we can deduce how one should reason with such statements: the appropriate logic is determined as soon as a clear understanding of the meaning of the statement with which it has to operate has been reached. Thus the problem of the interpretation of grades of membership cannot be considered independently of the problem of fuzzy reasoning itself: it is part and parcel of the latter; a solution of one is a solution of both.

Giles further pointed out that the "assertions" concept, instead of the statements themselves, is fundamental, and the meaning of an assertion can be identified by its payoff function in terms of the decision theory. For example, consider the assertion "Yao is tall". The meaning of the fuzzy statement "Yao is tall" is that information specific to this fuzzy statement which in conjunction with one's state of belief regarding the relative possibility of different world states (here the relevant is the height of Yao in these various possible states) will allow him to decide whether (and how willingly) to assert "Yao is tall".

Now, let us use this example, "Yao is tall", to discuss how to construct the degree of membership from the "assertions" via the payoff concept.

Usually, one who asserts "Yao is tall" will receive approval from his peers to an extent dependent primarily on the height of John. In other words, the payoff of the assertion in any world state is approximately a function $f(h)$ of the height h of Yao in that state. Insofar as this is the case, we can represent the payoff function of the assertion by the graph of the function f. Clearly, the utility will crease with height from some negative initial value to positive values. For example, the graph might be as shown in Figure. 1.23. In this figure the payoff value $f(h)$ may be described as the degree of willingness of an agent to assert "Yao is tall". In particular, a positive (negative) value for the payoff $f(h)$ indicates that the agent will (won't) choose to assert "Yao is tall" when he knows Yao's height h. So one can tell where the graph crosses the axis just by asking for the least height at which Yao would be described as tall.

We can then determine other points on the graph of $f(h)$ by offering bribes. For instance, the graph suggests that the agent would be willing to assert "Yao is tall" even if $h = 170$, provided he was offered as a reward a prize worth 5 units of payoff. Actually, this argument is not quiet correct, since then two events occur together their payoffs do not necessarily increase. Fortunately, one can flip a fair coin. If it shoes a head, he will get the reward. Otherwise, he will be obligated to assert "Yao is tall". Therefore, if he agrees to this proposal, provided that he was sure Yao's height was at least 170 cm, then $f(170) \rightarrow 5$ is obtained. By using a lottery instead of the coin, one can similarly determine all other negative ordinates on the graph. On the other hand, the positive ordinates are also found by using a penalty instead of a reward.

It is noted that the previous discussion of the payoff function for the assertion "Yao is tall" has been simplified in the following two respects:

(1) It is only approximately true that the payoff function of height, according to the following two reasons:

(a) In a normal society he term "tall" does not solely represent geometrical height. For instance, a child is more likely to be considered tall than an adult of the same height.

(b) Even the assertion that the payoff depends only on the nature of John must be qualified.

(2) The meaning of an assertion, as represented by its payoff function, is determined by the society in which the assertion used, and does not depend on the particular person who makes the assertion.

As a matter of fact, an assertion can be brought into standard form by a scaling and a shift. Here let us call the payoff function of the resulting standardized assertion "the truth function of the asserted fuzzy statement". The function is then the truth function of the sentence. Truth functions of fuzzy statements this coincide with payoff functions of payoff functions of standardized assertions except that the extreme values 0 and 1 are necessarily attained in the latter case. Therefore, the graph of the truth function of the asserted fuzzy statement may be obtained from the graph of the payoff function of the assertion simply by changing the calibration on the y-axis. From Figure. 1.23, the degree of the truth function for the fuzzy statement "Yao is tall" can be obtained by reading the right hand scale. The degree of membership of Yao in the fuzzy set of tall men is then identified with the truth of the fuzzy statement "Yao is tall".

For the payoff function, the following two significant points should be noted:

(a) It is clear that for any practical assertion, one value must be positive and the other negative. Otherwise, the assertion would either always or never be made, so that it would be useless for conveying the beliefs of a speaker.

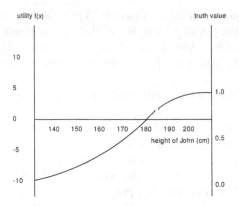

Fig. 1.23 Payoff function of the assertion "Yao is tall"

(b) A scaling of the payoff function (multiplying it by a positive scale factor) corresponds to a change in emphasize of the assertion. For example, the payoff function for "Yao is tall" may be taken to differ only by a scale factor (> 1) from that for the less emphatic "Yao is tall". Turing this observation around, one should regard any assertions a and a', whose payoff functions differ only by a positive scale factor, as assertion of the same fuzzy sentence except more (or less) emphasized.

(4) Other examples

Besides the above three presented approaches, there are several other considerations and approaches. Rapoport, Wallsten and Cox[18] dealt with the synthesis of subjective data for deriving membership functions. They provided direct (magnitude estimation) and indirect (grade pair-comparison) concepts to establish membership for probability phrase such as probable, rather likely, very likely, very unlikely, and so forth. Eliciting membership function is considered as eliciting utility function in the decision theory.

Bandemer[17] developed a fuzzy counterpart to regression analysis methods accounting for fuzzy observation. The fuzzy (functional) relations were then derived either directly. or via families of parametrized functions.

Pedryca[16] solved an identification problem in terms of fuzzy relational equations from input/output fuzzy sets, Based on the clustering technique, Pedrycz roposed a general methodological scheme which includes the following three steps: (i) structure determination, (ii) parameter determination, and (iii) model valuation.

Bezdek and Hathaway[15] proposed the relational hard c-means (HCM) algorithm for the classification of objects when information about pairwise relationship between these objects is available. Their approach can be used to derive membership functions.

1.6   Possibility Distribution

What is a possibility distribution? It is convenient to answer this question in terms of another concept, namely, that of a fuzzy restrictio [13, 19], to which the concept of a possibility distribution bears a close relation.

Let X be a variable which takes values in a universe of discourse U, with the generic element of U denoted by u and

$$X = u \tag{1.71}$$

signifying that X is assigned the value $u, u \in U$.

Let F be a fuzzy subset of U which is characterized by a membership function μ_F. Then F is a fuzzy restriction on X (or associated with X) if F acts as an elastic constraint on the values that may be assigned to X-in the sense that the assignment of a value μ to X has the form

$$X = u : \mu_F(u), \tag{1.72}$$

where $\mu_F(u)$ is interpreted as the degree to which the constraint represented by F is satisfied when μ is assigned to X. Equivalently, (1.72) implies that $1 - \mu_F(u)$ is the degree to which the constraint in question must be stretched in order to allow the assignment of u to X.

Let $R(X)$ denote a fuzzy restriction associated with X. Then, to express that F plays the role of a fuzzy restriction in relation to X, we write

$$R(X) = F. \tag{1.73}$$

An equation of this form is called a relational assignment equation because it represents the assignment of a fuzzy set (or a fuzzy relation) to the restriction associated with X.

To illustrate the concept of a fuzzy restriction, consider a proposition of the form $p\triangle = X$, where X is the name of an object, a variable or a proposition, and F is the name of a fuzzy subset of U, as in "Miller is very intelligent," "X is a small number," "Jeny is blonde is quite true," etc. As shown in [13] and [24], the translation of such a proposition may be expressed as

$$R(A(X)) = F, \tag{1.74}$$

where $A(X)$ is an implied attribute of X which takes values in U, and (1.74) signifies that the proposition $p\triangle = X$ is F has the effect of assigning F to the fuzzy restriction on the values of $A(X)$.

As a simple example of (1.74), let p be the proposition "Brown is young," in which young is a fuzzy subset of $U = [0, 100]$ characterized by the membership function

$$\mu_{\text{young}}(u) = 1 - S(u; 20, 30, 40), \tag{1.75}$$

where u is the numerical age and the S-function is defined by [13].

$$
\begin{aligned}
S(u; \alpha, \beta, \gamma) &= 0 && \text{for } u \leq \alpha \\
&= 2(\tfrac{u-\alpha}{\gamma-\alpha})^2 && \text{for } \alpha \leq u \leq \beta \\
&= 1 - 2(\tfrac{u-\gamma}{\gamma-\alpha})^2 && \text{for } \beta \leq u \leq \gamma \\
&= 1 && \text{for } u \geq \gamma,
\end{aligned}
\tag{1.76}
$$

in which the parameter $\beta\triangle = (\alpha + \gamma)/2$ is the crossover point, that is, $S(\beta; \alpha, \beta, \gamma) = 0.5$. In this case, the implied attribute $A(X)$ is Age(Brown) and the translation of "Li is young" assumes the form:

$$
\text{Brown is young} \rightarrow R(\text{Age(Brown)}) = \text{young}.
\tag{1.77}
$$

To relate the concept of a fuzzy restriction to that of a possibility distribution, we interpret the right-hand member of (1.76) in the following manner.

Consider a numerical age, say $u = 28$, whose grade of membership in the fuzzy set young (as defined by (1.75)) is approximately 0.7. First, we interpret 0.7 as the degree of compatibility of 28 with the concept labeled young. Then, we postulate that the proposition "Li is young" converts the meaning of 0.7 from the degree of compatibility of 28 with young to the degree of possibility that John is 28 given the proposition "Li is young." In short, the compatibility of a value of μ with young becomes converted into the possibility of that value of u given "Li is young."

Stated in more general terms, the concept of a possibility distribution may be defined as follows. (For simplicity, we assume that $A(X) = X$.)

Definition 1.7. (Zadeh [13]) Let F be a fuzzy subset of a universe of discourse U which is characterized by its membership function μ_F, with the grade of membership, $\mu_F(u)$, interpreted as the compatibility of u with the concept labeled F.

Let X be a variable taking values in U, and let F act as a fuzzy restriction, $R(X)$, associated with X. Then the proposition "X is F," which translates into

$$
R(X) = F
\tag{1.78}
$$

associates a possibility distribution, Π_X, with X which is postulated to be equal to $R(X)$, i.e.,

$$
\Pi_X = R(X).
\tag{1.79}
$$

Correspondingly, the possibility distribution function associated with X (or the possibility distribution function of Π_X,) is denoted by π_X and is defined to be numerically equal to the membership function of F, i.e.,

$$
\pi_X \triangle = \mu_F.
\tag{1.80}
$$

Thus, $\pi_X(u)$, the possibility that $X = u$, is postulated to be equal to $\mu_F(u)$.

In view of (1.79), the relational assignment equation (1.78) may be expressed equivalently in the form

$$\Pi_X = F \tag{1.81}$$

placing in evidence that the proposition $p\triangle = X$ is F has the effect of associating X with a possibility distribution Π_X, which, by (1.79), is equal to F. When expressed in the form of (1.81), a relational assignment equation will be referred to as a possibility assignment equation, with the understanding that Π_X is induced by p.

As a simple illustration, let U be the universe of positive integers and let F be the fuzzy set of small integers defined by ($+\triangle =$ union)

$$\text{small integer} = 1/1 + 1/2 + 0.8/3 + 0.6/4 + 0.4/5 + 0.2/6.$$

Then, the proposition "X is a small integer" associates with X the possibility distribution

$$\Pi_X = 1/1 + 1/2 + 0.8/3 + 0.6/4 + 0.4/5 + 0.2/6, \tag{1.82}$$

in which a term such as 0.8/3 signifies that the possibility that X is 3, given that X is a small integer, is 0.8.

There are several important points relating to the above definition which are in need of comment.

First, (1.79) implies that the possibility distribution Π_X, may be regarded as an interpretation of the concept of a fuzzy restriction and, consequently, that the mathematical apparatus of the theory of fuzzy sets-and, especially, the calculus of fuzzy restrictions -provides a basis for the manipulation of possibility distributions by the rules of this calculus.

Second, the definition implies the assumption that our intuitive perception of the ways in which possibilities combine is in accord with the rules of combination of fuzzy restrictions. Although the validity of this assumption cannot be proved at this juncture, it appears that there is a fairly close agreement between such basic operations as the union and intersection of fuzzy sets, on the one hand, and the possibility distributions associated with the disjunctions and conjunctions of propositions of the form "X is F." However, since our intuition concerning the behavior of possibilities is not very reliable, a great deal of empirical work would have to be done to provide us with a better understanding of the ways in which possibility distributions are manipulated by humans. Such an understanding would be enhanced by the development of an axiomatic approach to the definition of subjective possibilities-an approach which might be in the spirit of the axiomatic approaches to the definition of subjective probabilities [38, 39].

Third, the definition of $\pi_X(u)$ implies that the degree of possibility may be any number in the interval [0, 1] rather than just 0 or 1. In this connection, it should be noted that the existence of intermediate degrees of possibility is implicit in such commonly encountered propositions as "There is a slight possibility that Wilson is very rich," "It is quite possible that Davis will be promoted," etc.

Difference between probability and possibility

It could be argued, of course, that a characterization of an intermediate degree of
possibility by a label such as "slight possibility" is commonly meant to be interpreted as "slight probability." Unquestionably, this is frequently the case in everyday
discourse. Nevertheless, there is a fundamental difference between probability and
possibility which, once better understood, will lead to a more careful differentiation
between the characterizations of degrees of possibility vs. degrees of probability-
especially in legal discourse, medical diagnosis, synthetic languages and, more generally, those applications in which a high degree of precision of meaning is an
important desideratum.

To illustrate the difference between probability and possibility by a simple example, consider the statement "Gan ate X eggs for breakfast," with X taking values
in $U = \{1, 2, 3, 4, \cdots\}$. We may associate a possibility distribution with X by interpreting $\pi_X(u)$ as the degree of ease with which Gan can eat u eggs. We may also
associate a probability distribution with X by interpreting $p_X(u)$ as the probability
of Gan eating u eggs for breakfast. Assuming that we employ some explicit or implicit criterion for assessing the degree of ease with which Gan can eat u eggs for
breakfast, the values of $\pi_X(u)$ and $p_X(u)$ might be as shown in Table 1.28.

Table 1.28 The possibility and probability distributions associated with X

u	1	2	3	4	5	6	7	8
$\pi_X(u)$	1	1	1	1	0.8	0.6	0.4	0.2
$p_X(u)$	0.1	0.8	0.1	0	0	0	0	0

We observe that, whereas the possibility that Gan may eat 3 eggs for breakfast
is 1, the probability that he may do so might be quite small, e.g., 0.1. Thus, a high
degree of possibility does not imply a high degree of probability, nor does a low
degree of probability imply a low degree of possibility. However, if an event is
impossible, it is bound to be improbable. This heuristic connection between possibilities and probabilities may be stated in the form of what might be called the
possibility/probability consistency principle, namely:

If a variable X can take the values $u_1, \cdots, u_n$, with respective possibilities $\Pi =
(\pi_1, \cdots, \pi_n)$ and probabilities $P = (p_1, \cdots, p_n))$, then the degree of consistency of
the probability distribution P with the possibility distribution Π is expressed by
$(+\triangle = \text{arithmetic sum})$

$$\gamma = \pi_1 p_1 + \cdots + \pi_2 p_2. \tag{1.83}$$

It should be understood, of course, that the possibility/probability consistency
principle is not a precise law or a relationship that is intrinsic in the concepts of
possibility and probability. Rather it is an approximate formalization of the heuristic observation that a lessening of the possibility of an event tends to lessen its
probability-but not vice-versa. In this sense, the principle is of use in situations in

which what is known about a variable X is its possibility-rather than its probability distribution. In such cases-which occur far more frequently than those in which the reverse is true-the possibility/probability consistency principle provides a basis for the computation of the possibility distribution of the probability distribution of X. Such computations play a particularly important role in decision-making under uncertainty and in the theories of evidence and belief[40, 43].

In the example discussed above, the possibility of X assuming a value u is interpreted as the degree of ease with which u may be assigned to X, e.g., the degree of ease with which Gan may eat u eggs for breakfast. It should be understood, however, that this "degree of ease" may or may not have physical reality. Thus, the proposition "Li is young" induces a possibility distribution whose possibility distribution function is expressed by (1.75). In this case, the possibility that the variable Age(Li) may take the value 28 is 0.7, with 0.7 representing the degree of ease with which 28 may be assigned to Age(Li) given the elasticity of the fuzzy restriction labeled young. Thus, in this case "the degree of ease" has a figurative rather than physical significance.

Chapter 2
Fuzzy Multiple Objective Decision Making

The fuzzy programming approach [194, 410, 412] is useful and efficient for treating a programming problem under uncertainty. While a classical and stochastic programming approach may cost a lot to obtain the exact coefficient value or distribution, fuzzy programming approach does not [411]. From this fact, fuzzy programming approach can be very advantageous when the coefficients are not known exactly. Fuzzy programming offers a powerful means of handling optimization problems with fuzzy parameters. Fuzzy programming has been used in different ways in the past.

In this chapter, we introduce basic knowledge about the multi-objective decision making model under fuzzy environments, and we use fuzzy variables to describe these fuzzy coefficients. The first section is about fuzzy variable, we present a definition of a fuzzy variable, some useful fuzzy variables with good properties, three chances to measure the occurrence of fuzzy events: possibility, necessity and credibility, chance distribution of a fuzzy variable, expected value using three measures for fuzzy variables, and four ranking methods of fuzzy variables. We then introduce the general fuzzy multi-objective model and the a group of models which can deal with the fuzzy multi-objective model: fuzzy EVM, CCM and DCM. For each kind of model, an equivalence model based on some special fuzzy variables with good properties and the traditional method of the multi-objective decision making model are given to solve the crisp models. For the general fuzzy decision making model, it is usually hard to give the equivalent form, so three fuzzy simulations-based PSO are presented to handle. In the final section, an application to farm structure optimization problem is presented as illustration.

2.1 Farm Structure Optimization Problem under Fuzzy Environment

Fuzziness uncertainty exists in many. Consider when we ask people the following question: How far is it between two places? How long will it take to finish a job? How much will it take to produce a product? How many products do you demand?

J. Xu and X. Zhou: Fuzzy-Like Multiple Objective Decision Making, STUDFUZZ 263, pp. 57–133.
springerlink.com

How many resources will you need? How many products can you supply? What are the importance of different events? Because people have subjectivity, the answer will usually include these words: "about, more or less, around, between a and b," and so on. Actually these words delegate the fuzziness, and when we have to describe these words, we use fuzzy variables.

For the above multi-objective problems, we should use a fuzzy decision making model to clarify. Assume that x is a decision vector, ξ is a fuzzy vector, $f_i(x, \xi)$ is a return function, and $g_j(x, \xi)$ are constraint functions, $i = 1, 2, \cdots, m$; $j = 1, 2, \cdots, p$. The fuzzy decision making model is as follows:

$$\begin{cases} \max \ [f_1(x, \xi), f_2(x, \xi), \cdots, f_m(x, \xi)] \\ \text{s.t.} \begin{cases} g_r(x, \xi) \leq 0, \ r = 1, 2, \cdots, p \\ x \in X. \end{cases} \end{cases} \qquad (2.1)$$

The model (2.1) is not well-defined because we cannot maximize the fuzzy quantity $f_i(x, \xi)$, $i = 1, 2, \cdots, m$ (just like that we cannot maximize a random quantity), and the constraints $g_r(x, \xi)$, $r = 1, 2, \cdots, p$ do not produce a crisp feasible set.

Unfortunately, the form of fuzzy programming like (2.1) appears frequently in the literature. Fuzzy programming is a class of mathematical models. Different from fashion or building models, everyone should have the same understanding of the same mathematical model. In other words, a mathematical model must have an unambiguous explanation. The form (2.1) does not have mathematical meaning because it has different interpretations.

Let's consider a real-life farm structure optimization problem with imprecise input data, where the coefficients are fuzzy numbers. This problem is to search for the best structure of a typical Polish private farm [425] having 2 hectares (ha) of arable land and 4 ha of permanent grassland. The farmer possesses 6 sows and 4 cows. There are 26 considered activities which can be divided into the following six groups:

(1) plant production for sale:
  x_1: winter wheat,
  x_2: winter barley,
  x_3: triticale,
  x_4: spring wheat,
  x_5: spring barley,
  x_6: rape,
  x_7: peas,
  x_8: potatoes,
  x_9: sugar beets.
(2) plant production for fodder consumed in the farm:
  x_{10}: winter barley,
  x_{11}: spring barley,
  x_{12}: triticale,
  x_{13}: spring wheat,

x_{14}: potatoes,
x_{15}: fodder beet,
x_{16}: lucerne,
x_{17}: clover,
x_{18} corn;
(3) x_{19}: permanent grassland cultivation;
(4) purchase of fertilizers:
  x_{20}: phosphorus,
  x_{21}: nitrogen,
  x_{22}: potassium;
(5) x_{23}: purchase of a nutritive fodder;
(6) manpower hire:
  x_{24}: in the spring periods,
  x_{25}: in the summer periods,
  x_{26}: in the autumnal periods.

All these activities correspond to decision variables $x_1, x_2, \cdots, x_{26}$, respectively, with proper definition of their dimension, i.e., number of hectares of winter barley for sale, number of kilograms of phosphorus to be bought, and number of manpower hire in the spring period.

The problem here is how to decide in order to get the maximal gross profit, maximal structure-forming plants area and minimal manpower hire.

2.2 Fuzzy Variable

Let's introduce basic knowledge about fuzzy variable, which includes the measure, the definition and the properties of fuzzy variables.

2.2.1 Definition of Fuzzy Variable

Since its introduction in 1965 by Zadeh [9], fuzzy set theory has been well developed and applied in a wide variety of real problems. The term fuzzy variable was first introduced by Kaufmann [210], then it appeared in Zadeh [19, 22] and Nahmias [291]. Possibility theory was proposed by Zadeh [22], and developed by many researchers such as Dubois and Prade [71].

In order to provide an axiomatic theory to describe fuzziness, Nahmias [244] suggested a theoretical framework. Let us give the definition of possibility space (also called pattern space by Nahmias).

Definition 2.1. (Dubois and Prade [71]) Let Θ be a nonempty set, and $P(\Theta)$ be the power set of Θ. For each $A \subseteq P(\Theta)$, there is a nonnegative number $Pos\{A\}$, called its possibility, such that

(i) $Pos\{\emptyset\} = 0$;

(ii) $Pos\{\Theta\} = 1$;

(iii) $Pos\{\bigcup_k A_k\} = \sup_k Pos\{A_k\}$ for any arbitrary collection $\{A_k\}$ in $P(\Theta)$.

The triplet $(\Theta, P(\Theta), Pos)$ is called a possibility space, and the function Pos is referred to as a possibility measure.

It is easy to obtain the the following properties of *Pos* from the axioms above.

Property 2.1. The properties of *Pos* measure:

(i) $0 \le Pos\{A\} \le 1, \forall A \in P(\Theta)$;

(ii) $Pos\{A\} \le Pos\{B\}$, if $A \subseteq B$.

Several researchers have defined fuzzy variable in different ways, such as Kaufman [210], Zadah [19, 22] and Nahmias [291]. In this book we use the following definition of fuzzy variable.

Definition 2.2. (Nahmias [291]) A fuzzy variable is defined as a function from the possibility space $(\Theta, P(\Theta), Pos)$ to the real line **R**.

Definition 2.3. (Dubois and Prade [71]) Let ξ be a fuzzy variable on the possibility space $(\Theta, P(\Theta), Pos)$. Then its membership function $\mu : \mathbf{R} \mapsto [0, 1]$ is derived from the possibility measure *Pos* by

$$\mu(x) = Pos\{\theta \in \Theta | \xi(\theta) = x\}. \tag{2.2}$$

Remark 2.1. For any fuzzy variable ξ with membership function μ, we have $\sup_x \mu(x) = \sup_x Pos\{\theta \in \Theta | \xi(\theta) = x\} = Pos\{\Theta\} = 1$. That is, any fuzzy variables defined by Definition 2.2 are normalized.

Remark 2.2. Let ξ be a fuzzy variable with membership function μ. Then ξ may be regarded as a function from the possibility space $(\Theta, P(\Theta), Pos)$ to **R**, provided that $Pos\{A\} = \sup\{\mu(\xi(\theta)) | \theta \in A\}$ for any $A \in P(\Theta)$.

Remark 2.3. Since $\Theta = A \cup A^c$, we have $Pos\{A\} \vee Pos\{A^c\} = Pos\{\Theta\} = 1$ which implies that $Pos\{A\} \le 1$. On the other hand, since $A = A \cup \phi$, we have $Pos\{A\} \vee 0 = Pos\{A\}$ which implies that $Pos\{A\} \ge 0$. It follows that $0 \le Pos\{A\} \le 1$ for any $A \subseteq P$.

Remark 2.4. Let $A \subseteq B$. Then there exists a set C such that $B = A \cup C$. Thus we have $Pos\{A\} \vee Pos\{C\} = Pos\{B\}$ which gives that $Pos\{A\} \le Pos\{B\}$.

Theorem 2.1. *Zadeh[9] Let* $\tilde{a}_1, \tilde{a}_2, \cdots, \tilde{a}_n$ *be fuzzy variables, and* $f : R^n \to R$ *be continuous functions. Then the membership function* $\mu_{\tilde{a}}$ *of* $f(\tilde{a}_1, \tilde{a}_2, \cdots, \tilde{a}_n) \le 0$ *is derived from the membership functions* $\mu_{\tilde{a}_1}, \mu_{\tilde{a}_2}, \cdots, \mu_{\tilde{a}_n}$ *by*

$$\mu_{\tilde{a}} = \sup_{x_1, x_2, \cdots, x_n \in R} \left\{ \min_{1 \le i \le n} \mu_{\tilde{a}_i}(x_i) | x = f(x_1, x_2, \cdots, x_n) \right\}. \tag{2.3}$$

Now we introduce the LR fuzzy variable on which the fuzzy arithmetics have good results.

Definition 2.4. (Dubois and Prade [305]) Let f be a function from real numbers set $\mathbf{R}$ to $[0,1]$. If f satisfies $(1)f(x)=f(-x),(2)f(0)=1,(3)f(x)$ is decreasing on $[0,+\infty)$, then $f(x)$ is called reference function of a fuzzy variable.

The following reference functions are used usually in practical application:
$(1)f(x) = \max\{0, 1 - |x|^p\}(p \geq 0)$.
$(2)f(x) = \exp(-|x|^p)(p \geq 0)$.
$(3)f(x) = \frac{1}{1+|x|^p}(p \geq 0)$.
$(4)f(x) = \begin{cases} 1, x \in [-1,1] \\ 0, \text{ otherwise.} \end{cases}$

Definition 2.5. (Dubois and Prade [305]) Let $L(\cdot), R(\cdot)$ be two reference functions. If the membership function of fuzzy variable ξ has the following form

$$\mu_\xi(x) = \begin{cases} L(\frac{m-x}{\alpha}), \ x \leq m, \alpha > 0 \\ R(\frac{x-m}{\beta}), \ x \geq m, \beta > 0, \end{cases} \tag{2.4}$$

then ξ is called *LR* fuzzy variable, L, R are called left and right branch of ξ respectively, α, β are called left and right spread of ξ respectively, m is called the main value of ξ. Denote ξ by $(m, \alpha, \beta)_{LR}$. In addition, we assume *LR* fuzzy variable degenerate to a real number as $\alpha = \beta = 0$, i.e. $(m; 0, 0)_{LR} = m$.

Lemma 2.1. *[305] Let $\xi_1 = (m_1, \alpha_1, \beta_1)_{LR}, \xi_2 = (m_2, \alpha_2, \beta_2)_{LR}, k(\neq 0)$ a real number, then*
$(1)\xi_1 + \xi_2 = (m_1 + m_2, \alpha_1 + \alpha_2, \beta_1 + \beta_2)_{LR}$;
$(2)\xi_1 - \xi_2 = (m_1 - m_2, \alpha_1 + \beta_2, \beta_1 + \alpha_2)_{LR}$;
$(3)k\xi_1 = \begin{cases} (km_1, k\alpha_1, k\beta_1)_{LR}, \\ (km_1, -k\alpha_1, -k\beta_1)_{LR}, \end{cases} k > 0.$

Actually, $\xi_1 - \xi_2 = \xi_1 + (-1)\xi_2$.

Now let us introduce a kind of special *LR* fuzzy variables. By trapezoidal fuzzy variable we mean the fuzzy variables fully determined by quadruples (r_1, r_2, r_3, r_4) of crisp numbers with $r_1 < r_2 \leq r_3 < r_4$, whose membership functions can be denoted by

$$\mu(x) = \begin{cases} \frac{x-r_1}{r_2-r_1}, & \text{if } r_1 \leq x \leq r_2 \\ 1, & \text{if } r_2 \leq x \leq r_3 \\ \frac{x-r_3}{r_3-r_4}, & \text{if } r_3 \leq x \leq r_4 \\ 0, & \text{otherwise.} \end{cases}$$

The membership function curve of a trapezoidal fuzzy variable are shown by Figure 2.1.

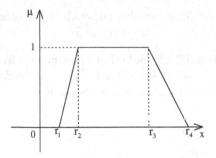

Fig. 2.1 The membership function curve of trapezoidal fuzzy variable (r_1, r_2, r_3, r_4)

We note that the trapezoidal fuzzy variable is a triangular fuzzy variable if $r_2 = r_3$, denoted by a triple (r_1, r_2, r_3). That is , the membership function of triangular fuzzy variable (r_1, r_2, r_3) is

$$\mu(x) = \begin{cases} \frac{x-r_1}{r_2-r_1}, & \text{if} \quad r_1 \leq x \leq r_2 \\ \frac{x-r_3}{r_2-r_3}, & \text{if} \quad r_2 \leq x \leq r_3 \\ 0, & \text{otherwise.} \end{cases}$$

The membership function curve of a triangular fuzzy variable are shown by Figure 2.2.

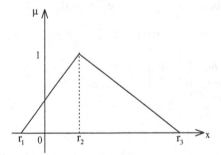

Fig. 2.2 The membership function curve of triangular fuzzy variable (r_1, r_2, r_3)

In some practical application, r_2 was in the center of interval $[r_1, r_3]$, i.e. $r_2 = \frac{r_1+r_3}{2}$. In this case, ξ is called centeral triangular fuzzy variable, denoted by $(a, \frac{a+b}{2}, b)(a < b)$. Its membership function is

$$\mu(x) = \begin{cases} \frac{2(x-b)}{b-a}, & \text{if} \quad a \leq x \leq \frac{a+b}{2} \\ \frac{2(x-b)}{a-b}, & \text{if} \quad \frac{a+b}{2} \leq x \leq a \\ 0, & \text{otherwise.} \end{cases}$$

The membership function curve of a triangular fuzzy variable are shown by Figure 2.3.

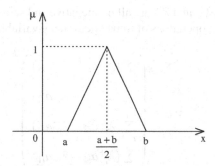

Fig. 2.3 The membership function curve of centeral triangular fuzzy variable $(a, \frac{a+b}{2}, b)(a < b)$

From the fuzzy arithmetic, we can obtain the sum of trapezoidal fuzzy variables $\tilde{a} = (a_1, a_2, a_3, a_4)$ and $\tilde{b} = (b_1, b_2, b_3, b_4)$ as

$$\mu_{\tilde{a}+\tilde{b}}(z) = \sup\{\min\{\mu_{\tilde{a}}(x), \mu_{\tilde{b}}(y)\}|z = x + y\}$$
$$= \begin{cases} \frac{z-(a_1+b_1)}{(a_2+b_2)-(a_1+b_1)}, & \text{if} \quad a_1 + b_1 \leq z \leq a_2 + b_2 \\ 1, & \text{if} \quad a_2 + b_2 \leq z \leq a_3 + b_3 \\ \frac{z-(a_4+b_4)}{(a_3+b_3)-(a_4+b_4)}, & \text{if} \quad a_3 + b_3 \leq z \leq a_4 + b_4 \\ 0, & \text{otherwise.} \end{cases}$$

That is, the sum of two trapezoidal fuzzy variables is also a trapezoidal fuzzy variable, and

$$\tilde{a} + \tilde{b} = (a_1 + b_1, a_2 + b_2, a_3 + b_3, a_4 + b_4).$$

Next we consider the product of a trapezoidal fuzzy variable and a scalar number λ. We have

$$\mu_{\lambda \cdot \tilde{a}}(z) = \sup\{\mu_{\tilde{a}}(x)|z = \lambda x\}$$

which yields that

$$\lambda \cdot \tilde{a} = \begin{cases} (\lambda \tilde{a}_1, \lambda \tilde{a}_2, \lambda \tilde{a}_3, \lambda \tilde{a}_4), & \text{if } \lambda \geq 0 \\ (\lambda \tilde{a}_4, \lambda \tilde{a}_3, \lambda \tilde{a}_2, \lambda \tilde{a}_1), & \text{if } \lambda < 0. \end{cases}$$

That is, the product of a trapezoidal fuzzy variable and a scalar number is also a trapezoidal fuzzy variable. Thus a weighted sum of trapezoidal fuzzy variables is also a trapezoidal fuzzy variable.

For example, we assume that $\tilde{a}_i$ are trapezoidal fuzzy variables $(a_{i1}, a_{i2}, a_{i3}, a_{i4})$, and λ_i are scalar numbers, $i = 1, 2, \cdots, n$, respectively. If we define

$$\lambda_i^+ = \begin{cases} \lambda_i, & \text{if } \lambda_i \leq 0 \\ 0, & \text{otherswise,} \end{cases} \qquad \lambda_i^- = \begin{cases} 0, & \text{if } \lambda_i \leq 0 \\ -\lambda_i, & \text{otherswise.} \end{cases}$$

for $i = 1, 2, \cdots, n$, then λ_i^+ and λ_i^- are all nonnegative and satisfy that $\lambda_i = \lambda_i^+ - \lambda_i^-$.
By the sum and product operations of trapezoidal fuzzy variables, we can obtain

$$
\tilde{a} = \sum_{i=1}^{n} \lambda_i \tilde{a}_i = \begin{pmatrix} \sum_{i=1}^{n} (\lambda_i^+ a_{i1} - \lambda_i^- a_{i4}) \\ \sum_{i=1}^{n} (\lambda_i^+ a_{i2} - \lambda_i^- a_{i3}) \\ \sum_{i=1}^{n} (\lambda_i^+ a_{i2} - \lambda_i^- a_{i2}) \\ \sum_{i=1}^{n} (\lambda_i^+ a_{i4} - \lambda_i^- a_{i1}) \end{pmatrix}^T
$$

As $\tilde{a}_i$ degenerate into triangular variables, i.e. $\tilde{a}_i = (a_{i1}, a_{i2}, a_{i3})$, we have

$$
\tilde{a} = \sum_{i=1}^{n} \lambda_i \tilde{a}_i = \begin{pmatrix} \sum_{i=1}^{n} (\lambda_i^+ a_{i1} - \lambda_i^- a_{i3}) \\ \sum_{i=1}^{n} (\lambda_i^+ a_{i2} - \lambda_i^- a_{i2}) \\ \sum_{i=1}^{n} (\lambda_i^+ a_{i3} - \lambda_i^- a_{i1}) \end{pmatrix}^T .
$$

2.2.2 Possibility, Necessity and General Fuzzy Measure

In order to measure the chances of occurrence of fuzzy events, three kinds of fuzzy measures of fuzzy events are introduced in this section. If decision making is performed optimistically, it is reasonable to use the possibility measure as follows:

More generally, we give the following theorem on possibility of fuzzy event.

Theorem 2.2. *[31] Let $\tilde{a}_1, \tilde{a}_2, \cdots, \tilde{a}_n$ be fuzzy variables, and $f : R^n \to R$ be continuous functions. Then the possibility of the fuzzy event characterized by $f(\tilde{a}_1, \tilde{a}_2, \cdots, \tilde{a}_n) \leq 0$ is*

$$
Pos\{f(\tilde{a}_1, \tilde{a}_2, \cdots, \tilde{a}_n) \leq 0\}
$$
$$
= \sup_{x_1, x_2, \cdots, x_n \in R} \left\{ \min_{1 \leq i \leq n} \mu_{\tilde{a}_i}(x_i) | f(x_1, x_2, \cdots, x_n) \leq 0 \right\}. \tag{2.5}
$$

Example 2.1. Let $\tilde{a}$ and $\tilde{b}$ be fuzzy variables on the possibility spaces $(\Theta_1, P(\Theta_1), Pos_1)$ and $(\Theta_2, P(\Theta_2), Pos_2)$, respectively. Then $\tilde{a} \leq \tilde{b}$ is a fuzzy event defined on the product possibility space $(\Theta, P(\Theta), Pos)$, whose possibility is

$$
Pos\{\tilde{a} \leq \tilde{b}\} = \sup_{x, y \in \mathbf{R}} \{\mu_{\tilde{a}}(x) \wedge \mu_{\tilde{b}}(y) | x \leq y\},
$$

where the abbreviation Pos represents possibility. This means that the possibility of $\tilde{a} \leq \tilde{b}$ is the largest possibility that there exists at least one pair of values $x, y \in \mathbf{R}$

such that $x \leq y$, and the values of $\tilde{a}$ and $\tilde{b}$ are x and y, respectively. Similarly, the possibility of $\tilde{a} = \tilde{a}$ is given by

$$Pos\{\tilde{a} = \tilde{b}\} = \sup_{x \in \mathbf{R}}\{\mu_{\tilde{a}}(x) \wedge \mu_{\tilde{b}}(x)\}.$$

If the decision maker prefers a pessimistic decision in order to avoid risk, it may be approximate to replace the possibility measure with the necessity measure.

A set function Nec defined on $P(\Theta)$ is said to be a necessity measure if it satisfies the following conditions:
(1) $Nec\{\emptyset\} = 0$, and $Nec\{\Theta\} = 1$;
(2) $Nec\{\bigcap_{i \in I} A_i\} = \inf_{i \in I}\{A_i\}$ for any subclass $\{A_i|i \in I\}$ of $P(\Theta)$, where I is an index set.

The necessity measure of a set A is defined as the impossibility of the opposite set A^c.

Definition 2.6. (Dubois [31]) Let $(\Theta, P(\Theta), Pos)$ be a possibility space, and A be a set in $P(\Theta)$. Then the necessity measure of A is

$$Nec\{A\} = 1 - Pos\{A^c\}.$$

Thus the necessity measure is the dual of possibility measure, that is, $Pos\{A\} + Nec\{A^c\} = 1$ for any $A \in P(\Theta)$.

Lemma 2.2. *[31] Let ξ_1 and ξ_1 be two fuzzy variables. Then we have*

$$Pos\{\xi_1 \geq \xi_2\} = \sup\{\mu_{\xi_1}(u) \wedge \mu_{\xi_1}(v)|u > v\}, \tag{2.6}$$

$$Pos\{\xi_1 > \xi_2\} = \sup\{\mu_{\xi_1}(u) \wedge \inf_{v}\{1 - \mu_{\xi_2}(v)|u \leq v\}\}. \tag{2.7}$$

If the decision maker prefers a pessimistic decision in order to avoid risk, it may be approximate to replace the possibility measure with the necessity measure.

$$Nec\{\xi_1 \geq \xi_2\} = \inf_{u}\{1 - \mu_{\xi}(u) \vee \sup_{v}\mu_{\xi_2}(v)|u \geq v\}, \tag{2.8}$$

$$Nec\{\xi_1 > \xi_2\} = \inf\{(1 - \mu_{\xi_1}(u)) \vee (1 - \mu_{\xi_2}(v))|u \leq v\}. \tag{2.9}$$

The conclusions of Lemma 2.2 can be shown by the following Figure 2.4.

By using the $\alpha-$level sets of fuzzy variable ξ_1 and ξ_2, $[m_\alpha^L, m_\alpha^R]$ and $[n_\alpha^L, n_\alpha^R]$, Lemma 2.2 can be rewritten as

$$Pos\{\xi_1 \geq \xi_2\} \geq \alpha \Longleftrightarrow m_\alpha^R \geq n_\alpha^L, \tag{2.10}$$

$$Pos\{\xi_1 > \xi_2\} \geq \alpha \Longleftrightarrow m_\alpha^R \geq n_{1-\alpha}^R, \tag{2.11}$$

$$Nec\{\xi_1 \geq \xi_2\} \geq \alpha \Longleftrightarrow m_{1-\alpha}^R \geq n_\alpha^L, \tag{2.12}$$

$$Nec\{\xi_1 > \xi_2\} \geq \alpha \Longleftrightarrow m_{1-\alpha}^L \geq n_{1-\alpha}^R. \tag{2.13}$$

Sometimes, the basic measures Pos and Nec proposed by Dubois and Prade [305, 31] can be limited in dealing with the realistic uncertain decision making problems, since

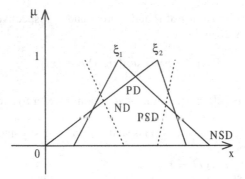

Fig. 2.4 Conclusions of Theorem 2.2

these two measures reflect the extremely optimistic and pessimistic attitudes, so we need a more general measure for fuzzy decision making problems. In fact, in realistic uncertain decision making process, decision makers often have different optimistic-pessimistic attitudes. They will decide the optimistic-pessimistic parameters according to their own judgement. When they have good forecasts, they will be optimistic about a fuzzy event and vice versa. This optimistic-pessimistic parameter of a fuzzy event should be considered to avoid extreme attitudes. Therefore, it is necessary to design a more flexible measure to measure fuzzy events in a decision making problem by introducing an optimistic-pessimistic parameter, and this measure should be adjustable according to the varying attitudes of the decision makers. So if we set the pessimistic measure Nec as the reference measure, and we can design a new measure Me by adding an optimistic-pessimistic adjusting factor $\lambda (Pos\{A\} - Nec\{A\})$, where $(Pos\{A\} - Nec\{A\})$ is the range in which the value of the measure can change from a pessimistic value to an optimistic value. Based on the above idea and discussion, a fuzzy measure Me is suitable for the background of decision making under a fuzzy environment and has practical significance.

Definition 2.7. (Xu and Zhou [402, 413]) Let $(\Theta, P(\Theta), Pos)$ be a possibility space, and A be a set in $P(\Theta)$. Then the fuzzy measure of A is

$$Me\{A\} = Nec\{A\} + \lambda (Pos\{A\} - Nec\{A\}),$$

where λ $(0 \leq \lambda \leq 1)$ is the optimistic-pessimistic parameter to determine the combined attitude of a decision maker.

Remark 2.5. Note that the fuzzy measure Me is to evaluate a degree that a fuzzy variable takes values in an interval with different optimistic-pessimistic attitudes. It is equal to the convex combination of Pos and Nec, i. e., $Me\{A\} = \lambda Pos\{A\} + (1 - \lambda)Nec\{A\}$.

When $\lambda = 1$, we have $Me = Pos$, it means the decision maker is optimistic, it's the measure of best case of that event, and it is the maximal chance of A holds;

When $\lambda = 0$, we have $Me = Nec$, it means the decision maker is pessimistic, it gives the measure of worst case of that event, and it is the minimal chance of A holds;

When $\lambda = 0.5$, we have $Me = Cr$, where Cr is the credibility measure introduced by Liu [220], it's a special case of Me, it means the decision maker takes compromise attitude. Cr and it is self dual, that is, $Cr\{A\} + Cr\{A^c\} = 1$ for any $A \in P(\Theta)$.

Theorem 2.3. *Let* $(\Theta, P(\Theta), Pos)$ *be a possibility space, and A be a set in $P(\Theta)$.*
(i) $Me\{\Theta\} = 1$, and $Me\{\Phi\} = 0$, where Θ is the collection and Φ is the empty set;
(ii) Since $\lambda(Pos\{A\} - Nec\{A\}) \geq 0$, so $Nec\{A\} \leq Me\{A\} \leq Pos\{A\}$ for any $A \in P(\Theta)$ and thus 0-1 boundedness holds for $Me\{\cdot\}$;
(iii) For any two sets which satisfy $A \subset B$, $Me\{A\} \leq Me\{B\}$, which means that monotonicity holds for $Me\{\cdot\}$;
(iv) For any $A \in P(\Theta)$, when the optimistic-pessimistic parameter $\lambda \geq 0.5$, we have $1 \leq Me\{A\} + Me\{A^c\} \leq 2$, and when $\lambda \leq 0.5$, we have $0 \leq Me\{A\} + Me\{A^c\} \leq 1$;
(v) For any $A, B \in P(\Theta)$ and $\lambda \geq 0.5$, we have $Me\{A \cup B\} \leq Me\{A\} + Me\{B\}$, which means the restricted subadditivity holds for $Me\{\cdot\}$.

Example 2.2. Let $\Theta = \{\theta_1, \theta_2\}$, with $Pos\{\theta_1\} = 1.0$, and $Pos\{\theta_2\} = 0.7$. Then we have
$$Nec\{\theta_1\} = 1 - Pos\{\theta_2\} = 0.3,$$
$$Nec\{\theta_2\} = 0.$$

So based on the above results, we have
$$Me\{\theta_1\} = Nec\{\theta_1\} + \lambda(Pos\{\theta_1\} - Nec\{\theta_1\}) = 0.3 + 0.7\lambda,$$
$$Me\{\theta_2\} = 0.7\lambda,$$
$$Me\{\theta_1\} + Me\{\theta_2\} = 0.3 + 1.4\lambda.$$

If we want to make $Me\{\theta_1\} + Me\{\theta_2\} = 1$, we need to set the optimistic-pessimistic parameter $\lambda = 0.5$.
If we set $\lambda = 0.2$, we have $Me\{\theta_1\} + Me\{\theta_2\} = 0.58$.
If we set $\lambda = 0.6$, we have $Me\{\theta_1\} + Me\{\theta_2\} = 1.14$.

Lemma 2.3. *[402, 413] Let $(\Theta, P(\Theta), Pos)$ be a possibility space, and A be a set in $P(\Theta)$. Then for any $\lambda_1 \geq \lambda_2$, we have*
$$Me^{\lambda_1}\{A\} \geq Me^{\lambda_2}\{A\}.$$

Proof. According to Definition 2.7, we have that
$$Me^{\lambda_1}\{A\} - Me^{\lambda_2}\{A\}$$
$$= (\lambda_1 Pos\{A\} + (1 - \lambda_1)Nec\{A\}) - (\lambda_2 Pos\{A\} + (1 - \lambda_2)Nec\{A\})$$
$$= (\lambda_1 - \lambda_2)Pos\{A\} - (\lambda_1 - \lambda_2)Nec\{A\}$$
$$= (\lambda_1 - \lambda_2)(Pos\{A\} - Nec\{A\}).$$

Because $\lambda_1 - \lambda_2 \geq 0$. So in order to prove $Me^{\lambda_1}\{A\} - Me^{\lambda_2}\{A\} \geq 0$, we just need to prove $Pos\{A\} \geq Nec\{A\}$.

If $Pos\{A\} = 1$, then it is obvious that $Pos\{A\} \geq Nec\{A\}$. Otherwise, we must have $Pos\{A^c\} = 1$, which implies that $Nec\{A\} = 1 - Pos\{A^c\} = 0$. Thus $Pos\{A\} \geq Nec\{A\}$ always holds, so we have

$$(\lambda_1 - \lambda_2)(Pos\{A\} - Nec\{A\}) \geq 0 \Rightarrow Me^{\lambda_1}\{A\} \geq Me^{\lambda_2}\{A\}.$$

The theorem is proved.   □

Remark 2.6. A fuzzy event may fail even though its possibility achieves 1, and hold even though its necessity is 0. However, the fuzzy event must hold if its credibility is 1, and fail if its credibility is 0.

Example 2.3. By a triangular fuzzy variable we mean the fuzzy variable ξ fully determined by the triplet (r_1, r_2, r_3) of crisp numbers with $r_1 < r_2 < r_3$, whose membership function is given by

$$\mu(x) = \begin{cases} \frac{x-r_1}{r_2-r_1}, & \text{if } r_1 \leq x \leq r_2 \\ \frac{x-r_3}{r_2-r_3}, & \text{if } r_2 \leq x \leq r_3 \\ 0, & \text{otherwise.} \end{cases}$$

From the definition of (2.2),(2.6) and (2.7), the possibility, necessity, and the general measure of $\xi \leq x$ are as follows respectively:

$$Pos\{\xi \leq x\} = \begin{cases} 0, & \text{if } x \leq r_1 \\ \frac{x-r_1}{r_2-r_1}, & \text{if } r_1 \leq x \leq r_2 \\ 1, & \text{if } x \geq r_2, \end{cases}$$

$$Nec\{\xi \leq x\} = \begin{cases} 0, & \text{if } x \leq r_2 \\ \frac{x-r_2}{r_3-r_2}, & \text{if } r_2 \leq x \leq r_3 \\ 1, & \text{if } x \geq r_3, \end{cases}$$

$$Me\{\xi \leq x\} = \begin{cases} 0, & \text{if } x \leq r_1 \\ \lambda\frac{x-r_1}{r_2-r_1}, & \text{if } r_1 \leq x \leq r_2 \\ \lambda + (1-\lambda)\frac{x-r_2}{r_3-r_2}, & \text{if } r_2 \leq x \leq r_3 \\ 1, & \text{if } x \geq r_3, \end{cases}$$

$$Cr\{\xi \leq x\} = \begin{cases} 0, & \text{if } x \leq r_1 \\ \frac{x-r_1}{2(r_2-r_1)}, & \text{if } r_1 \leq x \leq r_2 \\ \frac{x-2r_2+r_3}{2(r_3-r_2)}, & \text{if } r_2 \leq x \leq r_3 \\ 1, & \text{if } x \geq r_3. \end{cases}$$

Example 2.4. By a trapezoidal fuzzy variable, we mean the fuzzy variable ξ fully determined by quadruplet (r_1, r_2, r_3, r_4) of crisp numbers with $r_1 < r_2 < r_3 < r_4$, whose membership function is given by Figure 2.5.

$$\mu(x) = \begin{cases} \frac{x-r_1}{r_2-r_1}, & \text{if } r_1 \leq x \leq r_2 \\ 1, & \text{if } r_2 \leq x \leq r_3 \\ \frac{x-r_4}{r_3-r_4}, & \text{if } r_3 \leq x \leq r_4 \\ 0, & \text{otherwise.} \end{cases}$$

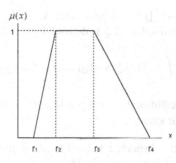

Fig. 2.5 A trapezoidal fuzzy variable $\xi = (r_1, r_2, r_3, r_4)$

From the definition of (2.2),(2.6) and (2.7), the possibility, necessity, and credibility of $\xi \leq x$ are as follows respectively:

$$Pos\{\xi \leq x\} = \begin{cases} 1, & \text{if } x \geq r_2 \\ \frac{x-r_1}{r_2-r_1}, & \text{if } r_1 \leq x \leq r_2 \\ 0, & \text{otherwise,} \end{cases}$$

$$Nec\{\xi \leq x\} = \begin{cases} 1, & \text{if } x \geq r_4 \\ \frac{x-r_3}{r_4-r_3}, & \text{if } r_3 \leq x \leq r_4 \\ 0, & \text{otherwise,} \end{cases}$$

$$Me\{\xi \leq x\} = \begin{cases} 0, & \text{if } x \leq r_1 \\ \lambda \frac{x-r_1}{r_2-r_1}, & \text{if } r_1 \leq x \leq r_2 \\ \lambda, & \text{if } r_2 \leq x \leq r_3 \\ \lambda + (1-\lambda)\frac{x-r_3}{r_4-r_3}, & \text{if } r_3 \leq x \leq r_4 \\ 1, & \text{if } x \geq r_4, \end{cases}$$

$$Cr\{\xi \leq x\} = \begin{cases} 0, & \text{if } x \leq r_1 \\ \frac{x-r_1}{2(r_2-r_1)}, & \text{if } r_1 \leq x \leq r_2 \\ \frac{1}{2}, & \text{if } r_2 \leq x \leq r_3 \\ \frac{x-2r_3+r_4}{2(r_4-r_3)}, & \text{if } r_3 \leq x \leq r_4 \\ 1, & \text{otherwise.} \end{cases}$$

Lemma 2.4. *[77] Let $\xi = (r_1, r_2, r_3, r_4)$ be a trapezoidal fuzzy variable. Then for any given confidence level α with $0 < \alpha \leq 1$, we have*
(1). *$Pos\{\xi \leq 0\}$ if and only if $(1-\alpha)r_1 + \alpha r_2 \leq 0$;*
(2). *$Nec\{\xi \leq 0\}$ if and only if $(1-\alpha_1)r_3 + \alpha r_4 \leq 0$;*
(3). *when $\alpha \leq 0.5, Cr\{\xi \leq 0\}$ if and only if $(1-2\alpha)r_1 + 2\alpha_2 \leq 0$;*
(4). *when $\alpha > 0.5, Cr\{\xi \leq 0\}$ if and only if $(2-2\alpha)r_3 + (2\alpha_2 - 1)r_4 \leq 0$.*

2.2.3 Expected Value Operator of Fuzzy Variables

In order to measure the mean of a fuzzy variable, several researchers defined different expected values for fuzzy variables, such as Dubois and Prade [30], Campos and Verdegay [33], González [34] and Yager [35, 36]. In this book, we define three kinds of expected values for fuzzy variables based on three kinds fuzzy measures:

Definition 2.8. (Xu, Zhou[413]) Let ξ be a fuzzy variable on the possibility space $(\Theta, P(\Theta), Pos)$. The expected value of ξ is defined by

$$E^{Me}[\xi] = \int_0^{+\infty} Me\{\xi \geq r\}dr - \int_{-\infty}^0 Me\{\xi \leq r\}dr, \qquad (2.14)$$

where integrals are defined through Lebesgue integral, *i.e.* , well defined in case either of the two takes finite value.

Similarly, we can define the expected value based on the Pos, Nec, Cr measure, which are the special case of fuzzy measure Me.

$$\begin{cases} E^{Pos}[\xi] = \int_0^{+\infty} Pos\{\xi \geq r\}dr - \int_{-\infty}^0 Pos\{\xi \leq r\}dr, \\ E^{Nec}[\xi] = \int_0^{+\infty} Nec\{\xi \geq r\}dr - \int_{-\infty}^0 Nec\{\xi \leq r\}dr, \qquad (2.15) \\ E^{Cr}[\xi] = \int_0^{+\infty} Cr\{\xi \geq r\}dr - \int_{-\infty}^0 Cr\{\xi \leq r\}dr. \end{cases}$$

Obviously, $E^{Me}[\xi] = \lambda E^{Pos}[\xi] + (1 - \lambda)E^{Nec}[\xi]$, and $E^{Cr}[\xi] = \frac{1}{2}(E^{Pos}[\xi] + E^{Nec}[\xi])$.

Remark 2.7. The Definition 2.8 is not only applicable to continuous fuzzy variable, but also discrete fuzzy variable and a function of multiple fuzzy variables.

Proposition 2.1. *Let ξ be a fuzzy variable with a membership function*

$$\mu_\xi(x) = \begin{cases} 1, & if \quad x \in [a, b] \\ 0, & otherwise. \end{cases}$$

Then the expected value of ξ is

$$E^{Me}[\xi] = \begin{cases} (1 - \lambda)a + \lambda b, & if \ 0 \leq a \leq b \\ \lambda(a + b), & if \ a \leq 0 \leq b \\ \lambda a + (1 - \lambda)b, & if \ a \leq b \leq 0. \end{cases}$$

Proof. First, we calculate $E^{Pos}[\xi]$ and $E^{Nec}[\xi]$. There are three cases. Let's discuss every case in turn.

   Case 1: $0 \leq a \leq b$

$$\begin{aligned} E^{Pos}[\xi] &= \int_0^{+\infty} Pos\{\xi \geq r\}dr - \int_{-\infty}^0 Pos\{\xi \leq r\}dr \\ &= \int_0^b Pos\{\xi \geq r\}dr \\ &= b. \end{aligned}$$
$$E^{Nec}[\xi] = \int_0^{+\infty} Nec\{\xi \geq r\}dr - \int_{-\infty}^0 Nec\{\xi \leq r\}dr$$

$$\begin{aligned} &= \int_0^{+\infty}(1 - Pos\{\xi < r\})dr - \int_{-\infty}^0 (1 - Pos\{\xi > r\})dr \\ &= \int_{-\infty}^0 Pos\{\xi > r\}dr - \int_0^{+\infty} Pos\{\xi < r\}dr \\ &= \int_{-\infty}^0 1dr - [\int_0^a Pos\{\xi < r\}dr + \int_a^b Pos\{\xi < r\}dr + \int_b^{+\infty} Pos\{\xi < r\}dr] \\ &= a. \end{aligned}$$

Then we have

$$E^{Me}[\xi] = \lambda E^{Pos}[\xi] + (1-\lambda)E^{Nec}[\xi]$$
$$= (1-\lambda)a + \lambda b.$$

Case 2: $a \leq 0 \leq b$

$$E^{Pos}[\xi] = \int_0^{+\infty} Pos\{\xi \geq r\}dr - \int_{-\infty}^0 Pos\{\xi \leq r\}dr$$
$$= \int_0^b Pos\{\xi \geq r\}dr - \int_a^0 Pos\{\xi \leq r\}dr$$
$$= b+a.$$

$$E^{Nec}[\xi] = \int_0^{+\infty} Nec\{\xi \geq r\}dr - \int_{-\infty}^0 Nec\{\xi \leq r\}dr$$
$$= \int_0^{+\infty}(1-Pos\{\xi < r\})dr - \int_{-\infty}^0(1-Pos\{\xi > r\})dr$$
$$= \int_{-\infty}^0 Pos\{\xi > r\}dr - \int_0^{+\infty} Pos\{\xi < r\}dr$$
$$= \int_{-\infty}^0 1dr - \int_0^{+\infty} 1dr$$
$$= 0.$$

Then we have

$$E^{Me}[\xi] = \lambda(a+b).$$

Case 3: $a \leq b \leq 0$

$$E^{Pos}[\xi] = \int_0^{+\infty} Pos\{\xi \geq r\}dr - \int_{-\infty}^0 Pos\{\xi \leq r\}dr$$
$$= -\int_a^0 Pos\{\xi \leq r\}dr$$
$$= a.$$

$$E^{Nec}[\xi] = \int_0^{+\infty} Nec\{\xi \geq r\}dr - \int_{-\infty}^0 Nec\{\xi \leq r\}dr$$
$$= \int_0^{+\infty}(1-Pos\{\xi < r\})dr - \int_{-\infty}^0(1-Pos\{\xi > r\})dr$$
$$= \int_{-\infty}^0 Pos\{\xi > r\}dr - \int_0^{+\infty} Pos\{\xi < r\}dr$$
$$= \int_{-\infty}^b 1dr + \int_b^0 0dr - \int_0^{+\infty} 1dr$$
$$= b.$$

Then we have

$$E^{Me}[\xi] = \lambda a + (1-\lambda)b.$$

Above all, for any cases we have

$$E^{Me}[\xi] = \begin{cases} (1-\lambda)a + \lambda b, & \text{if } 0 \leq a \leq b \\ \lambda(a+b), & \text{if } a \leq 0 \leq b \\ \lambda a + (1-\lambda)b, & \text{if } a \leq b \leq 0. \end{cases}$$

This theorem is proved.   □

Remark 2.8. If $\lambda = \frac{1}{2}$, i.e. Me is actually Cr, then $E^{Cr}[\xi] = \frac{a+b}{2}$.

Proposition 2.2. *Let* $\xi = (r_1, r_2, r_3, r_4)$ *be a trapezoidal fuzzy variable. Then its expected value is*

$$E^{Me}[\xi] = \begin{cases} \frac{\lambda}{2}(r_1 + r_2) + \frac{1-\lambda}{2}(r_3 + r_4), & \text{if } r_4 \leq 0 \\ \frac{\lambda}{2}(r_1 + r_2) + \frac{\lambda r_4^2 - (1-\lambda)r_3^2}{2(r_4 - r_3)}, & \text{if } r_3 \leq 0 \leq r_4 \\ \frac{\lambda}{2}(r_1 + r_2 + r_3 + r_4), & \text{if } r_2 \leq 0 \leq r_3 \\ \frac{(1-\lambda)r_2^2 - \lambda r_1^2}{2(r_2 - r_1)} + \frac{\lambda}{2}(r_3 + r_4), & \text{if } r_1 \leq 0 \leq r_2 \\ \frac{1-\lambda}{2}(r_1 + r_2) + \frac{\lambda}{2}(r_3 + r_4), & \text{if } 0 \leq r_1. \end{cases} \quad (2.16)$$

Proof. Since there 5 cases, let's discuss every case in turn.

Case 1: $r_4 \leq 0$

$$\begin{aligned} E^{Me}[\xi] &= \int_0^{+\infty} Me\{\xi \geq r\}dr - \int_{-\infty}^0 Me\{\xi \leq r\}dr \\ &= -(\int_{-\infty}^{r_1} Me\{\xi \leq r\}dr + \int_{r_1}^{r_2} Me\{\xi \leq r\}dr + \int_{r_2}^{r_3} Me\{\xi \leq r\}dr \\ &\quad + \int_{r_3}^{r_4} Me\{\xi \leq r\}dr + \int_{r_4}^0 Me\{\xi \leq r\}dr) \\ &= -(\int_{r_1}^{r_2} \lambda \frac{r - r_1}{r_2 - r_1}dr + \int_{r_2}^{r_3} \lambda dr + \int_{r_3}^{r_4}(\lambda + (1-\lambda)\frac{r - r_3}{r_4 - r_3})dr + \int_{r_4}^0 1dr) \\ &= -(\frac{\lambda r^2}{2(r_2 - r_1)} \mid_{r_1}^{r_2} - \frac{\lambda r_1 r}{r_2 - r_1} \mid_{r_1}^{r_2} + \lambda r \mid_{r_2}^{r_3} + \lambda r \mid_{r_3}^{r_4} + \frac{(1-\lambda)r^2}{2(r_4 - r_3)} \mid_{r_3}^{r_4} - \frac{(1-\lambda)r_3 r}{r_4 - r_3} \mid_{r_3}^{r_4} + r \mid_{r_4}^0) \\ &= \frac{\lambda}{2}(r_1 + r_2) + \frac{1-\lambda}{2}(r_3 + r_4). \end{aligned}$$

Case 2: $r_3 \leq 0 \leq r_4$

$$\begin{aligned} E^{Me}[\xi] &= \int_0^{+\infty} Me\{\xi \geq r\}dr - \int_{-\infty}^0 Me\{\xi \leq r\}dr \\ &= \int_0^{r_4} Me\{\xi \geq r\}dr + \int_{r_4}^{+\infty} \lambda Cr\{\xi \geq r\}dr - (\int_{-\infty}^{r_1} Me\{\xi \leq r\}dr \\ &\quad + \int_{r_1}^{r_2} Me\{\xi \leq r\}dr + \int_{r_2}^{r_3} \lambda Cr\{\xi \leq r\}dr + \int_{r_3}^0 Me\{\xi \leq r\}dr) \\ &= \int_0^{r_4} \lambda \frac{r_4 - r}{r_4 - r_3}dr - (\int_{r_1}^{r_2} \lambda \frac{r - r_1}{r_2 - r_1}dr + \int_{r_2}^{r_3} \lambda dr + \int_{r_3}^0(\lambda + (1-\lambda)\frac{r - r_3}{r_4 - r_3})dr) \\ &= \frac{\lambda r_4 r}{r_4 - r_3} \mid_0^{r_4} - \frac{\lambda r^2}{2(r_4 - r_3)} \mid_0^{r_4} - (\frac{\lambda r^2}{2(r_2 - r_1)} \mid_{r_1}^{r_2} - \frac{\lambda r_1 r}{r_2 - r_1} \mid_{r_1}^{r_2} + \lambda r \mid_{r_2}^{r_3} + \lambda r \mid_{r_3}^0 + \frac{(1-\lambda)r^2}{2(r_4 - r_3)} \mid_{r_3}^0 \\ &\quad - \frac{(1-\lambda)r_3 r}{r_4 - r_3} \mid_{r_3}^0) \\ &= \frac{\lambda}{2}(r_1 + r_2) + \frac{\lambda r_4^2 - (1-\lambda)r_3^2}{2(r_4 - r_3)}. \end{aligned}$$

Case 3: $r_2 \leq 0 \leq r_3$

$$\begin{aligned} E^{Me}[\xi] &= \int_0^{+\infty} Me\{\xi \geq r\}dr - \int_{-\infty}^0 Me\{\xi \leq r\}dr \\ &= \int_0^{r_3} Me\{\xi \geq r\}dr + \int_{r_3}^{r_4} Me\{\xi \geq r\}dr + \int_{r_4}^{+\infty} Me\{\xi \geq r\}dr \\ &\quad - (\int_{-\infty}^{r_1} Me\{\xi \leq r\}dr + \int_{r_1}^{r_2} Me\{\xi \leq r\}dr + \int_{r_2}^0 \lambda Cr\{\xi \leq r\}dr) \\ &= \int_0^{r_3} \lambda dr + \int_{r_3}^{r_4} \lambda \frac{r_4 - r}{r_4 - r_3}dr - (\int_{r_1}^{r_2} \lambda \frac{r - r_1}{r_2 - r_1}dr + \int_{r_2}^0 \lambda dr) \\ &= \lambda r \mid_0^{r_3} + \frac{\lambda r_4 r}{r_4 - r_3} \mid_{r_3}^{r_4} - \frac{\lambda r^2}{2(r_4 - r_3)} \mid_{r_3}^{r_4} - (\frac{\lambda r^2}{2(r_2 - r_1)} \mid_{r_1}^{r_2} - \frac{\lambda r_1 r}{r_2 - r_1} \mid_{r_1}^{r_2} + \lambda r \mid_{r_2}^0) \\ &= \frac{\lambda}{2}(r_1 + r_2 + r_3 + r_4). \end{aligned}$$

Case 4: $r_1 \leq 0 \leq r_2$

$$
\begin{aligned}
E^{Me}[\xi] &= \int_0^{+\infty} Me\{\xi \geq r\}dr - \int_{-\infty}^0 Me\{\xi \leq r\}dr \\
&= \int_0^{r_2} Me\{\xi \geq r\}dr + \int_{r_2}^{r_3} Me\{\xi \geq r\}dr + \int_{r_3}^{r_4} Me\{\xi \geq r\}dr \\
&\quad + \int_{r_4}^{+\infty} Me\{\xi \geq r\}dr - (\int_{-\infty}^{r_1} Me\{\xi \leq r\}dr + \int_{r_1}^0 Me\{\xi \leq r\}dr) \\
&= \int_0^{r_2}(\lambda + (1-\lambda)\frac{r_2-r}{r_2-r_1})dr + \int_{r_2}^{r_3} \lambda dr + \int_{r_3}^{r_4} \lambda \frac{r_4-r}{r_4-r_3}dr - (\int_{r_1}^0 \lambda \frac{r-r_1}{r_2-r_1}dr) \\
&= \lambda r \big|_0^{r_2} + \frac{(1-\lambda)r_2 r}{r_2-r_1} \big|_0^{r_2} - \frac{(1-\lambda)r^2}{2(r_2-r_1)} \big|_0^{r_2} + \lambda r \big|_{r_2}^{r_3} + \frac{\lambda r_4 r}{r_4-r_3} \big|_{r_3}^{r_4} - \frac{\lambda r^2}{2(r_4-r_3)} \big|_{r_3}^{r_4} \\
&\quad - (\frac{\lambda r^2}{2(r_2-r_1)} \big|_{r_1}^0 - \frac{\lambda r_1 r}{r_2-r_1} \big|_{r_1}^0) \\
&= \frac{(1-\lambda)r_2^2 - \lambda r_1^2}{2(r_2-r_1)} + \frac{\lambda}{2}(r_3 + r_4).
\end{aligned}
$$

Case 5: $0 \leq r_1$

$$
\begin{aligned}
E^{Me}[\xi] &= \int_0^{+\infty} Me\{\xi \geq r\}dr - \int_{-\infty}^0 Me\{\xi \leq r\}dr \\
&= \int_0^{r_1} Me\{\xi \geq r\}dr + \int_{r_1}^{r_2} Me\{\xi \geq r\}dr + \int_{r_2}^{r_3} Me\{\xi \geq r\}dr \\
&\quad + \int_{r_3}^{r_4} Me\{\xi \geq r\}dr + \int_{r_4}^{+\infty} Me\{\xi \leq r\}dr \\
&= \int_0^{r_1} 1 dr + \int_{r_1}^{r_2}(\lambda + (1-\lambda)\frac{r_2-r}{r_2-r_1})dr + \int_{r_2}^{r_3} \lambda dr + \int_{r_3}^{r_4} \lambda \frac{r_4-r}{r_4-r_3}dr + 0 \\
&= r \big|_0^{r_1} + \lambda r \big|_{r_1}^{r_2} + \frac{(1-\lambda)r_2 r}{r_2-r_1} \big|_{r_1}^{r_2} - \frac{(1-\lambda)r^2}{2(r_2-r_1)} \big|_{r_1}^{r_2} + \lambda r \big|_{r_2}^{r_3} + \frac{\lambda r_4 r}{r_4-r_3} \big|_{r_3}^{r_4} - \frac{\lambda r^2}{2(r_4-r_3)} \big|_{r_3}^{r_4} \\
&= \frac{1-\lambda}{2}(r_1 + r_2) + \frac{\lambda}{2}(r_3 + r_4).
\end{aligned}
$$

Then we have

$$
E^{Me}[\xi] = \begin{cases}
\frac{\lambda}{2}(r_1 + r_2) + \frac{1-\lambda}{2}(r_3 + r_4), & \text{if } r_4 \leq 0 \\
\frac{\lambda}{2}(r_1 + r_2) + \frac{\lambda r_4^2 - (1-\lambda)r_3^2}{2(r_4-r_3)}, & \text{if } r_3 \leq 0 \leq r_4 \\
\frac{\lambda}{2}(r_1 + r_2 + r_3 + r_4), & \text{if } r_2 \leq 0 \leq r_3 \\
\frac{(1-\lambda)r_2^2 - \lambda r_1^2}{2(r_2-r_1)} + \frac{\lambda}{2}(r_3 + r_4), & \text{if } r_1 \leq 0 \leq r_2 \\
\frac{1-\lambda}{2}(r_1 + r_2) + \frac{\lambda}{2}(r_3 + r_4), & \text{if } 0 \leq r_1.
\end{cases}
$$

This theorem is proved. $\qquad\qquad\qquad\qquad\qquad\qquad\qquad\qquad\qquad\qquad\qquad\square$

Remark 2.9. If $\lambda = 0.5$, then

$$
E^{Cr}[\xi] = \frac{r_1 + r_2 + r_3 + r_4}{2}.
$$

Remark 2.10. If $r_2 = r_3$, i.e. ξ degenerates to a triangular fuzzy variable, then

$$
E^{Cr}[\xi] = \frac{1}{4}(r_1 + 2r_2 + r_4).
$$

Remark 2.11. If LR fuzzy number $\xi = (z, \alpha, \beta)_{LR}$ degenerates to a triangular fuzzy number which means the reference function $L(x) = R(x) = 1 - x$, then

$$
E^{Cr}[\xi] = z + \frac{1}{4}(\alpha + \beta).
$$

Proposition 2.3. *Let ξ be a fuzzy variable with a membership function*

$$\mu_\xi(x) = e^{-k(x-a)^2} (k > 0)$$

Then the expected value of ξ is

$$E^{Me}[\xi] = a - \int_0^a e^{-k(x-a)^2} dx$$

Proof. There are five cases. Let's discuss every case in turn.
  Case 1: $a \geq 0$

$$
\begin{aligned}
E^{Pos}[\xi] &= \int_0^{+\infty} Pos\{\xi \geq r\}dx - \int_{-\infty}^0 Pos\{\xi \leq x\}dx \\
&= (\int_0^a 1dx + \int_a^{+\infty} e^{-k(x-a)^2}dx) - \int_{-\infty}^0 e^{-k(x-a)^2}dx \\
&= a - \int_0^a e^{-k(x-a)^2}dx. \\
E^{Nec}[\xi] &= \int_0^{+\infty} Nec\{\xi \geq x\}dx - \int_{-\infty}^0 Nec\{\xi \leq x\}dx \\
&= \int_{-\infty}^0 Pos\{\xi > x\}dx - \int_0^{+\infty} Pos\{\xi < x\}dx \\
&= \int_{-\infty}^0 1dx - (\int_0^a e^{-k(x-a)^2}dx + \int_a^{+\infty} 1dr \\
&= a - \int_0^a e^{-k(x-a)^2}dx.
\end{aligned}
$$

Then we have

$$
\begin{aligned}
E^{Me}[\xi] &= \lambda E^{Pos}[\xi] + (1-\lambda)E^{Nec}[\xi] \\
&= \lambda(a - \int_0^a e^{-k(x-a)^2}dx) + (1-\lambda)(a - \int_0^a e^{-k(x-a)^2}dx) \\
&= a - \int_0^a e^{-k(x-a)^2}dx.
\end{aligned}
$$

Case 2: $a < 0$

$$
\begin{aligned}
E^{Pos}[\xi] &= \int_0^{+\infty} Pos\{\xi \geq r\}dx - \int_{-\infty}^0 Pos\{\xi \leq x\}dx \\
&= \int_0^{+\infty} e^{-k(x-a)^2}dx - (\int_{-\infty}^a e^{-k(x-a)^2}dx + \int_a^0 1dx) \\
&= \int_a^0 e^{-k(x-a)^2}dx + a. \\
E^{Nec}[\xi] &= \int_0^{+\infty} Nec\{\xi \geq x\}dx - \int_{-\infty}^0 Nec\{\xi \leq x\}dx \\
&= \int_{-\infty}^0 Pos\{\xi > x\}dx - \int_0^{+\infty} Pos\{\xi < x\}dx \\
&= (\int_{-\infty}^a 1dx + \int_a^0 e^{-k(x-a)^2}dx) - \int_0^{+\infty} 1dr \\
&= a + \int_a^0 e^{-k(x-a)^2}dx.
\end{aligned}
$$

Then we have

$$
\begin{aligned}
E^{Me}[\xi] &= \lambda E^{Pos}[\xi] + (1+\lambda)E^{Nec}[\xi] \\
&= \lambda(\int_a^0 e^{-k(x-a)^2}dx + a) + (1+\lambda)(a + \int_a^0 e^{-k(x-a)^2}dx) \\
&= a - \int_0^a e^{-k(x-a)^2}dx.
\end{aligned}
$$

So we always have $E^{Me}[\xi] = a - \int_0^a e^{-k(x-a)^2} dx$. This theorem is proved.   □

Remark 2.12. As well known, it is hard to calculate integral like $\int_0^a e^{-k(x-a)^2} dx$. However, we can substitute $\sum_{i=0}^{n} \int_0^a \frac{[-k(x-a)^2]^i}{i!} dx$ for $\int_0^a e^{-k(x-a)^2} dx$ to obtain the approximate results.

The definition of expected value operator is not only applicable to continuous case but also discrete case.

Proposition 2.4. *Assume that ξ is a discrete fuzzy variable whose membership function is given by*

$$\mu_\xi(x) = \begin{cases} \mu_1, & if \quad x = a_1 \\ \mu_2, & if \quad x = a_2 \\ \cdots \\ \mu_n, & if \quad x = a_n. \end{cases}$$

Without loss of generality, we also assume that $a_1 \leq a_2 \leq \cdots \leq a_n$. Then the expected value of ξ is

$$E^{Me}[\xi] = \sum_{i=1}^{n} w_i a_i, \tag{2.17}$$

where the weights $w_i, i = 1, 2, \cdots, m$ are given by

$$w_1 = \lambda \mu_1 + (1 - \lambda)(\max_{1 \leq j \leq n} \mu_j - \max_{1 < j \leq n} \mu_j),$$
$$w_i = \lambda(\max_{1 \leq j \leq i} \mu_j - \max_{1 \leq j < n} \mu_j) + (1 - \lambda)(\max_{i \leq j \leq n} \mu_j - \max_{i < j \leq n} \mu_j), \ 2 \leq i \leq n-1,$$
$$w_n = \lambda \mu_m + (1 - \lambda)(\max_{1 \leq j \leq n} \mu_j - \max_{1 \leq j < n} \mu_j).$$

Proof. It follows from Definition 2.8 that

$$E^{Pos}[\xi] = \int_0^{+\infty} Pos\{\xi \geq r\}dr - \int_{-\infty}^0 Pos\{\xi \leq r\}dr$$
$$= (\int_0^{x_{k+1}} Pos\{\xi \geq r\}dr + \int_{x_{k+1}}^{x_{k+2}} Pos\{\xi \geq r\}dr + \cdots + \int_{x_{n-1}}^{x_n} Pos\{\xi \geq r\}dr$$
$$+ \int_{x_n}^{+\infty} Pos\{\xi \geq r\}dr) - (\int_{-\infty}^{x_1} Pos\{\xi \leq r\}dr + \int_{x_1}^{x_2} Pos\{\xi \leq r\}dr + \cdots$$
$$+ \int_{x_{k-1}}^{x_k} Pos\{\xi \leq r\}dr + \int_{x_k}^0 Pos\{\xi \leq r\}dr)$$
$$= (\max_{k+1 \leq i \leq n} \mu_i \cdot x_{k+1} + \max_{k+2 \leq i \leq n} \mu_i \cdot (x_{k+2} - x_{k+1}) + \cdots + \mu_n \cdot (x_n - x_{n-1}))$$
$$- (\mu_1 \cdot (x_2 - x_1) + \max_{1 \leq i \leq 2} \mu_i \cdot (x_2 - x_1) + \cdots + \max_{1 \leq i \leq k-1} \mu_i \cdot (x_k - x_{k-1})$$
$$+ \max_{1 \leq i \leq k} \mu_i \cdot (0 - x_k))$$
$$= \mu_1 \cdot x_1 + (\max_{1 \leq i \leq 2} \mu_i - \mu_1)x_2 + \cdots + (\max_{1 \leq i \leq k-1} \mu_i - \max_{1 \leq i \leq k-2} \mu_i)x_{k-1}$$
$$+ (\max_{1 \leq i \leq k} \mu_i - \max_{1 \leq i \leq k-1} \mu_i)x_k + (\max_{k+1 \leq i \leq n} \mu_i - \max_{k+2 \leq i \leq n} \mu_i)x_{k+1} + \cdots$$
$$+ (\max_{n-1 \leq i \leq n} \mu_i - \mu_n)x_{n-1} + \mu_n x_n$$
$$= \mu_1 \cdot x_1 + \sum_{j=2}^{k} (\max_{1 \leq i \leq j} \mu_i - \max_{1 \leq i \leq j-1} \mu_i)x_j + \sum_{j=k+1}^{n-1} (\max_{j \leq i \leq n} \mu_i - \max_{j+1 \leq i \leq n} \mu_i)x_j$$
$$+ \mu_n \cdot x_n.$$

If $\mu_1 \leq \mu_2 \leq \cdots \leq \mu_k$ and $\mu_{k+1} \geq \mu_{k+2} \geq \cdots \geq \mu_n$, then $E[\xi] = \sum_{i=1}^{n} \mu_i x_1$.

It follows from Definition 2.8 that

$$
\begin{aligned}
E^{Nec}[\xi] &= \int_0^{+\infty} Nec\{\xi \geq r\}dr - \int_{-\infty}^0 Nec\{\xi \leq r\}dr \\
&= -\int_0^{+\infty} Pos\{\xi < r\}dr + \int_{-\infty}^0 Pos\{\xi > r\}dr \\
&= -(\int_0^{x_{k+1}} Pos\{\xi < r\}dr + \int_{x_{k+1}}^{x_{k+2}} Pos\{\xi < r\}dr + \cdots + \int_{x_{n-1}}^{x_n} Pos\{\xi < r\}dr \\
&\quad + \int_{x_n}^{+\infty} Pos\{\xi < r\}dr) + (\int_{-\infty}^{x_1} Pos\{\xi > r\}dr + \int_{x_1}^{x_2} Pos\{\xi > r\}dr + \cdots \\
&\quad + \int_{x_{k-1}}^{x_k} Pos\{\xi > r\}dr + \int_{x_k}^{0} Pos\{\xi > r\}dr) \\
&= -(\max_{1 \leq i \leq k} \mu_i \cdot x_{k+1} + \max_{1 \leq i \leq k+1} \mu_i \cdot (x_{k+2} - x_{k+1}) + \cdots + \max_{1 \leq i \leq n-1} \mu_i \cdot (x_n \\
&\quad - x_{n-1}) + \int_{x_n}^{+\infty} 1 dr) + (\int_{-\infty}^{x_1} 1 dr + \max_{2 \leq i \leq n} \mu_i \cdot (x_2 - x_1) + \cdots + \max_{k \leq i \leq n} \mu_i \cdot (x_k \\
&\quad - x_{k-1}) + \max_{k+1 \leq i \leq n} \mu_i \cdot (0 - x_k)) \\
&= (-\max_{2 \leq i \leq n} \mu_i) \cdot x_1 + (\max_{2 \leq i \leq n} \mu_i - \max_{3 \leq i \leq n} \mu_i) \cdot x_2 + \cdots + (\max_{k-1 \leq i \leq n} \mu_i - \max_{k \leq i \leq n} \mu_i) \\
&\quad + \cdot x_{k-1} + (\max_{k \leq i \leq n} \mu_i - \max_{k+1 \leq i \leq n} \mu_i) \cdot x_k + (\max_{1 \leq i \leq k+1} \mu_i - \max_{1 \leq i \leq k} \mu_i) \cdot x_{k+1} \\
&\quad + (\max_{1 \leq i \leq k+2} \mu_i - \max_{1 \leq i \leq k+1} \mu_i) \cdot x_{k+2} + \cdots + (\max_{1 \leq i \leq n-1} \mu_i - \max_{1 \leq i \leq n-2} \mu_i) \cdot x_{n-1} \\
&\quad + (-\max_{1 \leq i \leq n-1} \mu_i) x_n + (\int_{-\infty}^{x_1} 1 dr - \int_{x_n}^{+\infty} 1 dr).
\end{aligned}
$$

Since

$$
\begin{aligned}
\int_{-\infty}^{x_1} 1 dr - \int_{x_n}^{+\infty} 1 dr &= \int_{-\infty}^{x_1} 1 dr + \int_{-x_1}^{x_n} 1 dr - (\int_{-x_1}^{x_n} 1 dr + \int_{x_n}^{+\infty} 1 dr) \\
&= (\int_{-\infty}^{x_1} 1 dr - \int_{-x_1}^{+\infty} 1 dr) + \int_{-x_1}^{x_n} 1 dr \\
&= x_n + x_1,
\end{aligned}
$$

we have

$$
E^{Nec}(\xi) = \sum_{j=1}^{k} (\max_{j \leq i \leq n} \mu_i - \max_{j+1 \leq i \leq n} \mu_i) x_j + \sum_{j=k+1}^{n} (\max_{1 \leq i \leq j} \mu_i - \max_{1 \leq i \leq j-1} \mu_i) x_j.
$$

if $\mu_1 \geq \mu_2 \geq \cdots \geq \mu_k$ and $\mu_{k+1} \leq \mu_{k+2} \leq \cdots \leq \mu_n$, then $E^{Nec}[\xi] = \sum_{i=1}^{n} \mu_i x_1$.

Thus we have

$$
E^{Me}[\xi] = \lambda E^{Pos}[\xi] + (1 - \lambda) E^{Nec}[\xi] = \sum_{i=1}^{n} w_i a_i
$$

where the weights $w_i, i = 1, 2, \cdots, m$ are given by

$$
\begin{aligned}
w_1 &= \lambda \mu_1 + (1 - \lambda)(\max_{1 \leq j \leq n} \mu_j - \max_{1 < j \leq n} \mu_j), \\
w_i &= \lambda(\max_{1 \leq j \leq i} \mu_j - \max_{1 \leq j < n} \mu_j) + (1 - \lambda)(\max_{i \leq j \leq n} \mu_j - \max_{i < j \leq n} \mu_j), \quad 2 \leq i \leq n-1, \\
w_n &= \lambda \mu_m + (1 - \lambda)(\max_{1 \leq j \leq n} \mu_j - \max_{1 \leq j < n} \mu_j).
\end{aligned}
$$

It is easy to verify that all $w_i \geq 0$ and $\sum_{i=1}^{n} w_i = \max_{1 \leq n} \mu_i = 1$ since any fuzzy variables defined on a possibility space are normalized. This theorem is completed. $\square$

Based on the Cr measure, we also have the linearity of fuzzy expected value operator given by the following theorem.

Theorem 2.4. *(Liu [220]) Assume that ξ and η are fuzzy variables with finite expected values. For any real numbers a and b, we have*

$$E^{Cr}[a\xi + b\eta] = aE^{Cr}[\xi] + bE^{Cr}[\eta]. \qquad (2.18)$$

Definition 2.9. Let ξ be a fuzzy variable with finite expected value $E[\xi]$. The variance of ξ is defined as

$$V[\xi] = E[(\xi - E[\xi])^2]. \qquad (2.19)$$

Definition 2.10. Let ξ and η be fuzzy variables such that $E^{[}\xi]$ and $E^{Cr}[\eta]$ are finite. Then the convariance of ξ and η is defined by

$$Cov[\xi, \eta] = E[(\xi - E^{Cr}[\xi])(\eta - E[\eta])]. \qquad (2.20)$$

2.3 Fuzzy EVM

We give the general form of the fuzzy multi-objective decision making model as follows, and the discussion of this section is based on this model,

$$\begin{cases} \max \ [f_1(x,\xi), f_2(x,\xi), \cdots, f_m(x,\xi)] \\ \text{s.t.} \begin{cases} g_r(x,\xi) \leq 0, \ r = 1,2,\cdots, p \\ x \in X, \end{cases} \end{cases} \qquad (2.21)$$

where ξ are fuzzy vectors.

It is necessary for us to know that the model (2.21) is a conceptual model rather than a mathematical model, because we cannot maximize an uncertain quantity. There does not exist a natural ordership in an uncertain world. So the fuzzy multi-objective model needs to be transformed into some approximate certain models to describe the uncertain model. In general, there are three kinds of models which could deal with the fuzzy multi-objective model as follows.

2.3.1 General Model for Fuzzy EVM

From basic knowledge, we know that the expected value operator is a tool that can transform fuzzy uncertainty into crisp. In this section, we use the expected value operator based on the Cr measure. The spectrum of fuzzy expected value model is as follows:

$$\begin{cases} \max \ E[f_1(x,\xi), f_2(x,\xi), \cdots, f_m(x,\xi)] \\ \text{s.t.} \begin{cases} E[g_r(x,\xi)] \leq 0, \ r = 1,2,\cdots, p \\ x \in X. \end{cases} \end{cases} \qquad (2.22)$$

Definition 2.11. A solution x^* of problem (2.22) satisfies $E[g_r(x,\xi)] \leq 0$, $r = 1,2,\cdots, p$ is called a feasible solution.

Definition 2.12. A feasible point, x^* is said to be an efficient solution(non-dominated solution, Pareto solution) for Problem (2.22) such that $E[f_i(x,\xi)] \geq E[f_i(x^*,\xi)]$ for $i = 1,2,\cdots,m$ with strict inequality holding for at least one i.

In order to balance the multiple conflicting objectives, we may employ the following fuzzy expected value goal programming model,

$$
\begin{cases}
\min \ \sum_{j=1}^{l} P_j \sum_{i=1}^{m} (u_{ij}d_i^+ + v_{ij}d_i^-) \\
\text{s.t.} \begin{cases}
E[f_i(x,\xi)] + d_i^- - d_i^+ = \overline{f}_i, \ i = 1,2,\cdots,m \\
E[g_r(x,\xi)] \leq 0, \ r = 1,2,\cdots,p \\
d_i^+, d_i^- \geq 0, \ i = 1,2,\cdots,m \\
x \in X,
\end{cases}
\end{cases} \tag{2.23}
$$

where P_j is the preemptive priority factor which expresses the relative importance of various goals, $P_j \gg P_{j+1}$, for all j, u_{ij} is the weighting factor corresponding to positive deviation for goal i with priority j assigned, v_{ij} is the weighting factor corresponding to negative deviation for goal i with priority j assigned, d_i^+ is the positive deviation from the target of goal i, defined as

$$
d_i^+ = [E[f_i(x,\xi)] - \overline{f}_i] \vee 0, \tag{2.24}
$$

d_i^- is the negative deviation from the target of goal i, defined as

$$
d_i^- = [\overline{f}_i - E[f_i(x,\xi)]] \vee 0, \tag{2.25}
$$

f_i is a function in goal constraints, g_j is a function in real constraints, $\overline{f}_i$ is the target value according to goal i, l is the number of priorities, m is the number of goal constraints, and p is the number of real constraints.

To help decision makers efficiently make the decision, the convexity of (2.22) will be discussed in this section. As we know, for the multi-objective programming problem, if the feasible set is convex and each objective function is convex function, then it is called convex multi-objective programming. For problem (2.22), we prove a convexity theorem of fuzzy expected value model by adding some convexity conditions on the objective and constraint functions.

Theorem 2.5. *Let ξ be a fuzzy vector. Suppose that, for any fixed u, the functions $f_i(x,u), (j = 1,2,\cdots,m)$ and $g_r(x,u)$, $(r = 1,2,\cdots,p)$ are convex in x. Then the fuzzy multi-objective EVM*

$$
\begin{cases}
\max \ E[f_1(x,\xi), f_2(x,\xi),\cdots, f_m(x,\xi)] \\
\text{s.t.} \begin{cases}
E[g_r(x,\xi)] \leq 0, \ r = 1,2,\cdots,p \\
x \in X
\end{cases}
\end{cases} \tag{2.26}
$$

is a convex programming.

Proof. Let x_1 and x_2 be two feasible solutions. Because $g_r(x, \xi)$ is a convex continuous function with respect to x, then

$$g_r(\rho x_1 + (1 - \rho)x_2, \xi) \le \rho g_r(x_1, \xi) + (1 - \rho)g_r(x_2, \xi),$$

where $0 \le \rho \le 1, r = 1, 2, \cdots, p$. We can get that

$$E[g_r(\rho x_1 + (1 - \rho)x_2, \xi)] \le E[\rho g_r(x_1, \xi) + (1 - \rho)g_r(x_2, \xi)].$$

Because for any $\theta \in \Theta$, $\xi(\theta)$ is a real variable. Then by the linearity of expected value of random variable, we have

$$E[g_r(x_1, \xi(\theta)) + (1 - \rho)g_r(x_2, \xi(\theta))] = \rho E[g_r(x_1, \xi(\theta))] + (1 - \rho)E[g_r(x_2, \xi(\theta))].$$

Following the linearity of the expected value operator of a real variable, we can obtain

$$\begin{aligned}
E[\rho g_r(x_1, \xi) &+ (1 - \rho)g_r(x_2, \xi)] \\
&= E[\rho E[g_r(x_1, \xi(\theta))] + (1 - \rho)E[g_r(x_2, \xi(\theta))]] \\
&= \rho E[E[g_r(x_1, \xi(\theta))]] + (1 - \rho)E[E[g_r(x_2, \xi(\theta))]] \\
&= \rho E[g_r(x_1, \xi)] + (1 - \rho)E[g_r(x_2, \xi)].
\end{aligned}$$

Then $E[g_r(\rho x_1 + (1 - \rho)x_2, \xi)] \le \rho E[g_r(x_1, \xi)] + (1 - \rho)E[g_r(x_2, \xi)] \le 0$. This means that $\rho x_1 + (1 - \rho)x_2$ is also a feasible solution. Then it is a convex feasible set.

Similarly, For every i, $f_i(x, \xi)$ is a convex continuous function with respect to x, it follows that

$$f_i(\rho x_1 + (1 - \rho)x_2, \xi) \le \rho f_i(x_1, \xi) + (1 - \rho)f_i(x_2, \xi),$$

then

$$E[f_i(\rho x_1 + (1 - \rho)x_2, \xi)] \le \rho E[f_i(x_1, \xi)] + (1 - \rho)E[f_i(x_2, \xi)],$$

then

$$\begin{aligned}
E[f(\rho x_1 &+ (1 - \rho)x_2, \xi)] \\
&= \sum_{i=1}^m w_i E[f_i(\rho x_1 + (1 - \rho)x_2, \xi)] \\
&\le \sum_{i=1}^m w_i \{\rho E[f_i(x_1, \xi)] + (1 - \rho)E[f_i(x_2, \xi)]\} \\
&= \rho \sum_{i=1}^m w_i E[f_i(x_1, \xi)] + (1 - \rho) \sum_{i=1}^m w_i E[f_i(x_2, \xi)] \\
&= \rho E[f(x_1, \xi)] + (1 - \rho)E[f(x_2, \xi)].
\end{aligned}$$

This means function $E[f(x, \xi)]$ is convex.

Above all, the expected value programming problem (2.22) is convex programming. And the proof is completed. $\square$

2.3.2 Linear Fuzzy EVM and Two-Stage Method

For regular linear programming problems with fuzzy coefficients, we can use the expected value operator,

$$\begin{cases} \max\ E[\sum\limits_{j=1}^{n}\tilde{c}_{ij}x_j, i=1,2,\cdots,m] \\ \text{s.t.}\ \begin{cases} E[\tilde{a}_{rj}x_j] \geq E[\tilde{b}_r],\ r=1,2,\cdots,p \\ x_j \geq 0,\ j=1,2,\cdots,n. \end{cases} \end{cases} \tag{2.27}$$

It is noted that for the linear fuzzy EVM, it is assumed that the combination of the fuzzy variables is linear, but not the decision making variables. In this book, if all the objective functions and constraints in a model are linear combinations of fuzzy-like variables, we call it a linear fuzzy-like model.

Because of the introduction of the expected value operator, the model (2.27) is crisp, and we can use the section 2.3.4 to obtain the expected value.

2.3.2.1 Crisp Equivalent Models

Theorem 2.6. *If the the coefficients of fuzzy EVM for multi-objective decision-making are trapezoidal fuzzy numbers, that is, $\tilde{c}_{ij} = (c_{ij}^1, c_{ij}^2, c_{ij}^3, c_{ij}^4)$, with $c_{ij}^1 > 0$, $\tilde{a}_{rj} = (a_{rj}^1, a_{rj}^2, a_{rj}^3, a_{rj}^4)$, with $a_{rj}^1 > 0$, $\tilde{b}_{rj} = (b_{rj}^1, b_{rj}^2, b_{rj}^3, b_{rj}^4)$, with $b_{rj}^1 > 0$, then the model (2.28)*

$$\begin{cases} \max\ \left[E[\sum\limits_{j=1}^{n}(c_{1j}^1, c_{1j}^2, c_{1j}^3, c_{1j}^4)x_j], \cdots, E[\sum\limits_{j=1}^{n}(c_{mj}^1, c_{mj}^2, c_{mj}^3, c_{mj}^4)x_j] \right] \\ \text{s.t.}\ \begin{cases} E[(a_{rj}^1, a_{rj}^2, a_{rj}^3, a_{rj}^4)x_j] \geq E[(b_{rj}^1, b_{rj}^2, b_{rj}^3, b_{rj}^4)],\ r=1,2,\cdots,p \\ x_j \geq 0,\ j=1,2,\cdots,n \end{cases} \end{cases} \tag{2.28}$$

is equivalent to (2.29) when the decision maker's optimistic-pessimistic index $\lambda = 1/2$.

$$\begin{cases} \max\ \left[\sum\limits_{j=1}^{n}\frac{(c_{1j}^1+c_{1j}^2+c_{1j}^3+c_{1j}^4)}{4}x_j, \cdots, \sum\limits_{j=1}^{n}\frac{(c_{mj}^1+c_{mj}^2+c_{mj}^3+c_{mj}^4)}{4}x_j \right] \\ \text{s.t.}\ \begin{cases} (a_{rj}^1+a_{rj}^2+a_{rj}^3+a_{rj}^4)x_j \geq b_{rj}^1+b_{rj}^2+b_{rj}^3+b_{rj}^4,\ r=1,2,\cdots,p \\ x_j \geq 0,\ j=1,2,\cdots,n. \end{cases} \end{cases} \tag{2.29}$$

Proof. By Theorem 2.18, we know that

$$E[\sum_{j=1}^{n}(c_{1j}^1, c_{1j}^2, c_{1j}^3, c_{1j}^4)x_j] = \sum_{j=1}^{n} E[(c_{1j}^1, c_{1j}^2, c_{1j}^3, c_{1j}^4)]x_j.$$

Since $c_{ij}^1, a_{rj}^1 > 0, b_{rj}^1 > 0$ and $\lambda = 1/4$, by Proposition 2.2, we have

$$E[(c_{1j}^1, c_{1j}^2, c_{1j}^3, c_{1j}^4)] = (c_{1j}^1 + c_{1j}^2 + c_{1j}^3 + c_{1j}^4)/4.$$

According to Theorem 2.4, we can get

$$E[\sum_{j=1}^{n}(c_{1j}^1, c_{1j}^2, c_{1j}^3, c_{1j}^4)x_j] = \sum_{j=1}^{n}\frac{(c_{1j}^1 + c_{1j}^2 + c_{1j}^3 + c_{1j}^4)}{4}x_j.$$

And similarly, we can get

$$E[(a_{rj}^1, a_{rj}^2, a_{rj}^3, a_{rj}^4)x_j] \geq E[(b_{rj}^1, b_{rj}^2, b_{rj}^3, b_{rj}^4)]$$
$$\Leftrightarrow (a_{rj}^1 + a_{rj}^2 + a_{rj}^3 + a_{rj}^4)x_j \geq b_{rj}^1 + b_{rj}^2 + b_{rj}^3 + b_{rj}^4.$$

For simplification, we have the following equivalent model of problem (2.28),

$$\begin{cases} \max \{\Psi_1^c x, \Psi_2^c x, \cdots, \Psi_m^c x\} \\ \text{s.t.} \begin{cases} \Psi_r^a x \leq \Psi_r^b, r = 1, 2, \cdots, p \\ x \in X, \end{cases} \end{cases} \qquad (2.30)$$

where

$$\Psi_i^c = (c_{ij}^1 + c_{ij}^2 + c_{ij}^3 + c_{ij}^4)/4, \ i = 1, 2, \cdots, m.$$

$$\Psi_r^a = (a_{rj}^1 + a_{rj}^2 + a_{rj}^3 + a_{rj}^4)/4, \Psi_i^b = (b_{rj}^1 + b_{rj}^2 + b_{rj}^3 + b_{rj}^4)/4, \ r = 1, 2, \cdots, p.$$

respectively denote the vectors.

Then the proof is completed.   □

2.3.2.2   Two-Stage Method

If an uncertain problem can be transformed into its crisp equivalent model, then we can solve it by classical multi-objective decision making. Many researchers have studied this for decades, Nakayama [112], Luc [113], Szidarovszky [115], Ijiri [114], Lee [126], Kendall [127], Teargny [125], Shimizu [121], Choo [120], Wierzbicki [119], Zimmermann [44], Leberling [117], Werners [116], Sakawa [128], Sakawa et al. [129, 130], Seo and Sakawa [131], Hu [110], Lin and Dong [111], Chen and Lee [108], Youness [109], Chen and Lu [94]. Recently, Xu and Li [92] summed up predecessors' work. Readers who want to study multi-objective decision making methods and theory systematically may refer to these researches.

In this section, we will use the two-stage method to seek an efficient solution to the crisp multi-objective programming problem (2.31). The two-stage method is proposed by Li and Lee [392] on the basis of the maximin method proposed by Zimmermann [393].

$$\begin{cases} \max \{H_1(x), \cdots, H_k(x), \cdots, H_m(x)\} \\ \text{s.t.} \ x \in X. \end{cases} \qquad (2.31)$$

First stage: apply Zimmermann's minimum operator to obtain the maximal satisfying degree α^0 of the objective set and the related feasible solution x^0, i.e.,

$$\begin{cases} \max \alpha \\ \text{s.t.} \begin{cases} \mu_k(x) = \dfrac{H_k(x) - H_k'}{H_k^* - H_k'} \geq \alpha, k = 1, 2, \cdots, , m \\ x \in X. \end{cases} \end{cases} \qquad (2.32)$$

Assume that the optimal solution of problem (2.32) is (x^0, α^0), where α^0 is the optimal satisfying degree of the whole objective sets. If the optimal solution of problem (2.32) is unique, x^0 is the efficient of the problem (2.31). However, we cannot usually know if the optimal solution of problem (2.32) is unique, then the efficiency of x^0 must be checked by the following stage.

Second stage: check the efficiency of efficiency of x^0 or seek the new efficient solution x^1. Construct a new model whose objective function is to maximize the average satisfying degree of all objects subject to the additional constraint $\alpha_k \geq \alpha^0 (k = 1, 2, \cdots, m)$. Since the compensatory of the arithmetic mean operator, the solution obtained in the second stage is efficient. The existence of the constraint $\alpha_k \geq \alpha^0 (k = 1, 2, \cdots, m)$ guarantees the mutual equilibrium of every objective functions.

$$\begin{cases} \max \dfrac{1}{m} \sum\limits_{k=1}^{m} \alpha_k \\ \text{s.t.} \begin{cases} \alpha^0 \leq \alpha_k \leq \mu_k(x), k = 1, 2, \cdots, m \\ 0 \leq \alpha_k \leq 1 \\ x \in X. \end{cases} \end{cases} \qquad (2.33)$$

Assume that the optimal solution of problem (2.33) is x^1. It's easy to prove that x^1 is also the solution of problem (2.32), thus we have $x^1 = x^0$ if the solution of problem (2.32) is unique. But if the solution of problem (2.32) is not unique, x^0 may be efficient solution or not and we can guarantee x^1 is definitely efficient. Thus in any case, the two-stage method can provide an efficient solution in the second stage.

2.3.2.3 Numerical Example

Example 2.5. Let us consider the following example with fuzzy variables,

$$\begin{cases} \max \xi_1 x_1 + \xi_2 x_2 + \xi_3 x_3 + \xi_4 x_4 \\ \min x_1 + x_2 + x_3 + x_4 \\ \text{s.t.} \begin{cases} \xi_5 x_1 + \xi_6 x_2 + \xi_7 x_3 + \xi_8 x_4 \leq \xi_9 \\ 3x_1 + 5x_2 + 10x_3 + 15x_4 \leq 100 \\ x_j \geq 0, j = 1, 2, 3, 4, \end{cases} \end{cases} \qquad (2.34)$$

where, the coefficients are triangular fuzzy variables,

$$\begin{aligned} &\xi_1 = (3.5, 4, 4.5), \quad \xi_2 = (4.5, 5, 5.5), \ \xi_3 = (8.5, 9, 9.5), \\ &\xi_4 = (10.5, 11, 11.5), \ \xi_5 = (6.5, 7, 7.5), \ \xi_6 = (4.5, 5, 5.5), \\ &\xi_7 = (2.5, 3, 3.5), \quad \xi_8 = (1.5, 2, 2.5), \ \xi_9 = (110, 120, 130). \end{aligned}$$

In order to solve it, we use the expected operator to deal with fuzzy objectives and fuzzy constraints, then we can obtain the model,

$$\begin{cases} \max E[\xi_1 x_1 + \xi_2 x_2 + \xi_3 x_3 + \xi_4 x_4] \\ \min x_1 + x_2 + x_3 + x_4 \\ \text{s.t.} \begin{cases} E[\xi_5 x_1 + \xi_6 x_2 + \xi_7 x_3 + \xi_8 x_4] \leq E[\xi_9] \\ 3x_1 + 5x_2 + 10x_3 + 15x_4 \leq 100 \\ x_j \geq 0, j = 1, 2, 3, 4. \end{cases} \end{cases} \tag{2.35}$$

Here we suppose that $\lambda = 1/2$, and by Theorem 2.6, we know that the problem (4.10) is equivalent to model (4.28),

$$\begin{cases} \max F_1 = 4x_1 + 5x_2 + 9x_3 + 11x_4 \\ \min F_2 = x_1 + x_2 + x_3 + x_4 \\ \text{s.t.} \begin{cases} 7x_1 + 5x_2 + 3x_3 + 2x_4 \leq 120 \\ 3x_1 + 5x_2 + 10x_3 + 15x_4 \leq 100 \\ x_j \geq 0, j = 1, 2, 3, 4. \end{cases} \end{cases} \tag{2.36}$$

Of course, we can use fuzzy simulation to compute the expected value, and use the genetic algorithm to solve it. But since the model (4.28) is crisp multi-objective model, so we can use the regular ways to solve it. Here we use the most simple weighted-sum method to transform the bi-objective model (4.28) into the single objective model (4.29) by introducing the preference w_1, w_2 of the decision maker.

$$\begin{cases} \max w_1(4x_1 + 5x_2 + 9x_3 + 11x_4) - w_2(x_1 + x_2 + x_3 + x_4) \\ \text{s.t.} \begin{cases} 7x_1 + 5x_2 + 3x_3 + 2x_4 \leq 120 \\ 3x_1 + 5x_2 + 10x_3 + 15x_4 \leq 100 \\ x_j \geq 0, j = 1, 2, 3, 4. \end{cases} \end{cases} \tag{2.37}$$

Suppose a decision maker give the weight $w_1 = 0.7, w_2 = 0.3$, and we solve this single-objective crisp objective model and gets the results as follows:

$$x^* = (14.7541, 0, 5.57377, 0), F_1 = 109.1803, F_2 = 20.32787$$

Also, we can use the ideal point method in the last subsection to solve model 4.10, before we use the interactive method, we give the standard formulation,

$$\begin{cases} \max F_1 = 4x_1 + 5x_2 + 9x_3 + 11x_4 \\ \max F_2 = -x_1 - x_2 - x_3 - x_4 \\ \text{s.t.} \begin{cases} 7x_1 + 5x_2 + 3x_3 + 2x_4 \leq 120 \\ 3x_1 + 5x_2 + 10x_3 + 15x_4 \leq 100 \\ x_j \geq 0, j = 1, 2, 3, 4. \end{cases} \end{cases}$$

$F_i^0, F_i^1 (i = 1, 2)$ are calculated as follows

$$F_1^0 = 0, \quad F_1^1 = 109.1803, \quad F_2^0 = -22, \quad F_2^1 = 0$$

According to the two-stage method, we calculate the weight as follows:

$$w_1 = 0.832, w_1 = 0.168.$$

By step 3, we obtain the following model

$$
\begin{cases}
\min \left[0.832 \times (4x_1 + 5x_2 + 9x_3 + 11x_4 - 109.1803)^2 \right. \\
\quad \left. +0.168 \times (-x_1 - x_2 - x_3 - x_4 - 22)^2 \right]^{\frac{1}{2}} \\
\text{s.t.} \begin{cases} 7x_1 + 5x_2 + 3x_3 + 2x_4 \leq 120 \\ 3x_1 + 5x_2 + 10x_3 + 15x_4 \leq 100 \\ x_j \geq 0, j - 1,2,3,4 \end{cases}
\end{cases}
$$

After solving the above model, we can obtain the efficient solution as follows:

$$
(x_1, x_2, x_3, x_4) = (11.41, 0, 6.58, 0).
$$

2.3.3 Non-linear Fuzzy EVM and Fuzzy Simulation-Based PSO

It is difficult to transform the non-linear fuzzy EVM into it's equivalent form, so we introduce the fuzzy simulation-based particle swarm optimization algorithm(PSO) to solve.

2.3.3.1 Fuzzy Simulation 1 for Expected Value Operator

Let f be a real-valued function, and ξ_i be a fuzzy variables with membership functions μ_i, $i = 1,2,\cdots,n$, respectively. We denote $\xi = (\xi_1, \xi_2, \cdots, \xi_n)$. Then $f(\xi)$ is also a fuzzy variable whose expected value is defined by

$$
E^{Me}[f(\xi)] = \int_0^\infty Me\{f(\xi) \geq r\}dr - \int_{-\infty}^0 Me\{f(\xi) \leq r\}dr. \tag{2.38}
$$

A fuzzy simulation will be designed to estimate $E^{Me}[f(\xi)]$. We randomly generate $u_{1j}, u_{2j}, \cdots, u_{nj}$ from the ε-level sets of $\xi_1, \xi_2, \cdots, \xi_n$, $j = 1,2,\cdots,m$, respectively, where ε is a sufficiently small number. Let $u_j = (u_{1j}, u_{2j}, \cdots, u_{nj})$ and $\mu_j = \mu_1(u_{1j}) \wedge \mu_2(u_{2j} \wedge \cdots \wedge \mu_n(u_{nj}))$ for $j = 1,2,\cdots,m$.

Then for any number $r \geq 0$, the fuzzy measure $Me\{f(\xi) \geq r\}$ can be estimated by

$$
\begin{aligned}
Me\{f(\xi) \geq r\} = \lambda &\left(\max_{j=1,2,\cdots,m} \{\mu_j | f(u_j) \geq r\} \right) \\
&+ (1-\lambda)\left(1 - \max_{j=1,2,\cdots,m} \{\mu_j | f(u_j) < r\} \right)
\end{aligned} \tag{2.39}
$$

and for any number $r < 0$, $Me\{f(\xi) \leq r\}$ can be estimated by

$$
\begin{aligned}
Me\{f(\xi) \leq r\} = \lambda &\left(\max_{j=1,2,\cdots,m} \{\mu_j | f(u_j) \leq r\} \right) \\
&+ (1-\lambda)\left(1 - \max_{j=1,2,\cdots,m} \{\mu_j | f(u_j) > r\} \right)
\end{aligned} \tag{2.40}
$$

provided that m is sufficiently large.

The procedure is as follows:

Step 1. Set $E = 0$.

Step 2. Randomly generate $u_{1j}, u_{2j}, \cdots, u_{nj}$ from the ε-level sets of $\xi_1, \xi_2, \cdots, \xi_n$, and denote $u_j = (u_{1j}, u_{2j}, \cdots, u_{nj})$, $j = 1, 2, \cdots, m$, respectively, where ε is a sufficiently small number.

Step 3. Set $a = f(u_1) \wedge f(u_2) \wedge \cdots \wedge f(u_m)$, $b = f(u_1) \vee f(u_2) \vee \cdots \vee f(u_m)$.

Step 4. Randomly generate r from $[a, b]$.

Step 5. If $r \geq 0$, then $E \leftarrow E + Me\{f(\xi) \geq r\}$.

Step 6. If $r < 0$, then $E \leftarrow E - Me\{f(\xi) \leq r\}$.

Step 7. Repeat the fourth to sixth steps for N times.

Step 8. $E[f(\xi)] = a \vee 0 + b \wedge 0 + E \cdot (b - a)/N$.

2.3.3.2 PSO

PSO which was first introduced by Kennedy and Eberhart [316, 315] is a population based random search method that imitates the physical movements of individuals in a swarm as a search mechanism. Within the swarm, the learning mechanisms are based on the cognitive and social behavior of particles.

In the PSO algorithm, a solution of a specific problem is represented by an n-dimensional position of a particle. A swarm of a fixed number of particles is generated and each particle is initialized with a random position in a multidimensional search space. Each particle flies through the multidimensional search space with a velocity. In each step of the iteration the velocity of each particle is adjusted based on three components. The first component is the current velocity of the particle which represents the inertia term or momentum of the particle, i.e. the tendency to continue to move in the same direction. The second component is based on a position corresponding to the best solution, usually referred to as the personal best. The third component is based on a position corresponding to the best solution achieved so far by all the particles, i.e. the global best. Once the velocity of each particle is updated, the particles are then moved to new positions. The cycle is repeated until the stopping criterion is met. The specific expressions used int the original particle swarm optimization algorithm are discussed as follows.

Basic Form of PSO

The PSO algorithm consistes of a population of particle initialized with random position and velocity. This population of particle is usually called a swarm. In one iteration step, each particle is first evaluated to find it's individual objective function value. For each particle, if a position is reached which has a better objective function than the previous best solution, the personal best position is updated. Also, if an objective function is found that is better than the previous best objective function of the swarm the global best position is updated. The velocity is then updated on the particle's personal best position and the global best position found so far by the swarm. Every particle is then moved from the current position to the new position based on its velocity. The precess repeats until the stopping criterion is met.

In PSO, a swarm of L particles served as searching agent for a specific problem solution. A particle's position (Θ_l), which consists of H dimensions, is representing (directly of indirectly) a solution of the problem. The ability of particle to search for solution is represented by its velocity vector (Ω), which drives the movement of particle. In each PSO iteration, every particle moves from one position to the next based on its velocity. By moving from one position to the next, a particle is reaching different prospective solution of the problem. The basic particle movement equation if presented below:

$$\theta_{lh}(t+1) = \theta_{lh}(t) + \omega_{lh}(t+1), \tag{2.41}$$

where:

$\theta_{lh}(t+1)$: position of the lth particle at the hth dimension in the $(t+1)^{\text{th}}$ iteration,

$\theta_{lh}(t)$: position of the lth particle at the hth dimension in the tth iteration,

$\omega_{lh}(t+1)$: velocity of the lth particle at the hth dimension in the $(t+1)^{\text{th}}$ iteration.

PSO also imitated swarm's cognitive and social behavior as local and global search abilities. In the basic version of PSO, the particle's personal best position (Ψ_l) and the global best position (Ψ_g) are always updated and maintained. The personal best position of a particle, which expresses the cognitive or self-learning behavior, is defined as the position that gives the best objective function among the positions that have been visited by that particle. Once a particle reaches a position that has a better objective function than the previous best objective function for this particle, i.e., $Z(\Theta_l) < Z(\Psi_l)$, the personal best position is updated, The global best position, which expresses the social behavior, is the position that gives the best objective function among the positions that have been visited by all particles in the swarm. Once a particle reaches a position that has a better objective function than the previous best objective function for whole swarm, i.e., $Z(\Psi_l) < Z(\Psi_g)$, the global best position is also updated.

The personal best and global best position are used as the basis to update velocity of particle. In each iteration step, the velocity Ω is updated based on three terms: inertia, cognitive learning and social learning terms.

The inertia term forces particle to move in the same direction as in previous iteration. This term is calculated as a product of current velocity with an inertia weight (w).

The cognitive term forces particle to go back to its personal best position. This term is calculated as a product of a random number (u), personal best acceleration constant (c_p), and the difference between personal best position Ψ_l and current position Θ_l.

The social term forces particle to move toward the global best position. This term is calculated as a product of random number (u), global best acceleration constant (c_g), and the difference between global best position Ψ_g and current position Θ_l. Specifically, the equation for velocity updated is expressed as follow:

$$\omega_{lh}(t+1) = w\omega_{lh}(t) + c_p u(\psi_{lh} - \theta_{lh}(t)) + c_g u(\psi_{gh} - \theta_{lh}(t)), \tag{2.42}$$

where:

$\omega_{lh}(t+1)$: velocity of the lth particle at the hth dimension in the tth iteration,

ψ_{lh}: personal best position of the lth particle at the hth dimension in the tth iteration,

ψ_{gh}: global best position at the hth dimension in the tth iteration.

In the velocity-updating formula, random numbers is incorporated in order to randomize particle movement. Hence, two different particles may move to different position in the subsequent iteration even though they have similar position, personal best, and the global best.

Notation

The notation used in the algorithm is given as follows:

t: iteration index, $t = 1 \cdots T$.

l: particle index, $l = 1 \cdots L$.

h: dimension index, $h = 1 \cdots H$.

u: uniform random number in the interval $[0, 1]$.

$w(t)$: inertia weight in the tth iteration.

$\omega_{lh}(t+1)$: velocity of the lth particle at the hth dimension in the tth iteration.

$\theta_{lh}(t)$: position of the lth particle at the hth dimension in the tth iteration.

ψ_{lh}: personal best position of the lth particle at the hth dimension in the tth iteration.

ψ_{gh}: global best position at the hth dimension in the tth iteration.

c_p: personal best position acceleration constant.

c_g: global best position acceleration constant.

$\theta^{\max}$: maximum position value.

$\theta^{\min}$: minimum position value.

Θ_l: vector position of the lth particle, $[\theta_{l1}, \theta_{l2}, \cdots, \theta_{lH}]$.

Ω_l: vector velocity of the lth particle, $[\omega_{l1}, \omega_{l2}, \cdots, \omega_{lH}]$.

Ψ_l: vector personal best position of the lth particle, $[\psi_{l1}, \psi_{l2}, \cdots, \psi_{lH}]$.

Ψ_g: vector global best position, $[\psi_{l1}, \psi_{l2}, \cdots, \psi_{lH}]$.

R_l: the lth set of solution.

$Z(\Theta_l)$: fitness value of Θ_l.

Procedure of PSO algorithm

Step 1. Initialize L particle as a swarm:

Generate the lth particle with random position Θ_l in the range $[\theta^{\min}, \theta^{\max}]$, velocity $\Omega_l = 0$ and personal best $\Psi_l = \Omega_l$ for $l = 1 \cdots L$. Set iteration $t = 1$.

Step 2. Decode particles into solutions:

For $l = 1 \cdots L$, decode $\Theta_l(t)$ to a solution R_l. (This step is only needed if the particles are not directly representing the solutions.)

Step 3. Evaluate the particles:

For $l = 1 \cdots L$, compute the performance measurement of R_l, and set this as the fitness value of Θ_l, represent by $Z(\Theta_l)$.

Step 4. Update pbest:
For $l = 1 \cdots L$, update $\Psi_l = \Theta_l$, if $Z(\Theta_l) < Z(\Psi_l)$.
Step 5. Update pbest:
For $l = 1 \cdots L$, update $\Psi_g = \Psi_l$, if $Z(\Psi_l) < Z(\Psi_g)$.
Step 6. update the velocity and the position of each lth particle:

$$w(t) = w(T) + \frac{t-T}{1-T}[w(1) \quad w(T)], \tag{? 43}$$

$$\omega_{lh}(t+1) = w\omega_{lh}(t) + c_p u(\psi_{lh} - \theta_{lh}(t)) + c_g u(\psi_{gh} - \theta_{lh}(t)), \tag{2.44}$$

$$\theta_{lh}(t+1) = \theta_{lh}(t) + \omega_{lh}(t+1). \tag{2.45}$$

If $\theta_{lh}(t+1) > \theta^{\max}$, then
$$\theta_{lh}(t+1) = \theta^{\max}, \tag{2.46}$$

$$\omega_{lh}(t+1) = 0. \tag{2.47}$$

If $\theta_{lh}(t+1) < \theta^{\max}$, then
$$\theta_{lh}(t+1) = \theta^{\min}, \tag{2.48}$$

$$\omega_{lh}(t+1) = 0. \tag{2.49}$$

Step. 7 if the stopping criteria is met, i.e., t=T, stop; otherwise, $t = t + 1$ and return to step 2.

The value $\theta^{\max}$ and $\theta^{\min}$ in equation (2.46) and (2.48) are the upper and lower bounds on the position of particles.

Kay Parameters of PSO

Let's discuss the possible qualifications and effects of each parameter on the performance of PSO. The parameters consist of the population size (L), two acceleration constants (c_p and c_g), and the inertia weight (w).

Population size (L): This parameter represents the number of particles in the system. It is one of the important parameters of PSO, because it affects the fitness value and computation time. Furthermore, increasing the size of the population always increases computation time, but might not improve the fitness value. Generally speaking, too small a population size can lead to poor convergence while too large a population size can yield good convergence at the expense of a long running time.

Acceleration constants (c_p and c_g): The constants c_p and c_g are the acceleration constants of the personal best position and the global best position, respectively. Each acceleration constant controls the maximum distance that a particle is allowed to move from the current position to each best position. The new velocity can be viewed as a vector which combines the current velocity, and the vectors of the best

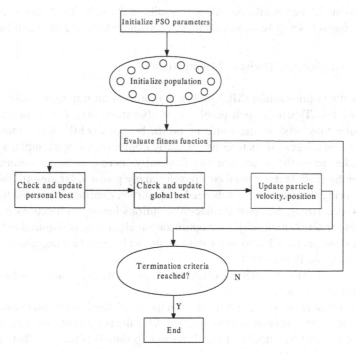

Fig. 2.6 Flow chart of PSO

positions. Each best positions's vector consists of the direction which is pointed from the particle's current position to the best position, and the magnitude of the movement can be between 0 to the acceleration constant of the best position times the distance between the best position and the current position.

Inertia weight (w): The new velocity is produced from the combination of vectors. One of these vectors is the current velocity. Inertia weight is a weight to control the magnitude of the current velocity on updating the new velocity. For $w = c$, it means that this vector has the same direction of the current velocity, and the parameters to control the search behavior of the swarm.

Velocity boundary (V_{max}) and position boundary (θ_{max}): Some PSO algorithms are implemented with a bound on velocity. For each dimension, the magnitude of velocity cannot be greater than V_{max}. This parameter is one of parameters that control the search behavior of the swarm. The smaller value of this parameter makes the particles in the population less aggressive in the search.

In the PSO particle movement mechanism, it is also common to limit the search space of particle location, i.e., the position value of particle dimension is bounded in the interval $[\theta^{min}, \theta^{max}]$. The use of position boundary θ^{max} is to force each particle to move within the feasible region to avoid solution divergence. Hence, the position value of certain particle dimension is being set at the minimum or maximum value whenever it moves beyond the boundary. In addition, the velocity of the corresponding dimension is reset to zero to avoid further movement beyond the boundary.

PSO for Multi-objective Decision-Making Model

Multi-objective optimization (MO) problems represent an important class of real-world problems. Typically such problems involve trade offs. For example, a car manufacturer may wish to maximize its profit, but meanwhile also minimize its production costs. These objectives are typically conflicting. For example, a higher profit could increase the production cost. Generally, there is no single optimal solution. Often the manufacturer needs to consider many possible "trade-off" solutions before choosing the one that suits its need. The curve or surface (for more than two objectives) describing the optimal trade-off solutions between objectives is known as the Pareto front. A multi-objective optimization algorithm is required to find solutions as close as possible to the Pareto front, while maintaining good solution diversity along the Pareto front.

To apply PSO to multi-objective optimization problems, several issues have to be taken into consideration:

1. How to choose pg (i.e., a leader) for each particle? The PSO needs to favor non-dominated particles over dominated ones, and drive the population towards different parts of the Pareto front, not just towards a single point. This requires that particles be allocated to different leaders.

2. How to identify non-dominated particles with respect to all particles current positions and personal best positions? And how to retain these solutions during the search process? One strategy is to combine all particles personal best positions and current positions, and then extract the non-dominated solutions from the combined population.

3. How to maintain particle diversity so that a set of well-distributed solutions can be found along the Pareto front? Some classic niching methods (e.g., crowding or sharing) can be adopted for this purpose.

The first PSO for solving multi-objective optimization was proposed by Moore and Chapman [317] in 1999. The main difference between single objective PSO and MOPSO is how to choose the global best. An *lbest* PSO was used, and p_g was chosen from a local neighborhood using a ring topology. All personal best positions were kept in an archive. At each particle update, the current position is compared with solutions in this archive to see if the current position represents a non-dominated solution. The archive is updated at each iteration to ensure it contains only non-dominated solutions.

Interestingly it was not until 2002 that the next publication on PSO for multi-objective optimization appeared. Coello and Lechuga [318] proposed MOPSO (Multi-objective PSO) which uses an external archive to store nondominated solutions. The diversity of solutions is maintained by keeping only one solution within each hypercube which is predefined by a user in the objective space. Parsopoulos and Vrahatis[319] adopted a more traditional weighted-sum approach. However, by using gradually changing weights, their approach was able to find a diverse set of solutions along the Pareto front. Fieldsend and Singh [320] proposed a PSO using a dominated tree structure to store non-dominated solutions found. The selection of leaders was also based on this structure. To maintain a better diversity, a turbulence operator was adopted to function as a 'mutation' operator in order to perturb the velocity value of a particle.

With the aim of increasing the efficiency of extracting non-dominated solutions from a swarm, Li [321] proposed NSPSO (Non-dominated Sorting PSO), which follows the principal idea of the well-known NSGA II algorithm [249]. In NSPSO, instead of comparing solely a particles personal best with its potential offspring, all particles personal best positions and offspring are first combined to form a temporary population. After this, domination comparisons for all individuals in this temporary population are carried out. This approach will ensure more non-dominated solutions can be discovered through the domination comparison operations rather than the above-mentioned multi-objective PSO algorithms.

Many more multi-objective PSO variants have been proposed in recent years. A survey conducted by Sierra [322] and Coello in 2006 shows that there are currently 25 different PSO algorithms for handling multi-objective optimization problems. Interested readers should refer to these for more information on different approaches.

Finally, we present the flow chart for fuzzy simulation-based PSO as follows, see Figure 2.7:

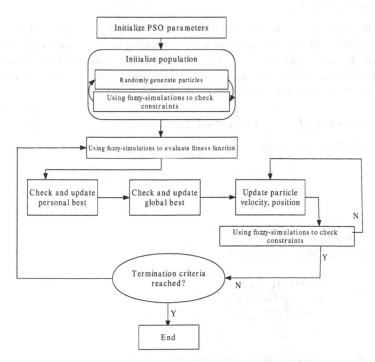

Fig. 2.7 Flow chart of Fu-Ra simulation-based GA

2.3.3.3   Numerical Example

Example 2.6. Consider the following programming problem,

$$
\begin{cases}
\max f_1(x,\xi) = \xi_1^2 x_1 + \xi_2^2 x_2 \\
\max f_2(x,\xi) = c_1 \xi_1^2 x_1 + c_2 \xi_2^2 x_2 \\
\text{s.t. } \begin{cases} x_1 + x_2 \le 30 \\ 3x_1 - 2x_2 \ge 8 \\ x_1 \ge 0, x_2 \ge 0, \end{cases}
\end{cases}
\tag{2.50}
$$

where $c = (c_1, c_2) = (1.2, -0.5)$, $\xi_1 = (1.8, 2, 2.2)$ and $\xi_2 = (1, 1.5, 2)$.

According to the constraints, we can obtain the interval of the two decision variables, that is,

$$
0 \le x_1 \le 30, 0 \le x_2 \le 17.
$$

Then we can initialize the particle by randomly generating the two dimensional particles from the intervals [0,30] and [0,17].

After using the fuzzy simulation-based PSO, we can get the optimal solution as follows:

$$
f_1 = 123.698, f_2 = 121.4493, x_1 = 13.694, x_2 = 0.5551.
$$

2.4   Fuzzy CCM

Besides the expected value operator which can be adopted to deal with the fuzzy multi-objective decision making model, we can also use the chance operator to transform the fuzzy uncertain into crisp, that we called fuzzy chance-constrained model. For the single fold uncertain problem-fuzzy problem, the chance operator is actually the *Pos* or *Nec* measure.

2.4.1   General Model for Fuzzy CCM

For the following fuzzy multi-objective model

$$
\begin{cases}
\max [f_1(x,\xi), f_2(x,\xi), \cdots, f_m(x,\xi)] \\
\text{s.t. } \begin{cases} g_r(x,\xi) \le 0, \ r = 1, 2, \cdots, p \\ x \in X, \end{cases}
\end{cases}
\tag{2.51}
$$

where ξ is fuzzy variable.

The general CCM is as following,

$$
\begin{cases}
\max [\bar{f}_1, \bar{f}_2, \cdots, \bar{f}_m] \\
\text{s.t. } \begin{cases} Ch\{f_i(x,\xi) \ge \bar{f}_i\} \ge \delta_i, \ i = 1, 2, \cdots, m \\ Ch\{g_r(x,\xi) \le 0\} \ge \theta_r, \ r = 1, 2, \cdots, p \\ x \in X, \end{cases}
\end{cases}
\tag{2.52}
$$

where δ_i, θ_r are the predetermined confidence levels.

We adopt Pos and Nec to measure the fuzzy event respectively, then we have two kinds of fuzzy CCM.

2.4.1.1   Fuzzy CCM Based on Pos Measure

The spectrum of chance-constrained model based on *Pos* measure is as follow:

$$
\begin{cases}
\max \; [\bar{f}_1, \bar{f}_2, \cdots, \bar{f}_m] \\
\text{s.t.} \begin{cases}
Pos\{f_i(x, \xi) \geq \bar{f}_i\} \geq \delta_i, \; i = 1, 2, \cdots, m \\
Pos\{g_r(x, \xi) \leq 0\} \geq \theta_r, \; r = 1, 2, \cdots, p \\
x \in X,
\end{cases}
\end{cases}
\tag{2.53}
$$

where δ_i, θ_r are the predetermined confidence levels.

Definition 2.13. A solution x^* of problem (2.53) satisfies $Pos\{g_r(x, \xi) \leq 0\} \geq \theta_r$, $r = 1, 2, \cdots, p$ is called a feasible solution at θ_r-possibility levels, $r = 1, 2, \cdots, p$.

Definition 2.14. A feasible solution at θ_r-possibility levels, x^*, is said to be a δ_i-efficient solution for Problem (2.53) if and only if there exists no other feasible solution at θ_r-possibility levels, such that $Pos\{f_i(x, \xi)\} \geq \delta_i$ with $f_i(x) \geq \bar{f}_i(x^*)$ for all i and $f_{i_0}(x) > \bar{f}_{i_0}(x^*)$ for at least one $i_0 \in \{1, 2, \cdots, m\}$.

The model (2.60) is called max max model since it is equivalent to

$$
\begin{cases}
\max\limits_{x} \max\limits_{\bar{f}_i} [\bar{f}_1, \bar{f}_2, \cdots, \bar{f}_m] \\
\text{s.t.} \begin{cases}
Pos\{f_i(x, \xi) \geq \bar{f}_i\} \geq \delta_i, \; i = 1, 2, \cdots, m \\
Pos\{g_r(x, \xi) \leq 0\} \geq \theta_r, \; r = 1, 2, \cdots, p \\
x \in X.
\end{cases}
\end{cases}
\tag{2.54}
$$

If we minimize objectives, the model may be formulated as follows,

$$
\begin{cases}
\min [\bar{f}_1, \bar{f}_2, \cdots, \bar{f}_m] \\
\text{s.t.} \begin{cases}
Pos\{f_i(x, \xi) \geq \bar{f}_i\} \geq \delta_i, \; i = 1, 2, \cdots, m \\
Pos\{g_r(x, \xi) \leq 0\} \geq \theta_r, \; r = 1, 2, \cdots, p \\
x \in X.
\end{cases}
\end{cases}
\tag{2.55}
$$

In addition, we have maxmin model and minmax model presented by (2.56) and (2.57) respectively,

$$
\begin{cases}
\max\limits_{x} \min\limits_{\bar{f}_i} [\bar{f}_1, \bar{f}_2, \cdots, \bar{f}_m] \\
\text{s.t.} \begin{cases}
Pos\{f_i(x, \xi) \geq \bar{f}_i\} \geq \delta_i, \; i = 1, 2, \cdots, m \\
Pos\{g_r(x, \xi) \leq 0\} \geq \theta_r, \; r = 1, 2, \cdots, p \\
x \in X,
\end{cases}
\end{cases}
\tag{2.56}
$$

$$
\begin{cases}
\min\limits_{x} \max\limits_{\bar{f}_i} [\bar{f}_1, \bar{f}_2, \cdots, \bar{f}_m] \\
\text{s.t.} \begin{cases}
Pos\{f_i(x, \xi) \geq \bar{f}_i\} \geq \delta_i, \; i = 1, 2, \cdots, m \\
Pos\{g_r(x, \xi) \leq 0\} \geq \theta_r, \; r = 1, 2, \cdots, p \\
x \in X.
\end{cases}
\end{cases}
\tag{2.57}
$$

2.4.1.2   Fuzzy CCM Based on Nec Measure

The measure "Pos" can be substituted by "Nec". The spectrum of chance-constrained model based on Nec measure is as follow:

$$
\begin{cases}
\max\ [\bar{f}_1, \bar{f}_2, \cdots, \bar{f}_m] \\
\text{s.t.}\ \begin{cases}
Nec\{f_i(x, \xi) \geq \bar{f}_i\} \geq \delta_i,\ i = 1, 2, \cdots, m \\
Nec\{g_r(x, \zeta) \leq 0\} \geq \theta_r,\ r = 1, 2, \cdots, p \\
x \in X,
\end{cases}
\end{cases}
\tag{2.58}
$$

where δ_i, θ_r are the predetermined confidence levels.

Definition 2.15. A solution x^* of problem (2.58) satisfies

$$
Nec\{g_r(x, \xi) \leq 0\} \geq \theta_r,\ r = 1, 2, \cdots, p
$$

is called a feasible solution at θ_r-necessity levels, $r = 1, 2, \cdots, p$.

Definition 2.16. A feasible solution at θ_r-necessity levels, x^*, is said to be a δ_i-efficient solution for Problem (2.58) if and only if there exists no other feasible solution at θ_r-necessity levels x, such that $Nec\{f_i(x, \xi)\} \geq \delta_i$ with $f_i(x) \geq \bar{f}_i(x^*)$ for all i and $f_{i_0}(x) > \bar{f}_{i_0}(x^*)$ for at least one $i_0 \in \{1, 2, \cdots, m\}$.

2.4.1.3   Fuzzy CCM Based on Cr Measure

We can also adopted the credibility measure. The spectrum of chance-constrained model based on Cr measure is as follows:

$$
\begin{cases}
\max\ [\bar{f}_1, \bar{f}_2, \cdots, \bar{f}_m] \\
\text{s.t.}\ \begin{cases}
Cr\{f_i(x, \xi) \geq \bar{f}_i\} \geq \delta_i,\ i = 1, 2, \cdots, m \\
Cr\{g_r(x, \xi) \leq 0\} \geq \theta_r,\ r = 1, 2, \cdots, p \\
x \in X,
\end{cases}
\end{cases}
\tag{2.59}
$$

where δ_i, θ_r are the predetermined confidence levels.

Definition 2.17. A solution x^* of problem (2.59) satisfies

$$
Cr\{g_r(x, \xi) \leq 0\} \geq \theta_r,\ r = 1, 2, \cdots, p
$$

is called a feasible solution at θ_r-credibility levels, $r = 1, 2, \cdots, p$.

Definition 2.18. A feasible solution at θ_r-credibility levels, x^*, is said to be a δ_i-efficient solution for Problem (2.59) if and only if there exists no other feasible solution x at θ_r-credibility levels, such that $Cr\{f_i(x, \xi)\} \geq \delta_i$ with $f_i(x) \geq \bar{f}_i(x^*)$ for all i and $f_{i_0}(x) > \bar{f}_{i_0}(x^*)$ for at least one $i_0 \in \{1, 2, \cdots, m\}$.

2.4.2 Linear Fuzzy CCM and Goal Programming Method

We consider the linear programming with fuzzy parameters $\tilde{c}_{ij}, \tilde{a}_{rj}, \tilde{b}_r$, $i = 1, 2, \cdots, n$, $j = 1, 2, \cdots, m$, $r = 1, 2, \cdots, p$, fuzzy constraints can no longer give a crisp feasible set. Naturally, decision maker requires constraints to hold at a pre-determined confidence level. The chance constraints with fuzzy parameters are expressed as follows:

$$Pos\{\sum_{j=1}^{n} \tilde{a}_{rj} x_j \leq \tilde{b}_r\} \geq \theta_r, \; r = 1, 2, \cdots, p. \qquad (2.60)$$

where $Pos\{\cdot\}$ denotes the possibility of the event $\{\cdot\}$.

If the decision maker hopes that the possibility of the objectives are not less than δ_i, and that the possibility of the constraints being no greater than $\tilde{b}_r$, is not less than θ_r, then the chance-constrained linear programming model with fuzzy parameters is formulated as follows:

$$\begin{cases} \max \; [\bar{f}_1, \bar{f}_2, \cdots, \bar{f}_m] \\ \text{s.t.} \begin{cases} Pos\{\sum_{j=1}^{n} \tilde{c}_{ij} x_j \geq \bar{f}_i\} \geq \delta_i, \; i = 1, 2, \cdots, m \\ Pos\{\sum_{j=1}^{n} \tilde{a}_{rj} x_j \leq \tilde{b}_r\} \geq \theta_r, \; r = 1, 2, \cdots, p \\ x_i \geq 0, \; i = 1, 2, \cdots, m, \end{cases} \end{cases} \qquad (2.61)$$

where $\max \bar{f}_i$ is the δ_i-return defined as

$$\max\{\bar{f}_i | Pos\{\sum_{j=1}^{n} \tilde{c}_{ij} x_j \geq \bar{f}_i\} \geq \delta_i\}.$$

And similarly, we have

$$\begin{cases} \max \; [\bar{f}_1, \bar{f}_2, \cdots, \bar{f}_m] \\ \text{s.t.} \begin{cases} Nec\{\sum_{j=1}^{n} \tilde{c}_{ij} x_j \geq \bar{f}_i\} \geq \delta_i, \; i = 1, 2, \cdots, m \\ Nec\{\sum_{j=1}^{n} \tilde{a}_{rj} x_j \leq \tilde{b}_r\} \geq \theta_r, \; r = 1, 2, \cdots, p \\ x_i \geq 0, \; i = 1, 2, \cdots, m, \end{cases} \end{cases} \qquad (2.62)$$

where $\max \bar{f}_i$ is the δ_i-return defined as

$$\max\{\bar{f}_i | Nec\{\sum_{j=1}^{n} \tilde{c}_{ij} x_j \geq \bar{f}_i\} \geq \delta_i\}.$$

Also, we have

$$
\begin{cases}
\max \ [\bar{f}_1, \bar{f}_2, \cdots, \bar{f}_m] \\
\text{s.t.} \begin{cases}
Cr\{\sum\limits_{j=1}^{n} \tilde{c}_{ij}x_j \geq \bar{f}_i\} \geq \delta_i, \ i = 1, 2, \cdots, m \\
Cr\{\sum\limits_{j=1}^{n} \tilde{a}_{rj}x_j \leq \tilde{b}_r\} \geq \theta_r, \ r = 1, 2, \cdots, p \\
x_i \gtrless 0, \ l = 1, 2, \cdots, m,
\end{cases}
\end{cases}
\tag{2.63}
$$

where $\max \bar{f}_i$ is the δ_i-return defined as

$$
\max\{\bar{f}_i | Cr\{\sum_{j=1}^{n} \tilde{c}_{ij}x_j \geq \bar{f}_i\} \geq \delta_i\}.
$$

2.4.2.1 Crisp Equivalent Model

The mathematical traditional solution methods require conversion of the chance constraints to their respective deterministic equivalents. However, this process is usually hard to perform and only successful for only some special cases.

First we present some useful results.

Lemma 2.5. *[77] Let ξ be the fuzzy variable with membership function μ, and the function $g(x,\xi)$ has the form $g(x,\xi) = h(x) - \xi$. Then we have:*
(1). $Pos\{g(x,\xi) \leq 0\} \geq \alpha$ if and only if $h(x) \leq K_\alpha$, where

$$
K_\alpha = \sup\{K | K = \mu^{-1}(\alpha)\},
\tag{2.64}
$$

(2). $Nec\{g(x,\xi) \leq 0\} \geq \alpha$ if and only if $h(x) \leq K_\alpha$, where

$$
K_\alpha = \inf\{K | K = \mu^{-1}(1 - \alpha)\},
\tag{2.65}
$$

(3). $Cr\{g(x,\xi) \leq 0\} \geq \alpha$ if and only if $h(x) \leq K_\alpha$, where

$$
K_\alpha = \begin{cases} \sup\{K | K = \mu^{-1}(2\alpha)\}, & \text{if } \alpha < 1/2 \\ K_\alpha = \inf\{K | K = \mu^{-1}(2(1 - \alpha))\}, & \text{if } \alpha \geq 1/2. \end{cases}
\tag{2.66}
$$

In the following content, we consider a special form, problem (2.61), in which model, $f_i, g_r, i = 1, 2, \cdots, m, r = 1, 2, \cdots, p$ are linear functions, and the fuzzy variables are LR fuzzy variables.

Before we give the theorem, let us recall the operations on trapezoidal fuzzy variables in advance.

Lemma 2.6. *[22] Let $\xi_1, \xi_2, \cdots, \xi_n$ be fuzzy variables, and $f : R^n \to R$ be a continuous function. Then the membership function μ_ξ of $\xi = f(\xi_1, \xi_2, \cdots, \xi_n)$ is derived from the membership functions $\mu_{\xi_1}, \mu_{\xi_2}, \cdots, \mu_{\xi_n}$ by*

$$
\mu_\xi(x) = \sup_{x_1, x_2, \ldots, x_n \in R} \left\{ \min_{1 \leq i \leq n} \mu_{\xi_i}(x_i) | x = f(x_1, x_2, \cdots, x_n) \right\}.
$$

Using Lemma 2.6, we can obtain the sum of trapezoidal fuzzy variables $\tilde{m} = (m_1, m_2, m_3, m_4)$ and $\tilde{n} = (n_1, n_2, n_3, n_4)$, that is,

$$\mu_{\tilde{m}+\tilde{n}}(z) = \sup\{\min\{\mu_{\tilde{m}}(x), \mu_{\tilde{n}}(y)\}|z = x+y\}$$
$$= \begin{cases} \frac{z-(m_1+n_1)}{(m_2+n_2)-(m_1+n_1)}, & \text{if } m_1 + n_1 \leq z \leq m_2 + n_2 \\ 1, & \text{if } m_2 + n_2 \leq z \leq m_3 + n_3 \\ \frac{z-(m_4+n_4)}{(m_3+n_3)-(m_4+n_4)}, & \text{if } m_3 + n_3 \leq z \leq m_4 + n_4 \\ 0, & \text{otherwise.} \end{cases}$$

That is, the sum of two trapezoidal fuzzy variables is also a trapezoidal fuzzy variable, and

$$\tilde{m} + \tilde{n} = (m_1 + n_1, m_2 + n_2, m_3 + n_3, m_4 + n_4). \tag{2.67}$$

Next we consider the product of a trapezoidal fuzzy variable $\tilde{m}$ and a scalar number λ. We have

$$\mu_{\lambda \cdot \tilde{m}}(z) = \sup\{\mu_{\tilde{m}}(x)|z = \lambda x\},$$

which yields that

$$\lambda \cdot \tilde{m} = \begin{cases} (\lambda m_1, \lambda m_2, \lambda m_3, \lambda m_4), & \text{if } \lambda \geq 0 \\ (\lambda m_4, \lambda m_3, \lambda m_2, \lambda m_1), & \text{if } \lambda < 0. \end{cases} \tag{2.68}$$

That is, the product of a trapezoidal fuzzy variable and a scalar number is also a trapezoidal fuzzy variable.

If the objective functions and the constraints are linear, the problem (2.61) can be converted to its crisp equivalent. The result is given by the following several theorems.

First, we give the crisp equivalent model for fuzzy CCM based on *Pos* measure.

Theorem 2.7. *Assume that $\tilde{c}_{ij}$ is LR fuzzy variable, the membership function of $\tilde{c}_{ij}$ is*

$$\mu_{\tilde{c}_{ij}}(t) = \begin{cases} L(\frac{c_{ij}-t}{\alpha_{ij}^c}), & t \leq c_{ij}, \alpha_{ij}^c > 0 \\ R(\frac{t-c_{ij}}{\beta_{ij}^c}), & t \geq c_{ij}, \beta_{ij}^c > 0, \end{cases} \tag{2.69}$$

where the vector $(c_{ij})_{n\times 1} = (c_{i1}, c_{i2}, \cdots, c_{in})^{\mathrm{T}}$ is real number, α_{ij}^c and β_{ij}^c are the left and right spread of $\tilde{c}_{ij}$, $i = 1, 2, \cdots, m$, $j = 1, 2, \cdots, n$, the reference function $L, R : [0, 1] \rightarrow [0, 1]$ satisfies that $L(1) = R(1) = 0$, $L(0) = R(0) = 1$, and it is monotone function. Then $\mathrm{Pos}\{\tilde{c}_i^{\mathrm{T}} x \geq f_i\} \geq \delta_i$ is equivalent to

$$f_i \leq c_i^T x + R^{-1}(\delta_i)\beta_i^{c\mathrm{T}} x, \quad i = 1, 2, \cdots, m, \tag{2.70}$$

Proof. Because $\tilde{c}_{ij}$ is LR fuzzy number, its membership function is $\mu_{\tilde{c}_{ij}}$. By extension principle[22], the membership function of fuzzy number $\tilde{c}_i^{\mathrm{T}} x$ is

$$\mu_{\tilde{c}_i^T x}(r) = \begin{cases} L(\frac{c_i^{\mathrm{T}} x-r}{\alpha_i^{c\mathrm{T}} x}), & r \leq c_i^{\mathrm{T}} x \\ R(\frac{r-c_i^{\mathrm{T}} x}{\beta_i^{c\mathrm{T}} x}), & r \geq c_i^{\mathrm{T}} x \end{cases} \quad i = 1, 2, \cdots, m. \tag{2.71}$$

For convenience, we denote $\tilde{c}_{ij} = (c_{ij}, \alpha_{ij}^c, \beta_{ij}^c)_{LR}$, $\tilde{c}_i^T x = (c_i^T x, \alpha_i^{cT} x, \beta_i^{cT} x)_{LR}$.

According to Lemma 2.2 we can get

$$\text{Pos}\{\tilde{c}_i^T x \geq f_i\} \geq \delta_i \Leftrightarrow c_i^T x + R^{-1}(\delta_i)\beta_i^{cT} x \geq f_i, \quad i = 1, 2, \cdots, m.$$

The proof is completed.   □

Remark 2.13. Especially, when the reference function of the variable $\tilde{c}_{ij}$ is $L(x) = R(x) = 1 - x$ ($x \in [0,1]$), then the LR fuzzy variable is specified as the triangular fuzzy variable, and $R^{-1}(\delta_i) = 1 - \delta_i$, so we have the following equivalent expressions:

$$\text{Pos}\{\tilde{c}_i^T x \geq f_i\} \geq \delta_i \Leftrightarrow c_i^T x + (1 - \delta_i)\beta_i^{cT} x \geq f_i, \quad i = 1, 2, \cdots, m.$$

Theorem 2.8. *Assume that $\tilde{a}_{rj}, \tilde{b}_r$ are LR fuzzy variables, the membership function of $\tilde{a}_{rj}$ and $\tilde{b}_r$ are*

$$\mu_{\tilde{a}_{rj}}(t) = \begin{cases} L(\frac{a_{rj}-t}{\alpha_{rj}^a}), & t \leq a_{rj}, \alpha_{rj}^a > 0 \\ R(\frac{t-a_{rj}}{\beta_{rj}^a}), & t \geq a_{rj}, \beta_{rj}^a > 0, \end{cases} \quad (2.72)$$

$$\mu_{\tilde{b}_r}(t) = \begin{cases} L(\frac{b_r-t}{\alpha_r^b}), & t \leq b_r, \alpha_r^b > 0 \\ R(\frac{t-b_r}{\beta_r^b}), & t \geq b_r, \beta_r^b > 0, \end{cases} \quad (2.73)$$

where the vector $(a_{rj})_{n\times 1} = (a_{r1}, a_{r2}, \cdots, a_{rn})^T$ is real number, α_{rj}^a and β_{rj}^a are the left and right spread of $\tilde{a}_{rj}$, α_r^b and β_r^b is the left and right spread of $\tilde{b}_r$, $r = 1, 2, \cdots, p$, $j = 1, 2, \cdots, n$, the reference function $L, R : [0, 1] \to [0, 1]$ satisfies that $L(1) = R(1) = 0, L(0) = R(0) = 1$, and it is monotone function. Suppose that a_{rj} and b_r are independent Then $\text{Pos}\{\tilde{a}_r^T x \leq \tilde{b}_r\} \geq \theta_r$ is equivalent to

$$b_r + R^{-1}(\theta_r)\beta_r^b \geq e_r^T x - L^{-1}(\theta_r)\alpha_r^{aT} x, \quad r = 1, 2, \cdots, p, \quad (2.74)$$

Proof. Because $\tilde{a}_{rj}$ are LR fuzzy number, its membership function is $\mu_{\tilde{a}_{rj}}$. By extension principle[22], the membership function of fuzzy number $\tilde{a}_r^T x$ is

$$\mu_{\tilde{a}_r^T x}(r) = \begin{cases} L(\frac{a_r^T x - r}{\alpha_r^{aT} x}), & r \leq a_r^T x \\ R(\frac{r - a_r^T x}{\beta_r^{aT} x}), & r \geq a_r^T x \end{cases} \quad i = 1, 2, \cdots, m. \quad (2.75)$$

And $\tilde{b}_r$ are also LR fuzzy number with membership function $\mu_{\tilde{a}_{rj}}$. According to Lemma 2.2, we can get

$$\begin{aligned} &\text{Pos}\{\tilde{a}_r^T x \leq \tilde{b}_r\} \geq \theta_r \\ &\Leftrightarrow b_r + R^{-1}(\theta_r)\beta_r^b \geq a_r^T x - L^{-1}(\theta_r)\alpha_r^{aT} x, \quad r = 1, 2, \cdots, p. \end{aligned} \quad (2.76)$$

The proof is completed.   □

Remark 2.14. Especially, when the reference function of the variables $\tilde{a}_{rj}, \tilde{b}_r$ are all $L(x) = R(x) = 1 - x$ $(x \in [0,1])$, then the LR fuzzy variables are specified as the triangular fuzzy variables, and $R^{-1}(\delta_i) = 1 - \delta_i$, so we have the following equivalent expressions:

$$\text{Pos}\{\tilde{a}_r^\text{T} x \le \tilde{b}_r\} \ge \theta_r$$
$$\Leftrightarrow b_r + (1 - \theta_r)\beta_r^b \ge a_r^\text{T} x - (1 - \theta_r)\alpha_r^{a\text{T}} x, \quad r = 1, 2, \cdots, p.$$

Then according to the Theorems 2.7 and 2.8, we derive the crisp equivalent model (2.77) of the models (2.61),

$$\begin{cases} \max \{f_1, f_2, \cdots, f_m\} \\ \text{s.t.} \begin{cases} f_i \le c_i^\text{T} x + R^{-1}(\delta_i)\beta_i^{c\text{T}} x, \quad i = 1, 2, \cdots, m \\ b_r + R^{-1}(\theta_r)\beta_r^b - a_r^\text{T} x + L^{-1}(\theta_r)\alpha_r^{a\text{T}} x \ge 0, \ r = 1, 2, \cdots, p \\ x \ge 0. \end{cases} \end{cases} \tag{2.77}$$

Then, we give the fuzzy CCM crisp equivalent model based on *Nec* measure.

Theorem 2.9. *Assume that the fuzzy variables are as the same as that in Theorem 2.7. Then* $\text{Nec}\{\tilde{c}_i^\text{T} x \ge f_i\} \ge \delta_i$ *is equivalent to*

$$f_i \le c_i^\text{T} x - L^{-1}(1 - \delta_i)\alpha_i^{c\text{T}} x, \quad i = 1, 2, \cdots, m. \tag{2.78}$$

Proof. The proof is similar as the proof of theorem 2.7. □

Remark 2.15. Especially, when the reference function of the variable $\tilde{c}_{ij}$ is $L(x) = R(x) = 1 - x$ $(x \in [0,1])$, then the LR fuzzy variable is specified as the triangular fuzzy variable, and $R^{-1}(\delta_i) = 1 - \delta_i$, so we have the following equivalent expressions:

$$\text{Nec}\{\tilde{c}_i^\text{T} x \ge f_i\} \ge \delta_i \Leftrightarrow f_i \le c_i^\text{T} x - \delta_i \alpha_i^{c\text{T}} x, \quad i = 1, 2, \cdots, m.$$

Theorem 2.10. *Assume that the fuzzy variables* $\tilde{a}_{rj}$ *and* $\tilde{b}_r$ *are as the same as that in Theorem 2.8. Then* $\text{Nec}\{\tilde{a}_r^\text{T} x \le \tilde{b}_r\} \ge \theta_r$ *is equivalent to*

$$b_r - L^{-1}(1 - \theta_r)\alpha_r^b \ge a_r^\text{T} x + R^{-1}(\theta_r)\beta_r^{a\text{T}} x, \quad r = 1, 2, \cdots, p. \tag{2.79}$$

Proof. The proof is similar as the proof of theorem 2.8. □

Remark 2.16. Especially, when the reference function of the variables $\tilde{a}_{rj}, \tilde{b}_r$ are all $L(x) = R(x) = 1 - x$ $(x \in [0,1])$, then the LR fuzzy variables are specified as the triangular fuzzy variables, and $R^{-1}(\delta_i) = 1 - \delta_i$, so we have the following equivalent expressions:

$$\text{Pos}\{\tilde{a}_r^\text{T} x \le \tilde{b}_r\} \ge \theta_r$$
$$\Leftrightarrow b_r - \theta_r \alpha_r^b \ge a_r^\text{T} x + (1 - \theta_r)\beta_r^{a\text{T}} x, \quad r = 1, 2, \cdots, p. \tag{2.80}$$

Then according to the Theorems 2.9 and 2.10,we derive the crisp equivalent model (2.81) of the model (2.62),

$$\begin{cases} \max \ \{f_1, f_2, \cdots, f_m\} \\ \text{s.t.} \begin{cases} f_i \leq c_i^T x - L^{-1}(1-\delta_i)\alpha_i^{cT}x, & i = 1,2,\cdots,m \\ b_r - L^{-1}(1-\theta_r)\alpha_r^b - a_r^T x - R^{-1}(\theta_r)\beta_r^{aT}x \geq 0, & r = 1,2,\cdots,p \\ x \geq 0. \end{cases} \end{cases} \quad (2.81)$$

Finally, we give the fuzzy CCM crisp equivalent model based on Cr measure.

In this study, we only consider the situation $\delta_r > 0.5$. Because in practice, the confidence level δ_r of the chance constraints in Model (2.63) should be more than 0.5, that is, the credibility of the fuzzy constraint is more than a number δ_r, and this number should be bigger than 0.5. If it is less than 0.5, the constraints cannot be deemed to be satisfied, and the model is meaningless. So it is not to be considered in this book when $\delta_r \leq 0.5$. We then propose the follow theorem which deals with the chance constraints.

Theorem 2.11. *For any given confidence level δ_r with $0.5 < \delta_r \leq 1, r = 1,2,\cdots,p$, we have*

$$Cr\left\{\sum_{j=1}^{n} \tilde{e}_{rj}x_j \leq \tilde{b}_r\right\} \geq \delta_r \Leftrightarrow (2-2\delta_r)(\sum_{j=1}^{n} e_{rj}^3 x_j - b_r^2) + (2\delta_r - 1)(\sum_{j=1}^{n} e_{rj}^4 x_j - b_r^1) \leq 0,$$

where $\tilde{e}_{rj}$ and $\tilde{b}_r$ are trapezoidal fuzzy variables, and x_j is a real number with $x_j > 0$.

Proof. In Model (2.63), $\tilde{e}_{rj}$ is a trapezoidal fuzzy variable, and x_j is a real number with $x_j > 0$. So by Equation (2.68) and (2.67), we know that $\tilde{e}_{rj}x_j = (e_{rj}^1 x_j, e_{rj}^2 x_j, e_{rj}^3 x_j, e_{rj}^4 x_j)$ is a trapezoidal fuzzy variable, and $\sum_{j=1}^{n} \tilde{e}_{rj}x_j = (\sum_{j=1}^{n} e_{rj}^1 x_j, \sum_{j=1}^{n} e_{rj}^2 x_j, \sum_{j=1}^{n} e_{rj}^3 x_j, \sum_{j=1}^{n} e_{rj}^4 x_j)$ is also a trapezoidal fuzzy variable.

Similarly, by (2.68), $-\tilde{b}_r = (-b_r^4, -b_r^3, -b_r^2, -b_r^1)$ is a trapezoidal fuzzy variable. Then we have

$$\sum_{j=1}^{n} \tilde{e}_{rj}x_j - \tilde{b}_r = (\sum_{j=1}^{n} e_{rj}^1 x_j - b_r^4, \sum_{j=1}^{n} e_{rj}^2 x_j - b_r^3, \sum_{j=1}^{n} e_{rj}^3 x_j - b_r^2, \sum_{j=1}^{n} e_{rj}^4 x_j - b_r^1).$$

For convenience, we denote

$$g_1 = \sum_{j=1}^{n} e_{rj}^1 x_j - b_r^4, \ g_2 = \sum_{j=1}^{n} e_{rj}^2 x_j - b_r^3, \ g_3 = \sum_{j=1}^{n} e_{rj}^3 x_j - b_r^2, \ g_4 = \sum_{j=1}^{n} e_{rj}^4 x_j - b_r^1.$$

From the definitions of credibility, it is easy to obtain

$$Cr\left\{\sum_{j=1}^{n}\tilde{e}_{rj}x_j - \tilde{b}_r \leq 0\right\} = \begin{cases} 1, & \text{if } g_4 \leq 0 \\ \frac{2g_3-g_4}{2(g_3-g_4)}, & \text{if } g_3 \leq 0 \leq g_4 \\ 0.5, & \text{if } g_2 \leq 0 \leq g_3 \\ \frac{g_1}{2(g_1-g_2)}, & \text{if } g_1 \leq 0 \leq g_2 \\ 0, & \text{otherwise.} \end{cases}$$

'$\Rightarrow$'

If $Cr\left\{\sum_{j=1}^{n}\tilde{e}_{rj}x_j - \tilde{b}_r \leq 0\right\} \geq \delta_r$, then we have either $g_4 \leq 0$, $\frac{2g_3-g_4}{2(g_3-g_4)} \geq \delta_r$ or $\frac{g_1}{2(g_1-g_2)} \geq \delta_r$.

When $g_4 \leq 0$, then $g_3 \leq g_4 \leq 0$, so that $(2-2\delta_r)g_3 + (2\delta_r - 1)g_4 \leq 0$;

When $\frac{2g_3-g_4)}{2(g_3-g_4)} \geq \delta_r$, then $(2-2\delta_r)g_3 + (2\delta_r - 1)r_4 \leq 0$ by the fact that $g_3 < g_4$;

When $\frac{g_1}{2(g_1-g_2)} \geq \delta_r$, then $g_1 \leq 2\delta_r(g_1 - g_2)$ by the fact that $g_1 < g_2$, so that $(1 - 2\delta_r)g_1 + 2\delta_r g_2 \leq 0$. However, since $(1 - 2\delta_r) < 0$, $g_1 \leq 0$, and $g_2 \geq 0$, it is obvious that $(1 - 2\delta_r)g_1 + 2\alpha g_2 \geq 0$. So that's conflicted.

So we have

$$Cr\left\{\sum_{j=1}^{n}\tilde{e}_{rj}x_j - \tilde{b}_r \leq 0\right\} \geq \delta_r \Rightarrow (2-2\delta_r)g_3 + (2\delta_r - 1)g_4 \leq 0,$$

for any cases.

'$\Leftarrow$'

Conversely, if $(2-2\delta_r)g_3 + (2\delta_r - 1)g_4 \leq 0$, the argument breaks down into two cases.

When $g_4 \leq 0$, we have $Cr\left\{\sum_{j=1}^{n}\tilde{e}_{rj}x_j - \tilde{b}_r \leq 0\right\} = 1$, which implies that

$$Cr\left\{\sum_{j=1}^{n}\tilde{e}_{rj}x_j - \tilde{b}_r \leq 0\right\} \geq \delta_r;$$

When $g_4 > 0$, we have $g_3 < 0$, we rearrange $(2-2\delta_r)g_3 + (2\delta_r - 1)g_4 \leq 0$ as $\frac{2g_3-g_4}{2(g_3-g_4)} \geq \delta_r$. Thus $Cr\left\{\sum_{j=1}^{n}\tilde{e}_{rj}x_j - \tilde{b}_r \leq 0\right\} \geq \delta_r$.

So we also have

$$(2-2\delta_r)g_3 + (2\delta_r - 1)g_4 \leq 0 \Rightarrow Cr\left\{\sum_{j=1}^{n}\tilde{e}_{rj}x_j - \tilde{b}_r \leq 0\right\} \geq \alpha.$$

Above all,

$$Cr\left\{\sum_{j=1}^{n}\tilde{e}_{rj}x_j - \tilde{b}_r \leq 0\right\} \geq \delta_r \Leftrightarrow (2-2\delta_r)g_3 + (2\delta_r - 1)g_4 \leq 0$$

or

$$Cr\left\{\sum_{j=1}^{n} \tilde{e}_{rj}x_j \leq \tilde{b}_r\right\} \geq \delta_r \Leftrightarrow (2-2\delta_r)(\sum_{j=1}^{n} e_{rj}^3 x_j - b_r^2) + (2\delta_r - 1)(\sum_{j=1}^{n} e_{rj}^4 x_j - b_r^1) \leq 0.$$

The proof is thus complete.   □

Remark 2.17. If $\tilde{e}_{rj}$ degenerates to a crisp number e_{rj}, then we have

$$Cr\left\{\sum_{j=1}^{n} e_{rj}x_j \leq \tilde{b}_r\right\} \geq \delta_r \Leftrightarrow (2-2\delta_r)(\sum_{j=1}^{n} e_{rj}x_j - b_r^2) + (2\delta_r - 1)(\sum_{j=1}^{n} e_{rj}x_j - b_r^1) \leq 0.$$

And if $\tilde{b}_r$ degenerates to a crisp number b_r, then we have

$$Cr\left\{\sum_{j=1}^{n} \tilde{e}_{rj}x_j \leq b_r\right\} \geq \delta_r \Leftrightarrow (2-2\delta_r)(\sum_{j=1}^{n} e_{rj}^3 x_j - b_r) + (2\delta_r - 1)(\sum_{j=1}^{n} e_{rj}^4 x_j - b_r) \leq 0,$$

for $0.5 < \delta_r \leq 1, r = 1, 2, \cdots, p.$

Theorem 2.12. *For any given confidence level δ_r with $0.5 < \delta_r \leq 1, r = 1, 2, \cdots, p$, we have*

$$Cr\left\{\sum_{j=1}^{n} \tilde{e}_{rj}x_j \geq \tilde{b}_r\right\} \geq \delta_r \Leftrightarrow (2-2\delta_r)(\sum_{j=1}^{n} e_{rj}^2 x_j - b_r^3) + (2\delta_r - 1)(\sum_{j=1}^{n} e_{rj}^1 x_j - b_r^4) \geq 0.$$

Proof. The proof is similar to that of Theorem 2.11, and thus omitted.   □

Remark 2.18. If $\tilde{e}_{rj}$ degenerates to a crisp number e_{rj}, then we have

$$Cr\left\{\sum_{j=1}^{n} e_{rj}x_j \geq \tilde{b}_r\right\} \geq \delta_r \Leftrightarrow (2-2\delta_r)(\sum_{j=1}^{n} e_{rj}x_j - b_r^3) + (2\delta_r - 1)(\sum_{j=1}^{n} e_{rj}x_j - b_r^4) \geq 0.$$

And if $\tilde{b}_r$ degenerates to a crisp number b_r, then we have

$$Cr\left\{\sum_{j=1}^{n} \tilde{e}_{rj}x_j \geq b_r\right\} \geq \delta_r \Leftrightarrow (2-2\delta_r)(\sum_{j=1}^{n} e_{rj}^2 x_j - b_r) + (2\delta_r - 1)(\sum_{j=1}^{n} e_{rj}^1 x_j - b_r) \geq 0,$$

for $0.5 < \delta_r \leq 1, r = 1, 2, \cdots, p.$

2.4.2.2 Goal Programming Method

The goal programming method was first proposed by Charnes and Cooper [161] in 1961. After that, Ijiri [394], Lee [126], Kendall and Lee [395], and Ignizio [396] researched and developed it. When dealing with many multi-objective decision making problems, it has been widely applied since it can provide a technique which is accepted by many decision makers, that is, it can point out preferential information and harmoniously inosculate it into the model.

The basic idea of goal programming method is that, for the objective function $f(x) = (f_1(x), f_2(x), \cdots, f_m(x))^T$, decision makers give a goal value $f^o = (f_1^o, f_2^o, \cdots, f_m^o)^T$ such that every objective function $f_i(x)$ approximates the goal value f_i^o as closely as possible. Let $d_p(f(x), f^o) \in R^m$ be the deviation between $f(x)$ and f^o, then consider the following problem,

$$\min_{x \in X} d_p(f(x), f^o), \tag{2.82}$$

where the goal value f^o and the weight vector w is predetermined by the decision maker. The weight w_i expresses the importance factor that the objective function $f_i(x)$ $(i = 1, 2, \cdots, m)$ approximates the goal value f_i^o, $1 \le p \le \infty$.

When $p = 1$, it is recalled the simple goal programming method which is most widely used. Then we have,

$$d_p(f(x), f^o) = \sum_{i=1}^{m} w_i |f(x) - f^o|.$$

Since there is the notation $|\cdot|$ in $d_p(f(x), f^o)$, it isn't a differentiable function any more. Therefore, denote that

$$d_i^+ = \frac{1}{2}(|f_i(x) - f_i^o| + (f_i(x) - f_i^o)),$$

$$d_i^- = \frac{1}{2}(|f_i(x) - f_i^o| - (f_i(x) - f_i^o)).$$

where d_i^+ expresses the quantity that $f_i(x)$ exceeds f_i^o and d_i^- expresses the quantity that $f_i(x)$ is less than f_i^o. It is easy to prove that, for any $i = 1, 2, \cdots, m$,

$$\begin{aligned} d_i^+ + d_i^- &= |f_i(x) - f_i^o|, \\ d_i^+ - d_i^- &= f_i(x) - f_i^o, \\ d_i^+ d_i^- &= 0,\ d_i^+, d_i^- \ge 0. \end{aligned} \tag{2.83}$$

When $p = 1$, problem (2.82) can be rewritten as,

$$\begin{cases} \min \sum\limits_{i=1}^{m} w_i(d_i^+ + d_i^-) \\ \text{s.t.} \begin{cases} f_i(x) + d_i^+ - d_i^- = f_i^o,\ i = 1, 2, \cdots, m \\ d_i^+ d_i^- = 0, d_i^+, d_i^- \ge 0,\ i = 1, 2, \cdots, m \\ x \in X. \end{cases} \end{cases} \tag{2.84}$$

In order to easily solve the problem (2.84), abandon the constraint $d_i^+ d_i^- = 0$ $(i = 1, 2, \cdots, m)$ and we have

$$\begin{cases} \min \sum\limits_{i=1}^{m} w_i(d_i^+ + d_i^-) \\ \text{s.t.} \begin{cases} f_i(x) + d_i^+ - d_i^- = f_i^o,\ i = 1, 2, \cdots, m \\ d_i^+, d_i^- \ge 0, \qquad\qquad i = 1, 2, \cdots, m \\ x \in X. \end{cases} \end{cases} \tag{2.85}$$

Theorem 2.13. *If $(x, \bar{d}^+, \bar{d}^-)$ is the optimal solution of problem (2.85), $\bar{x}$ is doubtlessly the optimal solution of problem (2.82), where $\bar{d}^+ = (\bar{d}_1^+, \bar{d}_2^+, \cdots, \bar{d}_m^+)$ and $\bar{d}^- = (\bar{d}_1^-, \bar{d}_2^-, \cdots, \bar{d}_m^-)$.*

Proof. Since $(x, \bar{d}^+, \bar{d}^-)$ is the optimal solution of problem (2.85), we have $x \in X$, $\bar{d}^+ \geq 0, \bar{d}^- \geq 0$ and

$$f_i(x) + \bar{d}_i^| - \bar{d}_i^- = f_i^o, \ l = 1, 2, \cdots, m. \tag{2.86}$$

(1) If $\bar{d}_i^+ = \bar{d}_i^- = 0$, we have $f_i(x) = f_i^o$, which means x is the optimal solution problem (2.82).

(2) If there exists $i_0 \in \{1, 2, \cdots, m\}$ such that $f_i(x) \neq f_i^o$, $\bar{d}_i^+ \bar{d}_i^- = 0$ doubtlessly holds. If not, we have $\bar{d}_i^+ > 0$ and $\bar{d}_i^- > 0$. We respectively discuss them as follows.

(i) If $\bar{d}_i^+ - \bar{d}_i^- > 0$, for $i \in \{1, 2, \cdots, m\}$, let

$$\tilde{d}_i^+ = \begin{cases} \bar{d}_i^+ - \bar{d}_i^-, & i = i_0 \\ \bar{d}_i^+, & i \neq i_0, \end{cases} \quad \tilde{d}_i^- = \begin{cases} 0, & i = i_0 \\ \bar{d}_i^-, & i \neq i_0. \end{cases} \tag{2.87}$$

Thus, $\tilde{d}_{i_0}^+ < \bar{d}_{i_0}^+$ and $\tilde{d}_{i_0}^- < \bar{d}_{i_0}^-$ both hold. It follows from equation (2.86) and (2.87) that,

$$f_i(x) + \tilde{d}_i^+ - \tilde{d}_i^- = \begin{cases} f_i(x) + 0 - (\bar{d}_i^+ - \bar{d}_i^-) = f_i^o, & i = i_0 \\ f_i(x) + \bar{d}_i^+ - \bar{d}_i^- = f_i^o, & i \neq i_0. \end{cases}$$

We also know $x \in X$, $\tilde{d}_i^+ \geq 0$ and $\tilde{d}_i^- \geq 0$. Denote $\tilde{d}^+ = (\tilde{d}_1^+, \tilde{d}_2^+, \cdots, \tilde{d}_m^+)$ and $\tilde{d}^- = (\tilde{d}_1^-, \tilde{d}_2^-, \cdots, \tilde{d}_m^-)$, then we have $(x, \tilde{d}^+, \tilde{d}^-)$ is a feasible solution of problem (2.85). If follows from $\tilde{d}_{i_0}^+ < \bar{d}_{i_0}^+$ and $\tilde{d}_{i_0}^- < \bar{d}_{i_0}^-$ that,

$$\sum_{i=1}^{m} (\tilde{d}_{i_0}^+ + \tilde{d}_{i_0}) < \sum_{i=1}^{m} (\bar{d}_{i_0}^+ + \bar{d}_{i_0}), \tag{2.88}$$

this conflict with the assumption that $(x, \bar{d}^+, \bar{d}^-)$ is the optimal solution of problem (2.85).

(ii) If $\bar{d}_i^+ - \bar{d}_i^- < 0$, for $i \in \{1, 2, \cdots, m\}$, let

$$\tilde{d}_i^+ = \begin{cases} 0, & i = i_0 \\ \bar{d}_i^+, & i \neq i_0, \end{cases} \quad \tilde{d}_i^- = \begin{cases} -(\bar{d}_i^+ - \bar{d}_i^-), & i = i_0 \\ \bar{d}_i^-, & i \neq i_0. \end{cases} \tag{2.89}$$

We can similarly prove that it conflicts with the assumption that $(x, \bar{d}^+, \bar{d}^-)$ is the optimal solution of problem (2.85).

So far, we have proved that $(x, \bar{d}^+, \bar{d}^-)$ is the optimal solution of problem (2.84). Since the feasible region of problem (2.84) is included in the one of problem (2.85), $(x, \bar{d}^+, \bar{d}^-)$ is the optimal solution of problem (2.85). Next, we will prove that $(x, \bar{d}^+, \bar{d}^-)$ is the optimal solution of problem (2.82). For any feasible solution (x, d^+, d^-), it follows from (2.83) that,

$$|f_i(x) - f_i^o| = d_i^+ + d_i^-, \ |f_i(\bar{x}) - f_i^o| = \bar{d}_i^+ + \bar{d}_i^-, \ i = 1, 2, \cdots, m.$$

For any $x \in X$, since

$$\sum_{i=1}^{m} |f_i(\bar{x}) - f_i^o| = \sum_{i=1}^{m} (\bar{d}_i^+ + \bar{d}_i^-) \leq \sum_{i=1}^{m} (d_i^+ + d_i^-) = \sum_{i=1}^{m} |f_i(x) - f_i^o|,$$

this means that $\bar{x}$ is the optimal solution of problem (2.82).   □

2.4.2.3 Numerical Example

We also use the ideal point method to solve the equivalent model of fuzzy CCM.

Example 2.7. Let consider the following problem,

$$
\begin{cases}
\max \ \xi_1 x_1 + \xi_2 x_2 + \xi_3 x_3 + \xi_4 x_4 \\
\min \ \xi_5 x_1 + \xi_6 x_2 + \xi_7 x_3 + \xi_8 x_4 \\
\text{s.t.} \begin{cases} 7x_1 + 5x_2 + 3x_3 + 2x_4 \leq 120 \\ \xi_9 x_1 + \xi_{10} x_2 + \xi_{11} x_3 + \xi_{12} x_4 \leq \xi_{13} \\ x_i \geq 0, \ i = 1, 2, \cdots, 4, \end{cases}
\end{cases}
\tag{2.90}
$$

where $\xi_j (j = 1, 2, \cdots, 13)$ are triangular LR fuzzy variables characterized as that in Theorem. 2.7,

$$
\begin{aligned}
&\xi_1 = (4, 0.5, 0.5)_{LR}, &&\xi_2 = (5, 0.5, 0.5)_{LR}, \\
&\xi_3 = (9, 0.5, 0.5)_{LR}, &&\xi_4 = (11, 0.5, 0.5)_{LR}, \\
&\xi_5 = (1, 0.1, 0.1)_{LR}, &&\xi_6 = (1, 0.1, 0.1)_{LR}, \\
&\xi_7 = (1, 0.1, 0.1)_{LR}, &&\xi_8 = (1, 0.1, 0.1)_{LR}, \\
&\xi_9 = (3, 0.5, 0.5)_{LR}, &&\xi_{10} = (5, 0.5, 0.5)_{LR}, \\
&\xi_{11} = (10, 0.5, 0.5)_{LR}, &&\xi_{12} = (15, 0.5, 0.5)_{LR}, \\
&\xi_{13} = (100, 5, 5)_{LR}.
\end{aligned}
$$

We use the chance operator to deal with the fuzzy multi-objective model, and here the decision maker is supposed to be comparatively optimistic, so we adopted the *Pos* measure to measure the chance,

$$
\begin{cases}
\max \ \{\overline{f}_1, \overline{f}_2\} \\
\text{s.t.} \begin{cases} Pos\{\xi_1 x_1 + \xi_2 x_2 + \xi_3 x_3 + \xi_4 x_4 \geq \overline{f}_1\} \geq \delta_1 \\ Pos\{-(\xi_5 x_1 + \xi_6 x_2 + \xi_7 x_3 + \xi_8 x_4) \geq \overline{f}_2\} \geq \delta_2 \\ 7x_1 + 5x_2 + 3x_3 + 2x_4 \leq 120 \\ Pos\{\xi_9 x_1 + \xi_{10} x_2 + \xi_{11} x_3 + \xi_{12} x_4 \leq \xi_{13}\} \geq \theta \\ x_i \geq 0, \ i = 1, 2, \cdots, 5. \end{cases}
\end{cases}
\tag{2.91}
$$

Since there are triangular LR fuzzy coefficients exist in the objective functions and a constraint, by Remarks 2.13, 2.13, for confidence level $\delta_1 = 0.9, \delta_2 = 0.9, \theta = 0.9$, we have the following equivalent model 2.92,

$$\begin{cases} \max \{\bar{f}_1, \bar{f}_2\} \\ \text{s.t.} \begin{cases} 4x_1 + 5x_2 + 9x_3 + 11x_4 + 0.1(0.5x_1 + 0.5x_2 + 0.5x_3 + 0.5x_4) \geq \bar{f}_1 \\ -(x_1 + x_2 + x_3 + x_4) + 0.1(0.1x_1 + 0.1x_2 + 0.1x_3 + 0.1x_4) \geq \bar{f}_2 \\ 7x_1 + 5x_2 + 3x_3 + 2x_4 \leq 120 \\ 3x_1 + 5x_2 + 10x_3 + 15x_4 - 0.1(0.5x_1 + 0.5x_2 + 0.5x_3 + 0.5x_4) \leq 100 + 0.1*5 \\ x_i \geq 0, \ i = 1, 2, \cdots, 5 \end{cases} \end{cases}$$

$$(2.92)$$

or

$$\begin{cases} \max F_1 = 4.05x_1 + 5.05x_2 + 9.05x_3 + 11.05x_4 \\ \max F_2 = -0.99(x_1 + x_2 + x_3 + x_4) \\ \text{s.t.} \begin{cases} 7x_1 + 5x_2 + 3x_3 + 2x_4 \leq 120 \\ 2.95x_1 + 4.95x_2 + 9.95x_3 + 14.95x_4 \leq 100.5 \\ x_i \geq 0, \ i = 1, 2, \cdots, 5. \end{cases} \end{cases}$$

$$(2.93)$$

For this problem, we could use the interactive solution method to solve it, $\bar{f}_i^0, \bar{f}_i^1 (i = 1, 2)$ are calculated as follows

$$\bar{f}_1^0 = 0, \quad \bar{f}_1^1 = 111.4737, \quad \bar{f}_2^0 = -21.9392, \quad \bar{f}_2^1 = 0.$$

Then we have that

$$\begin{cases} \min \lambda \\ \text{s.t.} \begin{cases} 4.05x_1 + 5.05x_2 + 9.05x_3 + 11.05x_4 \geq 111.4737(\bar{\mu}_1 - \lambda) \\ -0.99(x_1 + x_2 + x_3 + x_4) \geq 21.9392(\bar{\mu}_2 - \lambda) - 21.9392 \\ 7x_1 + 5x_2 + 3x_3 + 2x_4 \leq 120 \\ 2.95x_1 + 4.95x_2 + 9.95x_3 + 14.95x_4 \leq 100.5 \\ x_i \geq 0, \ i = 1, 2, \cdots, 5. \end{cases} \end{cases}$$

$$(2.94)$$

For the initial reference membership 1, each membership function value and the solution x as well as the objective function $F_i(x)$ are obtained (see the first row in Table 2.1). If the decision maker wishes to increase $\bar{f}_1(x)$ by sacrificing $\bar{f}_2(x)$, then the probability value $(\bar{\mu}_1, \bar{\mu}_2)$ needs to be updated, such as $(1, 0.98)$. Or else $(\bar{\mu}_1, \bar{\mu}_2)$ needs to be updated, such as $(0.98, 1)$. The results are listed in the second row. Suppose that the decision maker is satisfied with the solution when the probability is $(1, 0.90)$. Then the interactive process is stopped and the satisfactory solution is $x^* = ((0, 0, 4.299, 3.861)$ and $(F_1^*, F_2^*) = (81.162, 8.16)$. Moreover the decision maker can modify $\bar{f}_i^0, \bar{f}_i^1 (i = 1, 2)$ and build a new reference membership function to obtain his or her satisfactory solution.

Table 2.1 Results obtained from Interactive process

$\bar{\mu}_1$	$\bar{\mu}_2$	$x = (x_1, x_2, x_3, x_4)$	λ	$F_1(x)$	$F_2(x)$
1	1	(0,0,0.999,6.057)	0.318	75.971	6.985
1	0.98	(0,0,1.659,5.618)	0.308	72.116	7.204
1	0.96	(0,0,2.319,5.178)	0.298	78.204	7.422
1	0.94	(0,0,2.979,4.739)	0.288	79.326	7.641
1	0.92	(0,0,3.639,4.300)	0.278	80.448	7.592
1	0.90	(0,0,4.299,3.861)	0.268	81.57	8.078
0.98	1	(0,0,0.339,6.497)	0.308	74.86	6.768
0.96	1	(0,0,0,6.655)	0.300	73.538	6.588
0.94	1	(0,0,0,6.516)	0.294	72.0018	6.451
0.92	1	(0,0,0,6.378)	0.288	70.477	6.314
0.90	1	(0,0,0,6.239)	0.282	68.941	6.177

Suppose the risk tolerance given by decision maker is $\check{F}_1 = 78, \check{F}_2 = 7.5$. Then we can also use the goal programming method to handle the crisp linear model (2.93), and we can get the following goal programming model (2.95),

$$
\begin{cases}
\min \sum_{i=1}^{2} w_i(d_i^+ + d_i^-) \\
\text{s.t.}
\begin{cases}
4.05x_1 + 5.05x_2 + 9.05x_3 + 11.05x_4 + d_1^+ + d_1^- = \check{F}_1 \\
-0.99(x_1 + x_2 + x_3 + x_4) + d_2^+ + d_2^- = \check{F}_2 \\
7x_1 + 5x_2 + 3x_3 + 2x_4 \le 120 \\
2.95x_1 + 4.95x_2 + 9.95x_3 + 14.95x_4 \le 100.5 \\
x_i \ge 0, \ i = 1, 2, \cdots, 5. \\
d_i^+ d_i^- = 0, d_i^+, d_i^- \ge 0, i = 1, 2.
\end{cases}
\end{cases}
\tag{2.95}
$$

We take $w_1 = w_2 = 0.5$, and by solving the above model, we can obtain the efficient solution as follows:

$$(x_1, x_2, x_3, x_4) = (0, 0, 2.19, 5.26).$$

2.4.3 Non-linear Fuzzy CCM and Fuzzy Simulation-Based PSO with Preference Order

It is difficult to transform the non-linear fuzzy CCM into it's equivalent form, so we introduce fuzzy simulation-based particle swarm optimization with a preference order algorithm for solution.

2.4.3.1 Fuzzy Simulation 2 for Critical Value

Suppose that f is a real-valued function, and ξ_i are fuzzy variables with membership functions μ_i, $i = 1, 2, \cdots, n$, respectively. Let us find the maximal $\bar{f}$ such that the inequality

$$Pos\{f(\xi) \ge \bar{f}\} \ge \beta \tag{2.96}$$

holds, where $\xi = (\xi_1, \xi_2, \cdots, \xi_n)$. First we set $\bar{f} = -\infty$. Then we randomly generate $u_1, u_2, \cdots, u_n$ from the β-level sets of $\xi_1, \xi_2, \cdots, \xi_n$, respectively, and denote $u = (u_1, u_2, \cdots, u_n)$. We set $\bar{f} = f(u)$ provided that $\bar{f} < f(u)$. Repeat this process for N times. The value $\bar{f}$ is regarded as the estimation. We summarize this process as follows:

Step 1. Set $\bar{f} = -\infty$.

Step 2. Randomly generate $u_1, u_2, \cdots, u_n$ from the β-level sets of $\xi_1, \xi_2, \cdots, \xi_n$, respectively, and denote $u = (u_1, u_2, \cdots, u_n)$.

Step 3. If $\bar{f} < f(u)$, then we set $\bar{f} = f(u)$.

Step 4. Repeat the second and third steps N times.

Step 5. Return $\bar{f}$.

2.4.3.2 PSO with Preference Order

Preference order [389] is a generalization of Pareto optimality. It provides a way to designate some Pareto solutions superior to others when the size of the nondominated solution set is very large.

Definition 2.19. A point $x^* \in \Omega$ is considered efficiency of order k if $f(x^*)$ is not dominated by any of the k-element subsets of f_Ω, where Ω is the feasible region for an multi-objective optimization and f_Ω is the image of the feasible region in the objective space.

Remark 2.19. If x^* is efficiency of order k, then it is efficiency of order $k + 1$.

This claim has been proven by Das [389]. It is clear that the efficiency of order m is the ordinary concept of Pareto optimality, hence the efficiency of order k is an extension of Pareto optimality.

   Efficiency of order can be used to reduce the number of points in a nondominated set by retaining only those regarded as the "best compromise" [390]. In order to explain the "best compromise", consider a Pareto point P which has a near-minimum value for one objective but fairly high values for all the rest. Such a point lies at an extreme end of the Pareto surface and is not considered as a good compromise among all objectives. The point P is not qualified when the efficiency of order is $m - 1$, because it is not presented in one of the $(m - 1)$-element subsets of the objectives. For this subset, P is unlikely to be a Pareto optimal solution since it is dominated by other points in the subset. Therefore, it is concluded that the fewer extreme components a point has, the more likely it is to be efficiency of order.

   Then we describe the proposed MOPSO algorithm with a preference order ranking procedure. Firstly, PO ranking procedure is introduced, then the details of the proposed algorithm are given. In this study, the positions of Pareto nondominated particles are stored in the external archive. To find the "best compromise", the nondominated solutions are ranked according to PO. The ranking procedure can be summarized below:

(1) Identify the combinations of all subsets to m objectives;

(2) Assign the order to all nondominated solutions for each combination of all subsets based on PO;

(3) Identify the "best compromise" in all nondominated solutions according to their order.

On the basis of the above ranking procedure, the steps of the proposed algorithm are summarized as follows.

Step 1. Initialization. The population size, the archive size and the maximum number of iterations are initialized according to the problem concerned. The velocities and the positions are initialized randomly. The initial value of the personal best for each particle is equal to its initial position.

Step 2. Update the velocities and the positions. In order to control the balance of global and local search, a new updating equation for the velocity is adopted. Two parameters, namely r_1 and r_2, are independently and randomly generated, so there are cases in which two random parameters are both too large or too small. In the former case, both the personal and the social experiences accumulated so far are over used, and the result is that the particles are driven too far away from the local optimum. For the latter case, both the personal and the social experiences are not used fully, and the convergence speed of the algorithm is reduced. In the proposed algorithm the velocity is updated as follows:

$$v_{id}^{t+1} = w_{t+1}v_{id}^{t} + (1 - r_2)c_1r_1(p_{id}^{t} - x_{id}^{t}) + (1 - r_2)c_2(1 - r_1)(p_{gd}^{t} - x_{id}^{t}). \quad (2.97)$$

The inertia weight xtt1 in equation (2.97) is adjusted dynamically as follows:

$$w_{t+1} = w_t f_w, \quad (2.98)$$

where f_w is a constant between 0 and 1.For the personal best, it is easily defined by the concept of domination. But it is difficult to define the global best because there is a set of compromise solutions in MOPs. In order to identify the best compromise, the PO scheme is used to classify the solutions in this study. Neither the crowding distance[249] nor the density parameter[391] is needed to be computed. The superior solutions in the nondominated set are identified via PO scheme mentioned above. Then the best compromise is used as the global best to update the velocity according to equation (2.98). After updating the velocities of the particles, the positions are updated.

Step 3. Update archive. In each generation, the archive is updated by the nondominated solutions from the population. If the archive size exceeds the maximum size, the solutions in the archive are sorted in descending order according to their efficiency order. The solutions with lower efficiency order are deleted from the end of the archive.

Step 4. Termination. The algorithm is terminated if it reaches the maximum number of iterations. The final nondominated solutions are contained in the external archive.

Based on the above discussions, the pseudocode of the proposed algorithm is given as follows:

/*N is the population size, N_a is the external archive size, t_{max} is the maximum number of iterations, d is the dimensions of the particles.*/

Step 1. $t = 0$.

    (a) initialize x_{id}^0, $\forall i, i \in \{1, \cdots, N\}$ and $\forall d, d \in \{1, \cdots, D\}$.

    (b) initialize v_{id}^0, $\forall i, i \in \{1, \cdots, N\}$ and $\forall d, d \in \{1, \cdots, D\}$.

    (c) initialize p_{id}^0 for each particle, $p_{id}^0 \leftarrow x_{id}^0$.

    (d) initialize archive, $A_0 \leftarrow S_0$.

/*S_0: the nondominated solutions in the population*/

Step 2. for $t = 1$ to t_{max}.

    (a) Use PO scheme to identify the best compromise

(i) get combination.

/*returns the combination of all subsets for m objectives*/

(ii) get efficiency order.

/*compute the efficiency order for each nondominated solution from the archive on each subset*/

(iii) sort nondominated solutions.

/*sort nondominated solutions in descending order according to their efficiency order*/

(iv) get p_{gd}^t.

/*the solution with maximum efficiency order is defined as global best*/

    (b) update velocity of each particle according to Equation 2.97.

    (c) update position of each particle.

    (d) update p_{id}^t of each particle.

    (e) update w_t according to Equation 2.98.

    (f) update archive.

/*if the archive size N_a is exceeded, the solutions with lower efficiency order are deleted from the end of archive*/

Step 3. end.

2.4.3.3 Numerical Example

Example 2.8. Consider the following multi-objective programming problem with fuzzy parameters,

$$
\begin{cases}
\min \ \sqrt{(x_1 - \xi_1)^2 + (x_2 - \xi_2)^2 + (x_3 - \xi_3)^2} \\
\min \ \sqrt{(x_1 + \xi_1)^2 + (x_2 + \xi_2)^2 + (x_3 + \xi_3)^2} \\
\text{s.t.} \ \begin{cases} x_1^2 + x_2^2 + x_3^2 \leq 10 \\ x_1 \geq 0, x_2 \geq 0, x_3 \geq 0, \end{cases}
\end{cases} \tag{2.99}
$$

where $\xi_i, i = 1, 2, 3$ are triangular fuzzy numbers, that is,

$$
\begin{aligned}
\xi_1 &= (1, 1.5, 2), \\
\xi_2 &= (2.5, 3, 3.5), \\
\xi_3 &= (3.5, 4, 4.5).
\end{aligned}
$$

We use the fuzzy CCM to deal with the above nonlinear fuzzy multi-objective model, and we can get

$$
\begin{cases}
\min\ [\bar{f}_1, \bar{f}_2] \\
\text{s.t.}
\begin{cases}
Pos\{\sqrt{(x_1 - \xi_1)^2 + (x_2 - \xi_2)^2 + (x_3 - \xi_3)^2} \leq \bar{f}_1\} \geq \beta_1 \\
Pos\{\sqrt{(x_1 + \xi_1)^2 + (x_2 + \xi_2)^2 + (x_3 + \xi_3)^2} \leq \bar{f}_2\} \geq \beta_2 \\
x_1^2 + x_2^2 + x_3^2 \leq 10 \\
x_1 \geq 0, x_2 \geq 0, x_3 \geq 0.
\end{cases}
\end{cases}
\tag{2.100}
$$

As we know, it's difficult to convert the problem (2.100) into it's crisp inequivalent model. In order to solve this model, we have to make use of the fuzzy simulation-based PSO with preference order to get the efficient solution.

$$
\bar{f}_1 = 4.48, \bar{f}_2 = 5.95;
$$
$$
x_1 = 0.226, x_2 = 0.495, x_3 = 0.499.
$$

2.5 Fuzzy DCM

In this section, we will introduce two kinds of fuzzy dependent model based on two types of measure *Pos* and *Nec*.

2.5.1 General Model for Fuzzy DCM

We propose the general model of Fuzzy DCM.

$$
\begin{cases}
\max
\begin{bmatrix}
Ch\{f_1(x, \xi) \geq \bar{f}_1\}, \\
Ch\{f_2(x, \xi) \geq \bar{f}_2\}, \\
\cdots \\
Ch\{f_m(x, \xi) \geq \bar{f}_m\},
\end{bmatrix} \\
\text{s.t.}
\begin{cases}
g_r(x, \xi) \leq 0,\ r = 1, 2, \cdots, p \\
x \in X,
\end{cases}
\end{cases}
\tag{2.101}
$$

where $\xi = (\xi_1, \xi_2, \cdots, \xi_n)$ are fuzzy vector, Ch are the chance of the fuzzy event, and we could use possibility measure or necessity measure, $\bar{f}_i$, $i = 1, 2, \cdots, m$, are the predetermined ideal objective values for each objective functions.

2.5.1.1 Fuzzy DCP Based on Pos Measure

A typical formulation of fuzzy dependent-chance multi-objective model is given as follows,

$$
\begin{cases}
\max
\begin{bmatrix}
Pos\{f_1(x, \xi) \geq \bar{f}_1\}, \\
Pos\{f_2(x, \xi) \geq \bar{f}_2\}, \\
\cdots \\
Pos\{f_m(x, \xi) \geq \bar{f}_m\},
\end{bmatrix} \\
\text{s.t.}
\begin{cases}
g_r(x, \xi) \leq 0,\ r = 1, 2, \cdots, p \\
x \in X,
\end{cases}
\end{cases}
\tag{2.102}
$$

where $\xi = (\xi_1, \xi_2, \cdots, \xi_n)$ are fuzzy vector, Pos are the possibility measure of the fuzzy events, $\bar{f}_i, i = 1, 2, \cdots, m$, are the predetermined ideal objective values for each objective functions.

Thus, for a each given decision vector, we cannot judge whether or not a decision vector x is a feasible before the realization of the fuzzy vector ξ. Hence, the problem (2.102) is not well defined mathematically. Motivated by the ideas of EVM or CCM, we may consider the expect value of $g_r(x, \xi)$ or require the possibility measure of $g_r(x, \xi)$ are not less than a predetermined confidence level, i.e. we have the following models.

$$\begin{cases} \max \ [Pos\{f_1(x, \xi) \leq 0\}, Pos\{f_2(x, \xi) \leq 0\}, \cdots, Pos\{f_m(x, \xi) \leq 0\}] \\ \text{s.t.} \ \begin{cases} E[g_r(x, \xi)] \geq 0, r = 1, 2, \cdots, p \\ x \in X \end{cases} \end{cases} \quad (2.103)$$

and

$$\begin{cases} \max \ [Pos\{f_1(x, \xi) \leq 0\}, Pos\{f_2(x, \xi) \leq 0\}, \cdots, Pos\{f_m(x, \xi) \leq 0\}] \\ \text{s.t.} \ \begin{cases} Pos\{g_r(x, \xi) \leq 0\} \geq \theta_r, r = 1, 2, \cdots, p \\ x \in X, \end{cases} \end{cases} \quad (2.104)$$

where $\theta_r \in [0, 1], r = 1, 2, \cdots, p$ are the predetermined level value.

Obviously, Problem (2.103) and Problem (2.104) are mathematically meaningful decision problem. In other word, we can judge whether or not a decision vector is feasible. Thus, they can be formulated as

$$\max_{x \in X} [Pos\{f_1(x, \xi) \leq 0\}, Pos\{f_2(x, \xi) \leq 0\}, \cdots, Pos\{f_m(x, \xi) \leq 0\}], \quad (2.105)$$

where X is a fixed feasible set.

Definition 2.20. $x^* \in X$ is called a Pos fuzzy efficient solution of Problem (2.105) if and only if there exists no other fuzzy feasible solution x such that

$$Pos\{\theta \in \Theta | f_i(x, \xi)\} \geq Pos\{\theta \in \Theta | f_i(x^*, \xi)\}, i = 1, 2, \cdots, m,$$

for all i, and

$$Pos\{\theta \in \Theta | f_{i_0}(x, \xi)\} > Pos\{\theta \in \Theta | f_{i_0}(x^*, \xi)\}, i = 1, 2, \cdots, m,$$

for at least one $i_0 \in \{1, 2, \cdots, m\}$.

If the objective functions and the constraints are linear and with fuzzy coefficients, we propose the following dependent chance constraints models (2.106) and (2.107),

$$\begin{cases} \max \ [Pos\{\tilde{c}_i^T x \geq \bar{f}_i, i = 1, 2, \cdots, m\}] \\ \text{s.t.} \ \begin{cases} E[\tilde{a}_r^T]x \leq E[\tilde{b}_r], r = 1, 2, \cdots, p \\ x \geq 0 \end{cases} \end{cases} \quad (2.106)$$

and

$$\begin{cases} \max \ \left[Pos\{\tilde{c}_i^T x \geq \bar{f}_i, i = 1, 2, \cdots, m\}\right] \\ \text{s.t.} \ \begin{cases} Pos\{\tilde{a}_r^T x \leq \tilde{b}_r\} \geq \theta_r, \ r = 1, 2, \cdots, p \\ x \geq 0. \end{cases} \end{cases} \quad (2.107)$$

And the models (2.106) and (2.107) are equivalent to the following models (2.108) and (2.109) respectively,

$$\begin{cases} \max \ [\delta_1, \cdots, \delta_m] \\ \text{s.t.} \ \begin{cases} Pos\{\tilde{c}_i^T x \geq \bar{f}_i\} \geq \delta_i, \ i = 1, 2, \cdots, m \\ E[\tilde{a}_r^T] x \leq E[\tilde{b}_r], \ r = 1, 2, \cdots, p \\ x \geq 0 \end{cases} \end{cases} \quad (2.108)$$

and

$$\begin{cases} \max \ [\delta_1, \cdots, \delta_m] \\ \text{s.t.} \ \begin{cases} Pos\{\tilde{c}_i^T x \geq \bar{f}_i\} \geq \delta_i, \ i = 1, 2, \cdots, m \\ Pos\{\tilde{a}_r^T x \leq \tilde{b}_r\} \geq \theta_r, \ r = 1, 2, \cdots, p \\ x \geq 0. \end{cases} \end{cases} \quad (2.109)$$

2.5.1.2 Fuzzy DCP Based on Nec Measure

Another type of fuzzy dependent-chance multi-objective programming is also given as follows,

$$\begin{cases} \max \ \begin{bmatrix} Nec\{f_1(x, \xi) \geq \bar{f}_1\}, \\ Nec\{f_2(x, \xi) \geq \bar{f}_2\}, \\ \cdots \\ Nec\{f_m(x, \xi) \geq \bar{f}_m\}, \end{bmatrix} \\ \text{s.t.} \ \begin{cases} Nec\{g_j(x, \xi) \leq 0\} \geq \theta_r, \ r = 1, 2, \cdots, p \\ x \in X, \end{cases} \end{cases} \quad (2.110)$$

where Nec are the possibility measure of the fuzzy events.

And similarly to the CCM defined by Pos measure, motivated by the ideas of EVM or CCM, we have the following models.

$$\begin{cases} \max \ [Nec\{f_1(x, \xi) \leq 0\}, Nec\{f_2(x, \xi) \leq 0\}, \cdots, Nec\{f_m(x, \xi) \leq 0\}] \\ \text{s.t.} \ \begin{cases} E[g_r(x, \xi)] \geq 0, r = 1, 2, \cdots, p \\ x \in X \end{cases} \end{cases} \quad (2.111)$$

and

$$\begin{cases} \max \ [Nec\{f_1(x, \xi) \leq 0\}, Nec\{f_2(x, \xi) \leq 0\}, \cdots, Nec\{f_m(x, \xi) \leq 0\}] \\ \text{s.t.} \ \begin{cases} Nec\{g_r(x, \xi) \leq 0\} \geq \theta_r, \ r = 1, 2, \cdots, p \\ x \in X, \end{cases} \end{cases} \quad (2.112)$$

where $\theta_r \in [0, 1], r = 1, 2, \cdots, p$ are the predetermined levels.

We can judge whether or not a decision vector is feasible for model (2.111) and (2.112). Thus, they can be formulated as

$$\max_{x \in X} [Nec\{f_1(x,\xi) \leq 0\}, Nec\{f_2(x,\xi) \leq 0\}, \cdots, Nec\{f_m(x,\xi) \leq 0\}] \quad (2.113)$$

where X is a fixed feasible set.

Definition 2.21. $x^* \in X$ is called a *Nec* fuzzy efficient solution of Problem (2.113) if and only if there exists no other fuzzy feasible solution x such that

$$Nec\{\theta \in \Theta | f_i(x,\xi)\} \geq Nec\{\theta \in \Theta | f_i(x^*,\xi)\}, i = 1,2,\cdots,m,$$

for all i, and

$$Nec\{\theta \in \Theta | f_{i_0}(x,\xi)\} > Nec\{\theta \in \Theta | f_{i_0}(x^*,\xi)\}, i = 1,2,\cdots,m,$$

for at least one $i_0 \in \{1,2,\cdots,m\}$.

2.5.2 Linear Fuzzy DCM and Ideal Point Method

If the objective functions and the constraints are linear with fuzzy coefficients, we propose the following dependent chance models (2.114) and (2.115),

$$\begin{cases} \max & [Nec\{\tilde{c}^T x \geq \bar{f}_i, i = 1,2,\cdots,m\}] \\ \text{s.t.} & \begin{cases} E[\tilde{a}_r^T]x \leq E[\tilde{b}_r], \ r = 1,2,\cdots,p \\ x \geq 0 \end{cases} \end{cases} \quad (2.114)$$

and

$$\begin{cases} \max & [Nec\{\tilde{c}^T x \geq \bar{f}_i, i = 1,2,\cdots,m\}] \\ \text{s.t.} & \begin{cases} Nec\{\tilde{a}_r^T x \leq \tilde{b}_r\} \geq \theta_r, \ r = 1,2,\cdots,p \\ x \geq 0. \end{cases} \end{cases} \quad (2.115)$$

And the models (2.114) and (2.115) are equivalent to the following models (2.116) and (2.117) respectively,

$$\begin{cases} \max & [\delta_1,\cdots,\delta_m] \\ \text{s.t.} & \begin{cases} Nec\{\tilde{c}^T x \geq \bar{f}_i\} \geq \delta_i, \ i = 1,2,\cdots,m \\ E[\tilde{a}_r^T]x \leq E[\tilde{b}_r], \ r = 1,2,\cdots,p \\ x \geq 0 \end{cases} \end{cases} \quad (2.116)$$

and

$$\begin{cases} \max & [\delta_1,\cdots,\delta_m] \\ \text{s.t.} & \begin{cases} Nec\{\tilde{c}^T x \geq \bar{f}_i\} \geq \delta_i, \ i = 1,2,\cdots,m \\ Nec\{\tilde{a}_r^T x \leq \tilde{b}_r\} \geq \theta_r, \ r = 1,2,\cdots,p \\ x \geq 0. \end{cases} \end{cases} \quad (2.117)$$

2.5.2.1 Crisp Equivalent Models

When we use the expected value operator to deal with the fuzzy constraints, the theorems in subsection 2.2 can be used to transform some multi-objective models with special fuzzy variables into some certain type of models. So let's concentrate on the second form of CCM.

Fuzzy DCM Crisp Equivalent Model Based on Pos Measure

Theorem 2.14. *Assume that $\tilde{c}_{ij}$ is LR fuzzy variable, the membership function of $\tilde{c}_{ij}$ is*

$$\mu_{\tilde{c}_{ij}}(t) = \begin{cases} L(\frac{c_{ij}-t}{\alpha_{ij}^c}), & t \le c_{ij}, \alpha_{ij}^c > 0 \\ R(\frac{t-c_{ij}}{\beta_{ij}^c}), & t \ge c_{ij}, \beta_{ij}^c > 0, \end{cases} \tag{2.118}$$

where the vector $(c_{ij})_{n\times 1} = (c_{i1}, c_{i2}, \cdots, c_{in})^{\mathrm{T}}$ is real number, α_{ij}^c and β_{ij}^c are the left and right spread of $\tilde{c}_{ij}$, $i = 1, 2, \cdots, m$, $j = 1, 2, \cdots, n$, the reference function $L, R : [0,1] \to [0,1]$ satisfies that $L(1) = R(1) = 0, L(0) = R(0) = 1$, and it is monotone function. Then $\mathrm{Pos}\{\tilde{c}_i^{\mathrm{T}} x \ge f_i\} \ge \delta_i$ is equivalent to

$$R^{-1}(\delta_i) \ge \frac{f_i - c_i^{\mathrm{T}} x}{\beta_i^{c\mathrm{T}} x}, \quad i = 1, 2, \cdots, m. \tag{2.119}$$

Proof. The proof is the same as the proof of Theorem 2.7.

Since we assume that the function $R(\cdot)$ is a monotonically decreasing function, so max δ_i is equivalent to min $R^{-1}(\delta_i)$. By theorem 2.14, 2.8, it's easy for us to derive the crisp equivalent model (2.120) of the models (2.107, 2.109),

$$\begin{cases} \max \left\{ \frac{c_i^T x - f_i}{\beta_i^{c\mathrm{T}} x}, i = 1, 2, \cdots, m \right\} \\ \text{s.t.} \begin{cases} b_r + R^{-1}(\theta_r)\beta_r^b - a_r^{\mathrm{T}} x + L^{-1}(\theta_r)\alpha_r^{a\mathrm{T}} x \ge 0, r = 1, 2, \cdots, p \\ x \ge 0. \end{cases} \end{cases} \tag{2.120}$$

Fuzzy DCM Crisp Equivalent Model Based on *Nec* Measure

Theorem 2.15. *Assume that the fuzzy variable are as the same as that in Theorem 2.7. Then $Nec\{\tilde{c}_i^{\mathrm{T}} x \ge f_i\} \ge \delta_i$ is equivalent to*

$$L^{-1}(1 - \delta_i) \le \frac{c_i^T x - f_i}{\alpha_i^{c\mathrm{T}} x}, \quad i = 1, 2, \cdots, m. \tag{2.121}$$

Proof. The proof is similar as the proof of theorem 2.7.

Since we assume that the reference function $L(\cdot)$ are monotonically decreasing function, so max δ_i is equivalent to max $L^{-1}(1-\delta_i)$. By theorem 2.15, 2.10, it's easy for us to derive the crisp equivalent model (2.122) of the models (2.115,2.117),

$$\begin{cases} \max \left\{ \frac{c_i^T x - f_i}{\alpha_i^{cT} x}, i = 1,2,\cdots,m \right\} \\ \text{s.t.} \begin{cases} b_r + R^{-1}(\theta_r)\beta_r^b - a_r^T x + L^{-1}(\theta_r)\alpha_r^{aT} x \geq 0, r = 1,2,\cdots,p \\ x \geq 0. \end{cases} \end{cases} \quad (2.122)$$

2.5.2.2 Ideal Point Method

In this section, we make use of the ideal point method proposed in [92, 387, 388] to resolve the multiobjective problem with crisp parameters. If the decision maker can firstly propose an estimated value $\bar{F}_i$ for each objective function $\Psi_i^c x$ such that

$$\bar{F}_i \geq \max_{x \in X'} \Psi_1^c x, i = 1,2,\cdots,m, \quad (2.123)$$

where $X' = \{x \in X | \Psi_r^e x \leq \Psi_r^b, r = 1,2,\cdots,p, x \geq 0\}$, then $\bar{F}_i = (\bar{F}_1, \bar{F}_2, \cdots, \bar{F}_m)^T$ is called the ideal point, especially, if $\bar{F}_i \geq \max_{x \in X'} \Psi_1^c x$ for all i, we call $\bar{F}$ the most ideal point.

The basic theory of the ideal point method is to take a especial norm in the objective space R^m and obtain the feasible solution x that the objective value approaches the ideal point $\bar{F} = (\bar{F}_1, \bar{F}_2, \cdots, \bar{F}_m)^T$ under the norm distance, that is, to seek the feasible solution x satisfying

$$\min_{x \in X'} u(\Psi^c(x)) = \min_{x \in X'} ||\Psi^c(x) - \bar{F}||.$$

Usually, the following norm functions are used to describe the distance:
(1) p-mode function

$$d_p(\Psi^c(x), \bar{F}; \omega) = \left[\sum_{i=1}^m \omega_i |\Psi_i^c x - \bar{F}_i|^p \right]^{\frac{1}{p}}, 1 \leq p < +\infty. \quad (2.124)$$

(2)The maximal deviation function

$$d_{+\infty}(\Psi^c(x), \bar{F}; \omega) = \max_{1 \leq i \leq m} \omega_i |\Psi_i^c x - \bar{F}_i|. \quad (2.125)$$

(3) Geometric mean function

$$d(\Psi^c(x), \bar{F}) = \left[\prod_{i=1}^m |\Psi_i^c x - \bar{F}_i|^p \right]^{\frac{1}{m}}. \quad (2.126)$$

The weight parameter vector $\omega = (\omega_1, \omega_2, \cdots, \omega_m)^T > 0$ needs to be predetermined.

Theorem 2.16. *Assume that* $\bar{F}_i > \max_{x \in X'} \Psi_1^c x (i = 1, 2, \cdots, m)$. *If* x^* *is the optimal solution of the following problem*

$$\min_{x \in X'} d_p(\Psi^c(x), \bar{F}; \omega) = \left[\sum_{i=1}^m \omega_i |\Psi_i^c x - \bar{F}_i|^p \right]^{\frac{1}{p}}, \tag{2.127}$$

then x^* *is an efficient solution of problem (2.128). On the contrary, if* x^* *is an efficient solution of problem (2.128), then there exists a weight vector* ω *such that* x^* *is the optimal solution of problem (2.127).*

Proof. This result can be easily obtained, and we hereby don't prove it. $\qquad\square$

Next, we take the p-mode function to describe the procedure of solving the problem (2.128).

$$\begin{cases} \max \{\Psi_1^c x, \cdots \Psi_i^c x, \cdots, \Psi_m^c x\} \\ \text{s.t. } x \in X. \end{cases} \tag{2.128}$$

Step 1. Find the ideal point. If the decision maker can give the ideal objective value satisfying the condition (2.123), the value will be considered as the ideal point. However, decision makers themselves don't know how to give the objective value, so we can get the ideal point by solving the following programming problem,

$$\begin{cases} \max \Psi_i^c x \\ \text{s.t. } \begin{cases} \Psi_r^e x \leq \Psi_r^b, r = 1, 2, \cdots, p \\ x \in X. \end{cases} \end{cases} \tag{2.129}$$

Then the ideal point $\bar{F} = (\bar{F}_1, \bar{F}_2, \cdots, \bar{F}_m)^T$ can be fixed by $\bar{F}_i = \Psi_i^c x^*$, where x^* is the optimal solution of problem (2.129).

Step 2. Fix the weight. The method of selecting the weight is referred in research papers and interested readers can consult them. We usually use the following function to fix the weight,

$$\omega_i = \frac{\bar{F}_i}{\sum_{i=1}^m \bar{F}_i}.$$

Step 3. Construct the minimal distance problem. Solve the following single objective programming problem to obtain an efficient solution to problem (2.128),

$$\begin{cases} \min \left[\sum_{i=1}^m \omega_i |\Psi_i^c x - \bar{F}_i|^t \right]^{\frac{1}{t}} \\ \text{s.t. } \begin{cases} \Psi_r^e x \leq \Psi_r^b, r = 1, 2, \cdots, p \\ x \in X. \end{cases} \end{cases} \tag{2.130}$$

Usually, we take $t = 2$ to compute it.

2.5.2.3   Numerical Example

Example 2.9. Let consider the following problem ,

$$
\begin{cases}
\max \ \xi_1 x_1 + \xi_2 x_2 + \xi_3 x_3 + \xi_4 x_4 \\
\max \ \xi_5 x_1 + \xi_6 x_2 + \xi_7 x_3 + \xi_8 x_4 \\
\text{s.t.} \begin{cases}
7x_1 + 5x_2 + 3x_3 + 2x_4 \le 120 \\
3x_1 + 3x_2 + 10x_3 + 15x_4 \le \xi_{13} \\
x_i \ge 0, \ i = 1,2,\cdots,4,
\end{cases}
\end{cases}
\tag{2.131}
$$

where $\xi_j (j = 1,2,\cdots,8)$ are triangular LR fuzzy variables characterized as that in Theorem. 2.7,

$$
\begin{aligned}
&\xi_1 = (4,0.5,0.5)_{LR}, \quad \xi_2 = (5,0.5,0.5)_{LR}, \\
&\xi_3 = (9,0.5,0.5)_{LR}, \quad \xi_4 = (11,0.5,0.5)_{LR}, \\
&\xi_5 = (1,0.1,0.1)_{LR}, \quad \xi_6 = (1,0.1,0.1)_{LR}, \\
&\xi_7 = (1,0.1,0.1)_{LR}, \quad \xi_8 = (1,0.1,0.1)_{LR}.
\end{aligned}
$$

We use the chance operator and the expected value operator to deal with the objective functions and constraints of fuzzy multi-objective model respectively, and here the decision maker is supposed to be comparatively optimistic, so we adopt the *Pos* measure to measure the chance, so we can obtain the following DCM,

$$
\begin{cases}
\max \ Pos\{\xi_1 x_1 + \xi_2 x_2 + \xi_3 x_3 + \xi_4 x_4 \ge \bar{f}_1\} \\
\max \ Pos\{\xi_5 x_1 + \xi_6 x_2 + \xi_7 x_3 + \xi_8 x_4 \ge \bar{f}_2\} \\
\text{s.t.} \begin{cases}
7x_1 + 5x_2 + 3x_3 + 2x_4 \le 120 \\
3x_1 + 5x_2 + 10x_3 + 15x_4 \le 100 \\
x_i \ge 0, \ i = 1,2,\cdots,5
\end{cases}
\end{cases}
\tag{2.132}
$$

or

$$
\begin{cases}
\max \ \{\delta_1, \delta_2\} \\
\text{s.t.} \begin{cases}
Pos\{\xi_1 x_1 + \xi_2 x_2 + \xi_3 x_3 + \xi_4 x_4 \ge \bar{f}_1\} \ge \delta_1 \\
Pos\{\xi_5 x_1 + \xi_6 x_2 + \xi_7 x_3 + \xi_8 x_4 \ge \bar{f}_2\} \ge \delta_2 \\
7x_1 + 5x_2 + 3x_3 + 2x_4 \le 120 \\
3x_1 + 5x_2 + 10x_3 + 15x_4 \le 100 \\
x_i \ge 0, \ i = 1,2,\cdots,5.
\end{cases}
\end{cases}
\tag{2.133}
$$

Since triangular LR fuzzy coefficients exist in the objective functions and a constraint, by Theorem 2.14, if the decision maker gives the ideal value of the objective functions $\bar{f}_1, \bar{f}_2$, we have the following equivalent model 2.134,

$$
\begin{cases}
\max \ \left\{ \begin{aligned}
F_1 &= \frac{4x_1 + 5x_2 + 9x_3 + 11x_4 - \bar{f}_1}{0.5x_1 + 0.5x_2 + 0.5x_3 + 0.5x_4}, \\
F_2 &= \frac{x_1 + x_2 + x_3 + x_4 - \bar{f}_2}{0.1x_1 + 0.1x_2 + 0.1x_3 + 0.1x_4}
\end{aligned} \right\} \\
\text{s.t.} \begin{cases}
7x_1 + 5x_2 + 3x_3 + 2x_4 \le 120 \\
3x_1 + 5x_2 + 10x_3 + 15x_4 \le 100 \\
x_i \ge 0, \ i = 1,2,\cdots,5.
\end{cases}
\end{cases}
\tag{2.134}
$$

Then we use the ideal point method to solve, and we first calculate the following two single objective models (2.135) and (2.136),

$$\begin{cases} \max F_1 = \frac{4x_1+5x_2+9x_3+11x_4-\bar{f}_1}{0.5x_1+0.5x_2+0.5x_3+0.5x_4} \\ \text{s.t.} \begin{cases} 7x_1+5x_2+3x_3+2x_4 \leq 120 \\ 3x_1+5x_2+10x_3+15x_4 \leq 100 \\ x_i \geq 0, \ i=1,2,\cdots,5 \end{cases} \end{cases} \quad (2.135)$$

and

$$\begin{cases} \max F_2 = \frac{x_1+x_2+x_3+x_4-\bar{f}_2}{0.1x_1+0.1x_2+0.1x_3+0.1x_4} \\ \text{s.t.} \begin{cases} 7x_1+5x_2+3x_3+2x_4 \leq 120 \\ 3x_1+5x_2+10x_3+15x_4 \leq 100 \\ x_i \geq 0, \ i=1,2,\cdots,5. \end{cases} \end{cases} \quad (2.136)$$

We set $\bar{f}_1 = 100, \bar{f}_2 = 20$ and get the optimal solution as follows respectively:

$$(x_1,x_2,x_3,x_4) = (13.77,0,7.87,0), F_1 = 2.394,$$
$$(x_1,x_2,x_3,x_4) = (0,24,0,0), F_2 = 1.667.$$

Then we get the weights for each objective,

$$w_1 = 2.394/(2.394+1.667) = 0.59, w_2 = 0.41.$$

We can construct the model (2.137) according to the ideal point method,

$$\begin{cases} \max \sqrt{0.59(\frac{4x_1+5x_2+9x_3+11x_4-\bar{f}_1}{0.5x_1+0.5x_2+0.5x_3+0.5x_4})^2 + 0.41(\frac{x_1+x_2+x_3+x_4-\bar{f}_2}{0.1x_1+0.1x_2+0.1x_3+0.1x_4})^2} \\ \text{s.t.} \begin{cases} 7x_1+5x_2+3x_3+2x_4 \leq 120 \\ 3x_1+5x_2+10x_3+15x_4 \leq 100 \\ x_i \geq 0, \ i=1,2,\cdots,5. \end{cases} \end{cases} \quad (2.137)$$

By solving the above model, we could get a efficient solution of the original model (2.132) as follows,

$$(x_1,x_2,x_3,x_4) = (13.77,0,7.87,0).$$

2.5.3 Non-linear Fuzzy DCM and Fuzzy Simulation-Based PSO with Shrinkage Factor

It is difficult to transform the non-linear fuzzy DCM into it's equivalent form, so we introduce the fuzzy simulation 3-based PSO to solve.

2.5.3.1 Fuzzy Simulation 3 for Chance

Suppose that $g_1, g_2, \cdots, g_p$ are real-valued functions, and ξ_i are fuzzy variables with membership functions $\mu_i, i = 1, 2, \cdots, n$, respectively. We design a fuzzy simulation to compute the possibility

$$L = Pos\{g_j(\xi) \geq 0, j = 1, 2, \cdots, p\}, \quad (2.138)$$

where $\xi = (\xi_1, \xi_2, \cdots, \xi_n)$. In practice, we give a lower estimation of the possibility L, denoted by α. Then we randomly generate $u_1, u_2, \cdots, u_n$ from the α-level sets of $\xi_1, \xi_2, \cdots, \xi_n$, respectively, and denote $u = (u_1, u_2, \cdots, u_n)$. If the α-level set is not easy for a computer to describe, we can give a larger region, for example, a hypercube containing the α-level set. Certainly, the smaller the region, the more effective the fuzzy simulation. Now we set

$$\mu = \mu_1(u_1) \wedge \mu_2(u_2) \wedge \cdots \mu_n(u_n).$$

If $g_j(u) \leq 0, j = 1, 2, \cdots p$ and $L < \mu$, then we set $L = \mu$. Repeat this process N times. The value L is regarded as an estimation of the possibility. We now summarize it as follows:

Step 1. Set $L = \alpha$ as a lower estimation.

Step 2. Randomly generate u_i from the α-level sets of fuzzy variables $\xi_i, i = 1, 2, \ldots, n$, respectively, and denote $u = (u_1, u_2, \cdots, u_n)$.

Step 3. Set $\mu = \mu_1(u_1) \wedge \mu_2(u_2) \wedge \cdots \mu_n(u_n)$.

Step 4. If $g_j(u) \leq 0$ and $L < \mu$, then we set $L = \mu$.

Step 5. Repeat the second to fourth steps N times.

Step 6. Return L.

2.5.3.2   PSO with Shrinkage Factor

Particle swarm optimization (PSO) is one of the newest techniques within the family of evolutionary optimization algorithms. The algorithm is based on an analogy with the choreography of the flight of a flock of birds. Due to its fast convergence, PSO has been advocated to be especially suitable for multiobjective optimization. There are many variants of the single objective PSO but in most of them the movement of the particles towards the optimum is governed by equations similar to the following:

$$\mathbf{v}_i(t+1) = w\mathbf{v}_i(t) + c_1 \cdot r_1(\mathbf{P}_i(t) - \mathbf{x}_i(t)) + c_2 \cdot r_2(\mathbf{P}_g(t) - \mathbf{x}_i(t)), \qquad (2.139)$$

where w is an inertia coefficient that has an important role balancing the global (a large value of w) and local search (a small value of w), c_1 and c_2 are constants (usually $c_1 = c_2 = 2$), r_1 and r_2 are uniform random numbers in $[0,1]$, Pi is the best position vector of particle i so far, P_g is the best position vector of all particles so far, $\xi(t)$ is the current position vector of particle i, and $v_i(t)$ is the current velocity of particle i. Mendes et al. [128] suggests an inertia coefficient w of less than 1, while other authors recommend starting with larger values and decrease them with time, for example from a value of 1.4 to 0.5. Coello Coello et al. [130] highlighted the sensitivity of the standard PSO algorithm to the value of w and proposed the introduction of a mutation operator that assures an adequate global search while keeping a small value of w (suggested 0.4) which favors a refined local search.

Step 1. Give proper parameter setting, e.g. population size N, study factors c_1, c_2, part size M and the maximum generation *Gen*.

Step 2. Initialize all particle, $k = 0$.

Step 3. Evaluate $x(k)$.

Step 4. Judge whether the termination criterion is satisfied. If $k > Gen$, stop, otherwise refresh variables according to the following method: refresh $x(k)$ and $v(k)$ using the below formula:

$$\begin{cases} v_i(t+1) = wv_i(t) + c_1 \cdot r_1, \\ (P_i(t) - x_i(t)) + c_2 \cdot r_2(P_g(t) - x_i(t)). \end{cases}$$

2.5.3.3 Numerical Example

Example 2.10. Let us consider a multi-objective programming with fuzzy coefficients.

$$\begin{cases} \max F_1(x, \xi) = 2\xi_1^2 x_1 + 3\xi_2^2 x_2 - \xi_3 x_3 + \sqrt{\xi_4^2 + (3 - \xi_5 x_4)^2} \\ \max F_2(x, \xi) = \xi_6 x_1 + \xi_7 x_2 + \xi_8 x_3 \\ \text{s.t.} \begin{cases} 5x_1 - 3x_2^2 + 6\sqrt{x_3} + x_4 \le 80 \\ 4x_1 + 35x_2 - 4.5x_3 \le 20 \\ x_1 + x_2 + x_3 + x_4 \le 18 \\ x_1, x_2, x_3, x_4 \ge 0, \end{cases} \end{cases} \qquad (2.140)$$

where $\xi_i (i = 1, 2, \cdots, 8)$ are fuzzy variables as follows,

$$\begin{aligned} &\xi_1 = (0.4, 0.45, 0.5), \quad \xi_2 = (0.6, 0.65, 0.7), \\ &\xi_3 = (0.7, 0.75, 0.8), \quad \xi_4 = (0.8, 1, 1.2), \\ &\xi_5 = (3.8, 4, 4.2), \qquad \xi_6 = (-1.2, -1, -0.5), \\ &\xi_7 = (2.5, 3, 3.5), \qquad \xi_8 = (0.8, 1, 1.2). \end{aligned} \qquad (2.141)$$

From the mathematical view, the problem (2.140) is not well defined because of the uncertain parameters. Then we apply Fuzzy DCM to deal with this uncertain programming.

$$\begin{cases} \max \{\delta_1, \delta_2\} \\ \text{s.t.} \begin{cases} Ch\{2\xi_1^2 x_1 + 3\xi_2^2 x_2 - \xi_3 x_3 + \sqrt{\xi_4^2 + (3 - \xi_5 x_4)^2} \ge \bar{f}_1\}(\gamma_1) \ge \delta_1, \\ Ch\{\xi_6 x_1 + \xi_7 x_2 + \xi_8 x_3 \ge \bar{f}_2\}(\gamma_2) \ge \delta_2 \\ 5x_1 - 3x_2^2 + 6\sqrt{x_3} + x_4 \le 80 \\ 4x_1 + 35x_2 - 4.5x_3 \le 20 \\ x_1 + x_2 + x_3 + x_4 \le 18 \\ x_1, x_2, x_3, x_4 \ge 0. \end{cases} \end{cases} \qquad (2.142)$$

Since there exists non-linear objective functions, we cannot transform it into it's crisp equivalent model. In order to solve it, we use the fuzzy simulation based PSO to deal with it. After running, we get a solution as follows:

$$\begin{aligned} &x_1 = 11.32, x_2 = 3.44, x_3 = 3.24, x_4 = 0; \\ &\delta_1 = 0.63, \delta_2 = 0.55. \end{aligned}$$

2.6 Application of Farm Structure Optimization Problem

If we consider the farm structure optimization problem in section 2.1, the mathematical model as follows.

Constraints include:

(1) balance of arable land; since this farm has 20 hectares of arable land, so the land for plant production for sale and for fodder consumed in this farm should be no more than the total arable land;

$$\sum_{i=1}^{18} x_i \leq 20.$$

(2) two balances for crop succession for spring crops and rape;

$$-x_4 - x_5 + x_9 - x_{11} - x_{13} + x_{15} + x_{18} \geq 0,$$
$$x_5 + x_6 + x_{11} \geq 0.$$

(3) three constraints on the area of group of plants for crops, root crops and winter crops;

$$x_1 + x_2 + x_3 + x_4 + x_5 + x_{10} + x_{11} + x_{12} + x_{13} + x_{14} + x_{15} \leq 14,$$
$$x_9 + x_{10} + x_{15} \leq 5,$$
$$x_1 + x_2 + x_3 + x_{10} + x_{12} \leq 10.$$

(4) three balances for manpower for spring, summer and autumnal periods of extended manpower demand;

$$\tilde{a}_{7,1}x_1 + \tilde{a}_{7,2}x_2 + \cdots + \tilde{a}_{7,19}x_{19} - x_{24} \leq \tilde{b}_7,$$
$$\tilde{a}_{8,1}x_1 + \tilde{a}_{8,2}x_2 + \cdots + \tilde{a}_{8,19}x_{19} - x_{25} \leq \tilde{b}_8,$$
$$\tilde{a}_{9,1}x_1 + \tilde{a}_{9,2}x_2 + \cdots + \tilde{a}_{9,19}x_{19} - x_{26} \leq \tilde{b}_9.$$

(5) three balances for artificial fertilizers for phosphorus, nitrogen and potassium;

$$160x_1 + 140x_2 + 120x_3 + 120x_4 + 120x_5 + 180x_6 + 20x_7$$
$$+ 120x_8 + 160x_9 + 10x_{10} + 15x_{11} + 35x_{12} + 15x_{13} + 160x_{14} + 160x_{15}$$
$$+ 30x_{16} + 30x_{17} + 200x_{18} + 200x_{19} - x_{20} \leq 0,$$
$$120x_1 + 120x_2 + 120x_3 + 100x_4 + 100x_5 + 120x_6 + 100x_7$$
$$+ 120x_8 + 120x_9 + 120x_{10} + 100x_{11} + 120x_{12} + 100x_{13} + 120x_{14}$$
$$+ 120x_{15} + 120x_{16} + 120x_{17} + 140x_{18} + 120x_{19} - x_{21} \leq 0,$$
$$100x_1 + 100x_2 + 90x_3 + 80x_4 + 80x_5 + 180x_6 + 80x_7$$
$$+ 120x_8 + 120x_9 + 90x_{10} + 80x_{11} + 90x_{12} + 80x_{13} + 120x_{14} + 120x_{15}$$
$$+ 100x_{16} + 100x_{17} + 120x_{18} + 100x_{19} - x_{22} \leq 0.$$

(6) four balances of fodder for green fodder, potatoes, fodder beets and nutritive fodder;

$$300x_9 + 400x_{16} + 350x_{17} + 550x_{18} + 350x_{19} \geq 3000,$$
$$250x_{14} \geq 480,$$
$$900x_{15} \geq 450,$$
$$45x_{10} + 40x_{11} + 43x_{12} + 40x_{13} + x_{23} \geq 360.$$

(7) balance of permanent grassland.

$$x_{19} = 4.$$

The matrix of constraints is presented in Table 2.2, and a definition for imprecise coefficients is given in Tables 2.3 and 2.4; the imprecise coefficients are treated as triangular fuzzy numbers.

The following three objectives are used to evaluate the solutions/alternatives,
(1) gross profit:

$$f_1(x) = \tilde{c}_1 x_1 + \tilde{c}_2 x_2 + \cdots + \tilde{c}_9 x_9 - \tilde{c}_{10} x_{10} - \tilde{c}_{11} x_{11} - \cdots - \tilde{c}_{26} x_{26} + C$$

where the fuzzy cost coefficients are presented in Table 2.4 and constant C is equal to $(22000, 22000, 22000)$.
(2) structure-forming plants area:

$$f_2(x) = 0.5x_6 + x_7 + x_{16} + x_{17}.$$

(3) manpower hire:

$$f_3(x) = x_{24} + x_{25} + x_{26}.$$

It is obvious that the first objective and the constraints 7-9 are imprecise, so we need to transform them into a crisp to solve it. The simplest way is to transform the uncertain objective functions into crisp by an expected value operator, as follows,

$$f_1(x) = \tilde{c}_1 x_1 + \tilde{c}_2 x_2 + \cdots + \tilde{c}_9 x_9 - \tilde{c}_{10} x_{10} - \tilde{c}_{11} x_{11} - \cdots - \tilde{c}_{26} x_{26} + \sum_s T_s h_s$$
$$\Rightarrow E[f_1(x)] = E[\tilde{c}_1 x_1 + \tilde{c}_2 x_2 + \cdots + \tilde{c}_9 x_9 - \tilde{c}_{10} x_{10} - \tilde{c}_{11} x_{11} - \cdots - \tilde{c}_{26} x_{26} + \sum_s T_s h_s]$$
$$\Rightarrow E[f_1(x)] = E[\tilde{c}_1] x_1 + E[\tilde{c}_2] x_2 + \cdots + E[\tilde{c}_9] x_9 - E[\tilde{c}_{10}] x_{10} - E[\tilde{c}_{11}] x_{11} - \cdots$$
$$- E[\tilde{c}_{26}] x_{26} + E[\sum_s T_s h_s]$$
$$\Rightarrow E[f_1(x)] = 500x_1 + 400x_2 + 385x_3 + 350x_4 + 405x_5 + 565x_6 + 500x_7 + 1400x_8$$
$$+ 605x_9 - 170x_{10} - 170x_{11} - 170x_{12} - 175x_{13} - 760x_{14} - 300x_{15} - 135x_{16}$$
$$- 145x_{17} - 245x_{18} - 95x_{19} - 0.1x_{20} - 0.65x_{21} - 0.03x_{22} - 12x_{23} - 0.75x_{24}$$
$$- 0.8x_{25} - 0.9x_{26} + 22000.$$

$$(2.143)$$

Table 2.2 Constraints set

i	1	2	3	4	5	6	7	8	9	10	11	12	13	14	15	16	17
x_1	1			1		1	$\tilde{a}_{7,1}$	$\tilde{a}_{8,1}$	$\tilde{a}_{9,1}$	160	120	100					
x_2	1			1		1	$\tilde{a}_{7,2}$	$\tilde{a}_{8,2}$	$\tilde{a}_{9,2}$	140	120	100					
x_3	1			1		1	$\tilde{a}_{7,3}$	$\tilde{a}_{8,3}$	$\tilde{a}_{9,3}$	120	120	90					
x_4	1	-1		1			$\tilde{a}_{7,4}$	$\tilde{a}_{8,4}$	$\tilde{a}_{9,4}$	120	100	80					
x_5	1	-1	1	1			$\tilde{a}_{7,5}$	$\tilde{a}_{8,5}$	$\tilde{a}_{9,5}$	120	100	80					
x_6	1		1				$\tilde{a}_{7,6}$	$\tilde{a}_{8,6}$	$\tilde{a}_{9,6}$	180	120	180					
x_7	1						$\tilde{a}_{7,7}$	$\tilde{a}_{8,7}$	$\tilde{a}_{9,7}$	20	100	80					
x_8	1						$\tilde{a}_{7,8}$	$\tilde{a}_{8,8}$	$\tilde{a}_{9,8}$	120	120	120					
x_9	1	1			1		$\tilde{a}_{7,9}$	$\tilde{a}_{8,9}$	$\tilde{a}_{9,9}$	160	120	120	300				
x_{10}	1			1	1	1	$\tilde{a}_{7,10}$	$\tilde{a}_{8,10}$	$\tilde{a}_{9,10}$	10	120	190				45	
x_{11}	1	-1	1	1			$\tilde{a}_{7,11}$	$\tilde{a}_{8,11}$	$\tilde{a}_{9,11}$	15	100	80				40	
x_{12}	1			1		1	$\tilde{a}_{7,12}$	$\tilde{a}_{8,12}$	$\tilde{a}_{9,12}$	35	120	90				43	
x_{13}	1	-1		1			$\tilde{a}_{7,13}$	$\tilde{a}_{8,13}$	$\tilde{a}_{9,13}$	15	100	80				40	
x_{14}	1			1			$\tilde{a}_{7,14}$	$\tilde{a}_{8,14}$	$\tilde{a}_{9,14}$	160	120	120		250			
x_{15}	1	1		1	1		$\tilde{a}_{7,15}$	$\tilde{a}_{8,15}$	$\tilde{a}_{9,15}$	160	120	120			900		
x_{16}	1						$\tilde{a}_{7,16}$	$\tilde{a}_{8,16}$	$\tilde{a}_{9,16}$	30	120	100	400				
x_{17}	1						$\tilde{a}_{7,17}$	$\tilde{a}_{8,17}$	$\tilde{a}_{9,17}$	30	120	100	350				
x_{18}	1	1					$\tilde{a}_{7,18}$	$\tilde{a}_{8,18}$	$\tilde{a}_{9,18}$	200	140	120	550				
x_{19}							$\tilde{a}_{7,19}$	$\tilde{a}_{8,19}$	$\tilde{a}_{9,19}$	200	120	100	350				1
x_{20}									-1								
x_{21}										-1							
x_{22}											-1						
x_{23}													1				
x_{24}					-1												
x_{25}						-1											
x_{26}							-1										
sg.	$=$	$\geq$	$\geq$	$\leq$	$\leq$	$\leq$	$\leq$	$\leq$	$\leq$	$\leq$	$\leq$	$\leq$	$\geq$	$\geq$	$\geq$	$\geq$	$=$
	200	0	0	14	5	10	b_7	b_8	b_9	0	0	0	3000	480	450	360	4

$\tilde{b}_7 = (600, 615, 630); \tilde{b}_8 = (550, 580, 610); \tilde{b}_9 = (895, 955, 1015).$

Also we use the expected value operator to deal with the fuzzy constraints,

$$\tilde{a}_{7,1}x_1 + \tilde{a}_{7,2}x_2 + \cdots + \tilde{a}_{7,19}x_{19} - x_{24} \leq \tilde{b}_7$$
$$\Rightarrow E[\tilde{a}_{7,1}x_1 + \tilde{a}_{7,2}x_2 + \cdots + \tilde{a}_{7,19}x_{19} - x_{24}] \leq E[\tilde{b}_7]$$
$$\Rightarrow E[\tilde{a}_{7,1}]x_1 + E[\tilde{a}_{7,2}]x_2 + \cdots + E[\tilde{a}_{7,19}]x_{19} - x_{24} \leq E[\tilde{b}_7]$$
$$\Rightarrow 15x_1 + 15x_2 + 15x_3 + 25x_4 + 25x_5 + 30x_6 + 30x_7 + 110x_8 + 80x_9 + 30x_{10}$$
$$+ 25x_{11} + 15x_{12} + 25x_{13} + 110x_{14} + 80x_{15} + 30x_{16} + 30x_{17} + 25x_{18}$$
$$+ 10x_{19} - x_{24} \leq 615.$$

$$(2.144)$$

Table 2.3 Fuzzy coefficients of constraints

j	$\tilde{a}_{7,j}$	$\tilde{a}_{8,j}$	$\tilde{a}_{9,j}$
1	(14,15,16)	(38,40,42)	(32,35,38)
2	(14,15,16)	(38,40,42)	(9,10,11)
3	(14,15,16)	(38,40,42)	(32,35,38)
4	(22,25,28)	(38,40,42)	(14,15,16)
5	(22,25,28)	(38,40,42)	(14,15,16)
6	(27,30,33)	(86,90,94)	(9,10,11)
7	(27,30,33)	(72,75,78)	(17,20,23)
8	(107,110,113)	(14,15,16)	(116,120,124)
9	(76,80,84)	(18,20,22)	(155,160,165)
10	(27,30,33)	(38,40,42)	(9,10,11)
11	(23,25,27)	(38,40,42)	(13,15,17)
12	(14,15,16)	(38,40,42)	(32,35,38)
13	(22,25,28)	(38,40,42)	(14,15,16)
14	(105,110,115)	(14,15,16)	(116,120,124)
15	(78,80,82)	(18,20,22)	(155,160,165)
16	(28,30,32)	(31,34,37)	(9,10,11)
17	(28,30,32)	(31,34,37)	(9,10,11)
18	(23,25,27)	(9,10,11)	(57,60,63)
19	(9,10,11)	(38,40,42)	(38,40,42)

Table 2.4 Fuzzy cost coefficients

j	$\tilde{c}_j$	j	$\tilde{c}_j$
1	(480,500,520)	14	(740,760,780)
2	(380,400,420)	15	(280,300,320)
3	(370,385,395)	16	(130,135,140)
4	(330,350,370)	17	(140,145,150)
5	(400,405,410)	18	(240,245,250)
6	(560,565,570)	19	(90,95,100)
7	(470,500,530)	20	(0.99,0.10,0.11)
8	(1350,1400,1450)	21	(0.55,0.65,0.75)
9	(585,605,625)	22	(0.02,0.03,0.04)
10	(160,170,180)	23	(11,12,13)
11	(160,170,180)	24	(0.7,0.75,0.8)
12	(180,170,180)	25	(0.7,0.8,0.9)
13	(165,175,185)	26	(0.8,0.9,1.0)

And

$$40x_1 + 40x_2 + 40x_3 + 40x_4 + 40x_5 + 90x_6 + 75x_7 + 15x_8 + 20x_9 + 40x_{10}$$
$$+40x_{11} + 40x_{12} + 40x_{13} + 14x_{14} + 20x_{15} + 34x_{16} + 34x_{17} + 10x_{18}$$
$$+40x_{19} - x_{25} \leq 580$$
$$35x_1 + 10x_2 + 35x_3 + 15x_4 + 15x_5 + 10x_6 + 20x_7 + 120x_8 + 160x_9 + 10x_{10}$$
$$+15x_{11} + 35x_{12} + 15x_{13} + 120x_{14} + 160x_{15} + 10x_{16} + 10x_{17} + 60x_{18}$$
$$+40x_{19} - x_{26} \leq 955.$$

$$(2.145)$$

Also we could use *Pos* operator to turn the constraints (4) to chance constraints (2.146-2.148). For the predetermined confidence level $\beta_1, \beta_2, \beta_3$, the chance constraints are:

$$Pos\left\{\tilde{a}_{7,1}x_1 + \tilde{a}_{7,2}x_2 + \cdots + \tilde{a}_{7,19}x_{19} - x_{24} \leq \tilde{b}_7\right\} \geq \beta_1, \qquad (2.146)$$

$$Pos\left\{\tilde{a}_{8,1}x_1 + \tilde{a}_{8,2}x_2 + \cdots + \tilde{a}_{8,19}x_{19} - x_{25} \leq \tilde{b}_8\right\} \geq \beta_2, \qquad (2.147)$$

$$Pos\left\{\tilde{a}_{9,1}x_1 + \tilde{a}_{9,2}x_2 + \cdots + \tilde{a}_{9,19}x_{19} - x_{26} \leq \tilde{b}_9\right\} \geq \beta_3. \qquad (2.148)$$

According to the CCM, we can transform each *Pos*-chance constraint into it's crisp equivalent constraint:

$$15x_1 + 15x_2 + 15x_3 + 25x_4 + 25x_5 + 30x_6 + 30x_7 + 110x_8 + 80x_9 + 30x_{10} + 25x_{11}$$
$$+15x_{12} + 25x_{13} + 110x_{14} + 80x_{15} + 30x_{16} + 30x_{17} + 25x_{18} + 10x_{19}$$
$$-(1 - \beta_1)(x_1 + x_2 + x_3 + 3x_4 + 3x_5 + 3x_6 + 3x_7 + 3x_8 + 4x_9 + 3x_{10} + 2x_{11}$$
$$+x_{12} + 3x_{13} + 5x_{14} + 2x_{15} + 2x_{16} + 2x_{17} + 2x_{18} + x_{19}) - x_{24} \leq 615 + 15(1 - \beta_1),$$
$$40x_1 + 40x_2 + 40x_3 + 40x_4 + 40x_5 + 90x_6 + 75x_7 + 15x_8 + 20x_9 + 40x_{10} + 40x_{11}$$
$$+40x_{12} + 40x_{13} + 14x_{14} + 20x_{15} + 34x_{16} + 34x_{17} + 10x_{18} + 40x_{19}$$
$$-(1 - \beta_2)(2x_1 + 2x_2 + 2x_3 + 2x_4 + 2x_5 + 4x_6 + 3x_7 + x_8 + 2x_9 + 2x_{10} + 2x_{11}$$
$$+2x_{12} + 2x_{13} + x_{14} + 2x_{15} + 3x_{16} + 3x_{17} + x_{18} + 2x_{19}) - x_{25} \leq 580 + 30(1 - \beta_2),$$
$$35x_1 + 10x_2 + 35x_3 + 15x_4 + 15x_5 + 10x_6 + 20x_7 + 120x_8 + 160x_9 + 10x_{10} + 15x_{11}$$
$$+35x_{12} + 15x_{13} + 120x_{14} + 160x_{15} + 10x_{16} + 10x_{17} + 60x_{18} + 40x_{19}$$
$$-(1 - \beta_3)(3x_1 + x_2 + 3x_3 + x_4 + x_5 + x_6 + 3x_7 + 4x_8 + 5x_9 + x_{10} + 2x_{11}$$
$$+3x_{12} + x_{13} + 4x_{14} + 5x_{15} + x_{16} + x_{17} + 3x_{18} + 2x_{19}) - x_{26} \leq 955 + 60(1 - \beta_3).$$

$$(2.149)$$

If we adopt the expected value operator and the *Pos*-constraint method to deal with the objective functions and the fuzzy constraints, respectively, then we give the crisp equivalent model as follows:

$$\begin{cases} \max\ E[f_1(x)] = 500x_1 + 400x_2 + 385x_3 + 350x_4 + 405x_5 + 565x_6 + 500x_7 \\ \qquad +1400x_8 + 605x_9 - 170x_{10} - 170x_{11} - 170x_{12} - 175x_{13} - 760x_{14} - 300x_{15} \\ \qquad -135x_{16} - 145x_{17} - 245x_{18} - 95x_{19} - 0.1x_{20} - 0.65x_{21} - 0.03x_{22} - 12x_{23} \\ \qquad -0.75x_{24} - 0.8x_{25} - 0.9x_{26} + 22000 \\ \max\ f_2(x) = 0.5x_6 + x_7 + x_{16} + x_{17} \\ \min\ f_3(x) = x_{24} + x_{25} + x_{26} \\ \text{s.t.} \begin{cases} \sum_{i=1}^{18} x_i = 20 \\ -x_4 - x_5 + x_9 - x_{11} - x_{13} + x_{15} + x_{18} \geq 0 \\ x_5 + x_6 + x_{11} \geq 0 \\ x_1 + x_2 + x_3 + x_4 + x_5 + x_{10} + x_{11} + x_{12} + x_{13} + x_{14} + x_{15} \leq 14 \\ x_9 + x_{10} + x_{15} \leq 5 \\ x_1 + x_2 + x_3 + x_{10} + x_{12} \leq 10 \\ 15x_1 + 15x_2 + 15x_3 + 25x_4 + 25x_5 + 30x_6 + 30x_7 + 110x_8 + 80x_9 + 30x_{10} \\ \quad +25x_{11} + 15x_{12} + 25x_{13} + 110x_{14} + 80x_{15} + 30x_{16} + 30x_{17} + 25x_{18} \\ \quad +10x_{19} - (1-\beta_1)(x_1 + x_2 + x_3 + 3x_4 + 3x_5 + 3x_6 + 3x_7 + 3x_8 + 4x_9 \\ \quad +3x_{10} + 2x_{11} + x_{12} + 3x_{13} + 5x_{14} + 2x_{15} + 2x_{16} + 2x_{17} + 2x_{18} + x_{19}) \\ \quad -x_{24} \leq 615 + 15(1-\beta_1) \\ 40x_1 + 40x_2 + 40x_3 + 40x_4 + 40x_5 + 90x_6 + 75x_7 + 15x_8 + 20x_9 + 40x_{10} \\ \quad +40x_{11} + 40x_{12} + 40x_{13} + 14x_{14} + 20x_{15} + 34x_{16} + 34x_{17} + 10x_{18} \\ \quad +40x_{19} - (1-\beta_2)(2x_1 + 2x_2 + 2x_3 + 2x_4 + 2x_5 + 4x_6 + 3x_7 + x_8 \\ \quad +2x_9 + 2x_{10} + 2x_{11} + 2x_{12} + 2x_{13} + x_{14} + 2x_{15} + 3x_{16} + 3x_{17} + x_{18} \\ \quad +2x_{19}) - x_{25} \leq 580 + 30(1-\beta_2) \\ 35x_1 + 10x_2 + 35x_3 + 15x_4 + 15x_5 + 10x_6 + 20x_7 + 120x_8 + 160x_9 + 10x_{10} \\ \quad +15x_{11} + 35x_{12} + 15x_{13} + 120x_{14} + 160x_{15} + 10x_{16} + 10x_{17} + 60x_{18} \\ \quad +40x_{19} - (1-\beta_3)(3x_1 + x_2 + 3x_3 + x_4 + x_5 + x_6 + 3x_7 + 4x_8 \\ \quad +5x_9 + x_{10} + 2x_{11} + 3x_{12} + x_{13} + 4x_{14} + 5x_{15} + x_{16} + x_{17} + 3x_{18} + 2x_{19}) \\ \quad -x_{26} \leq 955 + 60(1-\beta_3) \\ 160x_1 + 140x_2 + 120x_3 + 120x_4 + 120x_5 + 180x_6 + 20x_7 \\ \quad +120x_8 + 160x_9 + 10x_{10} + 15x_{11} + 35x_{12} + 15x_{13} + 160x_{14} \\ \quad +160x_{15} + 30x_{16} + 30x_{17} + 200x_{18} + 200x_{19} - x_{20} \leq 0 \\ 120x_1 + 120x_2 + 120x_3 + 100x_4 + 100x_5 + 120x_6 + 100x_7 \\ \quad +120x_8 + 120x_9 + 120x_{10} + 100x_{11} + 120x_{12} + 100x_{13} + 120x_{14} + 120x_{15} \\ \quad +120x_{16} + 120x_{17} + 140x_{18} + 120x_{19} - x_{21} \leq 0 \\ 100x_1 + 100x_2 + 90x_3 + 80x_4 + 80x_5 + 180x_6 + 80x_7 + 120x_8 + 120x_9 + 90x_{10} \\ \quad +80x_{11} + 90x_{12} + 80x_{13} + 120x_{14} + 120x_{15} + 100x_{16} + 100x_{17} + 120x_{18} \\ \quad +100x_{19} - x_{22} \leq 0 \\ 300x_9 + 400x_{16} + 350x_{17} + 550x_{18} + 350x_{19} \geq 3000 \\ 250x_{14} \geq 480 \\ 900x_{15} \geq 450 \\ 45x_{10} + 40x_{11} + 43x_{12} + 40x_{13} + x_{23} \geq 360 \\ x_{19} = 4. \end{cases} \end{cases}$$

$$(2.150)$$

We set $\beta_1 = \beta_2 = \beta_3 = 1$, and the weights for each objective functions $w1 = 0.7, w2 = 0.2, w3 = 0.3$. We use the fuzzy simulation-based PSO to solve the above model (2.150), and the results are as follows:

$$f_1(x) = 33502,$$
$$f_2(x) = 43.574,$$
$$f_3(x) = 2251.2.$$

The results for the decision variables are:

$$x_1 \rightarrow x_7$$
6.3   18   15.1   3.1   14.3   16.8   15.7
$$x_8 \rightarrow x_{14}$$
20   11.7   7.2   7.5   19.7   17.7   5.3
$$x_{15} \rightarrow x_{21}$$
5.3   4.9   3.5   6.7   4   1624.9   1746.6
$$x_{22} \rightarrow x_{26}$$
3232.8   679.2   289.4   851.2   3095.8

If the decision maker can also change the values of the confidence levels and the weights for each objective functions, they will get different solutions with different parameters.

2.7   Multi-objective Decision Making with Fuzzy Relations

In this multi-objective decision making problems, the objectives and constraints in an imprecise situation assume that they can be represented by fuzzy sets. A decision, then, may be stated as the confluence of the fuzzy objectives and constraints, and may be defined by a max-min operator. That is, assume that we are given a fuzzy objective set F and a fuzzy constraints set C in a space of alternatives X. Then F and C combine to form decision D which is a fuzzy set resulting from the intersection of F and C, and corresponding $\mu_D = \mu_F \cap \mu_C$. This relationship between F, C and D is depicted in Figure 2.8.

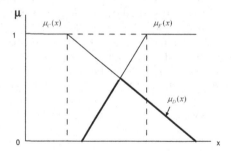

Fig. 2.8 The relationship of fuzzy sets C, F and D

2.7.1   Multi-objective Decision Making with Fuzzy Constraints

For constraints, there are available resources with fuzziness which are characterized by the membership function over a tolerance range (ambiguous range). The membership function of the optimal solution can then be constructed by the convolution of the membership functions of constraints.

The general model for this multi-objective decision making problem with fuzzy resources available is formulated as:

$$\begin{cases} \max F = [f_1(x), f_2(x), \cdots, f_n(x)] \\ \text{s.t.} \begin{cases} g_r(x) \leq \tilde{b}_r, \ r = 1, 2, \cdots, p \\ x \in X, \end{cases} \end{cases} \tag{2.151}$$

where $\tilde{b}_r$, $\forall r$ are in $[b_r, b_r + p_r]$ with given tolerance level p_r.

If the constraints are linear, we have the following model (2.152),

$$\begin{cases} \max F = [f_1(x), f_2(x), \cdots, f_n(x)] \\ \text{s.t.} \begin{cases} (Ax)_r \leq \tilde{b}_r, \ r = 1, 2, \cdots, p \\ x \geq 0. \end{cases} \end{cases} \tag{2.152}$$

We may also consider the following fuzzy inequality constraints (2.153),

$$\begin{cases} \max F = [f_1(x), f_2(x), \cdots, f_n(x)] \\ \text{s.t.} \begin{cases} (Ax)_r \widetilde{\leq} b_r, \ r = 1, 2, \cdots, p \\ x \geq 0, \end{cases} \end{cases} \tag{2.153}$$

where $\widetilde{\leq}$ is called fuzzy less than or equal to. Assume that tolerance p_i for each fuzzy constraint is known. Then $(Ax)_r \leq \tilde{b}_r$ will be equivalent to $(Ax)_r \leq (b_r + \theta p_r)$, $\forall r$, where θ is in $[0, 1]$. If the membership functions of both cases are the same, the Equations (2.152) and (2.153) will be the same problem[25]. Thus we will consider both problems as equivalent in this section.

Verdegay[26] first proved that the problems of (2.152) and (2.153) are equivalent to crisp parametric programming problems and we introduce his nonsymmetric method[27].

For Equation (2.152), let's consider the membership functions (2.154) of the constraints:

$$\mu_r(x) = \begin{cases} 1, & \text{if } (Ax)_r < b_r \\ 1 - \frac{(Ax)_r - b_r}{p_r}, & \text{if } b_r \leq (Ax)_r \leq b_r + p_r \\ 0, & \text{if } (Ax)_r > b_r + p_r. \end{cases} \tag{2.154}$$

If the membership functions are continuous and monotonic, and trade-off between those fuzzy constraints are allowed, then Equation (2.152) will be equivalent to:

$$\begin{cases} \max F = [f_1(x), f_2(x), \cdots, f_n(x)] \\ \text{s.t.} \ x \in X_\alpha, \end{cases} \tag{2.155}$$

where $X_\alpha = \{x_r | \mu_r(x) \geq \alpha, \forall r, x \geq 0\}$, for each $\alpha \in [0, 1]$. The α-level cut concept is based on the previous works of Tanaka et al.[275] and Orlovski[29].

The membership functions indicate that:

(1). If $(Ax)_r < b_r$ then the rth constraint is absolutely satisfied, $v_r(x) = 1$;

(2). If $(Ax)_r \geq b_r + p_r$ where p_r is the maximum tolerance from b_r and determined by the decision maker in any systematic or nonsystematic way, then the ith constraint is absolutely violated, $v_r(x) = 0$;

(3). If $(Ax)_r \in (b_r, b_r + p_r)$ then the membership functions are monotonically decreasing. That is, the less satisfaction the decision maker feels.

We can then substitute the membership functions of Equation (2.154) into Equation (2.155) and obtain the following problem (2.156):

$$\begin{cases} \max F = [f_1(x), f_2(x), \cdots, f_n(x)] \\ \text{s.t.} \begin{cases} (Ax)_r \lesssim b_r + (1-\alpha)p_r, \ \forall r \\ x \geq 0 \text{ and } \alpha \in [0,1], \end{cases} \end{cases} \tag{2.156}$$

which is equivalent to a parametric programming, while $\alpha = 1 - \theta$. Thus the fuzzy linear programming problem given by Equations (2.152) and (2.153) can be equivalent to a crisp parametric linear programming problem when some proper forms of membership functions of the fuzzy constraints are assumed. It is noted that for each α, we have an optimal solution, so the solution with α grade of membership id actually fuzzy.

For illustrating this approach, let us consider the following example.

Example 2.11. Let us consider the transportation problem with cost and time objectives and fuzzy supply and fuzzy demand constraints.

$$\begin{cases} \min cx \\ \min tx \\ \text{s.t.} \begin{cases} \sum_{i \in I} x_{ij} \geq \tilde{d}_i, \ j \in J \\ \sum_{j \in J} x_{ij} \leq \tilde{s}_i, \ i \in I \\ x_{ij} \geq 0 \ \forall i,j. \end{cases} \end{cases} \tag{2.157}$$

The membership functions of the fuzzy constraints, $\mu_j(\cdot), j \in J$, and $\mu_i(\cdot), i \in I$, are assumed to be continuous and strictly monotonic, respectively. Then it can be proved that:

$$\sum_{i \in I} \mu_i^{-1}(\alpha) \geq \sum_{j \in J} \mu_j^{-1}(\alpha), \ \alpha \in [0,1]. \tag{2.158}$$

Therefore, the fuzzy transportation problem can be solved by means of the crisp transportation problem. That is:

$$\begin{cases} \min cx \\ \min tx \\ \text{s.t.} \begin{cases} \sum_{i \in I} x_{ij} \geq \mu_j^{-1}(\alpha), \ j \in J \\ \sum_{j \in J} x_{ij} \leq \mu_i^{-1}(\alpha), \ i \in I \\ x_{ij} \geq 0 \ \forall i,j, \ \alpha \in [0,1]. \end{cases} \end{cases} \tag{2.159}$$

In order to get $\sum_i \mu_i^{-1}(\alpha) = \sum_j \mu_j^{-1}(\alpha)$, an n-th dummy destination is introduced. Thus, the following parametric linear programming problem is obtained:

$$
\begin{cases}
\min \ cx \\
\min \ tx \\
\text{s.t.} \begin{cases}
\sum\limits_{i \in I} x_{ij} = \mu_j^{-1}(\alpha), \ j \in J \\
\sum\limits_{j \in J} x_{ij} = \mu_i^{-1}(\alpha), \ i \in I \\
x_{ij} \geq 0 \ \forall i,j, \ \alpha \in [0,1].
\end{cases}
\end{cases}
\tag{2.160}
$$

2.7.2 Multi-objective Decision Making with Fuzzy Objectives

For objectives, there are some fuzzy objectives in a multi-objective decision making problem. The general model for this multi-objective decision making problem with fuzzy objectives is formulated as:

$$
\begin{cases}
\max \ [f_1(x), f_2(x), \cdots, f_k(x)] \\
\text{s.t.} \ x \in X = \{x | g_r(x) \{\geq, =, \leq\} 0, r = 1, 2, \cdots, p\},
\end{cases}
\tag{2.161}
$$

where $f_i(x), i \in I$ are the benefit (maximization) objectives, where $f_j(x), j \in J$ are the cost (minimization) objectives, and $I \cup J = \{1, 2, \cdots, K\}$.

Zimmermann[45] first used the max-min operator of Bellman and Zedeh to solve the multi-objective problems, and consider the following Equation (2.162) as:

$$
\begin{cases}
\text{Find} \quad x \\
\text{such that:} \begin{cases} f_k(x) \geq f_k^0, \ \forall k \\ x \in X, \end{cases}
\end{cases}
\tag{2.162}
$$

where f_k^0, $\forall k$ are corresponding goals, and all objective functions are assumed to be maximized. Here the objective functions of Equation (2.161) are considered as fuzzy constraints. If the tolerances of fuzzy constraints are given, we could establish their membership functions $\mu_k(x), \forall k$. Under the concept of min-operator, the feasible solution set is defined by the interaction of the fuzzy objective set. This feasible solution set is then characterized by its membership $\mu_D(x)$ which is:

$$
\mu_D(x) = \min(\mu_1(x), \cdots, \mu_k(x)).
$$

Further more, a decision maker makes a decision with a maximum μ_D value in the feasible decision set. The chosen solution can then be obtained by solving the problem of "maximize $\mu_D(x)$." That is:

$$
\begin{cases}
\max \ [\min_k \ \mu_k(x)] \\
\text{s.t.} \ x \in X.
\end{cases}
\tag{2.163}
$$

Now, let $\alpha = \min_k \mu_k(x)$ be the overall satisfactory level compromise. We obtain the following equivalent model ():

$$
\begin{cases}
\max\ \alpha \\
\text{s.t.}\ \begin{cases} \alpha \le \mu_k(x),\ \forall k \\ x \in X. \end{cases}
\end{cases}
\tag{2.164}
$$

To establish the membership functions of objective functions, we could first obtain the payoff table of positive ideal solutions (PIS) as shown in Table. 2.5 and assume membership functions are linear and non-decreasing between f_k^+ and f_K^-, $\forall k$.

Table 2.5 Payoff table

	f_1	f_2	f_3	$\cdots$	f_k	x
max f_1	f_1^+	$f_2(x^1)$	$f_3(x^1)$	$\cdots$	$f_k(x^1)$	x_1
max f_2	$f_1(x^2)$	f_2^+	$f_3(x^2)$	$\cdots$	$f_k(x^2)$	x_2
$\cdots$			$\cdots$			$\cdots$
$\cdots$			$\cdots$			$\cdots$
$\cdots$			$\cdots$			$\cdots$
max f_k	$f_1(x^k)$	$f_2(x^k)$	$f_3(x^k)$	$\cdots$	f_k^+	x_k
	f_1^-	f_2^-	f_3^-	$\cdots$	f_k^-	

Note: f_k^- is the minimum value in each column.

Then the membership functions would be (2.165), for $\forall k$,

$$
\mu_k(x) = \begin{cases}
1, & \text{if } f_k(x) > f_k^+ \\
\frac{f_k(x) - f_K^-}{f_k^+ - f_k^-}, & \text{if } f_k^- \le f_k(x) \le f_k^+ \\
0, & \text{if } f_k(x) < f_k^-.
\end{cases}
\tag{2.165}
$$

These membership functions are essentially based on the concept of preference or satisfaction. It is worth noting that the only feasible solution region of practical relevance includes those elements in the critical area, $\{x | f_k^- \le f_k(x) \le f_k^+, \forall k, \text{and } x \in X\}$. Finally we obtain the following problem (2.166):

$$
\begin{cases}
\max\ \alpha \\
\text{s.t.}\ \begin{cases} \mu_k(x) = \frac{f_k(x) - f_K^-}{f_k^+ - f_k^-} \ge \alpha \\ x \in X. \end{cases}
\end{cases}
\tag{2.166}
$$

However, the membership functions might sometimes be modelled by other types just as in chapter 1.

Let's consider the following example to illustrate the approach.

Example 2.12. Let's consider a trade balance problem. A company manufactures two products of given capacities. Product 1 yields a profit of \$2 per piece in foreign countries, and product 2 needs imported raw materials of \$1 per piece. Two objectives are established:

(i) maximum improvement of the balance of trade, that is, maximum difference of exports minus imports;
(ii) profit maximization.
   The problem can be formulated as:

$$
\begin{cases}
\max\ f_1(x) = -x_1 + 2x_2 & \text{effect on the balance trade} \\
\max\ f_2(x) = 2x_1 + x_2 & \text{profit} \\
\text{s.t.} \begin{cases}
-x_1 + 3x_2 \le 21 & \text{management} \\
x_1 + 3x_2 \le 27 & \text{space} \\
4x_1 + 3x_2 \le 45 & \text{material} \\
3x_1 + x_2 \le 21 & \text{labor hours} \\
x_1, x_2 \ge 0.
\end{cases}
\end{cases} \tag{2.167}
$$

To solve this problem, we first obtain the payoff table which is in Table 2.6,

Table 2.6 Payoff table of example 2.12

	f_1	f_2	x_1	x_2
max f_1	14	7	0	7
max f_2	-3	21	9	3

The membership functions for these two objectives are then obtained as follows:

$$
\mu_1(x) = \begin{cases}
1, & \text{if } 14 < -x_1 + 2x_2 \\
\frac{-x_1 + 2x_2 - (-3)}{17}, & \text{if } -3 \le -x_1 + 2x_2 \le 14 \\
0, & \text{if } -x_1 + 2x_2 < -3,
\end{cases} \tag{2.168}
$$

$$
\mu_2(x) = \begin{cases}
1, & \text{if } 21 < 2x_1 + x_2 \\
\frac{2x_1 + x_2 - 7}{14}, & \text{if } 7 \le 2x_1 + x_2 \le 21 \\
0, & \text{if } 2x_1 + x_2 < 7.
\end{cases} \tag{2.169}
$$

Finally, we compute the following linear programming problem:

$$
\begin{cases}
\max\ \alpha \\
\text{s.t.} \begin{cases}
\alpha \le \frac{-x_1 + 2x_2 - (-3)}{17} \\
\alpha \le \frac{2x_1 + x_2 - 7}{14} \\
\alpha \in [0,1] \text{ and } x \in X,
\end{cases}
\end{cases} \tag{2.170}
$$

where X is the original feasible capacity. The optimal solution is: $x^0 = (5.03, 7.32)$ with optimal effect of the balance of the balance of trade $f_1^0 = \$17.38$ and optimal profit $f_2^0 = \$4.58$ where the overall satisfactory level is $\alpha^0 = 0.74$.

Chapter 3
Fuzzy Random Multiple Objective Decision Making

Since the fuzzy set was initialized by Zadeh[9], the possibility theory has been developed Dubios and Prade[48], and it has been applied in many fields. Later, many scholars proposed the concept of two-fold uncertain variables, combined fuzzy variables with fuzzy and random variables. The concept of the fuzzy random variable was first defined by Kwakernaak [215]. There are further scholars who defined the concept of the fuzzy random variable, Puri and Ralescu [139], Kruse and Meyer [214], Wang and Qiao [281], López-Diaz and Gil [228],Colubi[166], Liu [226], which was widely extended to many fields. Li and Xu [1, 2], Xu and Liu [3] discussed the properties of fuzzy random variables, and introduced the fuzzy random(Fu-Ra) multi-objective decision making models and a way to deal with them; some crisp equivalent models are given and relative algorithms are proposed to solve the problem. Finally, these models and algorithms are applied to some realistic problems, such as, portfolio selection problems, inventory problems, and so on.

In this chapter, we first introduce the fuzzy random variable, the arithmetic, and the properties of the fuzzy random variable. Based on the expected value operator and chance operator of the fuzzy random variable, the three parts are presented respectively.

(1) Fuzzy random expected value decision-making model(Fu-Ra EVM). Usually, decision makers find it difficult to make a decision when they encounter fuzzy random parameters. A clear criteria must be determined to assist in the decision. The expected value operator of fuzzy random variables is introduced and the crisp equivalent model is deduced when the distribution is clear.

(2) Fuzzy random chance constraint decision-making model(Fu-Ra CCM). Sometimes, decision makers dont strictly require the objective value to be maximal benefit but only need to obtain the maximum benefit under a predetermined confidence level. Then the chance constrained model is proposed and the crisp equivalent model is deduced when the distribution is clear.

(3) Fuzzy random dependent chance decision-making model(Fu-Ra DCM). When decision makers predetermine an objective value and require the maximal probability that objective values exceed the predetermined one.

J. Xu and X. Zhou: Fuzzy-Like Multiple Objective Decision Making, STUDFUZZ 263, pp. 135–225.
springerlink.com

Finally, an application to portfolio problems is presented to show the decision making process under a fuzzy random environment.

3.1 Portfolio Selection Problem under a Fuzzy Random Environment

Portfolio selection problem is that the investors strike a balance between maximizing returns and minimizing the risks of their investment. In order to deal with this problem, the Mean-Variance model (3.1) was proposed by Markowitz [46, 47]. In this model, the return is quantified by the mean, and the risk is characterized by the variance of a portfolio of securities. After Markowitz's work, many scholars proposed different mathematical approaches to develop portfolio theory based on probability theory. In these portfolio selection models, security returns are still assumed to be random variables, and random uncertainty is considered the sole way of modeling uncertainty.

$$\begin{cases} \min \ x^{\mathrm{T}} V x \\ \text{s.t.} \begin{cases} E(r)^{\mathrm{T}} x \geq R_0 \\ \sum_{j=1}^{n} x_j = 1 \\ x_j \geq 0, j = 1, 2, \cdots, n \end{cases} \end{cases} \quad \text{or} \quad \begin{cases} \max \ E(r)^{\mathrm{T}} x \\ \text{s.t.} \begin{cases} x^{\mathrm{T}} V x \leq V_0 \\ \sum_{i=1}^{n} x_j = 1 \\ x_j \geq 0, j = 1, 2, \cdots, n, \end{cases} \end{cases} \quad (3.1)$$

where the future return r is random vector.

In 1952, Roy proposed the safety-first model for the portfolio selection problem. The decision rule is to minimize the probability that the return of the portfolio is less than the predetermined 'disaster level', then on the basis of the safety-first model, Kataoka proposed a different form of safety-first model [208]:

$$\begin{cases} \max \ R \\ \text{s.t.} \begin{cases} Pr\{\sum_{j=1}^{n} r_j x_j \leq R\} \leq \alpha \\ \sum_{j=1}^{n} x_j = 1 \\ x_j \geq 0, j = 1, 2, \cdots, n, \end{cases} \end{cases} \quad (3.2)$$

where the future return r_j is random variable, $j = 1, 2, \cdots, n$, α is the probability confidence level determined by the investor.

In the real world, there are many non-probabilistic factors that affect stock markets which should not be dealt with using probability approaches. With the introduction of fuzzy set theory and possibility theory [9, 22, 48], several scholars began to employ these theory to manage portfolios in a fuzzy environment. For example, Tanaka et al. [49] and Inuiguchi and Tanino [50] assumed security returns to be fuzzy variables with possibility distributions and proposed the possibilistic portfolio selection models respectively, Parra et al. [51] proposed a fuzzy goal programming approach for portfolio selection, and Zhang and Nie [52] proposed the admissible efficient portfolio model. More research on fuzzy portfolio selection may be found in [53, 54, 55, 56, 58] etc. More general reviews of the different portfolio models are available as [59].

The present portfolio selection models are based on either probability theory or fuzzy set theory, therefore only one kind of uncertainty, randomness or fuzziness is reflected. In fact, randomness and fuzziness are often mixed up together in real settings which requires taking them into account simultaneously in he portfolio selection process. Random uncertainty is the uncertainty of whether the event will happen or not, that is, it's hard to predict whether the event will happen or not, but the states of the event are clear, it's external uncertainty; Fuzzy uncertainty is the uncertainty of the states of that event itself, that is, the problem does not rest whether with the event will happen or not but rests with the fact that the states of the event itself aren't clear. Thus, different people will have different feelings when they observe the same event, so they could educe different conclusions, so fuzzy uncertainty is subjective uncertainty.

Let us explain the reasonableness of the state about fuzzy random security returns. We know the basic assumption behind the Markowitz's mean variance model is that the situation of the stock market in the future can be correctly reflected by securities data in the past, that is, the mean and covariance of a portfolio of securities in the future are similar to the past ones. However, there are so many uncertain factors that this assumption cannot be guaranteed for the real ever-changing stock markets, especially for new emerging stock markets without plenty of historical data such as the stock market in China. Since stock experts possess enough information and experience about the stock market, a good method is to let them provide their rough estimation about the future returns of securities, and the certain mean value could extend to a fuzzy number. In this case, the return rates of securities are fuzzy random variables, i.e., random variables whose actual values are fuzzy sets. Also, as Katagiri and Ishii proposed in [206], realistically random factors and fuzzy information often influences the security market simultaneously, because it's difficult for people to separate the randomness and the fuzziness, the fuzzy random future return is the assumption of his research.

When we use fuzzy random variables to describe the future return of the securities, and use the historical data and the advice about the historical returns from the experts, it is reasonable for people to believe that the fuzzy random portfolio selection problem is more realistic and proper.

Now, let's focus on the portfolio selection problem with fuzzy random returns. It is rational for people to consider that the future return of every security is a fuzzy variable which is around a value with left and right spreads, but here the middle value is usually not a certain number, but a random variable, so the future return is a fuzzy random variable, and in this situation, the portfolio selection problem is under a fuzzy random environment, see Figure 3.1.

And the general fuzzy random portfolio selection model is proposed as (3.3), (3.5) and (3.4).

$$\begin{cases} \max\ E\left(\sum_{j=1}^{n} \tilde{\bar{r}}_j x_j\right) \\ \min Var\left(\sum_{j=1}^{n} \tilde{\bar{r}}_j x_j\right) \\ \text{s.t.} \begin{cases} \sum_{j=1}^{n} x_j = 1 \\ x_j \geq 0, j = 1, 2, \cdots, n, \end{cases} \end{cases} \tag{3.3}$$

Fig. 3.1 Fuzzy random portfolio selection problem

$$\begin{cases} \max R \\ \text{s.t.} \begin{cases} Ch\{\sum_{j=1}^{n} \tilde{\bar{r}}_j x_j \geq R\}(\gamma) \geq \delta \\ \sum_{j=1}^{n} x_j = 1 \\ x_j \geq 0, j = 1, 2, \cdots, n, \end{cases} \end{cases} \quad (3.4)$$

$$\begin{cases} \max Ch\{\sum_{j=1}^{n} \tilde{\bar{r}}_j x_j \geq R\}(\gamma) \\ \text{s.t.} \begin{cases} \sum_{j=1}^{n} x_j = 1 \\ x_j \geq 0, j = 1, 2, \cdots, n, \end{cases} \end{cases} \quad (3.5)$$

where $\tilde{\bar{r}}_j$ are the fuzzy random future return of security j.

3.2 Fu-Ra Variable

Roughly speaking, a fuzzy random variable (Fu-Ra variable) is a measurable function from a probability space to a collection of fuzzy variables. For the better understanding of this chapter, let us first briefly review some necessary knowledge about fuzzy random variable.

3.2.1 Definition of Fu-Ra Variable

In order to describe the hybrid uncertainty which consists both of randomness and fuzziness-fuzzy random uncertainty, Kwakernaak [215] first introduced the concept of the fuzzy random variable in 1987. Then many scholars defined the fuzzy random variables from different perspectives, for example, Colubi et al. [166], Kruse and Meyer [214], López-Diaz and Gil [228], Puri and Ralescu [139], Wang and Qiao [281], and Liu [226]. Readers can refer to [178] for more history of the development of the fuzzy random variable.

We consider the fuzzy random variables which defined on the real number set, and in this situation, the above definitions are equivalent. Here we adopted the definition proposed by Puri and Ralescu [139].

In the following, we use R to denote the set of all real numbers, $\mathscr{F}_c(R)$ to denote the set of all fuzzy variables, $\mathscr{K}_c(R)$ to denote all of the non-empty bounded close interval.

Definition 3.1. (Puri and Ralescu [139]) In a given probability space $(\Omega, \mathscr{F}, P)$, a mapping $\xi : \Omega \to \mathscr{F}_c(R)$ is called a fuzzy random variable in $(\Omega, \mathscr{F}, P)$, if for $\forall \, \alpha \in (0,1]$, the set-valued function $\xi_\alpha : \Omega \to \mathscr{K}_c(R)$

$$\xi_\alpha(\omega) = (\xi(\omega))_\alpha = \{x | x \in R, \mu_{\xi(\omega)}(x) \geq \alpha\}, \quad \forall \, \omega \in \Omega,$$

is $\mathscr{F}$ measurable.

The following Theorem 3.1 gives a sufficient and necessary condition of real number fuzzy random variables.

Lemma 3.1. *[230] If ξ is a fuzzy random variable defined in the probability space $(\Omega, \mathscr{F}, P)$, then for $\omega \in \Omega, \forall \, \alpha \in [0,1], \xi_\alpha(\omega) = [\xi_\alpha^-(\omega), \xi_\alpha^+(\omega)]$ is a random interval, that is, $\xi_\alpha^-(\omega)$ and $\xi_\alpha^+(\omega)$ are real number random variable in the probability space $(\Omega, \mathscr{F}, P)$.*

Example 3.1. Suppose that a boy tosses a coin, if it is tail, he will lost 10RMB, the membership function could be

$$\mu(x) = \left[1 - \frac{(x+10)^2}{4}\right] \vee 0;$$

And if it is head, he will win some income which is between 100RMB and 1000RMB, the membership function could be

$$v(x) = \left[1 - \frac{(x-360)^2}{380^2}\right] \vee 0.$$

So, the income that this boy will get is a fuzzy random variable

$$\xi(\omega) = \begin{cases} \mu, & \text{if } \omega = \text{tail} \\ v, & \text{if } \omega = \text{head}. \end{cases}$$

Example 3.2. $\forall \, \omega \in \Omega$, if $\xi(\omega)$ is LR fuzzy numbers, then ξ is called LR fuzzy random variable, denoted by $\xi(\omega) = (a(\omega), l(\omega), r(\omega))_{LR}$, $\omega \in \Omega$, where $a(\omega), l(\omega), r(\omega)$ is a random variable defined in the probability space $(\Omega, \mathscr{F}, P)$.

Especially, if $\forall \, \omega \in \Omega$, $\xi(\omega)$ is triangular fuzzy variable, then ξ is triangular fuzzy random variable, denoted by $\xi(\omega) = (a(\omega) - l(\omega), a(\omega), a(\omega) + r(\omega))$, $\omega \in \Omega$. In this particular case, if $l(\omega), r(\omega)$ are constants, then the triangular fuzzy random variable ξ could be denoted by $\xi(\omega) = (a(\omega) - l, a(\omega), a(\omega) + r)$.

Definition 3.2. (Li and Xu [1]) If $\xi_1, \xi_2, \cdots, \xi_m$ are fuzzy random variables defined in the probability space $(\Omega, \mathscr{F}, P)$, then $\xi = (\xi_1, \xi_2, \cdots, \xi_n)$ is called fuzzy random vector.

3.2.2   Expected Value Operator of Fu-Ra Variables

The expected value of fuzzy random variables has been defined as a fuzzy number in several ways. We introduce the following two kinds of expected value operators of fuzzy variables by Puri and Ralescu [139] and Liu and Liu [220], respectively.

Puri and Ralescu used the Aumann integral based on the measurable set-valued functions to define the expected value of fuzzy random variables as fuzzy numbers[139]. The fuzzy expected value could describe the trend which the fuzzy random variable closes with the middle value.

Definition 3.3. (Puri and Ralescu [139]) On given probability space $(\Omega, \mathscr{F}, P)$, if $\forall \omega \in \Omega$, $\alpha \in [0, 1]$, the mappings $\omega \mapsto \xi_\alpha^-(\omega)$ and $\omega \mapsto \xi_\alpha^+(\omega)$ are integrable, then ξ is called the integrated bounded fuzzy random variable on the probability space $(\Omega, \mathscr{F}, P)$.

Definition 3.4. (Puri and Ralescu [139]) Let ξ be a integrated bounded fuzzy random variable on the probability space $(\Omega, \mathscr{F}, P)$, the expected value $E(\xi)$ of ξ is defined as a only fuzzy set in R, $\forall \alpha \in (0, 1]$, it satisfies

$$(E(\xi))_\alpha = \int_\Omega \xi_\alpha dP = \left\{ \int_\Omega f(\omega) dP(\omega) : f \in L^1(P), f(\omega) \in \xi_\alpha(\omega) \text{ a.s. } [P] \right\},$$

where $\int_\Omega \xi_\alpha dP$ is the Aumann integral of ξ_α about P, $L^1(P)$ denote all of the integrable function $f : \Omega \to R$ about the probability measure P.

Lemma 3.2. *[144] Let $(\Omega, \mathscr{F}, P)$ be complete probability space, $\xi : \Omega \to \mathscr{F}_c(R)$ is a integrated bounded fuzzy random variable. Then $\forall \alpha \in (0, 1]$, the α-set of $E(\xi)$ is the compact convex interval as follows,*

$$(E(\xi))_\alpha = [(E(\xi))_\alpha^-, (E(\xi))_\alpha^+] = \left[\int_\Omega (\xi(\omega))_\alpha^- dP(\omega), \int_\Omega (\xi(\omega))_\alpha^+ dP(\omega) \right].$$

Lemma 3.3. *[144] Let $(\Omega, \mathscr{F}, P)$ be complete probability space, ξ_1, ξ_2 are integrated bounded fuzzy random variables on $(\Omega, \mathscr{F}, P)$, $\lambda, \gamma \in R$, then*

$$E(\lambda \xi_1 + \gamma \xi_2) = \lambda E(\xi_1) + \gamma E(\xi_2).$$

The variance of a fuzzy random variable describes the spread that the fuzzy random variable spreads from the expected value. The covariance between two fuzzy random variables reflects the linear correlation degree between them. In [141], Feng et al gave the definition of variance and covariance of fuzzy random as crisp values. Readers can refer to [142] for more general forms about the variance of fuzzy random variables.

Definition 3.5. (Feng [141]) Let $(\Omega, \mathscr{F}, P)$ be complete probability space, ξ_1, ξ_2 are square integrable fuzzy random variable on the probability space $(\Omega, \mathscr{F}, P)$, the covariance of ξ_1 and ξ_2 is defined by

$$Cov(\xi_1, \xi_2) = \frac{1}{2} \int_0^1 \left[Cov((\xi_1)_\alpha^-, (\xi_2)_\alpha^-) + Cov((\xi_1)_\alpha^+, (\xi_1)_\alpha^+) \right] d\alpha, \qquad (3.6)$$

the variance of ξ_1 is defined by

$$Var(\xi_1) = Cov(\xi_1, \xi_1) = \frac{1}{2} \int_0^1 [Var((\xi_1)_\alpha^-) + Var(\xi_1)_\alpha^+)] d\alpha. \qquad (3.7)$$

Lemma 3.4. *[141, 142] Let $(\Omega, \mathscr{F}, P)$ be complete probability space, ξ_1, ξ_2 are square integrable fuzzy random variable on the probability space $(\Omega, \mathscr{F}, P)$, $\lambda, \gamma \in R$, then we have*
(1) $Var(\lambda \xi_1 + u) = \lambda^2 Var(\xi_1)$;
(2) $Var(\xi_1 + \xi_2) = Var(\xi_1) + Var(\xi_2) + 2Cov(\xi_1, \xi_2)$;
(3) $Cov(\lambda \xi_1 + u, \gamma \xi_2 + v) = \lambda \gamma Cov(\xi_1, \xi_2)$, here u, v are fuzzy numbers, $\lambda \gamma \geq 0$.

Similar as the definition of the expected value of random variables, Liu and Liu[220] give the following definition of the expected value of fuzzy random variables.

Definition 3.6. (Liu and Liu [220]) Let ξ be a fuzzy random variable defined on the probability space $(\Omega, \mathscr{F}, P)$. Then its expected value is defined by

$$E[\xi] = \int_0^{+\infty} Pr\{\omega \in \Omega | E[\xi(\omega)] \geq r\} dr - \int_{-\infty}^0 Pr\{\omega \in \Omega | E[\xi(\omega)] \leq r\} dr. \quad (3.8)$$

Definition 3.7. (Liu and Liu [220]) Let ξ be a fuzzy random variable with finite expected value $E[\xi]$. The variance of ξ is

$$V[\xi] = E[(\xi - E[\xi])^2].$$

Remark 3.1. The expected value operator E appears in both sides of the definitions of $E[\xi]$. In fact, the symbol E represents different meanings-it is overloaded. That is, the overloading allows us to use the same symbol E for different expected value operators, because we can deduce the meaning from the type of argument.

Remark 3.2. If the fuzzy random variable degenerates to a random variable, then the expected value operator degenerates to the form

$$E[\xi] = \int_0^{+\infty} Pr\{\xi \geq r\} dr - \int_{-\infty}^0 Pr\{\xi \leq r\} dr,$$

which is just the conventional expected value of random variable.

Remark 3.3. If the fuzzy random variable degenerates to a fuzzy variable, then the expected value operator degenerates to the form

$$E[\xi] = \int_0^{+\infty} Cr\{\xi \geq r\} dr - \int_{-\infty}^0 Cr\{\xi \leq r\} dr,$$

which is just the conventional expected value of the fuzzy variable.

Example 3.3. Assume that ξ is a fuzzy random variable defined as

$$\xi = (\rho, \rho + 1, \rho + 2), \text{ with } \rho \sim N(0,1).$$

Then for each $\omega \in \Omega$, we have

$$E[\xi(\omega)] = \frac{1}{4}[\rho(\omega) + 2(\rho(\omega) + 1) + (\rho(\omega) + 2)] = \rho(\omega) + 1.$$

Thus $E[\xi(\omega)] = E[\rho] + 1 = 1$.

3.2.3 Chance Operator of Fu-Ra Variables

Now let us consider the chance of a fuzzy random event. Recall that the probability of the random event and the possibility of the fuzzy event are defined as a real number. However, for a fuzzy random event, the primitive chance is defined as a function rather than a number.

Since there are three chance measures of the fuzzy variables, we introduce three kinds of chance of fuzzy random variable, we give the three types of primitive chance of fuzzy random event as follows.

Definition 3.8. Let $\xi = (\xi_1, \xi_2, \cdots \xi_n)$ be a fuzzy random vector defined on $(\Omega, \mathscr{A}, Pr)$, and $f : \mathbf{R}^n \to \mathbf{R}$ is real-valued continuous function. Then the primitive chance of a fuzzy random event characterized by $f(\xi) \leq 0$ is a function from $(0, 1]$ to $[0, 1]$, defined as the following,
(1) Pr-Pos chance,

$$Ch\{f(\xi) \leq 0\}(\alpha) = \sup_{\alpha \in [0,1]} \{\omega | Pr\{\omega \in \Omega | Pos\{f(\xi(\omega)) \leq 0\} \geq \beta\} \geq \alpha\}. \quad (3.9)$$

(2) Pr-Nec chance,

$$Ch\{f(\xi) \leq 0\}(\alpha) = \sup_{\alpha \in [0,1]} \{\omega | Pr\{\omega \in \Omega | Nec\{f(\xi(\omega)) \leq 0\} \geq \beta\} \geq \alpha\}.$$
$$(3.10)$$

(3) Pr-Cr chance,

$$Ch\{f(\xi) \leq 0\}(\alpha) = \sup_{\alpha \in [0,1]} \{\omega | Pr\{\omega \in \Omega | Cr\{f(\xi(\omega)) \leq 0\} \geq \beta\} \geq \alpha\}. \quad (3.11)$$

where $\alpha, \beta \in [0, 1]$ are predetermined confidence level.

Remark 3.4. According to the primitive chance of a fuzzy random event characterized by $f(\xi) \leq 0$, we have the equivalent forms respectively

$$Ch\{f(\xi) \leq 0\}(\alpha) \geq \beta$$
$$\Leftrightarrow Pr\{\omega \in \Omega | Pos\{f(\xi(\omega)) \leq 0\} \geq \beta\} \geq \alpha$$
$$\text{or} \Leftrightarrow Pr\{\omega \in \Omega | Nec\{f(\xi(\omega)) \leq 0\} \geq \beta\} \geq \alpha$$
$$\text{or} \Leftrightarrow Pr\{\omega \in \Omega | Cr\{f(\xi(\omega)) \leq 0\} \geq \beta\} \geq \alpha,$$

Remark 3.5. The primitive chance of a fuzzy random event characterized by $f(\xi) \leq 0$ defined as (3.8) have the equivalent forms respectively.

$$Ch\{f(\xi) \leq 0\}(\alpha) = \sup_{Pos\{A\} \geq \alpha} \inf_{\omega \in A} Pr\{f(\xi(\omega)) \leq 0\}, \qquad (3.12)$$

$$Ch\{f(\xi) \leq 0\}(\alpha) = \sup_{Nec\{A\} \geq \alpha} \inf_{\omega \in A} Pr\{f(\xi(\omega)) \leq 0\}, \qquad (3.13)$$

$$Ch\{f(\xi) \leq 0\}(\alpha) = \sup_{Cr\{A\} \geq \alpha} \inf_{\omega \in A} Pr\{f(\xi(\omega)) \leq 0\}. \qquad (3.14)$$

Remark 3.6. The primitive chance represents that the fuzzy random event holds with possibility $Ch\{f_j(\xi) \leq 0, j = 1, 2, \cdots, m\}(\alpha)$ at probability α.

Remark 3.7. It is obvious that $Ch\{f_j(\xi) \leq 0, j = 1, 2, \cdots, m\}(\alpha)$ is a decreasing function of α (see Figure 3.2).

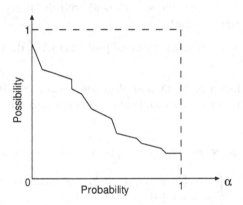

Fig. 3.2 Primitive chance curve $Ch\{f_j(\xi) \leq 0, j = 1, 2, \cdots, m\}$

Remark 3.8. If the fuzzy random vector ξ degenerates to a random vector, then the chance $Ch\{f_j(\xi) \leq 0, j = 1, 2, \cdots, m\}(\alpha)$ takes values either 0 or 1. That is,

$$Ch\{f_j(\xi) \leq 0, j = 1, 2, \cdots, m\}(\alpha) = \begin{cases} 1, & Pr\{f_j(\xi) \leq 0, j = 1, 2, \cdots, m\} \geq \alpha. \\ 0, & \text{otherwise.} \end{cases}$$

Remark 3.9. If the fuzzy random vector ξ degenerates to a fuzzy vector, then the chance $Ch\{f_j(\xi) \leq 0, j = 1, 2, \cdots, m\}(\alpha)$ (with $(\alpha) > 0$) is exactly the possibility of the event. That is,

$$Ch\{f_j(\xi) \leq 0, j = 1, 2, \cdots, m\}(\alpha) = Pos\{f_j(\xi) \leq 0, j = 1, 2, \cdots, m\}.$$

3.3 Fu-Ra EVM

In order to obtain the decision with optimizing the expected objective values subject to expected constraints, we may employ the following Fu-Ra EVM.

3.3.1 General Model for Fu-Ra EVM

Consider the following multi-objective programming problem with Fu-Ra coefficients

$$\begin{cases} \max \ [f_1(x,\xi), f_2(x,\xi), \cdots, f_m(x,\xi)] \\ \text{s.t.} \begin{cases} g_r(x,\xi) \le 0, r = 1, 2, \cdots, p \\ x \in X, \end{cases} \end{cases} \tag{3.15}$$

where x is a n-dimensional decision vector, $\xi = (\xi_1, \xi_2, \cdots, \xi_n)$ is a Fu-Ra vector, $f_i(x,\xi)$ are objective functions, $i = 1, 2, \cdots, m$. Because of the existence of Fu-Ra vector ξ, problem (3.16) is not well-defined. That is, the meaning of maximizing $f_i(x,\xi), i = 1, 2, \cdots, m$ is not clear and constraints $g_r(x,\xi) \le 0, r = 1, 2, \cdots, p$ do not define a deterministic feasible set. To deal with the fuzzy random events ξ, Fu-Ra EVM are brought forward.

Definition 3.9. If x is an efficient solution of problem (3.16), then it is called a fuzzy random efficient solution.

We can also formulate a fuzzy random decision system as a fuzzy random goal programming model according to the priority structure and target levels set by the decision maker:

$$\begin{cases} \min \ \sum_{j=1}^{l} P_j \sum_{i=1}^{m} (u_{ij}d_i^+ + v_{ij}d_i^-) \\ \text{s.t.} \begin{cases} f_i(x,\xi) + d_i^- - d_i^+ = b_i, & i = 1, 2, \cdots, m \\ g_r(x,\xi) \le 0, & r = 1, 2, \cdots, p \\ x \in X, \end{cases} \end{cases} \tag{3.16}$$

where P_j is the preemptive priority factor which expresses the relative importance of various goals, $P_j \gg P_{j+1}$, for all j, u_{ij} is the weighting factor corresponding to positive deviation for goal i with priority j assigned, v_{ij} is the weighting factor corresponding to negative deviation for goal i with priority j assigned, d_i^+ is the positive deviation from the target of goal i, defined as

$$d_i^+ = [f_i(x,\xi) - b_i] \vee 0,$$

d_i^- is the negative deviation from the target of goal i, defined as

$$d_i^- = [b_i - f_i(x,\xi)] \vee 0, \tag{3.17}$$

f_i is a function in goal constraints, g_r is a function in real constraints, b_i is the target value according to goal i, l is the number of priorities, m is the number of goal constraints, and p is the number of real constraints.

Based on the definition of the expected value of fuzzy random events f_i and g_j, the maximum expected value multi-objective decision making model (EVM) is proposed as follows,

$$\begin{cases} \max [E[f_1(x,\xi)], f_2(x,\xi), \cdots, f_m(x,\xi)]] \\ \text{s.t.} \begin{cases} E[g_r(x,\xi)] \le 0, r = 1, 2, \cdots, p \\ x \in X, \end{cases} \end{cases} \quad (3.18)$$

where x is n-dimensional decision vector and ξ is n-dimensional Fu-Ra variable.

Definition 3.10. If x^* is an efficient solution of problem (3.18), then it is called as a fuzzy random expected efficient solution.

Clearly, the problem (3.18) is multi-objective with crisp parameters. Then we can convert it into a single-objective programming by traditional weight sum method

$$\begin{cases} \max \sum_{i=1}^{m} w_i E[f_i(x,\xi)] \\ \text{s.t.} \begin{cases} E[g_r(x,\xi)] \le 0, & r = 1, 2, \cdots, p \\ w_1 + w_2 + \cdots + w_m = 1 \\ x \in X. \end{cases} \end{cases} \quad (3.19)$$

Theorem 3.1. *Problem (3.19) is equivalent to problem (3.18), i.e., the efficient solution of problem (3.18) is the optimal solution of problem (3.19) and the optimal solution of problem (3.19) is the efficient solution of problem (3.18).*

Proof. Apparently, the efficient solution of problem (3.18) is the optimal solution of problem (3.19). Let 's consider whether the optimal solution of problem (3.19) is the efficient solution of problem (3.18).

Suppose x^* is an optimal solution of problem (3.16). If x^* is not an efficient solution of problem (3.18), then there exists x_0 such that $E[f_i(x_0,\xi)] \ge E[f_i(x^*,\xi)](i = 1, 2, \cdots, m)$, and there at least exists a k such that $E[f_k(x_0,\xi)] > E[f_k(x^*,\xi)]$. Then,

$$\sum_{i=1}^{m} w_i E[f_i(x_0,\xi)] > \sum_{i=1}^{m} w_i E[f_i(x^*,\xi)].$$

This conflicts with the assumption that x^* is an optimal solution to problem (3.19). This completes the proof. □

Theorem 3.2. *Let $\xi = (\xi_1, \xi_2, \cdots, \xi_n)$ be a Fu-Ra vector on the probability space $(\Omega, \mathscr{F}, P)$, and f_i and $g_r : \mathscr{A}^n \to \mathscr{A}$ be convex continuous functions with respect to x, $i = 1, 2, \cdots, m; r = 1, 2, \cdots, p$. Then the expected value programming problem (3.19) is a convex programming.*

Proof. Let x_1 and x_2 be two feasible solutions. Because $g_r(x,\xi)$ is a convex continuous function with respect to x, then

$$g_r(\rho x_1 + (1-\rho)x_2, \xi) \le \rho g_r(x_1, \xi) + (1-\rho)g_r(x_2, \xi),$$

where $0 \leq \rho \leq 1, r = 1, 2, \cdots, p$. We can have

$$E[g_r(\rho x_1 + (1-\rho)x_2, \xi)] \leq E[\rho g_r(x_1, \xi) + (1-\rho)g_r(x_2, \xi)].$$

Because for any $\omega \in \Omega$, $\xi(\omega)$ is a fuzzy variable. Then by the linearity of expected value of fuzzy variable, we have

$$E[g_r(x_1, \zeta(\omega)) + (1-\rho)g_r(x_2, \xi(\omega))] = \rho E[g_r(x_1, \zeta(\omega))] + (1-\rho)E[g_r(x_2, \xi(\omega))].$$

Following the linearity of expected value operator of fuzzy variable, we can obtain

$$
\begin{aligned}
&E[\rho g_r(x_1, \xi) + (1-\rho)g_r(x_2, \xi)] \\
&= E[\rho E[g_r(x_1, \xi(\omega))] + (1-\rho)E[g_r(x_2, \xi(\omega))]] \\
&= \rho E[E[g_r(x_1, \xi(\omega))]] + 1-\rho)E[E[g_r(x_2, \xi(\omega))]] \\
&= E[g_r(x_1, \xi)] + (1-\rho)E[g_r(x_2, \xi)].
\end{aligned}
\tag{3.20}
$$

Then $E[g_r(\rho x_1 + (1-\rho)x_2, \xi)] \leq \rho E[g_r(x_1, \xi)] + (1-\rho)E[g_r(x_2, \xi)] \leq 0$. This means that $\rho x_1 + (1-\rho)x_2$ is also a feasible solution. Then $X(x \in X)$ is a convex feasible set.

For every i, $f_i(x, \xi)$ is a convex continuous function with respect to x, it follows that

$$f_i(\rho x_1 + (1-\rho)x_2, \xi) \leq \rho f_i(x_1, \xi) + (1-\rho)f_i(x_2, \xi),$$

then

$$E[f_i(\rho x_1 + (1-\rho)x_2, \xi)] \leq \rho E[f_i(x_1, \xi)] + (1-\rho)E[f_i(x_2, \xi)],$$

then

$$\sum_{i=1}^{m} w_i E[f_i(\rho x_1 + (1-\rho)x_2, \xi)] \leq \rho \sum_{i=1}^{m} w_i E[f_j(x_1, \xi)] + (1-\rho) \sum_{i=1}^{r} w_i E[f_j(x_2, \xi)].$$

This means function $\sum_{i=1}^{m} w_i E[f_i(x, \xi)]$ is convex. Above all, we can conclude that the expected value programming problem (3.19) is convex programming. $\square$

This guarantees that problem (3.19) has at least one feasible solution. Then we can also obtain the fuzzy random expected value goal of the programming model as follows:

$$
\begin{cases}
\min \sum_{i=1}^{m} P_i(u_i d_i^+ + v_i d_i^-) + \sum_{r=1}^{m} P_r(u_r d_r^+ + v_r d_r^-) \\
\text{s.t.} \begin{cases}
E[f_i(x, \xi)] + d_i^- - d_i^+ = q_i & i = 1, 2, \cdots, m \\
E[g_r(x, \xi)] + d_r^- - d_r^+ = 0, & r = 1, 2, \cdots, p \\
x \in X \\
d_i^-, d_i^+, d_r^-, d_r^+ \geq 0 \\
u_i, v_i, u_r, v_r = 0 \text{ or } 1,
\end{cases}
\end{cases}
\tag{3.21}
$$

where P_i, P_r are the priority coefficients that express the importance of goals.

Because of the introduction of the expected value operator, the model (3.18) is crisp, and we can use the regular way of solving the multi-objective model to deal with the Fu-Ra EVM.

3.3.2 Linear Fu-Ra EVM and Fuzzy Satisfied Method

In this section, we focus on the linear Fu-Ra EVM, we give the crisp equivalent model, and the interactive method which will be used to solve the crisp multi-objective model. Then a numerical example is given to illustrate the equivalent model and the solving method.

3.3.2.1 Crisp Equivalent Model

In order to solve the multi-objective decision making problem (3.18), we must compute the crisp expected value of ξ. However, as we know, this process is usually hard work most of the time. In this section, we will consider a special case-linear Fu-Ra EVM and present their results.

$$\begin{cases} \max \left[E[\tilde{\bar{c}}_1^T x], E[\tilde{\bar{c}}_2^T x], \cdots, E[\tilde{\bar{c}}_m^T x] \right] \\ \text{s.t.} \begin{cases} E[\tilde{\bar{a}}_r^T x] \le E[\tilde{\bar{b}}_r], r = 1, 2, \cdots, p \\ x \ge 0, \end{cases} \end{cases} \tag{3.22}$$

where $\tilde{\bar{c}}_i = (\tilde{\bar{c}}_{i1}, \tilde{\bar{c}}_{i1}, \cdots, \tilde{\bar{c}}_{in})^T, \tilde{\bar{a}}_r = (\tilde{\bar{a}}_{r1}, \tilde{\bar{a}}_{r1}, \cdots, \tilde{\bar{a}}_{rn})^T$ are Fu-Ra vectors, $\tilde{\bar{b}}_r$ are Fu-Ra variables, $i = 1, 2, \cdots, m, r = 1, 2, \cdots, p$. If these Fu-Ra vectors, as well as Fu-Ra variables have special forms, we have the following theorem.

Theorem 3.3. *If Fu-Ra variable* $\tilde{\bar{c}}_{ij}(\omega) = (\bar{c}_{ij1}(\omega), \bar{c}_{ij2}(\omega), \bar{c}_{ij3}(\omega), \bar{c}_{ij4}(\omega))$ *with* $\bar{c}_{ijt}(\omega) \sim \mathcal{N}(\mu_{ijt}, \delta_{ijt}^2), i = 1, 2, \cdots, m, j = 1, 2, \cdots, n, t = 1, 2, 3, 4$ *are trapezoidal Fu-Ra variables, then the objective functions*

$$E[\tilde{\bar{c}}_1^T x], \cdots, E[\tilde{\bar{c}}_m^T x]$$

are equivalent to

$$\frac{1}{4} \sum_{t=1}^{4} \sum_{j=1}^{n} \sum \mu_{1jt} x_j, \cdots, \frac{1}{4} \sum_{t=1}^{4} \sum_{j=1}^{n} \sum \mu_{mjt} x_j.$$

Proof. For any $i \in 1, 2, \cdots, m, \omega \in \Omega, \tilde{\bar{c}}_{ij}(\omega) = (\bar{c}_{ij1}(\omega), \bar{c}_{ij2}(\omega), \bar{c}_{ij3}(\omega), \bar{c}_{ij4}(\omega))$ is a trapezoidal fuzzy variable. It follows from Lemma 3.3 that

$$E[\sum_{j=1}^{n} \tilde{\bar{c}}_{ij}(\omega) x_j] = \frac{1}{4} \sum_{j=1}^{n} (\bar{c}_{ij1}(\omega) + \bar{c}_{ij2}(\omega) + \bar{c}_{ij3}(\omega) + \bar{c}_{ij4}(\omega)) x_j.$$

It follows from Definition (3.6) that

$$
\begin{aligned}
E[\tilde{\bar{c}}_i^T x] &= E[E[\tilde{\bar{c}}^T(\omega)_i x]] \\
&= E[\tfrac{1}{4} \sum_{j=1}^{n} (\bar{c}_{ij1}(\omega) + \bar{c}_{ij2}(\omega) + \bar{c}_{ij3}(\omega) + \bar{c}_{ij4}(\omega)) x_j] \\
&= \tfrac{1}{4} (\sum_{j=1}^{n} E[\bar{c}_{ij1}(\omega)] x_j + \sum_{j=1}^{n} E[\bar{c}_{ij2}(\omega)] x_j + \sum_{j=1}^{n} E[\bar{c}_{ij3}(\omega)] x_j + \sum_{j=1}^{n} E[\bar{c}_{ij4}(\omega)] x_j) \\
&= \tfrac{1}{4} (\sum_{j=1}^{n} \mu_{ij1} x_j + \sum_{j=1}^{n} \mu_{ij2} x_j + \sum_{j=1}^{n} \mu_{ij3} x_j + \sum_{j=1}^{n} \mu_{ij4} x_j) \\
&= \tfrac{1}{4} \sum_{t=1}^{4} \sum_{j=1}^{n} \sum \mu_{ijt} x_j.
\end{aligned}
$$

Then this theorem is proved.   □

Theorem 3.4. *If Fu-Ra variable $\tilde{\bar{a}}_{rj}, \tilde{\bar{b}}_r$ are trapezoidal Fu-Ra variables and defined as follows,*

$$
\tilde{\bar{a}}_{rj}(\omega) = (\bar{a}_{rj1}(\omega), \bar{c}_{rj2}(\omega), \bar{a}_{rj3}(\omega), \bar{a}_{rj4}(\omega)) \quad with \quad \bar{a}_{ijt}(\omega) \sim \mathcal{N}(\mu_{ijt}, \delta_{ijt}^2),
$$
$$
\tilde{\bar{b}}_r(\omega) = (\bar{b}_{r1}(\omega), \bar{b}_{r2}(\omega), \bar{b}_{r3}(\omega), \bar{c}_{r4}(\omega)) \quad with \quad \bar{b}_{rt}(\omega) \sim \mathcal{N}(\mu_{rt}, \delta_{rt}^2),
$$

for $r = 1, 2, \cdots, p, t = 1, 2, 3, 4$, then the constraints

$$
E[\tilde{\bar{a}}_r^T x] \le E[\tilde{\bar{b}}_r], r = 1, 2, \cdots, p
$$

is equivalent to

$$
\sum_{t=1}^{4} \sum_{j=1}^{n} \mu_{rjt} x_j \le \sum_{t=1}^{4} \mu_{rt}, r = 1, 2, \cdots, p.
$$

Proof. Similarly to the proof of Theorem 3.3, for any $r \in 1, 2, \cdots, p$, we have

$$
E[\tilde{\bar{a}}_r^T x] = \sum_{t=1}^{4} \sum_{j=1}^{n} \mu_{rjt} x_j
$$

and

$$
E[\tilde{\bar{b}}_r^T] = \sum_{t=1}^{4} \mu_{rt}.
$$

Then this theorem is proved.   □

According to Theorem 3.3-3.4, we can get the crisp equivalent model for Model (3.22) as follows:

$$
\begin{cases}
\max \left[\dfrac{1}{4} \sum_{t=1}^{4} \sum_{j=1}^{n} \sum \mu_{1jt} x_j, \dfrac{1}{4} \sum_{t=1}^{4} \sum_{j=1}^{n} \sum \mu_{2jt} x_j, \cdots, \dfrac{1}{4} \sum_{t=1}^{4} \sum_{j=1}^{n} \sum \mu_{mjt} x_j \right] \\
\text{s.t.} \begin{cases} \sum_{t=1}^{4} \sum_{j=1}^{n} \mu_{rjt} x_j \le \sum_{t=1}^{4} \mu_{rt}, r = 1, 2, \cdots, p \\ x_j \ge 0, j = 1, 2, \cdots, n. \end{cases}
\end{cases}
\tag{3.23}
$$

Here we give numerical examples to show the effectiveness of the algorithms we proposed for the fuzzy random expected value multi-objective decision making model problem.

3.3.2.2 Fuzzy Satisfied Method

In this subsection, we introduce the interactive fuzzy satisfied method proposed by Sakawa [257]. We consider the following multi-objective decision-making model,

$$\begin{cases} \max[H_i(x), \ i = 1, 2, \cdots, m] \\ \text{s.t. } x \in X. \end{cases} \tag{3.24}$$

The objective function of model (3.24) is to maximize $H_i(x)$, so for each objective we introduce the fuzzy objective "$H_i(x)$ approximately more than some value", and the membership function is

$$\mu_i(H_i(x)) = \begin{cases} 1, & H_i(x) > H_i^1 \\ \frac{H_i(x) - H_i^0}{H_i^1 - H_i^0}, & H_i^0 \leq H_i(x) \leq H_i^1 \\ 0, & H_i(x) < H_i^0. \end{cases} \tag{3.25}$$

In equation (3.25), the membership are 1 and 0 respectively when value of $H_i(x)$ are H_i^1 and H_i^0,

$$H_i^1 = \max_{x \in X} H_i(x), \quad H_i^0 = \min_{x \in X} H_i(x), \quad i = 1, 2, \cdots, m. \tag{3.26}$$

For model $\min_{x \in X} H_i(x)$, its optimal solution should be get at the boundary of the convex set X. If there exists no solution of $\max_{x \in X} H_i(x)$ or $\min_{x \in X} H_i(x)$, or $H_i^1 = \infty$, $H_i^0 = -\infty$, the decision maker may set the value of H_i^1, H_i^0 subjectively.

Hence, the model (3.24) could be transformed into the following form:

$$\begin{cases} \max \ [\mu_1(H_1(x)), \mu_2(H_2(x)), \cdots, \mu_m(H_m(x))] \\ \text{s.t. } x \in X. \end{cases} \tag{3.27}$$

For each objective function $\mu_i(H_i(x))$, let the decision maker give the reference value of membership function $\bar{\mu}_i$ to reflect the ideal value of membership function. Through solving the minmax problem (3.28) can get a efficient solution of model (3.24):

$$\begin{cases} \min \max_{i=1,2,\cdots,m} \{\bar{\mu}_i - \mu_i(H_i(x))\} \\ \text{s.t. } x \in X. \end{cases} \tag{3.28}$$

By introducing the assistant variable λ, model (3.28) is equivalent to

$$\begin{cases} \min \lambda \\ \text{s.t. } \begin{cases} \bar{\mu}_i - \mu_i(H_i(x)) \leq \lambda, & i = 1, 2, \cdots, m \\ 0 \leq \lambda \leq 1, & x \in X. \end{cases} \end{cases} \tag{3.29}$$

The following Theorem 3.5 present the relationship between the optimal solution of model (3.29) and the efficient model (3.24).

Theorem 3.5. *(i) For given* $\bar{\mu}_i$, $i = 1, 2, \cdots, m$, *if* $x^* \in X$ *is the single optimal solution of model (3.29), then* x^* *is the efficient solution of model (3.24).*

(ii) If x^* *is the efficient solution of model (3.24), and for any i, we have* $0 < \mu_i(H_i(x^*)) < 1$, *then exists* $\bar{\mu}_i$, $i = 1, 2, \cdots, m$ *make* x^* *the optimal solution of model (3.29).*

Proof. (i) Suppose that x^* is the single optimal solution of model (3.29) but not the efficient solution of model (3.24), then there exists $\bar{x} \in X$, such that $\forall i = 1, 2, \cdots, m$, we have $H_i(\bar{x}) \geq H_i(x^*)$, and at least there exists some i_0, $H_{i_0}(\bar{x}) > H_{i_0}(x^*)$ holds. Since $\mu_i(H_i(x))$ is monotone increasing function, then we have

$$\mu_i(H_i(x)) \geq \mu_i(H_i(x^*)), \quad \mu_{i_0}(H_i(x)) > \mu_{i_0}(H_i(x^*)), \quad \forall i = 1, 2, \cdots, m.$$

So, $\forall i = 1, 2, \cdots, m$,

$$\bar{\mu}_i - \mu_i(H_i(x)) \leq \bar{\mu}_i - \mu_i(H_i(x^*)), \quad \bar{\mu}_{i_0} - \mu_{i_0}(H_i(x)) < \bar{\mu}_{i_0} - \mu_{i_0}(H_i(x^*)).$$

Because

$$\lambda^* \geq \max_{i=1,2,\cdots,m} \{\bar{\mu}_i - \mu_i(H_i(x^*))\} \geq \max_{i=1,2,\cdots,m} \{\bar{\mu}_i - \mu_i(H_i(\bar{x}))\} = \bar{\lambda}.$$

It is conflict with the assumption, then x^* is the efficient solution of model (3.24).
(ii) If for at least one $\bar{\varepsilon}_i \geq 0$ and $\bar{x}$ is not the efficient solution of model (3.24), then there exists $x \in X$ such that for any $i = 1, 2, \cdots, m$, $\mu_i(H_i(x)) \geq \mu_i(H_i(\bar{x}))$, and there exists some i_0 such that $\mu_{i_0}(H_{i_0}(x)) \geq \mu_{i_0}(H_{i_0}(\bar{x}))$. So there exists $\varepsilon' \geq 0$ such that $\mu_i(H_i(x)) + \varepsilon_i' = \mu_i(H_i(\bar{x}))$, $i = 1, 2, \cdots, m$. It is conflict with the optimization.
The proof is completed.   □

According to Theorem 3.5, the optimal solution of model (3.29) is an efficient solution of model (3.27), thereby, it is a fuzzy random efficient solution of model (3.24).
In a word, we can get a satisfactory solution of model (4.41) for decision maker by employing the interactive fuzzy satisfied method. The steps are as follows:

Step 1. The decision maker set the reference value $\bar{\mu}_i$, $i = 1, 2, \cdots, m$ of membership function.

Step 2. Solve the model (3.29) and get the optimal solution x^*, and obtain a efficient solution of model (3.24).

Step 3. If the decision make is satisfied with the $\mu_i(H_i(x^*))$, then stops; Otherwise, the decision maker reset the reference value $\bar{\mu}_i$ of membership function, then turn to step 2.

3.3.2.3 Numerical Example

We will use the interactive fuzzy satisfied method to solve the crisp equivalent model of linear Fa-Ra EVM.

Example 3.4. Let us consider the following problem

$$\begin{cases} \max \left[E[\sum_{j=1}^{6} \xi_j x_j], E[\sum_{j=7}^{12} \xi_j x_j] \right] \\ \text{s.t.} \begin{cases} E[10\xi_1 x_1 + 8\xi_2 x_2 + 6\xi_3 x_3 + 4\xi_4 x_4 + 2\xi_5 x_5 + \xi_6 x_6] \leq E[100\xi_6] \\ E[9\xi_7 x_1 + 7\xi_8 x_2 + 5\xi_9 x_3 + 3\xi_{10} x_4 + \xi_{11} x_5 + 2\xi_{12} x_6] \leq E[120\xi_{12}] \\ x_1 + x_2 + x_3 + x_4 + x_5 + x_6 \leq 50 \\ x_1 + x_2 + x_3 + x_4 + x_5 + x_6 \geq 30 \\ 2x_1 + 3x_2 + 4x_3 + 5x_4 + 6x_5 + 7x_6 \leq 150 \\ 11x_1 + 9x_2 + 7x_3 + 5x_4 + 3x_5 + x_6 \geq 100 \\ x_j \geq 0, i = 1,2,3,4,5,6, \end{cases} \end{cases}$$

(3.30)

where $\xi_j (j = 1,2,\cdots,12)$ are fuzzy random variable characterized as

$$\xi_1 = (\rho_1 - 3, \rho_1 - 1, \rho_1 + 1, \rho_1 + 3), \text{ with } \rho_1 \sim \mathcal{N}(5,5),$$
$$\xi_2 = (\rho_2 - 3, \rho_2 - 1, \rho_2 + 1, \rho_2 + 3), \text{ with } \rho_2 \sim \mathcal{N}(8,6),$$
$$\xi_3 = (\rho_3 - 3, \rho_3 - 1, \rho_3 + 1, \rho_3 + 3), \text{ with } \rho_3 \sim \mathcal{N}(12,5),$$
$$\xi_4 = (\rho_4 - 3, \rho_4 - 1, \rho_4 + 1, \rho_4 + 3), \text{ with } \rho_4 \sim \mathcal{N}(18,6),$$
$$\xi_5 = (\rho_5 - 3, \rho_5 - 1, \rho_5 + 1, \rho_5 + 3), \text{ with } \rho_5 \sim \mathcal{N}(15,2),$$
$$\xi_6 = (\rho_6 - 3, \rho_6 - 1, \rho_6 + 1, \rho_6 + 3), \text{ with } \rho_6 \sim \mathcal{N}(28,3),$$
$$\xi_7 = (\rho_7 - 3, \rho_7 - 1, \rho_7 + 1, \rho_7 + 3), \text{ with } \rho_7 \sim \mathcal{N}(20,4),$$
$$\xi_8 = (\rho_8 - 3, \rho_8 - 1, \rho_8 + 1, \rho_8 + 3), \text{ with } \rho_8 \sim \mathcal{N}(15,6),$$
$$\xi_9 = (\rho_9 - 3, \rho_9 - 1, \rho_9 + 1, \rho_9 + 3), \text{ with } \rho_9 \sim \mathcal{N}(25,2),$$
$$\xi_{10} = (\rho_{10} - 3, \rho_{10} - 1, \rho_{10} + 1, \rho_{10} + 3), \text{ with } \rho_{10} \sim \mathcal{N}(8,1),$$
$$\xi_{11} = (\rho_{11} - 3, \rho_{11} - 1, \rho_{11} + 1, \rho_{11} + 3), \text{ with } \rho_{11} \sim \mathcal{N}(10,2),$$
$$\xi_{12} = (\rho_{12} - 3, \rho_{12} - 1, \rho_{12} + 1, \rho_{12} + 3), \text{ with } \rho_{12} \sim \mathcal{N}(30,6).$$

It follows from 3.4 that the problem is equivalent to

$$\begin{cases} \max F_1(x) = 5x_1 + 8x_2 + 12x_3 + 18x_4 + 15x_5 + 28x_6 \\ \max F_2(x) = 20x_1 + 15x_2 + 25x_3 + 8x_4 + 10x_5 + 30x_6 \\ \text{s.t.} \begin{cases} 50x_1 + 64_2 x_2 + 72x_3 + 72x_4 + 30x_5 + 28x_6 \leq 2800 \\ 180x_1 + 105x_2 + 125x_3 + 24x_4 + 10x_5 + 60x_6 \leq 3600 \\ x_1 + x_2 + x_3 + x_4 + x_5 + x_6 \leq 50 \\ x_1 + x_2 + x_3 + x_4 + x_5 + x_6 \geq 30 \\ 2x_1 + 3x_2 + 4x_3 + 5x_4 + 6x_5 + 7x_6 \leq 150 \\ 11x_1 + 9x_2 + 7x_3 + 5x_4 + 3x_5 + x_6 \geq 100 \\ x_j \geq 0, i = 1,2,3,4,5,6. \end{cases} \end{cases}$$

(3.31)

$F_i^0, F_i^1 (i = 1, 2)$ are calculated as follows

$$F_1^0 = 222, \quad F_1^1 = 564, \quad F_2^0 = 240, \quad F_2^1 = 843.3.$$

By the model , we have that

$$
\begin{cases}
\min \lambda \\
\text{s.t.} \begin{cases}
5x_1 + 8x_2 + 12x_3 + 18x_4 + 15x_5 + 28x_6 \geq 222 + 342(\bar{\mu}_1 - \lambda) \\
20x_1 + 15x_2 + 25x_3 + 8x_4 + 10x_5 + 30x_6 \geq 240 + 603.3(\bar{\mu}_2 - \lambda) \\
50x_1 + 64_2x_2 + 72x_3 + 72x_4 + 30x_5 + 28x_6 \leq 2800 \\
180x_1 + 105x_2 + 125x_3 + 24x_4 + 10x_5 + 60x_6 \leq 3600 \\
x_1 + x_2 + x_3 + x_4 + x_5 + x_6 \leq 50 \\
x_1 + x_2 + x_3 + x_4 + x_5 + x_6 \geq 30 \\
2x_1 + 3x_2 + 4x_3 + 5x_4 + 6x_5 + 7x_6 \leq 150 \\
11x_1 + 9x_2 + 7x_3 + 5x_4 + 3x_5 + x_6 \geq 100 \\
x_j \geq 0, i = 1, 2, 3, 4, 5, 6.
\end{cases}
\end{cases}
\tag{3.32}
$$

For the initial reference membership 1, each membership function value and the solution x as well as the objective function $F_i(x)$ are obtained (see the first row in Table 3.1). If the decision maker wishes to increase $F_1(x)$ by sacrificing $F_2(x)$, then the membership value $(\bar{\mu}_1, \bar{\mu}_2)$ needs to be updated, such as (1,0.98). Or else $(\bar{\mu}_1, \bar{\mu}_2)$ needs to be updated, such as (0.98,1). The results are listed in the second row. Suppose that the decision maker is satisfied with the solution when the probability is (1,0.90). Then the interactive process is stopped and the satisfactory solution is $x^* = (12.18, 0, 0, 0, 0, 17.95)$ and $(F_1^*, F_2^*) = (563.50, 782.10)$.

Moreover the decision maker can modify $F_i^0, F_i^1 (i = 1, 2)$ and build a new reference membership function to obtain his or her satisfactory solution.

Table 3.1 Results obtained from interactive process

$\bar{\mu}_1$	$\bar{\mu}_2$	$x = (x_1, x_2, x_3, x_4, x_5, x_6)$	λ	$F_1(x)$	$F_2(x)$
1	1	(11.59,0,4.7,0,0,15.43)	0.051	546.39	812.20
1	0.98	(12.79,0,2.55,0,0,16.32)	0.037	551.51	809.15
1	0.96	(171.11,0,0,0,0,208.17,80,0,0,0,0,191.83)	0.413	4454.37	5036.60
1	0.94	(171.11,0,0,0,0,212.79,80,0,0,0,0,187.21)	0.403	4519.05	4944.20
1	0.92	(171.11,0,0,0,0,217.42,80,0,0,0,0,182.58)	0.393	4583.87	4851.60
1	0.90	(171.11,0,0,0,0,222.04,80,0,0,0,0,177.96)	0.383	4508.55	4759.20
0.98	1	(12.79,0,6.86,0,0,14.54)	0.046	541.39	815.50
0.96	1	(13.99,0,0,0.39,0,0,17.20)	0.022	556.45	805.85
0.94	1	(13.62,0,0,0,0,17.54)	0.014	559.22	798.60
0.92	1	(12.90,0,0,0,0,17.74)	0.007	561.22	790.20
0.90	1	(12.18,0,0,0,0,17.95)	0.002	563.50	782.10

3.3.3 Non-linear Fu-Ra EVM and Fu-Ra Simulation-Based GA

For the non-linear Fu-Ra EVM, it is usually difficult to transform them into their equivalent forms, and it's is unnecessary to do this. In this case, simulation and an intelligent algorithm are the best way. So we introduce the Fu-Ra simulation-based weighted-sum GA first, and then we give a numerical example.

3.3.3.1 Fu-Ra Simulation 1 for Expected Value

In Fu-Ra EVM (3.18), one problem is to calculate the expected value $E[f(\xi)]$. Note that, for each $\omega \in \Omega$, we may calculate the expected value $E[f(\xi(\omega))]$ by fuzzy simulation. Since $E[f(\xi)]$ is essentially the expected value of stochastic variable $E[f(\xi(\omega))]$, we may combine stochastic simulation and fuzzy simulation to produce a fuzzy random simulation.

Firstly, we sample $\omega_1, \omega_2, \cdots, \omega_N$ from Ω according to Pr. For each $\omega_n (n = 1, 2, \cdots, N)$, $\xi(\omega_n)$ are all fuzzy variables, and $f(\xi(\omega_n))$ are also fuzzy variables. Then we can apply the fuzzy simulation 1 to get their expected values $E[f(\xi(\omega_n))]$, respectively.

In order to calculate the expected value $E[f(\xi)]$, we use the strong law of large numbers:

$$\frac{\sum\limits_{n=1}^{N} E[f(\xi(\omega_n))]}{N} \to E[f(\xi)], \tag{3.33}$$

as $N \to \infty$. Therefore, the value $E[f(\xi)]$ can be estimated by $\frac{1}{N} \sum\limits_{k=1}^{N} E[f(\xi(\omega_n))]$ provided that N is sufficiently large.

The procedure is as follows,

Step 1. Set $E = 0$.

Step 2. Sample ω from Ω according to the probability measure Pr.

Step 3. $E \leftarrow e + E[f(\xi(\omega))]$, where $E[f(\xi(\omega))]$ may be calculated by the fuzzy simulation.

Step 4. Repeat the second to fourth steps N times.

Step 5. $E[f(\xi)] = e/N$.

Example 3.5. We employ the fuzzy random simulation to calculate the expected value of $\xi_1 \xi_2$, where ξ_1 and ξ_2 are Fu-Ra variables defined as

$$\xi_1 = (\rho_1, \rho_1 + 1, \rho_1 + 2), \text{ with } \rho_1 \sim EXP(1),$$
$$\xi_2 = (\rho_2, \rho_2 + 1, \rho_2 + 2), \text{ with } \rho_2 \sim EXP(2).$$

After a run of the Fu-Ra simulation 1 with 5000 cycles shows that

$$E[\xi_1 \xi_2] = 6.34.$$

3.3.3.2 GA

Genetic algorithm (GA) is a stochastic search method for optimization problems based on the mechanics of natural selection and natural genetics-survival of the fittest. GAs have demonstrated considerable success in providing good solutions to many complex optimization problems and received more and more attentions during the past three decades. When the objective functions to be optimized in the optimization problems are multi-modal or the search spaces are particularly irregular, algorithms need to be highly robust in order to avoid getting stuck at a local optimal solution. The advantage of GAs is just able to obtain the global optimal solution fairly. In addition, GAs do not require the specific mathematical analysis of optimization problems, which makes GAs easily coded by users who are not necessarily good at mathematics on algorithms. GAs have been well-documented in the literature, such as in Holland, Goldberg, Michalewicz, and Fogel, etc., and have been applied to a wide variety of problems, such as optimal control problems, transportation problems, traveling salesman problems, scheduling, facility layout problems and network optimization and so on.

One of the important technical terms in GAs is chromosome, which is usually a string of symbols or numbers. A chromosome is a coding of a solution of an optimization problem, not necessarily the solution itself. GAs start with an initial set of random-generated chromosomes called population size. All chromosomes are evaluated by the so-called evaluation function, which is some measure of fitness. A new population will be formed by a selection process using a sampling mechanism based on the fitness values. The cycle from one population to the next one is called a generation. In each new generation, all chromosomes will be updated by the crossover and mutation operations. The revised chromosomes are also called offspring. The selection process enters a new generation. After performing the genetic system a given number of cycles, we decode the best chromosome into a solution which is regarded as the optimal solution of the optimization problem.

(1) Coding

How to encode a solution of the problem into a chromosome is a key issue when using GAs. The issue has been investigated from many aspects, such as mapping characters from genotype space to phenotype space when individuals are decoded into solutions, and metamorphosis properties when individuals are manipulated by genetic operators.

During the last 15 years, various encoding methods have been created for particular problems to provide effective implementation of GAs. According to what kind of symbol is used as the alleles of a gene, the encoding methods can be classified as follows:

    (i) Binary encoding,
    (ii) Real-number encoding,
    (iii) Integer or literal permutation encoding,
    (iv) General data structure encoding.

In holland' work, encoding is carried out using binary strings. Binary encoding for function optimization problems is known to have severe drawbacks due to the existence of Hamming cliffs, pairs of encodings having a large Hamming distance while belonging to points of minimal distance in phenotype space. For example, the pair 01111111111 and 1000000000 belong to neighboring points in phenotype space (points of minimal Euclidean distance) but have maximum Hamming distance simultaneously. The probability that crossover and mutation will occur can be very small. In this sense, the binary code does not preserve the locality of points in the phenotype space. Real-number encoding is best used for function optimization problems. It has been widely confirmed that real-number encoding performs better than binary of Gray encoding for function optimizations and constrained optimizations. Since the topological structure of the genotype space for real-number encoding is identical to that of the phenotype space, it is easy to form effective genetic operators by borrowing useful techniques from conventional methods. Integer of literal permutation encoding is best used for combinatorial optimization problems. Since the essence of combinatorial optimization problems is the search for a best permutation of combination of items subject to constraints, literal permutation encoding can be the best way to this type of problem. For more complex real-world problems, an appropriate data structure is suggested as the allele of a gene, to capture the nature of the problem. In such case, a gene may be an n-ary or more complex data structure.

According to the structure of encodings, the encoding methods can also be classified into the following two types: one-dimensional encoding and multi-dimensional encoding. In most practices, one-dimensional encoding is used. However, many real-world problems require solutions for multi-dimensional structures. It is nature to use a multi-dimensional encoding method to represent those solutions. For example, Vignaus and Michalewicz used an allocation matrix as encoding for the transportation problem. Cohoon and Paris used two-dimensional encoding for the VLSI circuit placement problem. Ono, Yamamura, and Kobayashi used a job-sequence matrix as encoding for job-shop scheduling peoblem. And a general discussion of multi-dimensional encoding and crossover was given by Bui and Moon.

According to the contents encoded, the following encoding methods can also be used: solution only and solution + parameters. In genetic algorithm practice, the first method is widely used to develop suitable encoding for a given problem. The second method is used in evolution strategies of Rechenberg and Schwefel. An individual consists of two pair: the first is the solution to a given problem, and the second, the strategy parameters, comprise variances and covariances of the normal distribution for mutation. The purpose of incorporating strategy parameters into the representation of individuals is to facilitate the evolutionary self-adaptation of these parameters by applying evolutionary operators to them. The search will then be performed in the space of solutions and strategy parameters together. In this way a suitable adjustment and diversity of mutation parameters should be provided under arbitrary circumstances.

Genetic algorithms work on two types of spaces alternatively: coding spaces and solution spaces, or in other words, genotype spaces and phenotype spaces. Genetic operators work on genotype space, and evaluation and selection work on the phenotype space. Natural selection is the link between chromosomes and the performance of decoded solutions. The mapping from genotype space to phenotype space has a considerable influence on the performance of genetic algorithms. One outstanding problem associated with mapping is that some individuals correspond to infeasible solutions to a given problem. This problem may become very severe for constrained optimization problems and combinatorial optimization problems.

We should distinguish between two basic concepts: infeasibility and illegality. Infeasibility refers to the phenomenon that a solution decoded from a chromosome lies outside the feasible region of a given problem; illegality refers to the phenomenon that a solution decoded from chromosome does not represent a solution to a given problem.

The infeasibility of chromosomes originates from the nature of the constrained optimization problem. Whichever technique is used, conventional methods or genetic algorithms, it must handle the constraints. For many optimization problems, the feasible region can be represented as a system of equalities or inequalities. For such cases, penalty methods can be used to handle infeasible chromosomes. In constrained optimization problems, the optimum typically occurs at the boundary between the feasible and infeasible areas. The penalty approach will force the genetic search to approach the optimum from both sides of the feasible and infeasible regions.

The illegality of chromosomes originates from the nature of encoding techniques. For many combinatorial optimization problems. problem -specific encodings are used, and such encodings usually yield illegal offsprings by simple one-cut point crossover operation. Because an illegal chromosome can not be decoded to a solution. Repair techniques are usually adopted to convert an illegal chromosome to a legal one. For example, the well-known PMX operator is essentially a two-cut point crossover for permutation representation, together with a repair procedure to resolve the illegitimacy caused by simple two-cut point crossover. Orvosh and Davis have shown that for many combinatorial optimization problems, it is relatively easy to repair an infeasible or illegal chromosome, and the repair strategy does indeed surpass other strategies, such as the rejecting or the penalizing strategy.

When a new encoding method is given, it is usually necessary to examine whether we can set up an effective genetic search using the encoding. Several principles have been proposed to evaluate an encoding as follows.

(i) Non-redundancy: The mapping between encodings and solutions must be 1 to 1.

The most desired case, 1 to 1 mapping, ensures that no trivial operations will occur when creating offspring. If n to 1 mapping occurs, genetic algorithms will waste time while searching. Because two individuals are duplicated in the phenotype space but not in the genotype space, distance measures in the genotype space cannot treat individuals as identical. It then becomes one reason for genetic algorithms to converge prematurely. The most undesirable case is 1 to n mapping, because we need another procedure performed on the phenotype space to determine the one solution among many possible solutions.

(ii) Legality: Any permutation of en encoding corresponds to a solution.
This principle guarantees that most existing genetic operators can easily be applied to the encoding.

(iii) Completeness: Any solution has a corresponding encoding.
This principle guarantees that any point in solution space is accessible for a genetic search.

(iv) Lamarckian property: The meaning of alleles for a gene is not context dependent.
The Lamarckian property for encoding concerns the issue of whether or not chromosome can pass on its merits to future populations through common genetic operations.

(v) Causality: Small variations on the genotype space due to mutation imply small variations in the phenotype space.

The principle suggested by Rechenberg in relation to evolution strategies. If focus on the conservation of neighborhood structures; that is, for the successful introduction of new information by mutation, the mutation operator should preserve the neighborhood structure in the corresponding phenotype space. The perspective is common among practitioners of genetic optimizations. Search processes that do not destroy the neighborhood structure are said to exhibit strong causality. The opposite extreme is no causality. Weak causality describes the case where small changes in the genotype space correspond to large changes in the phenotype space, and vice versa. Sendhoff, Kreuts, and Seelen suggested a condition to measure the causality associated with mapping from genotype to phenotype in combination with a mutation operator.

(2) Genetic operators
Search is one of the more universal problem-solving methods for problems in which one cannot determine a priori the sequence of steps leading to a solution. Typically, there are two types of search behaviors: random search and local search. Random search explores the entire solution and is capable of achieving escape from a local optimum. Local search exploits the best solution and is capable of climbing upward toward a local search optimum. The two types of search abilities from the mutual complementary components of a search. An ideal search should possess both types simultaneously. It is nearly impossible to design such a search method with conventional techniques. Genetic algorithms are a class of general-purpose search methods combining elements of directed and stochastic searches which can make a good balance between exploration and exploitation of the search space. In genetic algorithms, accumulated information is exploited by the selection mechanism, while mew regions of the search space are explored by means of genetic operators.

In conventional genetic algorithms, the crossover operator is used as the principle operator and the performance of a genetic system is heavily dependent on it. The mutation operator which produces spontaneous random changes in various chromosomes, is used as a background operator. In essence, genetic operators perform a random search and cannot guarantee to yield improved offspring. It has been discovered that the speed of convergence problems. There are many empirical studies on a comparison between crossover and mutation. It is confirmed that mutation can sometimes play a more important role than crossover.

There are two hypotheses for the explantation of how genetic algorithms exploit the distributed information to generate good solutions: the building-block hypothesis and the convergence-controlled variation hypothesis. The building-block hypothesis was proposed by Holland and refined by Goldberg. According to the hypothesis, crossover recombines features from two parents to produce offspring. Sometimes crossover will combines the best features form two parents, resulting in superior offspring. Since the fitness of an individual will often depend on complex patterns of simple features, it is important that the operator be able to propagate to offspring those patterns of features that contribute to the fitness of the parents. The concept of epistasis refers to strong interaction among genes in an encoding. In other words, epistasis measures the extent to which the contribution to fitness of one gene depends on the values of other genes. For a given problem, a high degree of epistasis means that building blocks cannot form.

The convergence-controlled variation hypothesis was given by Eshelman, Mathias and Schaffer. The hypothesis suggests using the convergence of a population to constrain the search. New points are sampled from a distribution that is a function of the population distribution at any point in time. As the population converges, the variation becomes more focused. Whereas the building-block hypothesis stresses recombining, and hence propagating, features that have survived in the parent population, the convergence-controlled variation hypothesis stresses randomly sampling from a distribution that is a function of the current population distribution.

How we conceptualize the genetic search will affect how we design genetic operators. From the point of view of search abilities, it is expected that a search provided by a method can possess the abilities of random search and directed search simultaneously. Cheng and Gen suggest the following approach for designing genetic operators. For the two genetic operators, crossover and mutation, one is used to perform a random search to try to explore the area beyond a local optimum, and the other is used to perform a local search to try to find an improved solution, The genetic search then possesses two types of the search abilities. With this approach, the mutation operator will play the same important role as that of the crossover operator in a genetic search.

(i) Selection
The principle behind genetic algorithms is essentially Darwinian natural selection. Selection provides the driving force in a genetic algorithm. With too much force, the genetic search will terminate prematurely; with too little force, the evolutionary progress will be slower than necessary. Typically, a lower selection pressure is indicated at the start of a genetic search in favor of a wide exploration of the search space, while a higher selection pressure is recommended at the end to narrow the search space. The selection directs the genetic search toward promising regions in the search space. During the past two decades, many selection methods have been proposed, examined, and compared. There are the following types.

(a) Roulette wheel selection: Proposed by Holland, is the best known selection type. The basic idea is to determine selection probability or survival probability for each chromosome proportional to the fitness value. Then a model roulette wheel can be made displaying these probabilities. The selection process is based on spinning the wheel the number of times equal to population size, each time selecting a single chromosome for the new population, The wheel features the selection method as a stochastic sampling method that uses a single wheel spin. The wheel is constructed in the same way as a standard roulette wheel, with a number of equally spaced markers equal to the population size. The basic strategy underlying this approach is to keep the expected number of copies of each chromosome in the next generation.

(b) $(\mu + \lambda)$-selection: In contrast with proportional selection, $(\mu + \lambda)$-selection and (μ, λ)-selection as proposed by Bäck are deterministic procedures that select the best chromosomes from parents and offspring. Note that both methods prohibit selection of duplicate chromosomes from the population, so many researchers prefer to use this method to deal with combinational optimization problems. Truncation selection and block selection are also deterministic procedures that rank all individuals according to their fitness and select the best as parents.

(c) Tournament selection: This type of selection contains random and deterministic features simultaneously. A special example is the tournament selection of Goldberg, Krob, and Deb. This method randomly chooses a set of chromosomes and picks out the best chromosome for reproduction. The number of chromosomes in the set is called the tournament size. A common tournament size is 2; this is called a binary tournament, Stochastic tournament selection was suggested by Wetzel. In this method, selection probabilities are calculated normally and successive pairs of chromosome are drawn using roulette wheel selection. After drawing a pair, the chromosome with higher fitness is inserted in the new population. The process continues until the population is full. Reminder stochastic sampling, proposed by Brindle, is a modified version of his deterministic sampling. In this method, each chromosome is allocated samples according to the fractional parts of the number expected.

(d) Steady-state reproduction: Generational replacement, replacing an entire set of parents by their offspring, can be viewed as another version of the deterministic approach. The steady-state reproduction of Whitely and Syswerdra belongs to this class, in which the n worst parents are replaced by offspring (n is the number of offspring).

(e) Ranking and scaling: The ranking and scaling mechanisms are proposed to mitigate these problems. The scaling method maps raw objective function values to positive real values, and the survival probability for each chromosome is determined according to these values. Fitness scaling has a twofold intention: to maintain a reasonable differential between the relative fitness ratings of chromosomes, and to prevent too-rapid a takeover by some superchromosomes to meet the requirement to limit competition early but to stimulate it later. Since De Jong's work, use of scaling objective functions has become widely accepted, and several scaling mechanisms have been proposed. According to the type of function used to transform the raw fitness into scaled fitness, scaling methods can be classified as linear scaling, sigma truncation, power law scaling, logarithmic scaling, an so on. If the transformation relation between scaled fitness and raw fitness is constant, it is called a static scaling method; if the transformation is variable with respect to some factors, it is called a dynamic scaling method. The windowing technique introduces a moving baseline into the fitness proposition selection to maintain a more constant selection pressure. The normalizing technique is also one type of dynamic scaling proposed by Cheng and Gen[69]. For most scaling methods, scaling parameters are problem dependent. Fitness ranking has an effect similar to that of fitness scaling but avoids the need for extra scaling parameters. Baker introduced the notion of ranking selection with genetic algorithms to overcome the scaling problems of the direct fitness-based approach. The ranking method ignores the actual object function values; instead, it uses a ranking of chromosomes to determine survival probability. The idea is straightforward: Sort the population according to the ranking but not its raw fitness, Two methods are in common use: linear ranking and exponential ranking.

(f) Sharing: The sharing techniques, introduced by Goldberg and Richardson[64] for multi-model function optimization, are used to maintain the diversity of population. A sharing function is a way of determining the degradation of an individual's fitness due to a neighbor at some distance. With the degradation, the reproduction probability of individuals in a crowd peak is restrained while other individuals are encouraged to give offspring.

(3) Procedure of genetic algorithm
We list the procedure of genetic algorithm as follows:

Step 1. Generate the input-output data for uncertain functions like

$$U_1 : x \to E[f(x, \xi)],$$
$$U_2 : x \to Ch\{g_j(x, \xi) \leq 0, j = 1, 2, \cdots, p\},$$
$$U_3 : x \to \max\{\bar{f} | Ch\{f(x, \xi) \geq \bar{f}\} \geq \alpha\},$$

by the fuzzy simulation.

Step 2. Initialize pop_size feasible chromosome.

Step 3. Update the chromosome by crossover and mutation operations, in which we may use the fuzzy random simulation.

Step 4. Calculate the objective values for all chromosome in which we may use the fuzzy random simulation.

Step 5. Compute the fitness of each chromosome according to the objective values.

Step 7. Select the chromosomes by spinning the roulette wheel.

Step 8. Repeat the fourth to seventh steps for a given number of cycles.

Step 9. Report the best chromosome as the optimal solution.

Weighted-Sum Approach

Conceptually, the weighted-sum approach can be viewed as an extension methods used in multi-objective optimization to genetic algorithms. It assigns weights to each objective function and combines the weighted objectives into a single objective function. In fact, the weighted-sum approaches used in the genetic algorithm are very different in nature from that in conventional multi-objective optimizations. In the multi-objective optimization, the weighted-sum approach is used to obtain a compromise solution. To make the method work, all that is needed a good weighting vector. It is usually very difficult to determine a set of appropriate weights for a given problem. In the genetic algorithms, the weighted-sum approach is used to primarily to adjust genetic search toward the Pareto frontier. Weights are readjusted adaptively along with the evolutionary process. Therefore, a good weighting vector is not a mandatory precondition to making genetic algorithms work. In addition, the drawbacks exhibited in the multi-objective optimization can be compensated by the powers of population-based search and evolutionary search.

Three weight-setting mechanisms have been proposed:

(1) fixed-weight approach;
(2) random-weight approach;
(3) adaptive weight approach.

The fixed-weight approach can be viewed as analogous to conventional scalarization techniques, it gives the genetic algorithms a tendency to sample the area toward a fixed point in the criterion space, while the random-weight and adaptive weight approaches are designed for genetic algorithms to fully utilize the power of genetic search which can only work due to the nature of population-based evolutionary search of genetic algorithms.

Random-Weight Approach

Murata, Ishibuchi, and Tanaka[65] proposed a random-weighted approach to obtaining a variable search direction toward the Pareto frontier. Typically, there are two types of search behavior in the objective space: fixed-direction search and multiple-direction search, as demonstrated in Figures 3.3 and 3.4. The random-weight approach gives the genetic algorithms a tendency to demonstrate a variable search direction, therefore, the ability to sample the area uniformly over the entire frontier.

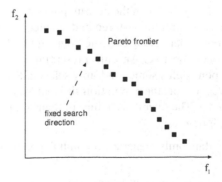

Fig. 3.3 Search in a fixed direction in criterion space

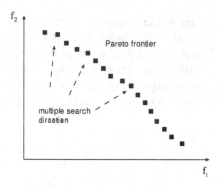

Fig. 3.4 Search in multiple directions in criterion space

Suppose that we are going to maximize q objective functions. The weighted-sum objective is given as follows:

$$z = \sum_{k=1}^{q} w_k f_k(x). \tag{3.34}$$

The random weights w_k are calculated by the equation

$$w_k = \frac{r_k}{\sum_{j=1}^{q} r_j}, k = 1, 2, \cdots, q, \tag{3.35}$$

where r_i are nonnegative random numbers.

Before selecting a pair of parents for crossover operation, a new set of random weights is specified by (3.35), and the fitness values for each individual are calculated by (3.34). The selection probability p_i for individual i is then defined by the following linear scaling function:

$$p_i = \frac{z_i - z_{min}}{\sum_{j=1}^{pop-size}(z_i - z_{min})}, \tag{3.36}$$

where z_{min} is the worst fitness value in the current population.

A tentative set of Pareto solutions is stored and updated at each generation. For a problem with q objectives, there are q extreme points int he Pareto solutions, each of which maximizes one objective. An elite preserving strategy is suggested for putting the n extreme points plus some randomly selected Pareto solutions into the next population. Let N_{pop} denote the population size and N_{elite} denote the number of elite solutions to preserve. The overall structure of their implementation of genetic algorithms is given as follows:

Step 1. Initialization. Randomly generate an initial population containing N_{pop} individuals.

Step 2. Evaluation. Calculation the values of q objective functions for each individual. Update a tentative set of Pareto solutions.

Step 3. Selection. Repeat the following steps to select $(N_{pop} - N_{elite})$ pairs of parent: Specify random weighted by (3.35), calculate fitness function by (3.34), calculate selection probability by (3.36), and select a pair of parent individuals for a crossover operation.

Step 4. Crossover. For each pair selected, apply a crossover operation to generate offspring.

Step 5. Mutation. Apply a mutation operation to each offspring generated by the crossover operation.

Step 6. Elitist strategy. Randomly select N_{pop} individuals from the tentative set of Pareto solutions. Add the selected solutions N_{elite} to $(N_{pop} - N_{elite})$ individuals generated in the foregoing steps to construct a population of N_{pop} of individuals.

Step 7. Termination test. If a prespecified stopping condition is satisfied, stop the run; otherwise return stop 1.

Random Weight GA

For the multi-objective optimization problems, more and more scholars have done some significant work and made some progress . Among them, Pareto [60] is widely regarded as one pioneer in this field. Because it is difficult to find a solution such that every objective get the optimization, Pareto introduced the non-dominated solutions or Pareto optimal solutions to obtain optimal objectives without sacrificing other objective functions. However, for a complex multi-objective optimization problem, it is also difficult to obtain its Pareto optimal solution. Recently, some scholars consider genetic algorithms as an efficient method to find its Pareto optimal solution. Such as, J. Xu, Q. Liu and R. Wang [61] apply spanning tree based on genetic algorithm to solve a class multi-objective programming problems with random fuzzy coefficients.

Since evolutionary computation was proposed, ingrowing researchers has been interested in simulating evolution to solve complex optimization problems. Among them, genetic algorithm introduced by Holland [62] is paid more and more attention to. As a kind of meta-heuristics, it could search the optimal solution without regard to the specific inner connections of the problem. Especially, the application of GA to multi-objective optimization problems has caused a theoretical and practical challenge to the mathematical community. In the past two decades, there are many approaches on GA developed by the scholars in all kinds of field. Goldberg [426, 427] firstly suggested the Pareto ranking based fitness assignment method to find the next set of non-dominated individuals. Then the multi-objective genetic algorithm in which the rank of individual corresponds to the number of current parent population was proposed by Fonseca and Fleming [174]. There are still two weighted sum genetic algorithms to solve multiobjective optimization problems. One is the random-weight genetic algorithm proposed by Ishibuchi et al. [63], the other is adaptive-weight genetic algorithm proposed by Gen and Cheng [70].

This section attempts to apply fuzzy random simulation to convert the uncertain multi-objective problem into deterministic one and make use of random-weight genetic algorithm to solve this multi-objective problem. For the following model,

$$\begin{cases} \max\ [E[f_1(x,\xi)], E[f_2(x,\xi)], \cdots, E[f_m(x,\xi)]] \\ \text{s.t.}\ \begin{cases} E[g_r(x,\xi)] \leq 0, & r = 1, 2, \cdots, p \\ x \in X. \end{cases} \end{cases}$$

No matter that the random rough variable is discrete or continuous, we can firstly simulate its expected value by random rough simulation and apply genetic algorithm to solve the multi-objective programming problem. It can be summarized as follows:

(1) Representation: A vector $x \in X$ is chosen as a chromosome to represent a solution to the optimization problem.

(2) Handling the objective and constraint function: To obtain a determined multi-objective programming problem, we can apply the technique of fuzzy random simulation to deal with them.

(3) Initializing process: Suppose that the decision maker is able to predetermine a region which contains the feasible set. Generate a random vector x from this region until a feasible one is accepted as a chromosome. Repeat the above process $N_{pop-size}$ times, then we have $N_{prop-size}$ initial feasible chromosomes $x^1, x^2, \cdots, x^{N_{pop-size}}$.

(4) Evaluation function: Decision maker's aim is to obtain the maximum expected value of every goal. Suppose $E[f(x,\xi)] = \sum_{i=1}^{r} w_i E[f_i(x,\xi)]$, where the weight coefficient w_i expresses the importance of $E[f_i(x,\xi)]$ to the decision-maker. Then the evaluation function could be given as follows:

$$eval(x) = \sum_{i=1}^{r} w_i E[f_i(x,\xi)],$$

where the random weight is generated as following formula,

$$w_k = \frac{r_k}{\sum_{j=1}^{r} r_j}, k = 1, 2, \cdots, r,$$

where r_i is a nonnegative random number.

(5) Selection process: We can apply the roulette wheel method to develop the selection process. Each time a single chromosome for a new population is selected in the following way: Compute the total probability q,

$$q = \sum_{j=1}^{N_{pop-size}} eval(x^j).$$

Then compute the probability of the ith chromosome q_i, $q_i = \frac{eval(x^i)}{q}$. Generate a random number r in $[0, 1]$ and select the ith chromosome x_i such that $q_{i-1} < r \leq q_i, 1 \leq i \leq N_{pop-size}$. Repeat the above process $N_{pop-size}$ times and we obtain $N_{pop-size}$

copies of chromosomes. The selection probability can be computed by the following function,

$$p_i = \frac{eval(x_i) - eval(x)_{min}}{\sum_{j=1}^{pop-size} eval(x_i) - eval(x)_{min}},$$

where $eval(x)_{min}$ is the minimum fitness value of current population.

(6) Crossover operation: Generate two random numbers λ_1, λ_2 from the open interval $(0, 1)$ satisfying $\lambda_1 + \lambda_2 = 1$ and the chromosome x^i is selected as a parent provided that $\lambda_i < P_{\lambda_i}$, where parameter P_{λ_i} is the probability of crossover operation. Repeat this process $N_{pop-size}$ times and $P_{\lambda_i} \cdot N_{pr-size}$ chromosomes are expected to be selected to undergo the crossover operation. The crossover operator on x^1 and x^2 will produce two children y^1 and y^2 as follows:

$$y^1 = \lambda_1 x^1 + \lambda_2 x^2, \quad y^2 = \lambda_1 x^2 + \lambda_2 x^1.$$

If both children are feasible, then we replace the parents with them, or else we keep the feasible one if it exists. Repeat the above operation until two feasible children are obtained or a given number of cycles is finished.

(7) Mutation operation: Similar to the crossover process, the chromosome x^i is selected as a parent to undergo the mutation operation provided that random number $m < P_m$, where parameter P_m as the probability of mutation operation. $P_{\lambda_i} \cdot N_{pr-size}$ are expected to be selected after repeating the process $N_{pr-size}$ times. Suppose that x^1 is chosen as a parent. Choose a mutation direction $d \in R^n$ randomly. Replace x with $x + M \cdot d$ if $x + M \cdot d$ is feasible, otherwise we set M as a random between 0 and M until it is feasible or a given number of cycle is finished. Here, M is a sufficiently large positive number.

We illustrate the fuzzy random simulation-based genetic algorithm procedure as follows:

Step 0. Input the parameters $N_{pop-size}, P_{\lambda_i}$ and P_m.

Step 1. Initialize $N_{pop-size}$ chromosomes whose feasibility may be checked by fuzzy random simulation.

Step 2. Update the chromosomes by crossover and mutation operations and random rough simulation is used to check the feasibility of offspring. Compute the fitness of each chromosome based on weight-sum objective.

Step 3. Select the chromosomes by spinning the roulette wheel.

Step 4. Make the crossover operation.

Step 5. Make the mutation operation for the chromosomes generated by crossover operation.

Step 6. Repeat the second to fourth steps for a given number of cycles.

Step 7. Report the best chromosome as the optimal solution.

Above all, we combine the Fu-Ra simulations and GA to obtain the Fu-Ra simulation based GA, see Figure 3.5.

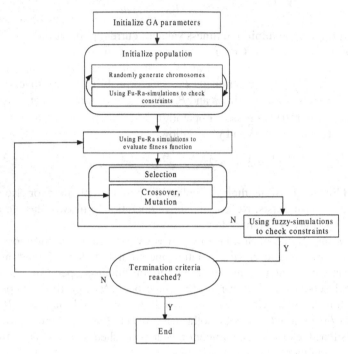

Fig. 3.5 Flow chart of Fu-Ra simulation-based GA

3.3.3.3 Numerical Example

We will use the Fu-Ra simulation 1 based random weight GA to deal with the non-linear Fu-Ra EVM.

Example 3.6. Let us consider a multi-objective programming with fuzzy random coefficients.

$$
\begin{cases}
\max F_1(x,\xi) = 2\xi_1 x_1^2 + 3\xi_2 x_2 - \xi_3 x_3 + \sqrt{(1-\xi_4)^2 x_4 + (3-\xi_5)^2} \\
\max F_2(x) = -x_1 + 2.5x_2 + 1.5x_3 \\
\text{s.t.} \begin{cases}
5x_1 - 3x_2^2 + 6\sqrt{x_3} + x_4 \le 50 \\
\xi_6\sqrt{x_1} + \xi_7 x_2 - \xi_8 x_3 \le \xi_9 \\
x_1 + x_2 + x_3 + x_4 \le 18 \\
x_1, x_2, x_3, x_4 \ge 0,
\end{cases}
\end{cases} \tag{3.37}
$$

where $\xi_i(i = 1, 2, \cdots, 9)$ are fuzzy random variables as follows,

$$
\begin{aligned}
\xi_1 &= (\rho_1 - 0.1, \rho_1, \rho_1 + 0.2), \text{ with } \rho_1 \sim U(0.4, 0.5), \\
\xi_2 &= (\rho_2 - 0.2, \rho_2, \rho_2 + 0.2), \text{ with } \rho_2 \sim U(0.6, 0.7), \\
\xi_3 &= (\rho_3 - 0.2, \rho_3, \rho_3 + 0.2), \text{ with } \rho_3 \sim U(0.7, 0.8), \\
\xi_4 &= (\rho_4 - 0.2, \rho_4, \rho_4 + 0.2), \text{ with } \rho_4 \sim N(2, 0.1), \\
\xi_5 &= (\rho_5 - 0.2, \rho_5, \rho_5 + 0.2), \text{ with } \rho_5 \sim N(4, 0.1), \\
\xi_6 &= (\rho_6 - 0.5, \rho_6, \rho_6 + 0.5), \text{ with } \rho_6 \sim N(4, 0.1), \\
\xi_7 &= (\rho_7 - 0.5, \rho_7, \rho_7 + 0.5), \text{ with } \rho_7 \sim N(6, 0.1), \\
\xi_8 &= (\rho_8 - 0.5, \rho_8, \rho_8 + 0.5), \text{ with } \rho_8 \sim N(4.5, 0.1), \\
\xi_9 &= (\rho_9 - 2, \rho_9, \rho_9 + 2), \text{ with } \rho_9 \sim N(20, 4).
\end{aligned}
\tag{3.38}
$$

From the mathematical view, the problem (3.37) is not well defined because of the uncertain parameters. Then we apply the expected value technique to deal with this uncertain programming.

$$
\begin{cases}
\max F_1(x, \xi) = E\left[2\xi_1 x_1^2 + 3\xi_2 x_2 - \xi_3 x_3 + \sqrt{(1 - \xi_4)^2 x_4 + (3 - \xi_5)^2}\right] \\
\max F_2(x) = -x_1 + 2.5x_2 + 1.5x_3 \\
\text{s.t.} \begin{cases}
5x_1 - 3x_2^2 + 6\sqrt{x_3} + x_4 \leq 50 \\
E\left[\xi_6 \sqrt{x_1} + \xi_7 x_2 - \xi_8 x_3\right] \leq E[\xi_9] \\
x_1 + x_2 + x_3 + x_4 \leq 18 \\
x_1, x_2, x_3, x_4 \geq 0.
\end{cases}
\end{cases}
\tag{3.39}
$$

Since there exists non-linear objective function and constraint, we cannot transform it into it's crisp equivalent model. In order to solve it, we use the Fu-Ra simulation based random weight GA to deal with it.

Step 1. For this model, we use real number encoding, so we initialize the chromosome, randomly generate a real number between 0 and 18.

Step 2. Then we check the constraints, in which, the fuzzy random simulation will be used to check the constraint $E\left[\xi_6 \sqrt{x_1} + \xi_7 x_2 - \xi_8 x_3\right] \leq E[\xi_9]$.

Step 3. For the feasible chromosomes, which is through the constraints checking, we compute the objectives value F_1 and F_2, in which the computation of the objective function will use the fuzzy random simulation, and according to the objectives values in one generation, we can obtain the weight for each chromosome.

Step 4. We use the weight to evaluate the fitness value of each chromosome, and the chromosome which has larger fitness value, the possibility which the chromosome be chosen is higher.

Step 5. Then we use the crossover and mutation operator for real number parent chromosomes, obtain the children chromosomes, then update a generation.

Step 6. When the results reach the determinant standard of convergence, stop.

After 3780 iterations, the results converged, and we could get the solutions, see Table 3.2. We can see that the solutions are stable. And we get the Figure 3.6 to see the process of convergency.

Table 3.2 Results obtained from Fu-Ra simulation based-random weight GA

w_1	w_2	$x = (x_1, x_2, x_3, x_4)$	$F_1(x)$	$F_2(x)$
0.6	0.4	(13.40,2.48,2.12,0)	165.85	-4.02
0.5	0.5	(13.39,2.48,2.11,0.03)	165.63	-4.03
0.4	0.6	(13.40,2.48,2.12,2.41)	166.71	-4.02

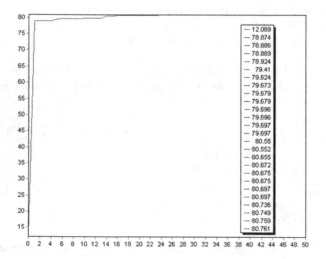

Fig. 3.6 Convergency curve

3.4 Fu-Ra CCM

For uncertain models, the chance operator is a useful tool to deal with it, so let's focus on this in this section.

3.4.1 General Model for Fu-Ra CCM

Let's introduce the general fuzzy random chance-constrained decision making model as follows.

$$
\begin{cases}
\max \ [\bar{f}_1, \bar{f}_2, \cdots, \bar{f}_m] \\
\text{s.t.} \begin{cases}
Ch\{f_i(x, \xi) \geq \bar{f}_i\}(\gamma_i) \geq \delta_i, \ i = 1, 2, \cdots, m \\
Ch\{g_r(x, \xi) \leq 0\}(\eta_r) \geq \theta_r, \ r = 1, 2, \cdots, p \\
x \in X,
\end{cases}
\end{cases}
\tag{3.40}
$$

where Ch is the chace measure of the Fu-Ra events, $\gamma_i, \delta_i, \eta_r, \theta_r$ are the predetermined confidence levels, f_i and x_i are the decision variables, $i = 1, 2, \cdots, m$.

Definition 3.11. If x^* is an efficient solution of problem (3.40), then it is called as a fuzzy random chance efficient solution.

According to the Definition 3.8 of the primitive chance measure:

$$Ch\{f_i(x,\xi) \geq \bar{f_i}\}(\gamma_i) \geq \delta_i \Leftrightarrow Pr\{\omega|Pos\{\{f_i(x,\xi) \geq \bar{f_i}\} \geq \delta_i\} \geq \gamma_i, \quad (3.41)$$

$$Ch\{g_r(x,\xi) \leq 0\}(\eta_r) \geq \theta_r \Leftrightarrow Pr\{\omega|Pos\{g_r(x,\xi) \leq 0\} \geq \theta_r\} \geq \eta_r. \quad (3.42)$$

So we can get Fu-Ra CCM based on Pr-Pos,

$$\begin{cases} \max [\bar{f_1}, \bar{f_2}, \cdots, \bar{f_n}] \\ \text{s.t.} \begin{cases} Pr\{\omega|Pos\{\{f_i(x,\xi) \geq \bar{f_i}\} \geq \delta_i\} \geq \gamma_i, & i = 1,2,\cdots,m \\ Pr\{\omega|Pos\{g_r(x,\xi) \leq 0\} \geq \theta_r\} \geq \eta_r, & r = 1,2,\cdots,p \\ x \in X, \end{cases} \end{cases} \quad (3.43)$$

where $\delta_i, \gamma_i, \theta_r, \eta_r \in [0,1]$ are the predetermined confidence levels, $Pos\{\cdot\}$ denotes the possibility of the fuzzy events in $\{\cdot\}$, and $Pr\{\cdot\}$ denotes the probability of the random events in $\{\cdot\}$.

Or we could also get Fu-Ra CCM based on Pr-Nec,

$$\begin{cases} \max [\bar{f_1}, \bar{f_2}, \cdots, \bar{f_n}] \\ \text{s.t.} \begin{cases} Pr\{\omega|Nec\{\{f_i(x,\xi) \geq \bar{f_i}\} \geq \delta_i\} \geq \gamma_i, & i = 1,2,\cdots,m \\ Pr\{\omega|Nec\{g_r(x,\xi) \leq 0\} \geq \theta_r\} \geq \eta_r, & r = 1,2,\cdots,p \\ x \in X, \end{cases} \end{cases} \quad (3.44)$$

where $Nec\{\cdot\}$ denotes the necessity of the fuzzy events in $\{\cdot\}$.

For simpleness, the parameters $\delta, \gamma, \theta, \eta$ can be the same confidence level, i.e. $\delta_i = \delta, \gamma_i = \gamma, \theta_r = \theta, \eta_r = \eta, i = 1,2,\cdots,m, r = 1,2,\cdots,p$.

Remark 3.10. If the fuzzy random vector ξ delegates to random vector, then $f_i(x,\xi) \geq \bar{f_i}$ is a random event. For $\omega \in \Omega$, $Pos\{f_i(x,\xi) \geq \bar{f_i}\} \geq \delta_i$ means $f_i(x,\xi) \geq \bar{f_i}$. So,

$$Pr\{\omega|Pos\{f_i(x,\xi) \geq \bar{f_i}\} \geq \delta_i\} \geq \gamma_i$$

is equivalent to $Pr\{\omega|f_i(x,\xi) \geq \bar{f_i}\} \geq \gamma_i, i = 1,2,\cdots,m$.

And similarly, the constraint

$$Pr\{\omega|Pos\{g_r(x,\xi) \leq 0\} \geq \theta_r\} \geq \eta_r$$

is equivalent to $Pr\{\omega|g_r(x,\xi) \leq 0\} \geq \eta_r, r = 1,2,\cdots,p$. So, the model (3.46) can be rewritten as

$$\begin{cases} \max [\bar{f_1}, \bar{f_2}, \cdots, \bar{f_m}] \\ \text{s.t.} \begin{cases} Pr\{\omega|f_i(x,\xi) \geq \bar{f_i}\} \geq \gamma_i, & i = 1,2,\cdots,m \\ Pr\{\omega|g_r(x,\xi) \leq 0\} \geq \eta_r, & r = 1,2,\cdots,p \\ x \in X. \end{cases} \end{cases}$$

This is coincident to the random CCM.

Remark 3.11. If the fuzzy random vector ξ delegates to fuzzy vector, then $Pos\{f_i(x,\xi) \geq \bar{f}_i\} \geq \delta_i$ is a crisp event. In order to satisfy $p_i := Pr\{\omega|Pos\{f_i(x,\xi) \geq \bar{f}_i\} \geq \delta_i\} \geq \gamma_i$, the probability p_i should be 1.

So the constraint

$$Pr\{\omega|Pos\{f_i(x,\xi) \geq \bar{f}_i\} \geq \delta_i\} = 1 \geq \gamma_i,$$

is equivalent to $Pos\{f_i(x,\xi) \geq \bar{f}_i\} \geq \delta_i$, $i = 1,2,\cdots,m$.

And similarly, the constraint

$$Pr\{\omega|Pos\{g_r(x,\xi) \leq 0\} \geq \theta_r\} \geq \eta_r$$

is equivalent to $Pos\{g_r(x,\xi) \leq 0\} \geq \theta_r$, $r = 1,2,\cdots,p$. So the model (3.46) is equivalent to

$$\begin{cases} \max\ [\bar{f}_1,\bar{f}_2,\cdots,\bar{f}_m] \\ \text{s.t.} \begin{cases} Pos\{f_i(x,\xi \geq \bar{f}_i\} \geq \delta_i, & i = 1,2,\cdots,m \\ Pos\{g_r(x,\xi) \leq 0\} \geq \theta_r, & r = 1,2,\cdots,p \\ x \in X. \end{cases} \end{cases}$$

This is coincident to the fuzzy CCM introduced in Chapter 2.

3.4.2 Linear Fu-Ra CCM and Surrogate Worth Trade-Off Method

Let's consider the linear Fu-Ra CCM:

$$\begin{cases} \max\ \{f_i,\ i = 1,2,\cdots,m\} \\ \text{s.t.} \begin{cases} Ch\{\tilde{\bar{c}}_i^T x \geq f_i\}(\gamma_i) \geq \delta_i,\ i = 1,2,\cdots,m \\ Ch\{\tilde{\bar{e}}_r^T x \leq \tilde{\bar{b}}_r\}(\eta_r) \geq \theta_r,\ r = 1,2,\cdots,p \\ x \geq 0, \end{cases} \end{cases} \tag{3.45}$$

where $\tilde{\bar{c}}_i, \tilde{\bar{e}}_r, \tilde{\bar{b}}_r$ are fuzzy random variables.

3.4.2.1 Crisp Equivalent Model

In order to solve the model (3.45), one feasible method is that we turn the chance-constraints to crisp equivalents.

Pr-Pos Constrained Multi-objective Linearity Model

Let's consider the linear Fu-Ra CCM based on Pr-Pos,

$$\begin{cases} \max\ \{f_1,f_2,\cdots,f_m\} \\ \text{s.t.} \begin{cases} Pr\{\omega|Pos\{\tilde{\bar{c}}_i(\omega)^T x \geq f_i\} \geq \delta_i\} \geq \gamma_i, & i = 1,2,\cdots,m \\ Pr\{\omega|Pos\{\tilde{\bar{e}}_r(\omega)^T x \leq \tilde{\bar{b}}_r(\omega)\} \geq \theta_r\} \geq \eta_r, & r = 1,2,\cdots,p \\ x \geq 0, \end{cases} \end{cases} \tag{3.46}$$

where $\delta_i, \gamma_i, \theta_r, \eta_r \in [0,1]$ are the predetermined confidence level, $Pos\{\cdot\}$ denotes the possibility of the fuzzy events in $\{\cdot\}$, and $Pr\{\cdot\}$ denotes the probability of the random events in $\{\cdot\}$.

Assume that $\tilde{c}_{ij}$, $\tilde{e}_{rj}$ and $\tilde{b}$ are LR fuzzy random variables, we give the following two theorems to transform the chance-constrained model into its crisp model.

Theorem 3.6. *Assume that $\tilde{c}_{ij}$ is LR fuzzy random variable, for any $\omega \in \Omega$, the membership function of $\tilde{c}_{ij}(\omega)$ is*

$$\mu_{\tilde{c}_{ij}(\omega)}(t) = \begin{cases} L(\frac{c_{ij}(\omega)-t}{\alpha_{ij}^c}), & t \le c_{ij}(\omega), \alpha_{ij}^c > 0 \\ R(\frac{t-c_{ij}(\omega)}{\beta_{ij}^c}), & t \ge c_{ij}(\omega), \beta_{ij}^c > 0 \end{cases} \quad \omega \in \Omega, \qquad (3.47)$$

where the random vector $(c_{ij}(\omega))_{n\times 1} = (c_{i1}(\omega), c_{i2}(\omega), \cdots, c_{in}(\omega))^{\mathrm{T}}$ is normally distributed, the mean vector is d_i^c, the covariance matrix is V_i^c, denoted by $(c_{ij}(\omega))_{n\times 1} \sim N(d_i^c, V_i^c)$, α_{ij}^c and β_{ij}^c are the left and right spread of $\tilde{c}_{ij}(\omega)$, $i = 1, 2, \cdots, m$, $j = 1, 2, \cdots, n$, the reference function $L, R : [0,1] \to [0,1]$ satisfies that $L(1) = R(1) = 0$, $L(0) = R(0) = 1$, and it is monotone function. Then $Pr\{\omega | Pos\{\tilde{c}_i(\omega)^{\mathrm{T}}x \ge f_i\} \ge \delta_i\} \ge \gamma_i$ is equivalent to

$$f_i \le R^{-1}(\delta_i)\beta_i^{c\mathrm{T}}x + d_i^{c\mathrm{T}}x + \Phi^{-1}(1-\gamma_i)\sqrt{x^{\mathrm{T}}V_i^c x}, \quad i = 1, 2, \cdots, m, \qquad (3.48)$$

where Φ are standard normally distributed, $\delta_i, \gamma_i \in [0,1]$ are predetermined confidence level.

Proof. For certain $w \in \Omega$, $\tilde{c}_{ij}(\omega)$ are fuzzy number, its membership function is $\mu_{\tilde{c}_{ij}(\omega)}(t)$. By extension principle, the membership function of fuzzy number $\tilde{c}_i(\omega)^{\mathrm{T}}x$ is

$$\mu_{\tilde{c}_i(\omega)^{\mathrm{T}}x}(r) = \begin{cases} L(\frac{c_i(\omega)^{\mathrm{T}}x-r}{\alpha_i^{c\mathrm{T}}x}), & r \le c_i(\omega)^{\mathrm{T}}x \\ R(\frac{r-c_i(\omega)^{\mathrm{T}}x}{\beta_i^{c\mathrm{T}}x}), & r \ge c_i(\omega)^{\mathrm{T}}x \end{cases} \quad i = 1, 2, \cdots, m. \qquad (3.49)$$

For convenience, denote $\tilde{c}_{ij}(\omega) = (c_{ij}(\omega), \alpha_{ij}^c, \beta_{ij}^c)_{\mathrm{LR}}$, $\tilde{c}_i(\omega)^{\mathrm{T}}x = (c_i(\omega)^{\mathrm{T}}x, \alpha_i^{c\mathrm{T}}x, \beta_i^{c\mathrm{T}}x)_{\mathrm{LR}}$.

Since $(c_{ij}(\omega))_{n\times 1} \sim N(d_i^c, V_i^c)$, so $c_i(\omega)^{\mathrm{T}}x \sim N(d_i^{c\mathrm{T}}x, x^{\mathrm{T}}V_i^c x)$. According to Lemma 2.2 we can get

$$Pos\{\tilde{c}_i(\omega)^{\mathrm{T}}x \ge f_i\} \ge \delta_i \Leftrightarrow c_i(\omega)^{\mathrm{T}}x + R^{-1}(\delta_i)\beta_i^{c\mathrm{T}}x \ge f_i, \quad i = 1, 2, \cdots, m.$$

So for predetermined level $\delta_i, \gamma_i \in [0,1]$,

$$Pr\{\omega|Pos\{\tilde{\bar{c}}_i(\omega)^{\mathrm{T}}x \geq f_i\} \geq \delta_i\} \geq \gamma_i$$
$$\Leftrightarrow Pr\{\omega|c_i(\omega)^{\mathrm{T}}x \geq f_i - R^{-1}(\delta_i)\beta_i^{c\mathrm{T}}x\} \geq \gamma_i$$
$$\Leftrightarrow Pr\{\omega|\frac{c_i(\omega)^{\mathrm{T}}x - d_i^{c\mathrm{T}}x}{\sqrt{x^{\mathrm{T}}V_i^c x}} \geq \frac{f_i - R^{-1}(\delta_i)\beta_i^{c\mathrm{T}}x - d_i^{c\mathrm{T}}x}{\sqrt{x^{\mathrm{T}}V_i^c x}}\} \geq \gamma_i$$
$$\Leftrightarrow \Phi\left(\frac{f_i - R^{-1}(\delta_i)\beta_i^{c\mathrm{T}}x - d_i^{c\mathrm{T}}x}{\sqrt{x^{\mathrm{T}}V_i^c x}}\right) \leq 1 - \gamma_i$$
$$\Leftrightarrow f_i \leq R^{-1}(\delta_i)\beta_i^{c\mathrm{T}}x + d_i^{c\mathrm{T}}x + \Phi^{-1}(1 - \gamma_i)\sqrt{x^{\mathrm{T}}V_i^c x}.$$

The proof is completed.   □

Similarly, the chance-constraint $Pr\{\omega|Pos\{\tilde{\bar{e}}_r(\omega)^{\mathrm{T}}x \leq \tilde{\bar{b}}_r(\omega)\} \geq \theta_r\} \geq \eta_r$ can also be transformed into crisp equivalent constraint.

Theorem 3.7. *Assume that $\tilde{\bar{e}}_{rj}$ and $\tilde{\bar{b}}_r$ are LR fuzzy random variables, for $\omega \in \Omega$, the membership function of $\tilde{\bar{e}}_{rj}(\omega)$ are $\tilde{\bar{b}}_r(\omega)$ are*

$$\mu_{\tilde{\bar{e}}_{rj}(\omega)}(t) = \begin{cases} L(\frac{e_{rj}(\omega)-t}{\alpha_{rj}^e}), & t \leq e_{rj}(\omega), \alpha_{rj}^e > 0 \\ R(\frac{t-e_{rj}(\omega)}{\beta_{rj}^e}), & t \geq e_{rj}(\omega), \beta_{rj}^e > 0, \end{cases} \quad (3.50)$$

$$\mu_{\tilde{\bar{b}}_r(\omega)}(t) = \begin{cases} L(\frac{b_r(\omega)-t}{\alpha_r^b}), & t \leq b_r(\omega), \alpha_r^b > 0 \\ R(\frac{t-b_r(\omega)}{\beta_r^b}), & t \geq b_r(\omega), \beta_r^b > 0, \end{cases} \quad (3.51)$$

where $(e_{rj}(\omega))_{n \times 1} = (e_{r1}(\omega), e_{r2}(\omega), \cdots, e_{rn}(\omega))^{\mathrm{T}} \sim N(d_r^e, V_r^e)$, $b_r(\omega) \sim N(d_r^b, (\sigma_r^b)^2)$, α_{rj}^e and β_{rj}^e are left and right spread of $\tilde{\bar{e}}_{rj}(\omega)$, α_r^b and β_r^b are the left and right spread of $\tilde{\bar{b}}_r(\omega)$, $r = 1, 2, \cdots, p$, $j = 1, 2, \cdots, n$, the reference function $L, R : [0,1] \to [0,1]$ are monotone decreasing continuous function, and it satisfies $L(1) = R(1) = 0, L(0) = R(0) = 1$.

For any $j = 1, 2, \cdots, n$, If $e_{rj}(\omega)$ and $b_r(\omega)$ are independent random variables. Then $Pr\{\omega|Pos\{\tilde{\bar{e}}_r(\omega)^{\mathrm{T}}x \leq \tilde{\bar{b}}_r(\omega)\} \geq \theta_r\} \geq \eta_r$ is equivalent to

$$R^{-1}(\theta_r)\beta_r^b + L^{-1}(\theta_r)\alpha_r^{e\mathrm{T}}x - (d_r^{e\mathrm{T}}x - d_r^b) - \Phi^{-1}(\eta_r)\sqrt{x^{\mathrm{T}}V_r^e x + (\sigma_r^b)^2} \geq 0.$$

When $\eta_r \geq 0.5$,

$$X := \{x \in \mathbf{R}^n | r\{\omega|Pos\{\tilde{\bar{e}}_r(\omega)^{\mathrm{T}}x \leq \tilde{\bar{b}}_r(\omega)\} \geq \theta_r\} \geq \eta_r, r = 1, 2, \cdots, p; \ x \geq 0\}$$

is a convex set.

Proof. According to Lemma 2.2, we have

$$Pos\{\tilde{\bar{e}}_r(\omega)^{\mathrm{T}}x \leq \tilde{\bar{b}}_r(\omega)\} \geq \theta_r \Leftrightarrow b_r(\omega) + R^{-1}(\theta_r)\beta_r^b \geq e_r(\omega)^{\mathrm{T}}x - L^{-1}(\theta_r)\alpha_r^{e\mathrm{T}}x.$$

Since $(e_{rj}(\omega))_{n\times 1} \sim N(d_r^e, V_r^e)$, $b_r(\omega) \sim N(d_r^b, (\sigma_r^b)^2)$, so

$$Pr\{\omega|Pos\{\tilde{\tilde{e}}_r(\omega)^{\mathrm{T}}x \leq \tilde{\tilde{b}}_r(\omega)\} \geq \theta_r\} \geq \eta_r$$
$$\Leftrightarrow Pr\{\omega|e_r(\omega)^{\mathrm{T}}x - b_r(\omega) \leq R^{-1}(\theta_r)\beta_r^b + L^{-1}(\theta_r)\alpha_r^{e\mathrm{T}}x\} \geq \eta_r$$
$$\Leftrightarrow \Phi\left(\frac{R^{-1}(\theta_r)\beta_r^b + L^{-1}(\theta_r)\alpha_r^{e\mathrm{T}}x - (d_r^{e\mathrm{T}}x - d_r^b)}{\sqrt{x^{\mathrm{T}}V_r^e x + (\sigma_r^b)^2}}\right) \geq \eta_r$$
$$\Leftrightarrow g_r(x) \geq 0,$$

Where

$$g_r(x) = R^{-1}(\theta_r)\beta_r^b + L^{-1}(\theta_r)\alpha_r^{e\mathrm{T}}x - (d_r^{e\mathrm{T}}x - d_r^b) - \Phi^{-1}(\eta_r)\sqrt{x^{\mathrm{T}}V_r^e x + (\sigma_r^b)^2}.$$

If $\eta_r \geq 0.5$, it is clear that $\Phi^{-1}(\eta_r) \geq 0$, $g_r(x)$ is a concave function. So X is a convex set.

The proof is completed. $\qquad\qquad\qquad\qquad\qquad\qquad\qquad\qquad\qquad\qquad$ □

By Theorems 3.6 and 3.7, the model (3.46) is equivalent to the following multi-objective model,

$$\begin{cases} \max\{f_1, f_2, \cdots, f_m\} \\ \text{s.t.} \begin{cases} f_i \leq R^{-1}(\delta_i)\beta_i^{c\mathrm{T}}x + d_i^{c\mathrm{T}}x + \Phi^{-1}(1-\gamma_i)\sqrt{x^{\mathrm{T}}V_i^c x}, \ i = 1,2,\cdots,m \\ x \in X \end{cases} \end{cases} \quad (3.52)$$

or

$$\begin{cases} \max \{H_1(x), H_2(x), \cdots, H_m(x)\} \\ \text{s.t. } x \in X, \end{cases} \qquad\qquad\qquad (3.53)$$

where

$$H_i(x) := R^{-1}(\delta_i)\beta_i^{c\mathrm{T}}x + d_i^{c\mathrm{T}}x + \Phi^{-1}(1-\gamma_i)\sqrt{x^{\mathrm{T}}V_i^c x}, \quad i = 1,2,\cdots,m. \quad (3.54)$$

Pr-Nec Constrained Multi-objective Linearity Model

Also, let's consider the linear Fu-Ra CCM based on Pr-Nec,

$$\begin{cases} \max [f_1, f_2, \cdots, f_m] \\ \text{s.t.} \begin{cases} Pr\{\omega|Nec\{\tilde{\tilde{c}}_i(\omega)^{\mathrm{T}}x \geq f_i\} \geq \delta_i\} \geq \gamma_i, \quad i = 1,2,\cdots,m \\ Pr\{\omega|Nec\{\tilde{\tilde{e}}_r(\omega)^{\mathrm{T}}x \leq \tilde{\tilde{b}}_r(\omega)\} \geq \theta_r\} \geq \eta_r, \quad r = 1,2,\cdots,p \\ x \geq 0, \end{cases} \end{cases} \quad (3.55)$$

where $\delta_i, \gamma_i, \theta_r, \eta_r \in [0,1]$ are the predetermined confidence level, $Nec\{\cdot\}$ denotes the necessity of the fuzzy events in $\{\cdot\}$, and $Pr\{\cdot\}$ denotes the probability of the random events in $\{\cdot\}$.

Similar to Theorems 3.6 and 3.7, The following theorems 3.8 and 3.9 presents the crisp equivalents of the chance constraints of model (3.55).

Theorem 3.8. *Assume that the fuzzy random variable* $\tilde{\bar{c}}_{ij}$ *is as same as the assumption in Theorem 3.6,* $i = 1, 2, \cdots, m$, $j = 1, 2, \cdots, n$. *For confidence level* $\delta_i, \gamma_i \in [0, 1]$, $i = 1, 2, \cdots, m$, *we have*

$$Pr\{\omega | Nec\{\tilde{\bar{c}}_i(\omega)^T x \geq f_i\} \geq \delta_i\} \geq \gamma_i$$
$$\Leftrightarrow f_i \leq d_i^{cT} x - L^{-1}(1 - \delta_i)\alpha_i^{cT} x + \Phi^{-1}(1 - \gamma_i)\sqrt{x^T V_i^c x}.$$

Proof. Fr certain $\omega \in \Omega$, by Lemma 2.2 we have

$$Nec\{\tilde{\bar{c}}_i(\omega)^T x \geq f_i\} \geq \delta_i \Leftrightarrow c_i(w)^T x - L^{-1}(1 - \delta_i)\alpha_i^{cT} x \geq f_i, \quad i = 1, 2, \cdots, m.$$

Since $c_i(\omega) \sim N(d_i^c, V_i^c)$, so $c_i(w)^T x \sim N(d_i^{cT} x, x^T V_i^c x)$.

$$Pr\{\omega | Nec\{\tilde{\bar{c}}_i(\omega)^T x \geq f_i\} \geq \delta_i\} \geq \gamma_i$$
$$\Leftrightarrow Pr\{\omega | c_i(\omega)^T x \geq f_i + L^{-1}(1 - \delta_i)\alpha_i^{cT} x\} \geq \gamma_i$$
$$\Leftrightarrow Pr\left\{w \Big| \frac{c_i(\omega)^T x - d_i^{cT} x}{\sqrt{x^T V_i^c x}} \geq \frac{f_i + L^{-1}(1 - \delta_i)\alpha_i^{cT} x - d_i^{cT} x}{\sqrt{x^T V_i^c x}}\right\} \geq \gamma_i$$
$$\Leftrightarrow \Phi\left(\frac{f_i + L^{-1}(1 - \delta_i)\alpha_i^{cT} x - d_i^{cT} x}{\sqrt{x^T V_i^c x}}\right) \leq 1 - \gamma_i$$
$$\Leftrightarrow f_i \leq d_i^{cT} x - L^{-1}(1 - \delta_i)\alpha_i^{cT} x + \Phi^{-1}(1 - \gamma_i)\sqrt{x^T V_i^c x}.$$

The proof is completed.   □

Theorem 3.9. *Assume that the fuzzy random variables* $\tilde{\bar{e}}_{rj}$ *and* $\tilde{\bar{b}}_r$ *are as same as the assumption in Theorem 3.7,* $j = 1, 2, \cdots, n$, $r = 1, 2, \cdots, p$. *Then for certain confidence level* $\theta_r, \eta_r \in [0, 1]$, $r = 1, 2, \cdots, p$, *we have*

$$Pr\{\omega | Nec\{\tilde{\bar{e}}_r(\omega)^T x \leq \tilde{\bar{b}}_r(\omega)\} \geq \theta_r\} \geq \eta_r$$
$$\Leftrightarrow \Phi^{-1}(1 - \eta_r)\sqrt{x^T V_r^e x + (\sigma_r^b)^2} - L^{-1}(1 - \theta_r)\alpha_r^b - R^{-1}(\theta_r)\beta_r^{eT} x + (d_r^b - d_r^{eT} x) \geq 0.$$

When $\eta_r \geq 0.5$,

$$X' := \{x \in \mathbf{R}^n | Pr\{\omega | Nec\{\tilde{\bar{e}}_r(\omega)^T x \leq \tilde{\bar{b}}_r(\omega)\} \geq \theta_r\} \geq \eta_r, r = 1, 2, \cdots, p; x \geq 0\}$$

is a convex set.

The proof is similar to Theorem 3.7, so here we skip the proof of Theorem 3.9 over.

By Theorems 3.8 and 3.9, the model (3.55) is equivalent to the following multi-objective problem,

$$\begin{cases} \max \ [f_1, f_2, \cdots, f_m] \\ \text{s.t.} \ \begin{cases} f_i \leq d_i^{cT} x - L^{-1}(1 - \delta_i)\alpha_i^{cT} x + \Phi^{-1}(1 - \gamma_i)\sqrt{x^T V_i^c x}, \ i = 1, 2, \cdots, m, \\ x \in X' \end{cases} \end{cases}$$

$$(3.56)$$

or

$$\begin{cases} \max \ [G_1(x), G_2(x), \cdots, G_m(x)] \\ \text{s.t.} \ x \in X', \end{cases} \quad (3.57)$$

where $G_i(x) := d_i^{cT} x - L^{-1}(1 - \delta_i)\alpha_i^{cT} x + \Phi^{-1}(1 - \gamma_i)\sqrt{x^T V_i^c x}$, $i = 1, 2, \cdots, m$.

3.4.2.2 Surrogate Worth Trade-Off Method

The surrogate worth trade-off method, which is called SWT method for short, was proposed by Haimes et al. [399] in 1974 to solve the multi-objective programming problem. It can be applied to continuous variables, objective functions and constraints which can be differentiated twice.

In its original version, SWT is, in principle, noninteractive and assumes continuous variables and twice differentiable objective functions and constraints. It consists of four steps: (1) generate a representative subset of efficient solutions, (2) obtain relevant trade-off information for each generated solution, (3) interact with DM t obtain information about preference expressed in terms of worth, and (4) retrieve the best-compromise solution from the information obtained.

Take the problem (3.53) as an example and list the detailed steps according to [400] as follows:

Step 1: Generation of a Representative Subset of Efficient Solutions. The ε-constraint method is recommended to obtain the representative subset of efficient solutions. Without loss of generality, we choose a reference objective H_1 and formulate the ε-constraint problem:

$$\begin{cases} \max H_1(x) \\ \text{s.t.} \begin{cases} H_i(x) \geq \varepsilon_i, i = 2, 3, \cdots, m \\ x \in X. \end{cases} \end{cases} \tag{3.58}$$

Although there is no rule to specify which objective should be chosen as a reference, the most important objective is recommended. To guarantee that the ε-constraint problem has feasible solution, a reasonable ε_i should be selected, usually, in the range $[a_i, b_i]$, where $a_i = \min_{x \in X} H_i(x)$ and $b_i = \max_{x \in X} H_i(x)$.

Step 2: Obtaining Trade-off Information. In the process of solving the problem (3.58), the trade-off information can easily be obtained merely by observing the optimal Kuhn-Tucker multipliers corresponding to the ε-constraints. Let these multipliers be denoted by $\lambda_{1i}(x(\varepsilon))$. If $\lambda_{1k}(x(\varepsilon)) > 0 (k = 1, 2, \cdots, m)$, then the efficient surface in the objective function space around the neighborhood of $H^\varepsilon = (H_1(x(\varepsilon)), H_2(x(\varepsilon)), \cdots, H_m(x(\varepsilon)))^T$ can be represented by $H_1 = (H_1, H_2, \cdots, H_m)$ and

$$\lambda_{1k}(x(\varepsilon)) = -\left.\frac{\partial H_1}{\partial H_k}\right|_H = H^\varepsilon, \ k = 2, 3, \cdots, m. \tag{3.59}$$

Thus each $\lambda_{1k}(x(\varepsilon))$ represents the efficient partial trade-off rate between H_1 and H_k at H^ε when all other objective are held fixed at their respective values at $x(\varepsilon)$. The adjective "efficient" is used to signify that after the trade-off is made the resulting point remains on the efficient surface. The detail can be referred to [400].

Step 3: Interacting with the Decision Maker to Elicit Preference. DM is supplied with trade-off information from *Step 2* and the levels of all criteria. He then expresses his ordinal preference on whether or not (and by how much) he would like to make such a trade at that level. Haimes et al. [400] constructed the following surrogate worth function: DM is asked "How much would you like to improve H_1 by $\lambda_{1k}(x(\varepsilon))$ units per one-unit degradation of H_k while all other objective remain fixed at $H_l(r(\varepsilon))$, $l \neq 1, k$? Indicate your preference on a scale of -10 to 10, where the values have the following meaning:

(1). +10 means you have the greatest desire to improve improve H_1 by $\lambda_{1k}(x(\varepsilon))$ units per one-unit degradation of H_k,

(2). 0 means you are indifferent about the trade,

(3). -10 means you have the greatest desire to degrade improve H_1 by $\lambda_{1k}(x(\varepsilon))$ units per one-unit improvement in H_k,

Values between -10 and 0, and 0 and 10 show proportional desire to make the trade."

DM's response is recorded as $w_{1k}(x(\varepsilon))$, called the *surrogate worth* of the trade-off between H_1 and H_k at the efficient solution $x(\varepsilon)$. At a particular efficient solution, there will be $m-1$ questions to obtain $w_{1k}(x(\varepsilon)), k = 2, 3, \cdots, m$.

Step 4: Retrieving the Best-Compromise Solution. If there exists an efficient solution $x(\varepsilon_0)$ such that

$$w_{1k}(x(\varepsilon_0)) = 0, \ k = 2, 3, \cdots, m, \tag{3.60}$$

the DM has obtained a best-compromise solution. Thus equation (3.60) is the best-compromise condition of $x(\varepsilon_0)$. If there is such $x(\varepsilon_0)$ in the representative set, then stop and output $x(\varepsilon_0)$. Otherwise we use multiple regression to construct the surrogate worth function as follows,

$$w_{1k} = w_{1k}(H_1, H_2, \cdots, H_m), \ k = 2, 3, \cdots, m.$$

Then the system of equations

$$w_{1k}(H_1, H_2, \cdots, H_m) = 0, \ k = 2, 3, \cdots, m,$$

is solved to determine $(H_2^*, \cdots, H_m^*)$.

Let $\varepsilon_{0k} = H_k^*(k = 2, \cdots, m)$, $\varepsilon_0 = (\varepsilon_{02}, \cdots, \varepsilon_{0m})^T$. The best-compromise solution $x(\varepsilon_0)$ is then found by solving the problem (3.58).

3.4.2.3 Numerical Example

We use the Interactive fuzzy satisfied method to solve the equivalent form (4.59) of the linear Fu-Ra CCM.

Usually, people will require that the value confidence level γ_i is no less than 0.5, so we will not consider the situation of $\gamma_i < 0.5$, and we only consider the situation of $\gamma_i \geq 0.5$. If $\gamma_i \geq 0.5$, then $\Phi^{-1}(1 - \gamma_i) \leq 0$. By equation (3.54), $H_i(x), i = 1, 2, \cdots, m$

are concave function. So $\max_{x \in X} H_i(x)$ is a convex programming, it's easy to find the maximal value point. For model $\min_{x \in X} H_i(x)$, its optimal solution should be get at the boundary of the convex set X. If there exists no solution of $\max_{x \in X} H_i(x)$ or $\min_{x \in X} H_i(x)$, or $H_i^1 = \infty$, $H_i^0 = -\infty$, the decision maker may set the value of H_i^1, H_i^0 subjectively.

For model (3.53), by model (3.29) in Interactive fuzzy satisfied method, we get

$$
\begin{cases}
\min \lambda \\
\text{s.t.} \begin{cases}
R^{-1}(\delta_i)\beta_i^{cT}x + d_i^{cT}x + \Phi^{-1}(1 - \gamma_i)\sqrt{x^T V_i^c x} \geq H_i^0 + \\
\qquad\qquad (\bar{\mu}_i - \lambda)(H_i^1 - H_i^0), \quad i = 1, 2, \cdots, m \\
0 \leq \lambda \leq 1, x \in X.
\end{cases}
\end{cases}
\tag{3.61}
$$

When $\gamma_i \geq 0.5$, $i = 1, 2, \cdots, m$, model (3.29) is convex programming, so it's easy to get the global optimal solution.

According to Theorem 3.5, the optimal solution of model (3.61) is a efficient solution of model (3.53), thereby, it is a $\gamma_i - Pr\ \delta_i - Pos$ fuzzy random efficient solution of model (3.46).

For model (3.57), by model (3.29) in Interactive fuzzy satisfied method, we get

$$
\begin{cases}
\min \lambda \\
\text{s.t.} \begin{cases}
d_i^{cT}x - L^{-1}(1 - \delta_i)\alpha_i^{cT}x + \Phi^{-1}(1 - \gamma_i)\sqrt{x^T V_i^c x} \geq G_i^0 + \\
\qquad\qquad (\bar{\mu}_i - \lambda)(G_i^1 - G_i^0), \quad i = 1, 2, \cdots, m \\
x \in X'.
\end{cases}
\end{cases}
\tag{3.62}
$$

When $\gamma_i \geq 0.5$, $\Phi^{-1}(1 - \gamma_i)\sqrt{x^T V_i^c x}$ is a concave function,

According to Theorem 3.5, the optimal solution of model (3.62) is a efficient solution of model (3.57), thereby, it is a $\gamma_i - Pr\ \delta_i - Nec$ fuzzy random efficient solution of model (3.55).

Example 3.7. We use the following example to illustrate this interactive fuzzy satisfied method. Consider the following Pr-Pos constrained multi-objective linear model:

$$
\begin{cases}
\max\ [f_1, f_2] \\
\text{s.t.} \begin{cases}
Pr\{\omega | Pos\{\tilde{\bar{\xi}}_1 x_1 + \tilde{\bar{\xi}}_2 x_2 + \tilde{\bar{\xi}}_3 x_3 + \tilde{\bar{\xi}}_4 x_4 + \tilde{\bar{\xi}}_5 x_5 \geq f_1\} \geq \delta_1\} \geq \gamma_1 \\
Pr\{\omega | Pos\{c_1\tilde{\bar{\xi}}_6 x_1 + c_2\tilde{\bar{\xi}}_7 x_2 + c_3\tilde{\bar{\xi}}_8 x_3 + c_4\tilde{\bar{\xi}}_9 x_4 + c_5\tilde{\bar{\xi}}_{10} x_5 \geq f_2\} \geq \delta_2\} \geq \gamma_2 \\
x_1 + x_2 + x_3 + x_4 + x_5 \leq 350 \\
x_1 + x_2 + x_3 + x_4 + x_5 \geq 300 \\
4x_1 + 2x_2 + 1.5x_3 + x_4 + 2x_5 \leq 1085 \\
x_1 + 4x_2 + 2x_3 + 5x_4 + 3x_5 \leq 660 \\
x_1 \geq 20, x_2 \geq 20, x_3 \geq 20, x_4 \geq 20, x_5 \geq 20,
\end{cases}
\end{cases}
\tag{3.63}
$$

where $c = (c_1, c_2, c_3, c_4, c_5) = (1.2, 0.5, 1.3, 0.8, 0.9)$,

$$
\begin{aligned}
\xi_1 &= (\rho_1, 3, 3)_{\mathrm{LR}}, \text{with } \rho_1 \sim N(113, 1), \\
\xi_2 &= (\rho_2, 8, 8)_{\mathrm{LR}}, \text{with } \rho_2 \sim N(241, 4), \\
\xi_3 &= (\rho_3, 3, 3)_{\mathrm{LR}}, \text{with } \rho_3 \sim N(87, 1), \\
\xi_4 &= (\rho_4, 7, 7)_{\mathrm{LR}}, \text{with } \rho_4 \sim N(56, 2), \\
\xi_5 &= (\rho_5, 5, 5)_{\mathrm{LR}}, \text{with } \rho_5 \sim N(92, 1), \\
\xi_6 &= (\rho_6, 10, 10)_{\mathrm{LR}}, \text{with } \rho_6 \sim N(628, 1), \\
\xi_7 &= (\rho_7, 7, 7)_{\mathrm{LR}}, \text{with } \rho_7 \sim N(143, 2), \\
\xi_8 &= (\rho_8, 12, 12)_{\mathrm{LR}}, \text{with } \rho_8 \sim N(476, 2), \\
\xi_9 &= (\rho_9, 5, 5)_{\mathrm{LR}}, \text{with } \rho_9 \sim N(324, 2), \\
\xi_{10} &= (\rho_{10}, 8, 8)_{\mathrm{LR}}, \text{with } \rho_{10} \sim N(539, 2),
\end{aligned}
$$

where ρ_i $(i = 1, 2, \cdots, 10)$ are independent random variables.

Let $\delta_i = \gamma_i = 0.9$, then $R^{-1}(\delta_i) = 0.1$, $\Phi^{-1}(1 - \gamma_i) = -1.28$, $i = 1, 2$. According to (3.53), we can get model (3.63) is equivalent to

$$
\begin{cases}
\max H_1(x) = 0.1(3x_1 + 8x_2 + 3x_3 + 7x_4 + 5x_5) + (113x_1 + 2411x_2 + 87x_3 + 56x_4 \\
\qquad\qquad + 92x_5) - 1.28\sqrt{x_1^2 + 4x_2^2 + x_3^2 + 2x_4^2 + x_5^2} \\
\max H_2(x) = 0.1(12x_1 + 3.5x_2 + 15.6x_3 + 4x_4 + 7.2x_5) + (753.6x_1 + 71.5x_2 + \\
\qquad\qquad 618.8x_3 + 259.2x_4 + 485.1x_5) - 1.28\sqrt{x_1^2 + 2x_2^2 + 2x_3^2 + 2x_4^2 + 2x_5^2} \\
\text{s.t.}
\begin{cases}
x_1 + x_2 + x_3 + x_4 + x_5 \le 350 \\
x_1 + x_2 + x_3 + x_4 + x_5 \ge 300 \\
4x_1 + 2x_2 + 1.5x_3 + x_4 + 2x_5 \le 1085 \\
x_1 + 4x_2 + 2x_3 + 5x_4 + 3x_5 \le 660 \\
x_1 \ge 20, x_2 \ge 20, x_3 \ge 20, x_4 \ge 20, x_5 \ge 20.
\end{cases}
\end{cases}
$$

$$(3.64)$$

The computation of H_i^0 and H_i^1 $(i = 1, 2)$ is as follows:

$$
H_1^1 = 43764.82, \quad H_1^0 = 30079.87, \quad H_2^1 = 225511.5, \quad H_2^0 = 158125.6.
$$

So we can get the membership function of H_1 and H_2 as follows,

$$
\mu_1(H_1(x)) = \begin{cases}
1, & H_1(x) > 43764.82 \\
\frac{H_1(x) - 30079.87}{13684.95}, & 30079.87 \le H_1(x) \le 43764.82 \\
0, & H_1(x) < 30079.87,
\end{cases}
$$

$$
\mu_2(H_2(x)) = \begin{cases}
1, & H_2(x) > 225511.5 \\
\frac{H_2(x) - 158125.6}{67385.9}, & 158125.6 \le H_2(x) \le 225511.5 \\
0, & H_2(x) < 158125.6.
\end{cases}
$$

Then we compute the following model to get the interactive satisfied solution,

$$
\begin{cases}
\min \lambda \\
\text{s.t.}
\begin{cases}
0.1(3x_1 + 8x_2 + 3x_3 + 7x_4 + 5x_5) + (113x_1 + 241x_2 + 87x_3 + 56x_4 \\
\quad + 92x_5) - 1.28\sqrt{x_1^2 + 4x_2^2 + x_3^2 + 2x_4^2 + x_5^2} \geq H_1^0 + (\bar{\mu}_1 - \lambda)(H_1^1 - H_1^0) \\
0.1(12x_1 + 3.5x_2 + 15.6x_3 + 4x_4 + 7.2x_5) + (753.6x_1 + 71.5x_2 \\
\quad + 618.8x_3 + 259.2x_4 + 485.1x_5) - 1.28\sqrt{x_1^2 + 2x_2^2 + 2x_3^2 + 2x_4^2 + 2x_5^2} \\
\quad \geq H_2^0 + (\bar{\mu}_2 - \lambda)(H_2^1 - H_2^0) \\
x_1 + x_2 + x_3 + x_4 + x_5 \leq 350 \\
x_1 + x_2 + x_3 + x_4 + x_5 \geq 300 \\
4x_1 + 2x_2 + 1.5x_3 + x_4 + 2x_5 \leq 1085 \\
x_1 + 4x_2 + 2x_3 + 5x_4 + 3x_5 \leq 660 \\
x_1 \geq 20, x_2 \geq 20, x_3 \geq 20, x_4 \geq 20, x_5 \geq 20 \\
0 \leq \lambda \leq 1.
\end{cases}
\end{cases}
$$

(3.65)

After solving the model (3.65), we can get the satisfied solution of model (3.63), which are listed in Table 3.3.

Table 3.3 Employ the interactive fuzzy satisfied method based on possibility($\gamma_i = 0.9, \delta_i = 0.9$)

$\bar{\mu}_1$	$\bar{\mu}_2$	H_1	H_2	$\mu_1(H_1)$	$\mu_2(H_2)$	x_1	x_2	x_3	x_4	x_5	λ
1	1	41476.7	214244.8	0.833	0.833	216.1	39.6	54.3	20.0	20.0	0.167
0.95	1	41093.2	215725.6	0.805	0.855	216.6	37.0	56.4	20.0	20.0	0.145
1	0.95	41860.2	212763.6	0.861	0.811	215.6	42.2	52.3	20.0	20.0	0.139
0.90	1	40709.6	217206.0	0.777	0.877	217.1	34.4	58.4	20.0	20.0	0.123
0.85	1	40325.9	218686.0	0.749	0.899	217.6	31.9	60.5	20.0	20.0	0.101
0.80	1	39942.1	220165.5	0.721	0.921	218.1	29.3	62.5	20.0	20.0	0.079

The first line of Table 3.3 lists each reference value value of membership function $\mu_i(H_1)$, when the initialized membership function is 1, the value of objective function $H_i(x)$, and its corresponding solution x. If the decision maker hopes that improve $H_2(x)$ on the basis of sacrifice $H_1(x)$. We may consider reset the reference value of membership function $(\bar{\mu}_1, \bar{\mu}_2)$, e.g., we set $(\bar{\mu}_1, \bar{\mu}_2) = (0.95, 1)$, or $(\bar{\mu}_1, \bar{\mu}_2) = (1, 0.95)$. The corresponding result are listed in the second and third lines. Suppose that when the reference value of membership function is $(\bar{\mu}_1, \bar{\mu}_2) = (0.80, 1)$, the decision maker is satisfied, then the interactive process is stopped, so we obtain the 0.9-Pr 0.9-Pos satisfied solution is $x^* = (218.1, 29.3, 62.5, 20.0, 20.0)^T$, and the corresponding value of objective function is $(f_1^*, f_2^*) = (39942.1, 220165.5)$.

Furthermore, the decision maker can also verify the value of H_i^1 and H_i^0 ($i = 1, 2$), and built new membership function and obtain other satisfied solution.

If the decision maker is comparatively pessimistic, we use *Nec* measure to substitute the *Pos*.

Let $\delta_i = \gamma_i = 0.9$, then $L^{-1}(1 - \delta_i) = 0.9$, $\Phi^{-1}(1 - \gamma_i) = -1.28, i = 1, 2$. According to (3.57), we can get model (3.63) is equivalent to

$$
\begin{cases}
\max \ G_1(x) = (113x_1 + 241x_2 + 87x_3 + 56x_4 + 92x_5) - 0.9(3x_1 + 8x_2 + 3x_3 \\
\qquad\qquad + 7x_4 + 5x_5) - 1.28\sqrt{x_1^2 + 4x_2^2 + x_3^2 + 2x_4^2 + x_5^2} \\
\max \ G_2(x) = +(753.6x_1 + 71.5x_2618.8x_3 + 259.2x_4 + 485.1x_5) - 0.9(12x_1 \\
\qquad\qquad + 3.5x_2 \ | \ 15.6x_3 \ | \ 1x_4 \ | \ 7.2x_5) \\
\qquad\qquad - 1.28\sqrt{x_1^2 + 2x_2^2 + 2x_3^2 + 2x_4^2 + 2x_5^2} \\
\text{s.t.} \begin{cases}
x_1 + x_2 + x_3 + x_4 + x_5 \leq 350 \\
x_1 + x_2 + x_3 + x_4 + x_5 \geq 300 \\
4x_1 + 2x_2 + 1.5x_3 + x_4 + 2x_5 \leq 1085 \\
x_1 + 4x_2 + 2x_3 + 5x_4 + 3x_5 \leq 660 \\
x_1 \geq 20, x_2 \geq 20, x_3 \geq 20, x_4 \geq 20, x_5 \geq 20.
\end{cases}
\end{cases}
$$

$$(3.66)$$

And suppose the value of G_i^1 and G_i^0 $(i = 1, 2)$ are

$$G_1^1 = 43477.1, \quad G_1^0 = 29855.9, \quad G_2^1 = 224706.3, \quad G_2^0 = 157567.1.$$

The corresponding satisfied results list in Table 3.4.

Table 3.4 Employ the interactive fuzzy satisfied method based on necessity($\gamma_i = 0.9, \delta_i = 0.9$)

$\bar{\mu}_1$	$\bar{\mu}_2$	G_1	G_2	$\mu_1(G_1)$	$\mu_2(G_2)$	x_1	x_2	x_3	x_4	x_5	λ
1	1	41201.6	213490.2	0.833	0.833	216.1	39.6	54.3	20.0	20.0	0.167
0.95	1	40820.2	214967.1	0.805	0.855	216.6	37.0	56.4	20.0	20.0	0.145
1	0.95	41582.9	212013.0	0.861	0.811	215.6	42.2	52.3	20.0	20.0	0.139
0.90	1	40438.6	216443.5	0.777	0.877	217.1	34.4	58.5	20.0	20.0	0.123
0.85	1	40057.0	217919.5	0.749	0.899	217.6	31.9	60.5	20.0	20.0	0.101
0.80	1	39675.3	219395.1	0.721	0.921	218.1	29.3	62.6	20.0	20.0	0.079

3.4.3 Non-linear Fu-Ra CCM and Fu-Ra Simulation-Based Adaptive Weight GA

For Fu-Ra CCM, we use Fu-Ra simulation-based adaptive weight GA to deal with.

3.4.3.1 Fu-Ra Simulation 2 for Critical Value

Let's introduce the simulation for Fu-Ra CCM, that is, we need to find the critical value $\bar{f}$ under the predetermined confidence levels. In the following, we explain how to find the maximal value $\bar{f}$ for any given confidence levels γ and δ,

$$Ch\{f(\xi) \geq \bar{f}\} \geq \beta.$$

If we use the Pr-Pos based chance, then the critical value $\bar{f}$ should satisfy

$$Pr\{\omega|Pos\{\{f(\xi) \geq \bar{f}\} \geq \delta\} \geq \gamma, \tag{3.67}$$

which means

$$\bar{f} = \max\{\bar{f}|Pr\{\omega|Pos\{\{f(\xi) \geq \bar{f}\} \geq \delta\} \geq \gamma\}.$$

If ξ ia s continuous Fu-Ra vector, then the maximal value $\bar{f}$ must be achieved at the equality case

$$Pr\{\omega|Pos\{\{f(\xi) \geq \bar{f}\} \geq \delta\} = \gamma.$$

First, we sample N independent random vector $\omega_1, \omega_2, \cdots, \omega_N$ from Ω according to the probability measure Pr, and define

$$h(\omega_n) = \begin{cases} 1, & \text{if } Pos\{f(\xi(\omega_n)) \geq \bar{f}\} \geq \delta \\ 0, & \text{otherwise}, \end{cases} \tag{3.68}$$

for $n = 1, 2, \cdots, N$, which are a sequence of random variables. And if for all n, $\bar{f}$ meets (3.67), then we have $E[h(\omega_n)] = \gamma$.

By the strong law of large numbers, we have

$$\frac{\sum_{n=1}^{N} h(\omega_n)}{N} \to \gamma \tag{3.69}$$

as $N \to \infty$. Note that the sum $\sum_{n=1}^{N} h(\omega_n)$ is just the number of ω_n satisfying $Pos\{f(\xi(\omega_n)) \geq \bar{f}\} \geq \delta$.

Let N' be the integer part of γN. Then the value $\bar{f}$ is equal to the N'_ith largest element in the sequence $\{\bar{f}_1, \bar{f}_2, \cdots, \bar{f}_N\}$, with

$$\bar{f}_n = \sup\{f_n|Pos\{f(\xi(\omega_n)) \geq f_n\} \geq \beta\}, \tag{3.70}$$

for $n = 1, 2, \cdots, N$, which can be obtained by fuzzy simulations.

We conclude the procedure as follows:

Step 1. Let $\bar{f} = -\infty$.

Step 2. Generate ω from Ω according to the probability measure Pr.

Step 3. Generate a determined vector $f(\xi(\omega))$ uniformly from the δ-cut of fuzzy vector $f(\xi(\omega))$.

Step 4. If $f(\xi(\omega)) \geq \bar{f}$, then let $\bar{f} = f(\xi(\omega))$.

Step 5. Return step 3, and repeat M times.

Step 6. Return step 2, and repeat N times.

Step 7. Set $N' = \gamma_i N$.

Step 8. Return the N'th largest element in $\{\bar{f}_1, \bar{f}_2, \cdots, f_N\}$.

If we use the Pr-Nec based chance, then we have

$$Pr\{\omega|Nec\{\{f(\xi) \geq \bar{f}\} \geq \delta\} \geq \gamma. \tag{3.71}$$

We could adopted the above procedure similarly to estimate $\bar{f}$, because that

$$Nec\{f(\xi) \geq \bar{f}\} \geq \delta \Leftrightarrow Pos\{f(\xi) < \bar{f}\} \leq (1-\delta).$$

We make the changes to step 3 and step 4 respectively:

Step 3'. Generate a determined vector $f(\xi(\omega))$ uniformly from the $(1-\delta)$-cut of fuzzy vector $f(\xi(\omega))$.

Step 4'. If $f(x,\xi(\omega)) < \bar{f}$, then let $\bar{f} = f(\xi(\omega))$.

Example 3.8. We employ the Fu-Ra simulation 2 to find the maximal value $\bar{f}$ such that $Ch\{\xi_1^2 + \xi_2^2 \geq \bar{f}\}(0.9) \geq 0.9$, where ξ_1 and ξ_2 are Fu-Ra variables defined as

$$\xi_1 = (\rho_1, \rho_1 + 1, \rho_1 + 2), \text{ with } \rho_1 \sim U(0,1),$$
$$\xi_2 = (\rho_2, \rho_2 + 1, \rho_2 + 2), \text{ with } \rho_1 \sim U(1,2).$$

After a run of Fu-Ra simulation 2 with 5000 cycles, we get that $\bar{f} = 16.39$.

3.4.3.2 Adaptive Weight GA

Gen and Cheng [66] proposed an adaptive weight approach which utilizes some useful information from the current population to readjust weights to obtain a search pressure toward a positive ideal point. Without loss of generality, consider the maximization problem with q objectives. For the solutions examined in each generation, we define two extreme points: the maximum extreme point z^+ and the minimum point z^- in criteria space as follows:

$$z^+ = \{z_1^{max}, z_2^{max}, \cdots, z_q^{max}\},$$
$$z^- = \{z_1^{min}, z_2^{min}, \cdots, z_q^{min}\},$$

where z_k^{min} and z_k^{max} are the maximal value and minimal value for objective k in the current population. Let P denote the set of the current population. For a given individual x, the maximal value and minimal value for each objective are defined as follows:

$$z_k^{max} = \max\{f_k(x)|x \in P\}, k = 1, 2, \cdots, q,$$
$$z_k^{min} = \min\{f_k(x)|x \in P\}, k = 1, 2, \cdots, q.$$

The hyperparallelogram defined by the two extreme points is a minimal hyperparallelogram containing all current solutions. The two extreme points are renewed at each generation. The maximum extreme point will gradually approximate the positive ideal point. The adaptive weight for objective k is calculated by the following equation:

$$w_k = \frac{1}{z_k^{\max} - z_k^{\min}}, k = 1, 2, \cdots, q.$$

For a given individual x, the weighted-sum objective function is given by the following equation:

$$
\begin{aligned}
z(x) &= \sum_{k=1}^{q} w_k(z_k - z_k^{\min}) \\
&= \sum_{k=1}^{q} \frac{z_k - z_k^{\min}}{z_k^{\max} - z_k^{\min}} \\
&= \sum_{k=1}^{q} \frac{f_k(x) - z_k^{\min}}{z_k^{\max} - z_k^{\min}}.
\end{aligned}
\tag{3.72}
$$

As the extreme points are renewed at each generation, the weights are renewed accordingly. Equation (3.72) is hyperplane defined by the following extreme point in current solutions:

$$
\begin{aligned}
&(z_1^{\max}, z_2^{\min}, \cdots, z_k^{\min}, \cdots, z_q^{\min}) \\
&(z_1^{\min}, z_2^{\max}, \cdots, z_k^{\min}, \cdots, z_q^{\min}) \\
&\cdots \\
&(z_1^{\min}, z_2^{\min}, \cdots, z_k^{\max}, \cdots, z_q^{\min}) \\
&\cdots \\
&(z_1^{\min}, z_2^{\min}, \cdots, z_k^{\min}, \cdots, z_q^{\max}).
\end{aligned}
$$

The hyperplane divides the criteria space Z into two half-space: One half-space contains the positive ideal point, denoted as Z^+, and the other half-space contains the negative ideal point, denoted as Z^-. All Pareto solutions examined lie in the space Z^+, and all points lying in Z^+ have larger fitness values than those in the points in the space Z^-. As the maximum extreme point approximates the positive ideal point along with the evolutionary progress, the hyperplane will gradually approach the positive ideal point. Therefore, the adaptive weight method can readjust its weights according to the current population to obtain a search pressure toward the positive ideal point.

For the minimization case, we just need to transform the original problem into its equivalent maximization problem and then apply (3.72). For a maximization problem, (3.72) can be simplified as follows:

$$
\begin{aligned}
z(x) &= \sum_{k=1}^{q} w_k z_k \\
&= \sum_{k=1}^{q} \frac{z_k}{z_k^{\max} - z_k^{\min}} \\
&= \sum_{k=1}^{q} \frac{f_k(x)}{z_k^{\max} - z_k^{\min}}.
\end{aligned}
$$

Let us look at an example of a bicriteria maximization problem:

$$
\begin{cases}
\max\{z_1 = f_1(x), z_2 = f_2(x)\} \\
\text{s.t.} \quad g_i(x) \le 0, i = 1, 2, \cdots, m.
\end{cases}
$$

For a given generation, two extreme points are identified as

$$z_1^{\max} = \max\{z_1(x^j), j = 1,2,\cdots,pop-size\},$$
$$z_2^{\max} = \max\{z_2(x^j), j = 1,2,\cdots,pop-size\},$$
$$z_1^{\min} = \min\{z_1(x^j), j = 1,2,\cdots,pop-size\},$$
$$z_2^{\min} = \min\{z_2(x^j), j = 1,2,\cdots,pop-size\},$$

and the adaptive weights are calculated as

$$w_1 = \frac{1}{z_1^{\max} - z_1^{\min}}, \ w_2 = \frac{1}{z_2^{\max} - z_2^{\min}}.$$

The weighted-sum objective function is then given by

$$z(x) = w_1 z_1 + w_2 z_2 = w_1 f_1(x) + w_2 f_2(x).$$

It is an adaptive moving line defined by the extreme points $(z_1^{\max}, z_2^{\min})$ and $(z_1^{\min}, z_2^{\max})$, as shown in Figure 3.7. The rectangle defined by the extreme point $(z_1^{\max}, z_2^{\max})$ and $(z_1^{\min}, z_2^{\min})$ is the minimal rectangle containing all current solutions.

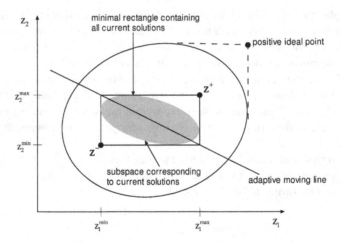

Fig. 3.7 Adaptive weights and adaptive hyperplane

One of important issues in the genetic multi-objective optimization is how to handle constraints because genetic operators used to manipulate chromosomes often yield infeasible offspring. Gen and Cheng[68] suggested an adaptive penalty method to deal with infeasible individuals. Given an individual x in current population $P(t)$, the adaptive penalty function is constructed as follows:

$$p(x) = 1 - \frac{1}{m} \sum_{i=1}^{m} \left(\frac{\Delta b_i(x)}{\Delta b_i^{\max}} \right)^a .$$

where $\Delta b_i(x) = \max\{0, g_i(x) - b_i\}$, $\Delta b_i^{\max} = \max\{\varepsilon, \Delta b_i(x) | x \in P(t)\}$.

Where $\Delta b_i(x)$ is the value of violation for constraint i for the ith chromosome, $\Delta b_i^{\max}$ the maximum of violation for constraint i among current population, and ε a small positive number used to have penalty avoid from zero division. For highly constrained optimization problems, the infeasible solutions take a relative big portion among population at each generation. The penalty approach adjusts the ratio of penalties adaptively at each generation to make a balance between the preservation of information and the pressure for infeasibility and avoid overpenalty. With the penalty function, the fitness function then takes the following form:

$$eval(x) = z(x)p(x).$$

The overall procedure of adaptive weight approach is summarized as follows:

Step 1. Initialization. Generate the initial population randomly.

Step 2. Evaluation. Calculate the objective value, penalty value, and fitness value for each individual.

Step 3. Pareto set. Update the set of Pareto solutions.

Step 4. Selection. Select the next generation using the roulette wheel method.

Step 5. Production. Produce offspring with crossover and mutation.

Step 6. Termination. If the maximal generation is reached, stop; otherwise, go to *step 2*.

3.4.3.3   Numerical Example

We will use the Fu-Ra simulation-based adaptive weight GA to deal with the nonlinear Fa-Ra CCM.

Example 3.9. Let us consider a multi-objective programming with fuzzy random coefficients.

$$\begin{cases} \max F_1(x, \xi) = 2\xi_1^2 x_1 + 3\xi_2^2 x_2 - \xi_3 x_3 + \sqrt{\xi_4^2 + (3 - \xi_5 x_4)^2} \\ \max F_2(x) = -x_1 + 2.5x_2 + 1.5x_3 \\ \text{s.t.} \begin{cases} 5x_1 - 3x_2^2 + 6\sqrt{x_3} + x_4 \leq 80 \\ \xi_6 x_1 + \xi_7^2 x_2 - \xi_8 x_3 \leq \xi_9 \\ x_1 + x_2 + x_3 + x_4 \leq 18 \\ x_1, x_2, x_3, x_4 \geq 0, \end{cases} \end{cases} \tag{3.73}$$

where $\xi_i (i = 1, 2, \cdots, 9)$ are fuzzy random variables as follows,

$$
\begin{aligned}
\xi_1 &= (\rho_1 - 0.1, \rho_1, \rho_1 + 0.2), \text{ with } \rho_1 \sim U(0.4, 0.5), \\
\xi_2 &= (\rho_2 - 0.2, \rho_2, \rho_2 + 0.2), \text{ with } \rho_2 \sim U(0.6, 0.7), \\
\xi_3 &= (\rho_3 - 0.2, \rho_3, \rho_3 + 0.2), \text{ with } \rho_3 \sim U(0.7, 0.8), \\
\xi_4 &= (\rho_4 - 0.2, \rho_4, \rho_4 + 0.2), \text{ with } \rho_4 \sim N(1, 0.1), \\
\xi_5 &= (\rho_5 - 0.2, \rho_5, \rho_5 + 0.2), \text{ with } \rho_5 \sim N(4, 0.1), \\
\xi_6 &= (\rho_6 - 0.5, \rho_6, \rho_6 + 0.5), \text{ with } \rho_6 \sim N(4, 0.1), \\
\xi_7 &= (\rho_7 - 0.5, \rho_7, \rho_7 + 0.5), \text{ with } \rho_7 \sim N(6, 0.1), \\
\xi_8 &= (\rho_8 - 0.5, \rho_8, \rho_8 + 0.5), \text{ with } \rho_8 \sim N(4.5, 0.1), \\
\xi_9 &= (\rho_9 - 2, \rho_9, \rho_9 + 2), \text{ with } \rho_9 \sim N(20, 4).
\end{aligned}
\tag{3.74}
$$

From the mathematical view, the problem (3.73) is not well defined because of the uncertain parameters. Then we apply the chance operator to deal with this uncertain programming.

$$
\begin{cases}
\max \{f_1, f_2\} \\
\text{s.t.} \begin{cases}
Ch\{2\xi_1 x_1 + 3\xi_2^2 x_2 - \xi_3 x_3 + \sqrt{\xi_4^2 + (3 - \xi_5 x_4)^2} \geq f_1\}(0.9) \geq 0.9 \\
-x_1 + 2.5x_2 + 1.5x_3 \geq f_2 \\
5x_1 - 3x_2^2 + 6\sqrt{x_3} + x_4 \leq 80 \\
Ch\{\xi_6 x_1 + \xi_7 x_2 - \xi_8 x_3 \leq \xi_9\}(0.9) \geq 0.9 \\
x_1 + x_2 + x_3 + x_4 \leq 18 \\
x_1, x_2, x_3, x_4 \geq 0.
\end{cases}
\end{cases}
\tag{3.75}
$$

Since there exists non-linear objective function and constraint, we cannot transform it into it's crisp equivalent model. In order to solve it, we use the Fu-Ra simulation based adaptive weight GA to deal with it.

After running, we get a solution as follows:

$$
x_1 = 11.32, x_2 = 3.44, x_3 = 3.24, x_4 = 0;
$$
$$
f_1 = 120.64, f_2 = 2.14.
$$

3.5 Fu-Ra DCM

Fu-Ra DCM is based on selecting the decision with maximize the chance to meet the event.

Uncertain environment, event, and chance function are key elements in dependent-chance model. So let define them first.

By uncertain environment (in this case the fuzzy random environment), we mean the fuzzy random constraints represented by

$$
g_r(x, \xi) \leq 0, \quad r = 1, 2, \cdots, p,
\tag{3.76}
$$

where x is a decision vector, and ξ is a fuzzy random vector.

By event we mean the system of inequalities

$$h_k(x, \xi) \leq 0, \quad k = 1, 2, \cdots, q. \tag{3.77}$$

The chance function of an event ε is characterized by (3.77) is defined as the chance measure of the event ε,

$$f(x) = Ch\{h_k(x, \xi) \leq 0, k = 1, 2, \cdots, q\} \tag{3.78}$$

subject to the uncertain environment (3.76).

3.5.1 General Model for Fu-Ra DCM

The decision maker determine the ideal objective values $\bar{f}_i$ for each objective $f_i(x, \xi)$, and maximize the chance measure of the fuzzy random events $f_i(x, \xi) \geq \bar{f}_i$ under a confidence level γ_i.

$$\begin{cases} \max & \begin{bmatrix} Ch\{f_1(x, \xi) \geq \bar{f}_1(\gamma_1)\} \\ Ch\{f_2(x, \xi) \geq \bar{f}_2(\gamma_2)\} \\ \cdots \\ Ch\{f_m(x, \xi) \geq \bar{f}_m(\gamma_m)\} \end{bmatrix} \\ \text{s.t.} & \begin{cases} g_r(x, \xi) \leq 0, & r = 1, 2, \cdots, p \\ x \in X, \end{cases} \end{cases} \tag{3.79}$$

where $\xi = (\xi_1, \xi_2, \cdots, \xi_n)$ is fuzzy random vector, $Ch\{\cdot\}$ denotes the chance measure of the fuzzy random event in $\{\cdot\}$, $\gamma_i, \eta_r, \theta_r \in [0, 1]$ are the predetermined confidence level, $i = 1, 2, \cdots, m, r = 1, 2, \cdots, p$.

If we introduce the new variables $\delta_i, i = 1, 2, \cdots, m$, then the model (3.86) can be written as the following equivalent form,

$$\begin{cases} \max & [\delta_1, \delta_2, \cdots, \delta_m] \\ \text{s.t.} & \begin{cases} Ch\{f_i(x, \xi) \geq \bar{f}_i\}(\gamma_i) \geq \delta_i, & i = 1, 2, \cdots, m \\ Ch\{g_r(x, \xi) \leq 0\}(\eta_r) \geq \theta_r, & r = 1, 2, \cdots, p \\ x \in X. \end{cases} \end{cases} \tag{3.80}$$

Definition 3.12. If x^* is an efficient solution of problem (3.80), we call it as a fuzzy random dependent chance efficient solution.

According to the definition of the chance measure, model (3.80) is also can be written as:

$$\begin{cases} \max & [\delta_1, \delta_2, \cdots, \delta_m] \\ \text{s.t.} & \begin{cases} Pr\{\omega | Pos\{f_i(x, \xi) \geq \bar{f}_i\} \geq \delta_i\} \geq \gamma_i, & i = 1, 2, \cdots, m \\ Pr\{\omega | Pos\{g_r(x, \xi) \leq 0\} \geq \theta_r\} \geq \eta_r, & r = 1, 2, \cdots, p \\ x \in X. \end{cases} \end{cases} \tag{3.81}$$

Let $x \in R^n$, for determined confidence level η_r, θ_r, if $x \geq 0$ and $Pr\{w|Pos\{g_r(x,\xi) \leq 0\} \geq \theta_r\} \geq \eta_r$ are tenable, then x is a feasible solution of model (3.80) or (3.81). And let X be these set of the whole feasible solution of model (3.86) or (3.81).

Remark 3.12. IF the fuzzy random vector ξ delegates to random vector, then we have

$$Ch\{f_i(x,\xi) \geq \bar{f}_i\}(\gamma_i) = \begin{cases} 1, & \text{if } Pr\{\omega|f_i(x,\xi) \geq \bar{f}_i\} \geq \gamma_i \\ 0, & \text{if } Pr\{\omega|f_i(x,\xi) \geq \bar{f}_i\} < \gamma_i \end{cases} \quad i = 1,2,\cdots,m.$$

So $\max Ch\{f_i(x,\xi) \geq \bar{f}_i\}(\gamma_i)$ is equivalent to $\max Pr\{f_i(x,\xi) \geq \bar{f}_i\}, i = 1,2,\cdots,m$. Similarly, for $\forall r = 1,2,\cdots,p$, we have

$$Pr\{\omega|Pos\{g_r(x,\xi) \leq 0\} \geq \theta_r\} \geq \eta_r \Leftrightarrow Pr\{\omega|g_r(x,\xi) \leq 0\} \geq \eta_r.$$

So model (3.86) is equivalent to:

$$\begin{cases} \max \; [Pr\{f_i(x,\xi) \geq \bar{f}_i\}, i = 1,2,\cdots,m] \\ \text{s.t.} \begin{cases} Pr\{\omega|g_r(x,\xi) \leq 0\} \geq \eta_r, & r = 1,2,\cdots,p \\ x \in X. \end{cases} \end{cases} \quad (3.82)$$

The model (3.82) is the extension of the CCM in random programming. When there is not constraint in model (3.82), then it is coincident to the CCM in random programming.

Remark 3.13. If the fuzzy random vectors ξ delegate to fuzzy vectors, then we have

$$Pr\{\omega|Pos\{f_i(x,\xi) \geq \bar{f}_i\} \geq \delta_i\} \geq \gamma_i \Leftrightarrow Pos\{f_i(x,\xi) \geq \bar{f}_i\} \geq \delta_i, \quad i = 1,2,\cdots,m$$
$$Pr\{\omega|Pos\{g_r(x,\xi) \leq 0\} \geq \theta_r\} \geq \eta_r \Leftrightarrow Pos\{g_r(x,\xi) \leq 0\} \geq \theta_r, \quad r = 1,2,\cdots,p.$$

So the model(3.86)is equivalent to

$$\begin{cases} \max \; [Pos\{f_i(x,\xi) \geq \bar{f}_i\}, i = 1,2,\cdots,m] \\ \text{s.t.} \begin{cases} Pos\{g_r(x,\xi) \leq 0\} \geq \theta_r, & r = 1,2,\cdots,p \\ x \in X. \end{cases} \end{cases} \quad (3.83)$$

The model (4.80) is the extension of the Modality model proposed by Inuiguchi[188], If there is no possibility constraint in (4.80), it is coincident to the Modality model.

In models (3.86) or (3.81), We can also use the chance measure based on the necessary measure, then we have the following model

$$\begin{cases} \max \; [\delta_1, \delta_2, \cdots, \delta_m] \\ \text{s.t.} \begin{cases} Pr\{\omega|Nec\{f_i(x,\xi) \geq \bar{f}_i\} \geq \delta_i\} \geq \gamma_i, & i = 1,2,\cdots,m \\ Pr\{\omega|Nec\{g_r(x,\xi) \leq 0\} \geq \theta_r\} \geq \eta_r, & r = 1,2,\cdots,p \\ x \in X. \end{cases} \end{cases} \quad (3.84)$$

Denote the feasible region of (3.84) as,

$$X^N := \{x \in \mathbf{R}^n | Pr\{\omega | Nec\{g_r(x,\xi) \leq 0\} \geq \theta_r\} \geq \eta_r, r = 1,2,\cdots,p\}. \quad (3.85)$$

3.5.2 Linear Fu-Ra DCM and Satisfying Trade-Off Method

In the following content we discuss how to transform the models (3.81) and (3.84) to crisp equivalent model when the fuzzy random coefficients are some special fuzzy variables, and we give the method to obtain the weak efficient solution.

We consider when the objective function and constraints function are linear, let $f_i(x,\xi) = \tilde{\bar{c}}_i^T x$, $g_r(x,\xi) = \tilde{\bar{e}}_r^T x - \tilde{b}_r$, then the model (3.79)can be written as,

$$\begin{cases} \max \; [Ch\{\tilde{\bar{c}}_1^T x \geq \tilde{f}_1\}(\gamma_1),\cdots,Ch\{\tilde{\bar{c}}_m^T x \geq \tilde{f}_m\}(\gamma_m)] \\ \text{s.t.} \begin{cases} Ch\{\tilde{\bar{e}}_r^T x \leq \tilde{b}_r\}(\eta_r) \geq \theta_r, & r = 1,2,\cdots,p \\ x \geq 0. \end{cases} \end{cases} \quad (3.86)$$

3.5.2.1 Crisp Equivalent Model

We introduce the crisp model of linear Fu-Ra DCM based on Pr-Pos and Pr-Nec, respectively.

Crisp Equivalent Model Based on Pr-Pos

First we assume that for $x \in X$, $\forall i = 1,2,\cdots,m$, $\beta_i^{cT} x > 0$ holds. Here X denote the feasible region of problem (3.86) and (3.81).

Theorem 3.10. *Assume that $\tilde{\bar{c}}_{ij}$ is LR fuzzy random variable, for certain $\omega \in \Omega$, the membership function of $\tilde{c}_{ij}(\omega)$ is,*

$$\mu_{\tilde{c}_{ij}(\omega)}(t) = \begin{cases} L(\frac{c_{ij}(\omega)-t}{\alpha_{ij}^c}), & t \leq c_{ij}(\omega), \alpha_{ij}^c > 0 \\ R(\frac{t-c_{ij}(\omega)}{\beta_{ij}^c}), & t \geq c_{ij}(\omega), \beta_{ij}^c > 0 \end{cases} \quad \omega \in \Omega, \quad (3.87)$$

where the random vector $(c_{ij}(\omega))_{n\times1} = (c_{i1}(\omega),c_{i2}(\omega),\cdots,c_{in}(\omega))^T$ obeys multiple dimensional normally distribution, mean vector is d_i^c, covariance matrix is V_i^c, denoted by $(c_{ij}(\omega))_{n\times1} \sim N(d_i^c,V_i^c)$, α_{ij}^c and β_{ij}^c are the left and right spread of $\tilde{c}_{ij}(\omega)$, $i = 1,2,\cdots,m$, $j = 1,2,\cdots,n$, Suppose that the reference function L,R : $[0,1] \rightarrow [0,1]$ satisfy $L(1) = R(1) = 0, L(0) = R(0) = 1$, and it is monotone decreasing function. Then $Pr\{\omega | Pos\{\tilde{c}_i(\omega)^T x \geq \tilde{f}_i\} \geq \delta_i\} \geq \gamma_i$ is equivalent to

$$R^{-1}(\delta_i) \geq \frac{\bar{f}_i - d_i^{cT} x - \Phi^{-1}(1-\gamma_i)\sqrt{x^T V_i^c x}}{\beta_i^{cT} x}, \quad i = 1,2,\cdots,m,$$

Here Φ is standard normally distributed function, $\gamma_i \in [0,1]$ are predetermined confidence level.

Proof. The proof is as the same as the proof of Theorem 3.6.   □

Because the reference function $R(\cdot)$ is monotone decreasing function, $\max \delta_i$ is equivalent to $\min R^{-1}(\delta_i)$. By Theorem 3.10 and Theorem 3.7, we can transform the model (3.81) to its crisp equivalent model (3.88),

$$
\begin{cases}
\max & \left[\dfrac{\Phi^{-1}(1-\gamma_i)\sqrt{x^{\mathrm{T}}V_i^c x}+d_i^{c\mathrm{T}}x-\bar{f}_i}{\beta_i^{c\mathrm{T}}x}, i=1,2,\cdots,m \right] \\
\text{s.t.} & \begin{cases} R^{-1}(\theta_r)\beta_r^b + L^{-1}(\theta_r)\alpha_r^{e\mathrm{T}}x - (d_r^{e\mathrm{T}}x - d_r^b) \\ \qquad - \Phi^{-1}(\eta_r)\sqrt{x^{\mathrm{T}}V_r^e x + (\sigma_r^b)^2} \geq 0, \quad r=1,2,\cdots,p \\ x \geq 0. \end{cases}
\end{cases}
$$

(3.88)

For convenience, we denote the objective function as

$$
F_i(x) := \frac{\Phi^{-1}(1-\gamma_i)\sqrt{x^{\mathrm{T}}V_i^c x}+d_i^{c\mathrm{T}}x-\bar{f}_i}{\beta_i^{c\mathrm{T}}x}, \quad i=1,2,\cdots,m. \tag{3.89}
$$

Actually, the decision maker usually require $\eta_r, \gamma_i \geq 0.5$, So we suppose $\eta_r \geq 0.5, \gamma_i \geq 0.5$ holds. By theorem 3.7, when $\eta_r \geq 0.5, r=1,2,\cdots,p$, the feasible region X of the model (3.88) is a convex set. So the model (3.88) is a multi-objective nonlinear fractional programming with convex set feasible region.

In order to solve (3.88), we adopt the reference point produced by Wierzbicki[283, 284] to transform the model (3.88) to the following nonlinear programming problem:

$$
\begin{cases}
\min D \\
\text{s.t.} \begin{cases} D \geq b_i(r_i - F_i(x)), \quad i=1,2,\cdots,m \\ x \in X, \end{cases}
\end{cases}
\tag{3.90}
$$

where r_i is the reference point of each objective decided by the decision maker, $b_i > 0$, different value of b_i will not influence the final optimal solution of model (3.90), $i=1,2,\cdots,m$.

Theorem 3.11. *The optimal solution of model (3.90) is a weak efficient solution of model (3.88).*

Proof. (proof in contrapositive form) Suppose that (x^*, D^*) is the optimal solution of model (3.90), but not the weak efficient solution of model (3.88), then there exists $x' \in X$, such that $F_i(x^*) < F_i(x'), \forall i=1,2,\cdots,m$.

So, since $b_i > 0$, we have

$$
b_i(r_i - F_i(x^*)) > b_i(r_i - F_i(x')), \quad i=1,2,\cdots,m.
$$

Set $D' = \max_i b_i(r_i - F_i(x'))$, It is clear that (x', D') is the feasible solution of model (3.90), However,

$$D' < b_i(r_i - F_i(x^*)) \le D^*.$$

It is conflict with that (x^*, D^*) is the optimal solution of model (3.90), so (x^*, D^*) is a weak efficient solution of model (3.88).

The prof is completed. $\square$

According to Theorem 3.11, we can obtain a weak efficient solution of model (3.88) by solving the nonlinear programming model (3.90). For regular nonlinear programming problem, the existed method can only obtain the local optimal solution, unless this nonlinear programming problem has some convexity, like quasiconvexity. In the following text we proof that the model (3.90) has this quasiconvexity.

Definition 3.13. (Bazaraa and Shetty [132]) Set function $f : S \to R^1$, S is a nonempty set in R^n. If for any $x^1, x^2 \in R$, $f(x^1) \ne f(x^2)$, and $\lambda \in (0, 1)$, we have

$$f(\lambda x^1 + (1 - \lambda)x^2) < \max\{f(x^1), f(x^2)\},$$

then $f(x)$ is called strictly quasiconvex function.

If $-f(x)$ is strictly quasiconvex function, then $f(x)$ is called strictly quasiconcave.

Theorem 3.12. *For any $i = 1, 2, \cdots, m$, when $\gamma_i \ge 0.5$, $F_i(x)$ is strictly quasiconcave function, $r_i - F_i(x)$ is strictly quasiconvex function, $x \in X$.*

Proof. For any $x^1, x^2 \in X$, $F_i(x^1) \ne F_i(x^2)$ and $\lambda \in (0, 1)$, We will prove that for any $i = 1, 2, \cdots, m$,

$$F_i(\lambda x^1 + (1 - \lambda)x^2) > \min\{F_i(x^1), F_i(x^2)\}, \quad \forall i = 1, 2, \cdots, m. \tag{3.91}$$

We may suppose $F_i(x^1) < F_i(x^2)$, i.e.,

$$\frac{\Phi^{-1}(1 - \gamma_i)\sqrt{x^{1\mathrm{T}}V_i^c x^1} + d_i^{c\mathrm{T}}x^1 - \bar{f}_i}{\beta_i^{c\mathrm{T}}x^1} < \frac{\Phi^{-1}(1 - \gamma_i)\sqrt{x^{2\mathrm{T}}V_i^c x^2} + d_i^{c\mathrm{T}}x^2 - \bar{f}_i}{\beta_i^{c\mathrm{T}}x^2}.$$

Since $\beta_i^{c\mathrm{T}}x^1, \beta_i^{c\mathrm{T}}x^2 > 0$, we have

$$\begin{aligned} P_i(x) := & \beta_i^{c\mathrm{T}}x^1 \left[\Phi^{-1}(1 - \gamma_i)\sqrt{x^{2\mathrm{T}}V_i^c x^2} + d_i^{c\mathrm{T}}x^2 - \bar{f}_i \right] \\ & - \beta_i^{c\mathrm{T}}x^2 \left[\Phi^{-1}(1 - \gamma_i)\sqrt{x^{1\mathrm{T}}V_i^c x^1} + d_i^{c\mathrm{T}}x^1 - \bar{f}_i \right] > 0. \end{aligned}$$

Denote $F_i(\lambda x^1 + (1 - \lambda)x^2) - F_i(x^1) = FP_i^1(x)/FP_i^2(x)$, it's easy to compute that

$$\begin{aligned} FP_i^2(x) = & \beta_i^{c\mathrm{T}}x^1(\lambda \beta_i^{c\mathrm{T}}x^1 + (1 - \lambda)\beta_i^{c\mathrm{T}}x^2) > 0, \\ FP_i^1(x) = & (1 - \lambda)\beta_i^{c\mathrm{T}}x^1(d_i^{c\mathrm{T}}x^2 - \bar{f}_i) - (1 - \lambda)\beta_i^{c\mathrm{T}}x^2(d_i^{c\mathrm{T}}x^1 - \bar{f}_i) \\ & + \Phi^{-1}(1 - \gamma_i)\beta_i^{c\mathrm{T}}x^1 \sqrt{\lambda^2 x^{1\mathrm{T}}V_i^c x^1 + (1 - \lambda)^2 x^{2\mathrm{T}}V_i^c x^2 + 2\lambda(1 - \lambda)x^{1\mathrm{T}}V_i^c x^2} \\ & - \Phi^{-1}(1 - \gamma_i)\beta_i^{c\mathrm{T}}x^1 \lambda \sqrt{x^{1\mathrm{T}}V_i^c x^1} - \Phi^{-1}(1 - \gamma_i)\beta_i^{c\mathrm{T}}x^2(1 - \lambda)\sqrt{x^{1\mathrm{T}}V_i^c x^1}. \end{aligned}$$

Let $\psi_i(x) = \sqrt{x^{\mathrm{T}}V_i^c x}$, it is easy to prove that $\psi_i(x)$ is convex function, So, for certain $\lambda \in (0, 1)$, x^1 and x^2, we have

$$\sqrt{\lambda^2 x^{1\mathrm{T}} V_i^c x^1 + (1-\lambda)^2 x^{2\mathrm{T}} V_i^c x^2 + 2\lambda(1-\lambda) x^{1\mathrm{T}} V_i^c x^2} = \psi(\lambda x^1 + (1-\lambda)x^2)$$
$$\leq \lambda \psi(x^1) + (1-\lambda)\psi(x^2) = \lambda \sqrt{x^{1\mathrm{T}} V_i^c x^1} + (1-\lambda)\sqrt{x^{2\mathrm{T}} V_i^c x^2}.$$

And because when $\gamma_i \geq 0.5$, we have $\Phi^{-1}(1-\gamma_i) \leq 0$, so

$$\frac{1}{1-\lambda} FP_i^1(x) \geq \beta_i^{c\mathrm{T}} x^1 (d_i^{c\mathrm{T}} x^2 - \bar{f}_i) - \beta_i^{c\mathrm{T}} x^2 (d_i^{c\mathrm{T}} x^1 - \bar{f}_i)$$
$$+ \beta_i^{c\mathrm{T}} x^1 \Phi^{-1}(1-\gamma_i)\sqrt{x^{2\mathrm{T}} V_i^c x^2} - \beta_i^{c\mathrm{T}} x^2 \Phi^{-1}(1-\gamma_i)\sqrt{x^{1\mathrm{T}} V_i^c x^1}$$
$$= \beta_i^{c\mathrm{T}} x^1 \left[\Phi^{-1}(1-\gamma_i)\sqrt{x^{2\mathrm{T}} V_i^c x^2} + d_i^{c\mathrm{T}} x^2 - \bar{f}_i\right]$$
$$- \beta_i^{c\mathrm{T}} x^2 \left[\Phi^{-1}(1-\gamma_i)\sqrt{x^{1\mathrm{T}} V_i^c x^1} + d_i^{c\mathrm{T}} x^1 - \bar{f}_i\right] = P_i(x) > 0.$$

Hence we have $FP_i^1(x) > 0$, $F_i(\lambda x^1 + (1-\lambda)x^2) - F_i(x^1) > 0$, i.e., equation (3.91) holds, $F_i(x)$ is strictly quasiconcave function, $b_i(r_i - F_i(x))$ is strictly quasiconvex function.

The proof is completed.   □

Actually, the objective function of model (3.90) that need to minimize is,

$$\varphi(x) := \max_{1 \leq i \leq m} b_i(r_i - F_i(x)).$$

So model (3.90) can rewrite as the following forms:

$$\min_{x \in X} \max_{1 \leq i \leq m} b_i(r_i - F_i(x)) = \min_{x \in X} \varphi(x). \qquad (3.92)$$

We will prove that the function $\varphi(x)$ is strictly quasiconvex function in Theorem 3.13.

Theorem 3.13. *For any $x \in X$, the function $\varphi(x) = \max\limits_{1 \leq i \leq m} b_i(r_i - F_i(x))$ is strictly quasiconvex function.*

Proof. For any $i = 1, 2, \cdots, m$, Let $q_i(x) = b_i(r_i - F_i(x))$. By Theorem 3.12, function $q_i(x)$ is strictly quasiconvex function.

Set $\lambda \in (0,1)$, for any $x^1, x^2 \in X$, since X is a convex set, so $\lambda x^1 + (1-\lambda)x^2 \in X$,

$$\varphi(\lambda x^1 + (1-\lambda)x^2) = \max_{1 \leq i \leq m} q_i(\lambda x^1 + (1-\lambda)x^2)$$
$$< \max_{1 \leq i \leq m} \left[\max(g_i(x^1), g_i(x^2))\right]$$
$$= \max\left[\max_{1 \leq i \leq m} g_i(x^1), \max_{1 \leq i \leq m} g_i(x^2)\right]$$
$$= \max\left[\varphi(x^1), \varphi(x^2)\right].$$

By Definition 3.13, we can get that $\varphi(x)$ is strictly quasiconvex function.

The proof is completed.   □

Lemma 3.5. *(Bazaraa and Shetty [132]) Set S is a convex set, $f(x)$ is strictly quasiconvex function, then any of the local optimal solution of model $\min_{x \in S} f(x)$ is global optimal solution.*

Since the feasible region X of model (3.92) is a convex set, the objective function $\varphi(x)$ is strictly quasiconvex function, So by Lemma 3.5, any of the local optimal solution of model (3.90) or (3.92) is global optimal solution, it means that we can use the existed nonlinear programming algorithm to get the global solution of model (3.90) or (3.92), By Theorem 3.11, we know that if we obtain a weak efficient solution of model (3.88), then we also get the fuzzy random weak efficient solution of model (3.86) and (3.81) which make the chance maximal on the confidence level γ.

Further more, when we set different reference points r_i, $i = 1, 2, \cdots, m$, we can get different weak efficient solutions of model (3.88).

Crisp Equivalent Model Based on Pr-Nec

In the following we talk about the crisp equivalent model of model(3.84) when the fuzzy random coefficients are some special fuzzy random variables. Assume that $\forall x \in X^N$, $\alpha_i^{cT} x > 0$ holds, $i = 1, 2, \cdots, m$.

Theorem 3.14. *Let the fuzzy random variable $\tilde{\bar{c}}_{ij}$ is as the same as the assumption in Theorem 3.10, $i = 1, 2, \cdots, m$, $j = 1, 2, \cdots, n$. For certain confidence level $\gamma_i \in [0, 1]$ and certain risk tolerance level $\bar{f}_i$, $Pr\{\omega | Nec\{\tilde{\bar{c}}_i(\omega)^T x \geq \bar{f}_i\} \geq \delta_i\} \geq \gamma_i$ is equivalent to*

$$L^{-1}(1 - \delta_i) \leq \frac{\Phi^{-1}(1 - \gamma_i)\sqrt{x^T V_i^c x} - \bar{f}_i + d_i^{cT} x}{\alpha_i^{cT} x}, \quad i = 1, 2, \cdots, m,$$

where Φ is standard normally distributed.

Proof. Similar to the proof of Theorem 3.6, For certain $\omega \in \Omega$, $\tilde{\bar{c}}_i(\omega)^T x$ is L-R fuzzy number, denoted by $\tilde{\bar{c}}_i(\omega)^T x = (c_i(\omega)^T x, \alpha_i^{cT} x, \beta_i^{cT} x)_{LR}$, and $c_i(\omega)^T x \sim N(d_i^{cT} x, x^T V_i^c x)$.

For given confidence level $\gamma_i \in [0, 1]$ and predetermined risk tolerance level $\bar{f}_i$, we have

$$Pr\{\omega | Nec\{\tilde{\bar{c}}_i(\omega)^T x \geq \bar{f}_i\} \geq \delta_i\} \geq \gamma_i$$
$$\Leftrightarrow Pr\{\omega | c_i(\omega)^T x \geq \bar{f}_i + L^{-1}(1 - \delta_i)\alpha_i^{cT} x\} \geq \gamma_i$$
$$\Leftrightarrow Pr\{\omega | \frac{c_i(\omega)^T x - d_i^{cT} x}{\sqrt{x^T V_i^c x}} \geq \frac{\bar{f}_i + L^{-1}(1 - \delta_i)\alpha_i^{cT} x - d_i^{cT} x}{\sqrt{x^T V_i^c x}}\} \geq \gamma_i$$
$$\Leftrightarrow \Phi\left(\frac{\bar{f}_i + L^{-1}(1 - \delta_i)\alpha_i^{cT} x - d_i^{cT} x}{\sqrt{x^T V_i^c x}}\right) \leq 1 - \gamma_i$$
$$\Leftrightarrow L^{-1}(1 - \delta_i)\alpha_i^{cT} x \leq \Phi^{-1}(1 - \gamma_i)\sqrt{x^T V_i^c x} - \bar{f}_i + d_i^{cT} x$$
$$\Leftrightarrow L^{-1}(1 - \delta_i) \leq \frac{\Phi^{-1}(1 - \gamma_i)\sqrt{x^T V_i^c x} - \bar{f}_i + d_i^{cT} x}{\alpha_i^{cT} x}.$$

The proof is completed.   □

Because the reference function $L(\cdot)$ is monotone decreasing function, $\max \delta_i$ is equivalent to $\max L^{-1}(1 - \delta_i)$, so, by theorem 3.14 and theorem 3.9, the crisp equivalent model of model (3.84) is

$$\begin{cases} \max & \left[\dfrac{\Phi^{-1}(1-\gamma_i)\sqrt{x^T V_i^c x} - \bar{f}_i + d_i^{cT} x}{\alpha_i^{cT} x}, i = 1, 2, \cdots, m \right] \\ \text{s.t.} & \begin{cases} \Phi^{-1}(1-\eta_r)\sqrt{x^T V_r^e x + (\sigma_r^b)^2} - L^{-1}(1-\theta_r)\alpha_r^b - R^{-1}(\theta_r)\beta_r^{eT} x \\ \qquad\qquad\qquad\qquad + (d_r^b - d_r^{eT} x) \geq 0, \quad r = 1, 2, \cdots, p \\ x \geq 0. \end{cases} \end{cases}$$

(3.93)

When $\eta_i \geq 0.5$, Through Theorem 3.9 we know that the feasible region X^N of model (3.93) is a convex set, $r = 1, 2, \cdots, p$. Denote the objective function of model (3.93) by

$$G_i(x) := \frac{\Phi^{-1}(1-\gamma_i)\sqrt{x^T V_i^c x} - \bar{f}_i + d_i^{cT} x}{\alpha_i^{cT} x}, \quad i = 1, 2, \cdots, m. \tag{3.94}$$

Since the difference between function $G_i(x)$ and $F_i(x)$ in equation (3.89) is the difference of denominator $\alpha_i^{cT} x$ and $\beta_i^{cT} x$. So the following discussion is as similar as the above "Crisp equivalent model based on Pr-Pos".

Still we adopt the reference point method to transform the multi-objective fractional programming problem(3.93) to nonlinear programming problem,

$$\begin{cases} \min D \\ \text{s.t.} \begin{cases} D \geq b_i(r_i - G_i(x)), & i = 1, 2, \cdots, m \\ x \in X^N. \end{cases} \end{cases} \tag{3.95}$$

Similar to the proof of Theorem 3.11, 3.12 and 3.13, we have the following conclusion:

Theorem 3.15. *The optimal solution of model (3.95 is a weak efficient solution of model (3.93).*

Proof. The proof is as the same as the proof of Theorem 3.11 and Theorem 3.13. □

Theorem 3.16. *(i) For any $i = 1, 2, \cdots, m$, when $\gamma_i \geq 0.5$, $G_i(x)$ is strictly quasiconcave function, $r_i - G_i(x)$ is strictly quasiconvex function, $x \in X^N$.*
(ii) $\psi(x) := \max_{1 \leq i \leq m} b_i(r_i - G_i(x))$ is strictly quasiconvex function, $x \in X^N$.

Proof. The proof is as the same as the proof of Theorem 3.12 and Theorem. □

After introducing $\psi(x) = \max_{i=1,2,\cdots,m} b_i(r_i - G_i(x))$, model (3.95) can be written as the following form,

$$\min_{x \in X^N} \max_{i=1,2,\cdots,m} b_i(r_i - G_i(x)) = \min_{x \in X^N} \psi(x). \tag{3.96}$$

Because of the feasible region X^N of model (3.96) is a convex set, the objective function $\psi(x)$ is strictly quasiconvex function, by Lemma 3.5, any of the local optimal solution of (3.96) is the global optimal solution, so we can directly use the nonlinear programming algorithm to get the global optimal solution of model (3.96) or (3.95), thereby, we get a weak efficient solution of model (3.93).

3.5.2.2   Satisfying Trade-Off Method

The satisfying trade-off method for multi-objective programming problems was proposed by Nakayama [368, 397]. It is an interactive method combining the satisfying level method with the ideal point method. This method can be applied to not only the linear multi-objective but also the nonlinear multi-objective programming.

$$\begin{cases} \max \ \{H_1(x,\xi), H_2(x,\xi), \cdots, H_m(x,\xi)\} \\ \text{s.t. } x \in X. \end{cases} \tag{3.97}$$

In the beginning, let's briefly introduce the simple satisfying level method which in mainly referred in [398]. In some real decision making problems, DM usually provides a reference objective values $\bar{H} = (\bar{H}_1, \bar{H}_2, \cdots, \bar{H}_m)^T$. If the solution satisfies the reference value, take it. The simple satisfying level method can be summarized as follows:

Step 1. DM gives the reference objective values $\bar{H}$.

Step 2. Solve the following programming problem,

$$\begin{cases} \max \sum_{i=1}^{m} H_i(x) \\ \text{s.t. } \begin{cases} H_i(x) \geq \bar{H}_i, \ i = 1, 2, \cdots, m \\ x \in X. \end{cases} \end{cases} \tag{3.98}$$

Step 3. If the problem (3.98) doesn't have the feasible solution, turn to *Step 4*. If the problem (3.98) has the optimal solution $\bar{x}$, output $\bar{x}$.

Step 4. DM re-gives the reference objective values $\bar{H}$ and turn to *Step 2*.
The satisfying trade-off method can be summarized as follows:

Step 1. Take the ideal point $H^* = (H_1^*, H_2^*, \cdots, H_m^*)^T$ such that $H_i^* > \max_{x \in X} f_i(x)$ $(i = 1, 2, \cdots, m)$.

Step 2. DM gives the objective level $\bar{H}^k = (\bar{H}_1^k, \bar{H}_2^k, \cdots, \bar{H}_m^k)^T$ and $\bar{H}_i^k < H_i^*(i = 1, 2, \cdots, m)$. Let $k = 1$.

Step 3. Compute the weight and solve the following problem to get the efficient solution.

$$w_i^k = \frac{1}{H_i^* - \bar{H}_i^k}, \ i = 1, 2, \cdots, m. \tag{3.99}$$

$$\max_{x \in X} \ \max_{1 \leq i \leq m} w_i^k |H_i^* - H_i(x)|, \tag{3.100}$$

or the equivalent problem,

$$\begin{cases} \min \lambda \\ \text{s.t.} \begin{cases} w_i^k(H_i^* - H_i(x)) \leq \lambda, \ i = 1, 2, \cdots, m \\ x \in X. \end{cases} \end{cases} \tag{3.101}$$

Suppose that the optimal solution is x^k.

Step 4. According to the objective value $H(x^k) = (H_1(x^k), H_2(x^k), \cdots, H_m(x^k))^T$, DM divide them into three classes: (1) which needs to improve, denote the related subscript set I_I^k, (2) which is permitted to release, denote the related subscript set I_R^k, (3) which is accepted, denote the related subscript set I_A^k. If $I_I^k = \Phi$, stop the iteration and output x^k. Otherwise, DM gives the new reference objective values $\tilde{H}_i^k$, $i \in I_I^k \cup I_R^k$ and let $\tilde{H}_i^k = H_i(x^k)$, $i \in I_A^k$.

Step 5. Let $u_i (i = 1, 2, \cdots, m)$ be the optimal Kuhn-Tucker operator of the first constraints. If there exists a minimal nonnegative number ε such that

$$\sum_{i=1}^m u_i w_i^k (\tilde{H}_i^k - H_i(x^k)) \geq -\varepsilon,$$

then we deem that $\tilde{H}_i^k$ passes the check for feasibility. Let $\tilde{H}_{i+1} = \tilde{H}_i^k$ ($i = 1, 2, \cdots, m$), turn to *Step 3*. Otherwise, $\tilde{H}_i^k$ isn't feasible. The detail can be referred in [368]. DM should re-give $\tilde{H}_i^k$, $i \in I_I^k \cup I_R^k$ and recheck it.

3.5.2.3 Numerical Example

We use the following example to illustrate the crisp equivalent model of linear Fu-Ra DCM.

Example 3.10. Suppose that the fuzzy random variable $\xi_1, \xi_2, \cdots, \xi_{10}$ and constant $c_1, c_2, \cdots, c_5$ are as the same as the assumption in Example 3.7.

Consider the following linear Fu-Ra DCM:

$$\begin{cases} \max \ [\delta_1, \delta_2] \\ \text{s.t.} \begin{cases} Ch\{\bar{\bar{\xi}}_1 x_1 + \bar{\bar{\xi}}_2 x_2 + \bar{\bar{\xi}}_3 x_3 + \bar{\bar{\xi}}_4 x_4 + \bar{\bar{\xi}}_5 x_5 \geq \bar{f}_1\}(0.9) \geq \delta_1 \\ Ch\{c_1 \bar{\bar{\xi}}_6 x_1 + c_2 \bar{\bar{\xi}}_7 x_2 + c_3 \bar{\bar{\xi}}_8 x_3 + c_4 \bar{\bar{\xi}}_9 x_4 + c_5 \bar{\bar{\xi}}_{10} x_5 \geq \bar{f}_2\}(0.9) \geq \delta_2 \\ x_1 + x_2 + x_3 + x_4 + x_5 \leq 350 \\ x_1 + x_2 + x_3 + x_4 + x_5 \geq 300 \\ 4x_1 + 2x_2 + 1.5x_3 + x_4 + 2x_5 \leq 1085 \\ x_1 + 4x_2 + 2x_3 + 5x_4 + 3x_5 \leq 660 \\ x_1 \geq 20, x_2 \geq 20, x_3 \geq 20, x_4 \geq 20, x_5 \geq 20, \end{cases} \end{cases}$$

$$\tag{3.102}$$

where the risk tolerance given by decision maker is $\bar{f}_1 = 35000$, $\bar{f}_2 = 170000$. Denote the feasible region of model (3.102) is X.

We adopt the Pr-Pos measure. By (3.88), model (3.102) ia equivalent to the following two objective model:

$$\max_{x \in X}\{F_1(x), F_2(x)\}, \tag{3.103}$$

where

$$F_i(x) = \frac{\Phi^{-1}(1 - \gamma_i)\sqrt{x^T V_i^c x} + d_i^{cT} x - \bar{f}_i}{\beta_i^{cT} x}, \quad i = 1, 2,$$

where $\beta_1^c = (3, 8, 3, 7, 5)$, $\beta_2^c = (12, 3.5, 15.6, 4, 7.2)$, $d_1^c = (113, 241, 87, 56, 92)$, $d_2^c = (753.6, \ 71.5, 618.8, 259.2, 485.1)$, $V_1^c = \text{diag}\{1, 4, 1, 2, 2\}$, $V_2^c = \text{diag}\{1, 2, 2, 2, 2\}$, $\Phi^{-1}(0.1) = -1.28$.

Let $\delta_i = \gamma_i = 0.9$, then $R^{-1}(\delta_i) = 0.1$, $\Phi^{-1}(1 - \gamma_i) = -1.28$, $i = 1, 2$. In order to help the decision maker to determine the reference point r_i of fractional objective function $F_i(x)$, first we figure out the maximal and minimal value of function $F_i(x)$, i.e., $\max_{x \in X} F_i(x)$ and $\min_{x \in X} F_i(x)$, $i = 1, 2$. Since $F_i(x)$ is strictly quasiconcave function, the local optimal solution of model $\max_{x \in X} F_i(x)$ is global optimal solution. It is easy (We can use the software, like LINGO, MATLAB, et,al.) to obtain the global maximal value of $F_i(x)$ as,

$$F_1^{\max}(x) = -14.87454, \quad F_2^{\max}(x) = 4.144418.$$

when we solve the model $\min_{x \in X} F_i(x)$, we need to use the Theorem 7.3.2 in literature [79]: when $-f(x)$ is continuous strictly quasiconvex function and the set of constrains is $X = \{x | Ax \le b, x \ge 0\}$ then the optimal solution of $\max_{x \in X} -f(x)$ is obtained at the vertex. The feasible region of model (3.102) is convex Polyhedra, $-F_i(x)$ is strictly quasiconvex function, so we can directly use the algorithm proposed in literature [79] to get the minimal value of $F_i(x)$,

$$F_1^{\min}(x) = -26.43929, \quad F_2^{\min}(x) = -13.64001.$$

Assume that the decision maker decide the reference points as $\bar{F}^k(\bar{F}_1^k, \bar{F}_2^k) = (-18, 1.5), k = 1$. Then according to the satisfying trade-off method, we calculate the weight as follows when $k = 1$,

$$w_1^k = \frac{1}{F_1^{\max}(x) - \bar{F}_1^k} = \frac{1}{-14.87454 + 18} = 0.32,$$
$$w_2^k = \frac{1}{F_2^{\max}(x) - \bar{F}_2^k} = \frac{1}{4.144418 - 1.5} = 0.378.$$

Follow Equation (3.101), we construct the following model (3.104),

$$\begin{cases} \min \ \lambda \\ \text{s.t.} \begin{cases} w_1^k(-14.87454 - F_1(x)) \le \lambda \\ w_2^k(4.144418 - F_2(x)) \le \lambda \\ x \in X \end{cases} \end{cases} \tag{3.104}$$

or

$$\begin{cases} \min \lambda \\ \begin{cases} 0.32(-14.87454- \\ \dfrac{-1.28\sqrt{x_1^2+4x_2^2+x_3^2+2x_4^2+x_5^2}+(113x_1+241x_2+87x_3+56x_4+92x_5)-35000}{3x_1+8x_2+3x_3+7x_4+5x_5}) \leq \lambda \\ 0.378(4.144418- \\ \dfrac{-1.28\sqrt{x_1^2+2x_2^2+2x_3^2+2x_4^2+2x_5^2}+(753.6x_1+71.5x_2+618.8x_3+239.2x_4+483.1x_5)-170000}{12x_1+3.5x_2+15.6x_3+4x_4+7.2x_5}) \leq \lambda \\ x_1+x_2+x_3+x_4+x_5 \leq 350 \\ x_1+x_2+x_3+x_4+x_5 \geq 300 \\ 4x_1+2x_2+1.5x_3+x_4+2x_5 \leq 1085 \\ x_1+4x_2+2x_3+5x_4+3x_5 \leq 660 \\ x_1 \geq 20, x_2 \geq 20, x_3 \geq 20, x_4 \geq 20, x_5 \geq 20. \end{cases} \end{cases}$$

(3.105)

Though solving the nonlinear model (3.106) we can get a weak efficient solution of model (3.104).

$$x^w = (169.9160, 23.36279, 76.50380, 20.00000, 47.87508),$$

If the decision maker is not satisfied with the current value of F_i^w, we can set a group of new reference points to figure out the new weak efficient solution and new value of F_i^w, we will stop until the decision maker is satisfied.

Similarly, if wo adopt the Pr-Nec measure to consider the model (3.104), by(3.93), model (3.102) is equivalent to the following two objective model:

$$\max_{x \in X}[G_1(x), G_2(x)].$$

(3.106)

The definition of $G_i(x)$ can refer to equation (3.94). Since $\alpha_i^c = \beta_i^c$, then $F_i(x) = G_i(x)$, $i = 1, 2$. So the crisp model of (3.102), (3.104) and (3.106) are coincident.

3.5.3 Non-linear Fu-Ra DCM and Fu-Ra Simulation-Based Compromise GA

Let's introduce the simulation for the a-chance which is critical in Fu-Ra DCM and the compromise approach in GA.

3.5.3.1 Fu-Ra Simulation 3 for Chance

In order to solve the Fu-Ra DCM, we use the Fu-Ra simulation to compute the objective function of Fu-Ra DCM $Ch\{f(\xi) \geq \bar{f}\}(\gamma)$, the confidence level γ is predetermined.

According to the definition of the Pr-Pos chance of fuzzy random variable which is,

$$Ch\{f(\xi) \geq \bar{f}\}(\gamma) = \sup\{\delta | Pr\{\omega \in \Omega | Pos\{f(\xi(\omega)) \geq \bar{f}\} \geq \delta\} \geq \gamma\}, \quad (3.107)$$

where $\omega = (\omega_1, \omega_2, \cdots, \omega_n)$, $\xi(\omega) = (\xi_1(\omega), \xi_2(\omega), \cdots, \xi_n(\omega))$.

Since the supremum is definitely reached when the confidence level is equal to γ, that is,

$$Pr\{\omega \in \Omega | Pos\{f(\xi(\omega)) \geq \bar{f}\} \geq \delta\} = \gamma,$$

so we can employ the following way to estimate $Ch\{f(\xi) \geq \bar{f}\}(\gamma)$.

First of all, according to the probability measure Pr, we generate N independent random vector $\omega_k = (\omega_{k1}, \omega_{k2}, \cdots, \omega_{kn})^T$, $k = 1, 2, \cdots, N$. Then for given sample $\omega_k \in \Omega$, we use the Fu simulation to compute the possibility of the fuzzy event $Pos\{f(\xi(\omega)) \geq \bar{f}\}$. The details are as follows: Generate an achieve value of fuzzy vector $\xi(\omega_k)$, and obtain the value of $f(\xi(\omega_k))$. If $f(\xi(\omega_k)) \geq \bar{f}$, then let the minimal value of membership function of each component be δ. Repeat the above process many times, keep the largest value of membership function and consider it as the possibility $Pos\{f(\xi(\omega_k)) \geq \bar{f}\}$.

For any $k = 1, 2, \cdots, N$, we define

$$h(\omega_k) = \begin{cases} 1, & \text{if } Pos\{f(\xi(\omega_k)) \geq \bar{f}\} \geq \delta \\ 0, & \text{otherwise.} \end{cases} \tag{3.108}$$

According to the Large Number Law, when $N \to \infty$, we have that

$$\frac{\sum_{k=1}^{N} h(\omega_k)}{N} \to \gamma. \tag{3.109}$$

Here we can see that $\sum_{k=1}^{N} h(\omega_k)$ denote the time when $Pos\{f(\xi(\omega_k)) \geq \bar{f}_i\} \geq \delta$. Let N_i' be the integer part of γN, then δ is the N_i'th largest element in the sequence $\{\delta_1, \delta_2, \cdots, \delta_N\}$.

We conclude the procedure as follows,

Step 1. Generate vector $\omega_k = (\omega_{k1}, \omega_{k2}, \cdots, \omega_{kn})^T$ randomly from Ω according to the probability measure Pr, $k = 1, 2, \cdots, N$.

Step 2. Compute the possibilities of fuzzy events $\delta_k = Pos\{f(\xi(\omega_k)) \geq \bar{f}\}$ for $k = 1, 2, \cdots, N$ by fuzzy simulation.

Step 3. Set N' be the integer part of γN.

Step 4. Return the N'th largest element in the sequence $\{\delta_1, \delta_2, \cdots, \delta_N\}$.

If we want to adopt the Pr-Nec chance of fuzzy random variable which is,

$$Ch\{f(\xi) \geq \bar{f}\}(\gamma) = \sup\{\delta | Pr\{\omega \in \Omega | Nec\{f(\xi(\omega)) \geq \bar{f}\} \geq \delta\} \geq \gamma\}.$$

The simulation procedure is similar as the above, expect that the following change:

Step 2'. Compute the possibilities of fuzzy events $\delta_k = Nec\{f(\xi(\omega_k)) \geq \bar{f}\}$ for $k = 1, 2, \cdots, N$ by fuzzy simulation.

Example 3.11. We employ the Fu-Ra simulation 3 to estimate the chance of the event $\xi_1 + \xi_2 \geq 2$ under the confidence level 0.9, and ξ_1, ξ_2 are described as follows:

$$\xi_1 = (\rho_1, \rho_1 + 1, \rho_1 + 2), \text{ with } \rho_1 \sim N(0, 1),$$
$$\xi_2 = (\rho_2, \rho_2 + 1, \rho_2 + 2), \text{ with } \rho_1 \sim N(1, 2).$$

After a run of Fu-Ra simulation 3 with 5000 cycles, we can get that

$$Ch\{\xi_1 + \xi_2 \geq 2\}(0.9) = 0.365.$$

3.5.3.2 Compromise GA

Compromise solution-based fitness assignment has been proposed by Cheng and Gen [67] as a means to obtain a compromised solution instead of generating all Pareto solutions. For many real-world problems, the set of Pareto solutions may be very large, possibly exponential in size. Thus the computational effort required to solve it can increase exponentially with problem size in the worst case. Having to evaluate a large set of Pareto solutions in order to select the best one poses a considerable cognitive burden on the decision maker. Therefore, in such cases, obtaining the entire set of Pareto solutions is of little interest to decision makers. In contrast to generating methods, the compromise approach searches for compromised solutions to overcome such difficulty.

The compromise approach identifies solutions that are closest to the ideal solution as determined by some measure of distance. The weighted L_p-norm is useful as a distance measure.

$$r(z; p, w) = ||z - z^*||_{p,w} = \left(\sum_{j=1}^{q} w_j^p |z_j - z_j^*|^p \right)^{1/p},$$

where $(z_1^*, z_2^*, \cdots, z_q^*)$ is the positive ideal point in the criterion space Z and wights $(w_1, w_2, \cdots, w_q)$ are assigned to objectives to emphasize different degrees of importance. As we know, the ideal solution is not usually attainable. However, it can serve as a good standard for evaluation of the nondominated solutions attainable. For many complex problems, to find an ideal point is also a difficult task. To overcome the difficulty, the concept of a proxy ideal point is suggested to replace the ideal point. The proxy ideal point is the ideal point corresponding to the current generation but not to a given problem. Let P denote the set of the current population, then the proxy ideal point $(z_1^{\min}, z_2^{\min}, \cdots, z_q^{\min})$ is calculated as follows:

$$z_1^{\min} = \min\{z_1(x) | x \in P\},$$
$$z_2^{\min} = \min\{z_2(x) | x \in P\},$$
$$\cdots$$
$$z_q^{\min} = \min\{z_q(x) | x \in P\}.$$

The proxy ideal point is easy to obtain at each generation. Along evolutionary progress, the proxy ideal point will gradually approximate the real ideal point.

Consider a minimization problem. Since the smaller the regret value, the better the individual, we have to convert the regret value into a fitness value to ensure that a fitter individual has a larger fitness value. Let $r(x)$ denote the regret value of individual x, r_{max} the biggest regret value, and r_{min} the smallest regret value in the current generation. The transformation is given as follows:

$$eval(x) = \frac{r_{max} - r(x) + \gamma}{r_{max} - r_{min} + \gamma},$$
(3.110)

where γ is a positive real number usually restricted within the open interval $(0,1)$. Its purpose is twofold:
(1) to present equation (3.110) from zero division;
(2) to make it possible to adjust the selection behavior from fitness proportional selection to pure random selection.

3.5.3.3 Numerical Example

We will use the Fu-Ra simulation-based compromise GA to deal with the non-linear Fu-Ra DCM.

Example 3.12. Let us consider a multi-objective programming with fuzzy random coefficients.

$$\begin{cases} \max F_1(x,\xi) = 2\xi_1^2 x_1 + 3\xi_2^2 x_2 - \xi_3 x_3 + \sqrt{\xi_4^2 + (3 - \xi_5 x_4)^2} \\ \max F_2(x,\xi) = \xi_6 x_1 + \xi_7 x_2 + \xi_8 x_3 \\ \text{s.t.} \begin{cases} 5x_1 - 3x_2^2 + 6\sqrt{x_3} + x_4 \leq 80 \\ 4x_1 + 35x_2 - 4.5x_3 \leq 20 \\ x_1 + x_2 + x_3 + x_4 \leq 18 \\ x_1, x_2, x_3, x_4 \geq 0, \end{cases} \end{cases}$$
(3.111)

where $\xi_i (i = 1, 2, \cdots, 8)$ are fuzzy random variables as follows,

$$\begin{aligned} \xi_1 &= (\rho_1 - 0.1, \rho_1, \rho_1 + 0.2), \text{ with } \rho_1 \sim U(0.4, 0.5), \\ \xi_2 &= (\rho_2 - 0.2, \rho_2, \rho_2 + 0.2), \text{ with } \rho_2 \sim U(0.6, 0.7), \\ \xi_3 &= (\rho_3 - 0.2, \rho_3, \rho_3 + 0.2), \text{ with } \rho_3 \sim U(0.7, 0.8), \\ \xi_4 &= (\rho_4 - 0.2, \rho_4, \rho_4 + 0.2), \text{ with } \rho_4 \sim N(1, 0.1), \\ \xi_5 &= (\rho_5 - 0.2, \rho_5, \rho_5 + 0.2), \text{ with } \rho_5 \sim N(4, 0.1), \\ \xi_6 &= (\rho_6 - 0.1, \rho_6, \rho_6 + 0.1), \text{ with } \rho_6 \sim N(-1, 0.1), \\ \xi_7 &= (\rho_7 - 0.5, \rho_7, \rho_7 + 0.1), \text{ with } \rho_7 \sim N(3, 0.1), \\ \xi_8 &= (\rho_8 - 0.1, \rho_8, \rho_8 + 0.5), \text{ with } \rho_8 \sim N(1, 0.1). \end{aligned}$$
(3.112)

From the mathematical view, the problem (3.111) is not well defined because of the uncertain parameters. Then we apply the chance operator to deal with this uncertain programming.

$$\begin{cases} \max \ [\delta_1, \delta_2] \\ \text{s.t.} \begin{cases} Ch\{2\xi_1^2 x_1 + 3\xi_2^2 x_2 - \xi_3 x_3 + \sqrt{\xi_4^2 + (3 - \xi_5 x_4)^2} \geq \bar{f}_1\}(\gamma_1) \geq \delta_1 \\ Ch\{\xi_6 x_1 + \xi_7 x_2 + \xi_8 x_3 \geq \bar{f}_2\}(\gamma_2) \geq \delta_2 \\ 5x_1 - 3x_2^2 + 6\sqrt{x_3} + x_4 \leq 80 \\ 4x_1 + 35x_2 - 4.5x_3 \leq 20 \\ x_1 + x_2 + x_3 + x_4 \leq 18 \\ x_1, x_2, x_3, x_4 \not\geq 0. \end{cases} \end{cases} \quad (3.113)$$

Since there exists non-linear objective functions, we cannot transform it into it's crisp equivalent model. In order to solve it, we use the Fu-Ra simulation based compromise GA to deal with it.

After running, we get a solution as follows:

$$x_1 = 11.32, x_2 = 3.44, x_3 = 3.24, x_4 = 0, \delta_1 = 0.63, \delta_2 = 0.55.$$

3.6 Application to the Portfolio Selection Problem

In this section, the future return is regarded as a fuzzy random variable, and we apply the Fu-Ra CCM and Fu-Ra EVM to the portfolio selection problem as an illustration.

3.6.1 Assumptions

In this section, we list the assumptions for portfolio selection problems under fuzzy random environments, then we use the λ-mean and the variance of the fuzzy random variable to measure the return and the risk of a portfolio respectively, that is we called λ-mean variance portfolio selection model.

First we explain the reasonableness of the assumption about fuzzy random security returns. We know a basic assumption behind the Markowitz's mean variance model is that the situation of the stock market in the future can be correctly reflected by securities data in the past, that is, the mean and covariance of a portfolio of securities in the future are similar to the past ones. However, there are so many uncertain factors that this assumption cannot be guaranteed for the real ever-changing stock markets, especially for new emerging stock markets without plenty of historical data such as the stock market in China. Since stock experts possess enough information and experience about the stock market, a good method is to let them provide their rough estimation about the future returns of securities. In this case, the return rates of securities are fuzzy random variables, i.e., random variables whose actual values are fuzzy sets. In this paper, following the idea of the mean variance model, we propose a new portfolio selection model which combines the statistical techniques with the experts' judgement based on fuzzy random theory. Moreover, we also notice that the assumption called homogeneous expectations in Markowitz's mean variance model that all investors share the same expected returns, predicted variances, and predicted covariances about the future is unrealistic in the real world. In fact, it is almost the hallmark of investors to specialize in different prognostications [251]. In our proposed fuzzy random portfolio selection model, the investors' subjective opinions as to the estimation of the return rates of each of the securities are also reflected by introducing a parameter vector λ.

Let r_j denote the return of the jth stock, x_j denote the investment scale of the jth stock, $j = 1, 2, \cdots, n$. Suppose that the return of the jth stock is L-R fuzzy random variable, denoted by $\tilde{\tilde{r}}_j$, and described by the following membership function, for $w \in \Omega$,

$$\mu_{\tilde{r}_j(\omega)}(x) = \begin{cases} L\left(\frac{a_j(\omega)-x}{\alpha_j}\right), & \text{if } a_j(\omega) - \alpha_j < x_j < a_j(\omega) \\ 1, & \text{if } x = a_j(\omega) \\ R\left(\frac{x-a_j(\omega)}{\beta_j}\right), & \text{if } a_j(\omega) < x_j < a_j(\omega) + \beta_j, \end{cases} \tag{3.114}$$

where the benchmark function $L(\cdot)$ and $R(\cdot)$ are monotonously decreasing left continuous function, and $L, R : [0,1] \to [0,1]$ satisfy that $R(0) = L(0) = 1, R(1) = L(1) = 0$. a_j is the normally distributed random variable, the mean is $E(a_j)$, and the variance is σ_j^2, denoted as $a_j \sim N(E(a_j), \sigma_j^2)$, the covariance of a_i and a_j is $Cov(a_i, a_j) = \sigma_{ij}$, we denote the covariance matrix as $V = (Cov(a_i, a_j))_{n \times n}$, $\alpha_j, \beta_j (> 0)$ are the left and the right spread of the LR fuzzy random variable $\tilde{\tilde{r}}_j(\omega)$. For convenience, in the following we denote the fuzzy random variable $\tilde{\tilde{r}}_j$ as $\tilde{\tilde{r}}_j(\omega) = (a_j(\omega), \alpha_j, \beta_j)_{LR}$, $\forall \omega \in \Omega$ (see Figure 3.8).

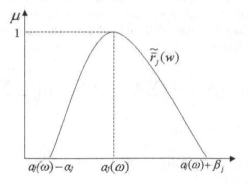

Fig. 3.8 LR-type fuzzy random return $\tilde{\tilde{r}}_j$

The parameters of the fuzzy random variable $\tilde{\tilde{r}}_j$ needs to determine are α_j, β_j, $E(a_j)$ and σ_{ij}. The left spread α_j and the right spread β_j, $j = 1, 2, \cdots, n$ can be determined by the experts' experience. For instance, choose a group of experts, let them give the left and the right spread, α_j^k, β_j^k, usually the left and the right spreads given by each expert are not the same, we use the arithmetical mean to determine the α_j and β_j, i.e., $\alpha_j = \sum_{k=1}^K \alpha_j^k / K$, $\beta_j = \sum_{k=1}^K \beta_j^k / K$. How to determine $E(a_j)$ and σ_{ij}? we can use the traditional method, that is, we collet the historical return and historical covariance to compute $E(a_j)$ and σ_{ij}, $i, j = 1, 2, \cdots, n$.

3.6.2 Chance Constrained Portfolio Selection Model

Similar to the safety-first model and the Fractile model proposed by Kataoka[208] and Inuiguchi&Tanino[50]. In this chapter, we assume the future returns as fuzzy random variables, like the ideas of the above two models, we can get the fuzzy random chance-constrained portfolio selection model.

Suppose the future return of every stock are triangular fuzzy random variables, denoted by $\tilde{\bar{r}}_j(\omega) = (a_j(\omega) - \alpha_j, a_j(\omega), a_j(\omega) + \beta_j)$, $\omega \in \Omega$, a_j is the random variables which obeys normal distribution, $a_j \sim N(E(a_j), \sigma_j^2)$, The covariance between a_i and a_j is $V = (\sigma_{ij})_{n \times n}$, α_j and $\beta_j (> 0)$ are the left and the right width of the triangular fuzzy number $\tilde{\bar{r}}_j(\omega)$, $\omega \in \Omega$.

The fuzzy random chance-constrained model for the portfolio selection problem is as follow,

$$\begin{cases} \max R \\ \text{s.t.} \begin{cases} Ch\{\sum_{j=1}^n \tilde{\bar{r}}_j x_j \geq R\}(\gamma) \geq \delta \\ \sum_{j=1}^n x_j = 1 \\ x_j \geq 0, \quad j = 1, 2, \cdots, n. \end{cases} \end{cases} \quad (3.115)$$

The parameters $\gamma, \delta \in [0,1]$ are the predetermined confidence level, R and x_j are the decision variables, $\gamma, \delta \in [0,1]$ are the predetermined confidence level, $j = 1, 2, \cdots, n$.

$X = \{x \in R | \sum_{j=1}^n x_j = 1, x_j \geq 0, j = 1, 2, \cdots, n\}$. Ch denote the chance measure, If the decision maker is comparatively optimistic, we can adopt the chance measure defined by the possibility measure Pos; If the decision maker is comparatively pessimistic, we can adopt the chance measure defined by the necessary measure Nec.

If we adopt the chance measure defined by Pos, by Theorem 3.6, $Ch\{\sum_{j=1}^n \tilde{\bar{r}}_j x_j \geq R\}(\gamma) \geq \delta$ is equivalent to

$$R \leq \sum_{j=1}^n E(a_j) x_j + (1 - \delta) \sum_{j=1}^n \beta_j x_j + \Phi^{-1}(1 - \gamma) \sqrt{\sum_{i=1}^n \sum_{j=1}^n \sigma_{ij} x_i x_j}.$$

So the model (3.115) is equivalent to

$$\begin{cases} \max \sum_{j=1}^n [E(a_j) + (1 - \delta)\beta_j] x_j + \Phi^{-1}(1 - \gamma) \sqrt{\sum_{i=1}^n \sum_{j=1}^n \sigma_{ij} x_i x_j} \\ \text{s.t.} \begin{cases} \sum_{j=1}^n x_j = 1 \\ x_j \geq 0, \quad j = 1, 2, \cdots, n. \end{cases} \end{cases} \quad (3.116)$$

Because the decision maker determine the confidence level $\gamma \geq 0.5$, then $\Phi^{-1}(1 - \gamma) \leq 0$. It is easily proved that the function $\sqrt{\sum_{i=1}^n \sum_{j=1}^n \sigma_{ij} x_i x_j} = x^T V x$ is convex function, so the modle (3.116) is convex programming.

Similarly, if we adopt the chance measure defined by Nec, by Theorem 3.8, model (3.115) is equivalent to

$$\begin{cases} \max \sum_{j=1}^{n}[E(a_j) - \delta\alpha_j]x_j + \Phi^{-1}(1-\gamma)\sqrt{\sum_{i=1}^{n}\sum_{j=1}^{n}\sigma_{ij}x_ix_j}, \\ \text{s.t.} \begin{cases} \sum_{j=1}^{n}x_j = 1 \\ x_j \geq 0, \ j = 1,2,\cdots,n. \end{cases} \end{cases} \tag{3.117}$$

When $\gamma \geq 0.5$, it's easily know that the model (3.117) is also a convex programming.

Let's compare the two equivalent model (3.116) and (3.117) of the chance constraint model (3.115),

$$F_1(x) := \sum_{j=1}^{n}[E(a_j) + (1-\delta)\beta_j]x_j + \Phi^{-1}(1-\gamma)\sqrt{\sum_{i=1}^{n}\sum_{j=1}^{n}\sigma_{ij}x_ix_j}$$

and

$$F_2(x) := \sum_{j=1}^{n}[E(a_j) - \delta\alpha_j]x_j + \Phi^{-1}(1-\gamma)\sqrt{\sum_{i=1}^{n}\sum_{j=1}^{n}\sigma_{ij}x_ix_j}.$$

Since the difference between the adopted definition of the chance measure. $\forall x \in X$, $F_1(x) \geq F_2(x)$ always holds. Let's use x^{1*} and x^{2*} to denote the optimal solution of the model (3.116) and (3.117), use $F_1^*(x^{1*})$ and $F_2^*(x^{2*})$ to denote the best value of model (3.116) and (3.117). So we have

$$F_1^*(x^{1*}) \geq F_2^*(x^{2*}).$$

The analytical result shows that, when the investors face the completely same set of investment scenarios, the return level of the portfolio adopted by the comparatively optimistic investors is higher than the return level of the portfolio adopted by the comparatively pessimistic investors.

3.6.2.1 Application to the Chinese Stock Market

30 different stocks are selected from the Shanghai Stock 180 Index for his/her investment. We assume that the return rate of each of the securities is a triangular fuzzy random variable, denoted by $r_j = (a_j - \alpha_j, a_j, a_j + \beta_j)$, where a_j is a normally distributed random variable, α_j and β_j are the left and right spread respectively, $j = 1,2,\cdots,30$.

First, we collect the historical data of the 30 stocks from January 2003 to January 2006, and use one month as a period to obtain the historical rates of returns for 36 periods. With the historical data, the expected return rates of the stocks and the covariance matrix $V = (Cov(a_i, a_j))_{n \times n}$ are estimated. Then we select K stock experts and let them give their estimation of α_j and β_j. Denote the estimation of the left and right spreads as α_j^k and β_j^k respectively, $k = 1,2,\cdots,K$. We let $\alpha_j = \sum_{k=1}^{K}\alpha_j^k$ and $\beta_j = \sum_{k=1}^{K}\beta_j^k$, $j = 1,2,\cdots,n$. The expected return rates of stocks and the left and right spreads are listed in Table 3.5. The covariance matrix V is omit here. From Table 3.5 we know that the fuzzy expected value of fuzzy random return rates of stocks, i.e. $E(r_j) = (E(a_j) - \alpha_j, E(a_j), E(a_j) + \beta_j)$ are obtained.

Table 3.5 The expected return rates, left and right spreads of 30 stocks (%)

Code	600000	600030	600026	600050	600036	600717	600642	600688	600104	600009
$E(a_j)$	0.45	1.42	1.73	0.61	1.39	1.16	1.01	1.87	0.44	2.35
α_j	0.15	0.22	0.23	0.1	0.19	0.16	0.21	0.17	0.04	0.35
β_j	0.15	0.08	0.27	0.39	0.71	0.64	0.09	0.53	0.36	0.25
Code	600008	600016	600808	600795	600033	600832	600011	600018	600019	600519
$E(a_j)$	0.68	1.41	0.46	1.38	1.03	1.65	0.71	1.96	2.07	3.6
α_j	0.08	0.11	0.06	0.18	0.13	0.15	0.11	0.16	0.07	0.1
β_j	0.22	0.39	0.14	0.12	0.37	0.1	0.19	0.54	0.33	0.3
Code	600660	600879	600277	600188	600270	600205	600177	600309	600428	600320
$E(a_j)$	1.66	2.14	2.09	0.91	2.32	2.51	0.41	3.61	2.42	3.19
α_j	0.26	0.14	0.39	0.21	0.22	0.51	0.11	0.61	0.42	0.19
β_j	0.24	0.26	0.11	0.09	0.28	0.09	0.39	0.29	0.58	0.61

Suppose that an investor determined that the confidence level is $\gamma = 0.6$, and he want to maximize the return when the possibility of that the return of portfolio is more than R is not less than $\delta = 0.8$ under the confidence level γ.

$$\begin{cases} \max R \\ \text{s.t.} \begin{cases} Ch\{\sum_{j=1}^{30} \tilde{\bar{r}}_j x_j \geq R\}(0.6) \geq 0.8 \\ \sum_{j=1}^{30} x_j = 1 \\ x_j \geq 0, \quad j = 1, 2, \cdots, 30. \end{cases} \end{cases} \tag{3.118}$$

We can get two concretely convex programming according to model (3.118) using the data. Since the model (3.116) and (3.117) are the crisp equivalent model of (3.118), we can use mathematical software to solve them, like LINGO, MATLAB, and so on. Also we can use the genetic algorithm to solve them.

Table 3.6 lists the optimal solution according to model (3.116) in which the chance is defined by Pos, that is, the investor should choose the best proportion as that $x_{19} = 0.2133$, $x_{20} = 0.4717$, $x_{28} = 0.2277$, $x_{30} = 0.0873$, the optimal return is 0.0149.

Table 3.6 Investment scale for optimistic investor ($\gamma = 0.6, \delta = 0.6$)

Code	600000	600030	600026	600050	600036	600717	600642	600688	600104	600009
Scale	0	0	0	0	0	0	0	0	0	0
Code	600008	600016	600808	600795	600033	600832	600011	600018	600019	600519
Scale	0	0	0	0	0	0	0	0	0.2133	0.4717
Code	600660	600879	600277	600188	600270	600205	600177	600309	600428	600320
Scale	0	0	0	0	0	0	0	0.2277	0	0.0873

Table 3.7 lists the optimal solution according to model (3.117) in which the chance is defined by Nec, the optimal solution for the investor is that $x_{19} = 0.2437$,

$x_{20} = 0.6309$, $x_{28} = 0.0429$, $x_{30} = 0.0824$, and he could get the optimal return 0.0126.

Table 3.7 Investment scale for pessimistic investor ($\gamma = 0.6, \delta = 0.6$)

Code	600000	600030	600026	600050	600036	600717	600642	600688	600104	600009
Scale	0	0	0	0	0	0	0	0	0	0
Code	600008	600016	600808	600795	600033	600832	600011	600018	600019	600519
Scale	0	0	0	0	0	0	0	0	0.2437	0.6309
Code	600660	600879	600277	600188	600270	600205	600177	600309	600428	600320
Scale	0	0	0	0	0	0	0	0.0429	0	0.0824

Also, the investor can determine a different confidence level γ, δ to obtain the corresponding optimal solution.

3.6.3 λ mean-Variance Portfolio Selection Model

Let recall the traditional mean variance model first, which can be described by the following bi-objective mathematical programming problem:

$$\begin{cases} \max E\left(\sum_{j=1}^{n} r_j x_j\right) \\ \min Var\left(\sum_{j=1}^{n} r_j x_j\right) \\ \text{s.t.} \begin{cases} \sum_{j=1}^{n} x_j = 1 \\ x_j \geq 0, \ j = 1, 2, \cdots, n, \end{cases} \end{cases}$$

where r_j are assumed to be random variables.

It is not the most ideal method to estimate the future returns of securities by merely utilizing historical data for real ever-changing stock markets especially in a small sample situation. Therefore, it may be a good choice to incorporate experts' experience and judgements about the stock market into the estimations of security returns.

3.6.3.1 Expected Return and Variance Risk for Portfolio

We assume the jth security return r_j to be a fuzzy random variable which is characterized by the following membership function

$$\mu_{r_j(\omega)}(x) = \begin{cases} L\left(\frac{a_j(\omega)-x}{\alpha_j}\right), & \text{if } a_j(\omega) - \alpha_j < x_j < a_j(w) \\ 1, & \text{if } x = a_j(\omega) \\ R\left(\frac{x-a_j(\omega)}{\beta_j}\right), & \text{if } a_j(\omega) < x_j < a_j(\omega) + \beta_j \end{cases} \quad \forall \, \omega \in \Omega,$$

where $L(\cdot)$ and $R(\cdot)$ are left-continuous and non-increasing functions, $L, R : [0, 1] \rightarrow [0, 1]$ with $R(0) = L(0) = 1$ and $R(1) = L(1) = 0$, a_j is a normally distributed random variable, denoted by $a_j \sim N(E(a_j), \sigma_j^2)$, the covariance of a_i and a_j is denoted by $Cov(a_i, a_j) = \sigma_{ij}$, the covariance matrix is denoted by $V = (Cov(a_i, a_j))_{n \times n}$, and $\alpha_j, \beta_j (> 0)$ are the left and right spreads of LR fuzzy number $r_j(w)$, respectively. For simplicity, we denote $r_j = (a_j(w), \alpha_j, \beta_j)_{LR}, \forall w \in \Omega$.

With the given historical data on the stocks over m period, the expected return rates vector and the covariance matrix of the stocks can be obtained. Spreads α_j and β_j are provided by stock experts according to their experience and judgements.

For given $\omega \in \Omega$, it can be easily shown that $(r_j(\omega))_\alpha^- = a_j(\omega) - L^{-1}(\alpha)\alpha_j$ and $(r_j(\omega))_\alpha^+ = a_j(\omega) + R^{-1}(\alpha)\beta_j$ for any $\alpha \in [0,1]$. From (3.4) we have that

$$(E(r_j))_\alpha = [E((r_j(\omega))_\alpha^-), E((r_j(\omega))_\alpha^+)] = [E(a_j) - L^{-1}(\alpha)\alpha_j, E(a_j) + R^{-1}(\alpha)\beta_j].$$

The expectation of fuzzy random variable r_j, $E(r_j)$ is also a LR fuzzy number denoted by $E(r_j) = (E(a_j), \alpha_j, \beta_j)_{LR}$. From (3.5) we have that

$$Var((r_j)_\alpha^-) = Var((r_j)_\alpha^+) = Var(a_j), \quad j = 1, 2, \cdots, n,$$

which implies that the variance of fuzzy random variable r_j is $Var(r_j) = Var(a_j)$.

It is easy to know that the total return rate of portfolio $\sum_{j=1}^n r_j x_j$ is also a fuzzy random variable. From property (i), we obtain that the expectation of fuzzy random variable. $\sum_{j=1}^n r_j x_j$ is given by

$$E(\sum_{j=1}^n r_j x_j) = \sum_{j=1}^n E(r_j) x_j = (\sum_{j=1}^n E(a_j) x_j, \sum_{j=1}^n \alpha_j x_j, \sum_{j=1}^n \beta_j x_j)_{LR}.$$

Because

$$Cov(r_i, r_j) = \frac{1}{2} \int_0^1 \left[Cov((r_i)_\alpha^-, (r_j)_\alpha^-) + Cov((r_i)_\alpha^+, (r_j)_\alpha^+) \right] d\alpha$$

$$= \frac{1}{2} \int_0^1 \left[Cov(a_i - L^{-1}(\alpha)\alpha_i, a_j - L^{-1}(\alpha)\alpha_j) \right.$$

$$\left. + Cov(a_i + R^{-1}(\alpha)\beta_i, a_j + R^{-1}(\alpha)\beta_j) \right] d\alpha$$

$$= Cov(a_i, a_j),$$

from property (iii) we obtain that the variance of $\sum_{j=1}^n r_j x_j$ is given by

$$Var(\sum_{j=1}^n r_j x_j) = \sum_{j=1}^n Var(r_j x_j) + \sum_{i=1}^n \sum_{j=1, j \neq i}^n Cov(r_i x_i, r_j x_j)$$

$$= \sum_{j=1}^n \sigma_j^2 x_j^2 + \sum_{i=1}^n \sum_{j=1, j \neq i}^n Cov(a_i, a_j) x_i x_j = x^T V x. \qquad (3.119)$$

3.6.3.2 λ-Mean Ranking Method for Fuzzy Expected Value

Note that the expected value of the fuzzy random variable $\sum_{j=1}^{n} r_j x_j$ is defined as a fuzzy number. Before introducing the fuzzy random portfolio selection model we need to employ a fuzzy ranking method to convert the fuzzy number $E(\sum_{j=1}^{n} r_j x_j)$ into a crisp number.

Up to now many fuzzy ranking methods have been proposed to rank fuzzy numbers. However, a optimal fuzzy ranking method which does not exist, nor does there exist a simple comparison principle for fuzzy numbers which can be accepted by all decision makers. In practice, the choice of a fuzzy ranking method depends on the properties of the problem and the decision makers' attitudes. As discussed in the introduction, homogeneous expectation is not a realistic assumption for the portfolio selection problem, and investors can be expected to hold differing forecasts about the future returns of securities. This is important because if everyone has different forecasts, everyone will have different feasible regions and hence different efficient frontiers.

Campos and Gonzalez [143] proposed the λ-average ranking method which is very useful especially in decision making, since it is able to incorporate the decision maker's subjective attitude into the decision making process. In this paper we adopt the λ-average ranking method to incorporate the investors' different forecasts about the future returns of securities into the portfolio selection model.

Let M be a fuzzy number with α-level sets $[M_\alpha^-, M_\alpha^+]$. The λ-average value of M is defined by [143]

$$V_S^\lambda(M) = \int_0^1 \left[\lambda M_\alpha^+ + (1-\lambda) M_\alpha^- \right] dS(\alpha), \tag{3.120}$$

where parameter $\lambda \in [0,1]$ is a subjective degree of investor's optimism-pessimism, and S is an additive measure on $(0,1]$ which determines the weight or importance associated with different α-level sets. In the following all α-level sets are assumed to have the same importance. In the case of continuous membership functions the integral in (3.120) is calculated with respect to $d\alpha$.

By (3.120) the crisp λ-average value of the expectation of the total return rate of portfolio is represented by

$$V^\lambda \left(E\left(\sum_{j=1}^{n} r_j x_j \right) \right) = \sum_{j=1}^{n} \left[E(a_j) + \lambda_j R^* \beta_j - (1-\lambda_j) L^* \alpha_j \right] x_j, \tag{3.121}$$

where $R^* = \int_0^1 R^{-1}(\alpha) d\alpha$, $L^* = \int_0^1 L^{-1}(\alpha) d\alpha$. The experts' judgement as well as the investors' subjective attitudes of optimism are reflected in (3.121). In this study, we use $V^\lambda(E(\sum_{j=1}^{n} r_j x_j))$ to denote the expected return rate of portfolio.

3.6.3.3 λ-Mean Variance Portfolio Selection Model

Assume an investor wants to maximize the λ-mean of expectation and minimize the risk of the portfolio. Based on the above discussion, a new portfolio selection model is proposed as follows:

$$\begin{cases} \max V^\lambda(E(\sum_{j=1}^n r_j x_j)) \\ \min Var(\sum_{j=1}^n r_j x_j) \\ \text{s.t.} \begin{cases} \sum_{j=1}^n x_j = 1 \\ x_j \geq 0, \quad j = 1, 2, \cdots, n \end{cases} \end{cases} \tag{3.122}$$

or equivalently

$$\begin{cases} \max \sum_{j=1}^n \left[E(a_j) + \lambda_j R^* \beta_j - (1 - \lambda_j) L^* \alpha_j \right] x_j \\ \min \sum_{i=1}^n \sum_{j=1}^n \sigma_{ij} x_i x_j \\ \text{s.t.} \begin{cases} \sum_{j=1}^n x_j = 1 \\ x_j \geq 0, \quad j = 1, 2, \cdots, n. \end{cases} \end{cases}$$

In this paper (3.122) is called as λ-mean variance portfolio selection model.

In model (3.122), parameter λ_j reflects the investor's subjective degree of optimism for the return rates of securities j, $j = 1, 2, \cdots, n$. For an aggressive and completely optimistic investor, λ_j should be chosen as 1, while for a conservative and completely pessimistic investor, λ_j may be chosen as 0. By varying the value of λ_j, the investor's optimism-pessimism opinion which may arise e.g. from having some additional information can be reflected in model (3.122).

We introduce the concepts of λ-mean variance efficient portfolios and λ-efficient frontiers.

Definition 3.14. (λ-mean variance efficient portfolio) For given parameter vector $\lambda = (\lambda_1, \lambda_2, \cdots, \lambda_n)$, feasible solution $(x_1^*, x_2^*, \cdots, x_n^*)$ is a λ-mean variance efficient portfolio if there does not exist a feasible solution $(x_1, x_2, \cdots, x_n)$ such that

$$V^\lambda(E(\sum_{j=1}^n r_j x_j)) \geq V^\lambda(E(\sum_{j=1}^n r_j x_j^*)), \quad Var(\sum_{j=1}^n r_j x_j) \leq Var(\sum_{j=1}^n r_j x_j^*)$$

with at least one strict inequality holding.

Definition 3.15. (λ-efficient frontier) The curve obtained from the bi-objective programming problem (3.122) is called a λ-efficient frontier.

When $\lambda = (1, 1, \cdots, 1)_{1 \times n}$, the λ-efficient frontier is called as an optimistic efficient frontier.

When $\lambda = (0, 0, \cdots, 0)_{1 \times n}$, the λ-efficient frontier is called as a pessimistic efficient frontier.

When $\lambda = (0.5, 0.5, \cdots, 0.5)_{1 \times n}$, the λ-efficient frontier is called as a neutral efficient frontier.

Remark 3.14. If $\alpha_j = \beta_j = 0$, $j = 1, 2, \cdots, n$, then the λ-efficient frontier is the traditional mean variance efficient frontier. It is obvious that the proposed model (3.122) is an extension of Markowitz's mean variance portfolio selection model.

The λ-mean variance efficient portfolios of model (3.122) can be obtained by solving the following parametric programming

$$\begin{cases} \max\ V^\lambda(E(\sum_{j=1}^n r_j x_j)) \\ \text{s.t.} \begin{cases} Var(\sum_{j=1}^n r_j x_j) \leq p_V \\ \sum_{j=1}^n x_j = 1, \quad x_j \geq 0, \quad j = 1, 2, \cdots, n \end{cases} \end{cases} \tag{3.123}$$

or

$$\begin{cases} \min\ Var(\sum_{j=1}^n r_j x_j) \\ \text{s.t.} \begin{cases} V^\lambda(E(\sum_{j=1}^n r_j x_j)) \geq p_E \\ \sum_{j=1}^n x_j = 1, \quad x_j \geq 0, \quad j = 1, 2, \cdots, n, \end{cases} \end{cases} \tag{3.124}$$

where p_V and p_E are the prescribed return rate value and risk level respectively.

Consider the following two mathematical programming problems:

$$\min_{x \in X} Var(\sum_{j=1}^n r_j x_j) \tag{3.125}$$

and

$$\max_{x \in X} V^\lambda(E(\sum_{j=1}^n r_j x_j)), \tag{3.126}$$

where $X = \{x \in R^n | \sum_{j=1}^n x_j = 1, x_j \geq 0, j = 1, 2, \cdots, n\}$. Denote the optimistic values of problem (3.125) and problem (3.126) are $V_{\min}$ and $E_{\max}$ respectively. Denote the optimal solution of problem (3.123) is $x^{\min}$ with optimal value $E_{\min}$ when $p_V = V_{\min}$. Denote the optimal solution of problem (3.124) is $x^{\max}$ with optimal value $V_{\max}$ when $p_E = E_{\max}$. If p_V ranges over $[V_{\min}, V_{\max}]$, all optimal solutions of problem (3.123) compose the λ-mean variance efficient portfolios of problem (3.122). Similarly, if p_E ranges over $[E_{\min}, E_{\max}]$, all optimal solutions of problem (3.124) also compose the λ-mean variance efficient portfolios of problem (3.122).

For given parameter vector $\lambda^i = (\lambda_1^i, \lambda_2^i, \cdots, \lambda_n^i)$, we denote the optimal solutions of problem (3.123) and (3.124) by x^{i*} and x^{i**} respectively, $i = 1, 2$.

The relationship of the λ-mean variance efficient portfolios located on different λ-efficient frontiers is presented in the following theorem.

Theorem 3.17. Suppose that $\lambda^1 \leq \lambda^2$, i.e. $\lambda_j^1 \leq \lambda_j^2$ for all $j = 1, 2, \cdots, n$. The optimal values of problem (3.123) and problem (3.124) satisfy

(1) $V^{\lambda^1}(E(\sum_{j=1}^n r_j x_j^{1*})) \leq V^{\lambda^2}(E(\sum_{j=1}^n r_j x_j^{2*}))$.

(2) $Var(\sum_{j=1}^n r_j x_j^{1**}) \geq Var(\sum_{j=1}^n r_j x_j^{2**})$.

Proof. (1) Since $\lambda_j^1 \leq \lambda_j^2$ for all $j = 1, 2, \cdots, n$, it follows from (3.121) that $V^{\lambda^1}(E(\sum_{j=1}^n r_j x_j)) \leq V^{\lambda^2}(E(\sum_{j=1}^n r_j x_j))$, where $x = (x_j)$ is feasible solution of problem (3.123). Therefore,

$$V^{\lambda^2}(E(\sum_{j=1}^{n} r_j x_j^{2*})) \ge V^{\lambda^2}(E(\sum_{j=1}^{n} r_j x_j)) \ge V^{\lambda^1}(E(\sum_{j=1}^{n} r_j x_j)),$$

which implies that $V^{\lambda^2}(E(\sum_{j=1}^{n} r_j x_j^{2*})) \ge V^{\lambda^1}(E(\sum_{j=1}^{n} r_j x_j^{1*}))$.

(2) Since $\lambda_j^1 \le \lambda_j^2$ for all $j = 1, 2, \cdots, n$, we have that

$$V^{\lambda^1}(E(\sum_{j=1}^{n} r_j x_j)) \ge p_E \Rightarrow V^{\lambda^2}(E(\sum_{j=1}^{n} r_j x_j)) \ge V^{\lambda^1}(E(\sum_{j=1}^{n} r_j x_j)) \ge p_E,$$

where $\sum_{j=1}^{n} x_j = 1, x_j \ge 0$. That is, the feasible solution of (3.124) for $\lambda = \lambda^1$ is also feasible for $\lambda = \lambda^2$. This implies that $Var(\sum_{j=1}^{n} r_j x_j^{1**}) \ge Var(\sum_{j=1}^{n} r_j x_j^{2**})$. □

The result of Theorem 3.17 is illustrated by Fig. 3.9. Different investors have different expectations for the future return of securities, therefore they may choose investment strategies from different λ-man variance efficient frontiers. Consider three mean variance efficient portfolios A, B and C, where A is located on the λ^2-mean variance efficient frontier, B and C are located on the λ^1-mean variance efficient frontier. From Fig. 3.9 we know that the return rate of portfolio A is better than that of B when the risk levels of the two portfolios are p_V; On the other hand, to reach the same return rate value p_E the risk level at portfolio A is less than that of B. For investment strategies B and C located on the same λ^1-mean variance efficient frontier, the risk level and the expected return rate of portfolio C are larger than those of portfolio B. In conclusion, if $\lambda^2 > \lambda^1$, the λ^2-efficient frontier is above the λ^1-efficient frontier.

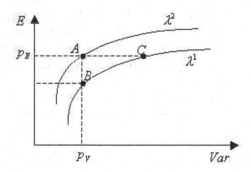

Fig. 3.9 The relationship of different λ-mean variance efficient portfolios

3.6.3.4   Application to the Chinese Stock Market

The data in the above subsection are also used to illustrate the proposed λ-mean variance portfolio selection model. Also we assume that the return rate of each security is a triangular fuzzy random variable, denoted by $r_j = (a_j - \alpha_j, a_j, a_j + \beta_j)$, where a_j is a normally distributed random variable, α_j and β_j are the left and right spread respectively, $j = 1, 2, \cdots, 30$.

To facilitate the determination of parameter λ_j for stock j, we propose to divide the investor into the following five classes according to his degree of optimism: (1) very optimistic: $\lambda_j \in (0.8, 1]$; (2) optimistic: $\lambda_j \in (0.6, 0.8]$; (3) neutral: $\lambda_j \in (0.4, 0.6]$; (4) pessimistic: $\lambda_j \in (0.2, 0.4]$; (5) very pessimistic: $\lambda_j \in [0, 0.2]$. For example, if an investor has an optimistic attitude for the future return rate of stock j, they can choose λ_j as the average value of 0.6 and 0.8, i.e. $\lambda_j = (0.6 + 0.8)/2 = 0.7$.

Suppose that an investor determines the parameter λ_j for all stocks (seen in Table 3.8). The corresponding λ-efficient frontier and the traditional mean variance efficient frontier are presented in Fig. 3.10.

From Fig. 3.10 we know that the λ-efficient frontier is above the traditional mean variance efficient frontier. Actually, the relationship between the λ efficient frontier and the traditional mean variance efficient frontier depends on the choice of parameter vector $\lambda = (\lambda_j)_{1 \times n}$. If an investor chooses a different subjective degree of optimism λ_j (seen in Table 3.9), then the corresponding λ-efficient frontier is under the mean variance efficient frontier (see Fig. 3.11); If an investor chooses different subjective degrees of optimism λ_j (seen in Table 3.10), then the two efficient frontiers intersect each other (see Fig. 3.12).

Moreover, three special λ-efficient frontiers are also presented in Fig. 3.13. If an investor has a conservative and pessimistic mind for the return rates of all stocks, the optimistic parameter vector λ should be set $(0, 0, \cdots, 0)_{1 \times 30}$. If an investor has an aggressive and optimistic mind, the optimistic parameter λ should be set as

Table 3.8 Investor's subjective optimistic degrees for the returns of 30 stocks (I)

Stock Code	600000	600030	600026	600050	600036	600717	600642	600688	600104	600009
λ_j	0.4	0.4	0.6	0.7	0.8	0.4	0.6	0.3	0.5	0.6
Stock Code	600008	600016	600808	600795	600033	600832	600011	600018	600019	600519
λ_j	0.5	0.3	0.8	0.9	0.5	0.4	0.6	0.5	0.7	0.3
Stock Code	600660	600879	600277	600188	600270	600205	600177	600309	600428	600320
λ_j	0.4	0.5	0.6	0.5	0.4	0.7	0.7	0.8	0.7	0.4

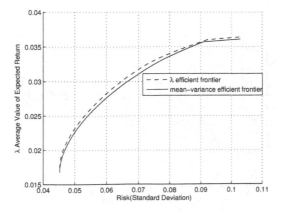

Fig. 3.10 λ efficient frontier and mean-variance efficient frontier (I)

Table 3.9 Investor's subjective optimistic degrees for the returns of 30 stocks (II)

Stock Code	600000	600030	600026	600050	600036	600717	600642	600688	600104	600009
λ_j	0.1	0.4	0.1	0.2	0.4	0.1	0.3	0.3	0.1	0.2
Stock Code	600008	600016	600808	600795	600033	600832	600011	600018	600019	600519
λ_j	0.2	0.3	0.4	0.3	0.2	0.4	0.3	0.2	0.5	0.3
Stock Code	600660	600879	600277	600188	600270	600205	600177	600309	600428	600320
λ_j	0.3	0.4	0.2	0.3	0.3	0.2	0.5	0.5	0.4	0.2

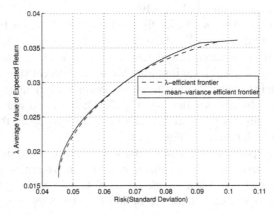

Fig. 3.11 λ efficient frontier and mean-variance efficient frontier (II)

Table 3.10 Investor's subjective optimistic degrees for the returns of 30 stocks (III)

Stock Code	600000	600030	600026	600050	600036	600717	600642	600688	600104	600009
λ_j	0.1	0.3	0.1	0.5	0.4	0.7	0.2	0.6	0.4	0.1
Stock Code	600008	600016	600808	600795	600033	600832	600011	600018	600019	600519
λ_j	0.2	0.3	0.8	0.2	0.1	0.1	0.4	0.2	0.1	0.3
Stock Code	600660	600879	600277	600188	600270	600205	600177	600309	600428	600320
λ_j	0.3	0.4	0.2	0.5	0.3	0.7	0.6	0.8	0.4	0.1

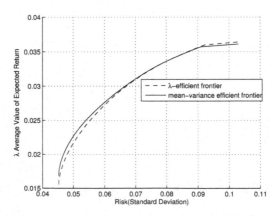

Fig. 3.12 λ efficient frontier and mean-variance efficient frontier (III)

$(1, 1, \cdots, 1)_{1 \times 30}$. If an investor has a neutral mind, the optimistic parameter λ should be set $(0.5, 0.5, \cdots, 0.5)_{1 \times 30}$. The pessimistic, optimistic and neutral efficient frontiers are given in Fig. 3.13.

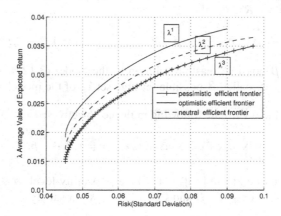

Fig. 3.13 λ efficient frontiers

From Theorem 3.17 we know that, if $\lambda^2 > \lambda^1$, the λ^2-efficient frontier is above the λ^1-efficient frontier. Because

$$\lambda^1 = (1, 1, \cdots, 1)_{1 \times 30} > \lambda^2 = (0.5, 0.5, \cdots, 0.5)_{1 \times 30} > \lambda^3 = (0, 0, \cdots, 0)_{1 \times 30},$$

the optimistic efficient frontier should be above the neutral efficient frontier and the neutral efficient frontier should be above the pessimistic efficient frontier. From Fig. 3.13 we know that the position relationships of the three efficient frontiers are accorded with the results of Theorem 3.17.

3.7 Another Way to Deal with Fuzzy Random Multi-objective Decision Making Model

In this section, we introduce another way to deal with the fuzzy random multi-objective decision making model, that is, we transform the fuzzy random variable into fuzzy variable. Here, we consider that fuzzy random multi-objective problem, i.e.,

$$\begin{cases} \max \left[\tilde{\bar{c}}_1 x, \tilde{\bar{c}}_2 x, \cdots, \tilde{\bar{c}}_m x \right] \\ \text{s.t. } x \in X = \{ x \in R^n | \tilde{\bar{A}} x \leq b, x \geq 0 \}. \end{cases} \tag{3.127}$$

where $\tilde{\bar{c}}_i (i = 1, 2, \cdots, m)$ and $\tilde{\bar{A}} = (\tilde{\bar{A}}_1 \tilde{\bar{A}}_2, \cdots, \tilde{\bar{A}}_r, \cdots, \tilde{\bar{A}}_p)^T$ are fuzzy random vectors, and $b = (b_1, b_2, \cdots, b_p)^T$.

3.7.1 (α, β)-Satisfied Solution for Fu-Ra Multi-objective Decision Making Model

Definition 3.16. (Xu and Liu [76]) Let $\beta = (\beta_1, \beta_2, \cdots, \beta_p)^T$ be possibility level vector, $\beta \in [0,1], x \in R^n$, and if

$$Pos(\tilde{\bar{A}}_{i}x \leq h_{r}, \tilde{\bar{r}}_{i}x) > \beta_r, \quad r = 1, 2, \cdots, p, i = 1, 2, \cdots, m,$$

then x is called β-possible feasible solution to the model (3.127). All β-possible feasible solutions are called β-possible feasible set X_β of the model (3.127).

Consider the problem with the following multiple objectives:

$$\max_{x \in D} [\tilde{\bar{c}}_1 x, \tilde{\bar{c}}_2 x, \cdots, \tilde{\bar{c}}_m x, \tilde{\bar{A}}_r x], \quad r = 1, 2, \cdots, p. \tag{3.128}$$

Definition 3.17. (Xu and Liu [76]) Let α be a possibility level, $\alpha \in [0,1], D \in R^n$ and $x_0 \in D$. if do not exist $x \in D$ and $k \in \{1, 2, \cdots, K\}$, x satisfy

$$Pos(\tilde{\bar{c}}_1 x \geq \tilde{\bar{c}}_1 x_0, \cdots, \tilde{\bar{c}}_{i-1} x \geq \tilde{\bar{c}}_{i-1} x_0, \tilde{\bar{c}}_i x > \tilde{\bar{c}}_i x_0, \tilde{\bar{c}}_{i+1} x \geq \tilde{\bar{c}}_{i+1} x_0, \cdots, \tilde{\bar{c}}_m x \geq \tilde{\bar{c}}_m x_0,$$
$$\tilde{\bar{A}}_{m+r} x \geq \tilde{\bar{A}}_{m+r} x_0) \geq \alpha, \quad r = 1, 2, \cdots, p, i = 1, 2, \cdots, m,$$
$$\tag{3.129}$$

then x_0 is called α-possible efficient solution of the model (3.128).

Definition 3.18. (Xu and Liu [76]) Let $x^0 \in X$, if x^0 be the α-possible efficient solution of problem

$$\begin{cases} \max [\tilde{\bar{c}}_1 x, \tilde{\bar{c}}_2 x, \cdots, \tilde{\bar{c}}_m x, \tilde{\bar{A}}_r x] \\ \text{s.t. } x \in X_\beta, \end{cases} \tag{3.130}$$

then x^0 is called (α, β)-satisfied solution of the model (3.127).

In fact, to solve the model (3.127) and find its (α, β)-satisfied solution, we may consider the multi-objective problem as follows:

$$\begin{cases} \max [(\tilde{\bar{c}}_1)_\alpha x, (\tilde{\bar{c}}_2)_\alpha x, \cdots, (\tilde{\bar{c}}_m)_\alpha x, (\tilde{\bar{A}}_r)_\alpha x] \\ \text{s.t. } x \in X_\beta, \end{cases} \tag{3.131}$$

where $(\tilde{\bar{c}}_i)_\alpha, (\tilde{\bar{A}}_r)_\alpha$ is α-level set of fuzzy random variables $\tilde{\bar{c}}_i, \tilde{\bar{A}}_r$ ($r = 1, 2, \cdots, p, i = 1, 2, \cdots, m$) respectively.

Theorem 3.18. *[76] x^0 is the (α, β)-satisfied solution of the model (3.127) if and only if x^0 is the efficient solution of the model (3.131).*

Proof. Let x^0 be the (α, β)-satisfied solution of the model (3.127). Following from the Definition 3.18, x^0 is the β-possible feasible solution and α-possible efficient solution to the model (3.130). If x^0 is not the efficient solution of the model

(3.131), then exist $x^1 \in X_\beta$ and row-vector $q_i, q_i \in (\tilde{\tilde{c}}_k, \tilde{\tilde{A}}_{m+r})_\alpha (i = 1, 2, \cdots, m, r = 1, 2, \cdots, p$, and $i_0 \in \{m+1, \cdots, m+p\})$, we can derive that

$$q_i x^1 \geq q_i x^0, \text{and } q_{i_0} x^1 > q_{i_0} x^0, \text{when } \forall i \in \{1, 2, \cdots, m, m+r\} \backslash \{i_0\}.$$

Due to $q_i \in (\tilde{\tilde{c}}_i, \tilde{\tilde{A}}_{m+r})_\alpha$, then $\min\{\pi_{\tilde{\tilde{c}}_1}(q_1), \pi_{\tilde{\tilde{c}}_2}(q_2), \cdots, \pi_{\tilde{\tilde{c}}_m}(q_m), \pi_{\tilde{\tilde{A}}_{m+r}}(q_{m+r})\} \geq \alpha, (r = 1, 2, \cdots, p$). Following from the Definition 3.16, based on the expand principle, we have

$$\sup_{(t_1, t_2, \cdots, t_m, t_{m+r}) \in T_{i_0}} \min[\pi_{\tilde{\tilde{c}}_1}(t_1), \cdots, \pi_{\tilde{\tilde{c}}_{i_0-1}}(t_{i_0-1}), \pi_{\tilde{\tilde{c}}_{i_0}}(t_{i_0}),$$
$$\pi_{\tilde{\tilde{c}}_{i_0+1}}(t_{i_0+1}), \cdots, \pi_{\tilde{\tilde{c}}_m}(t_m), \pi_{\tilde{\tilde{A}}_{m+r}}(t_{m+r})]$$
$$= Pos(\tilde{\tilde{c}}_1 x^1 \geq \tilde{\tilde{c}}_1 x^0, \cdots, \tilde{\tilde{c}}_{i_0-1} x^1 \geq \tilde{\tilde{c}}_{i_0-1} x^0, \tilde{\tilde{c}}_{i_0} x^1 > \tilde{\tilde{c}}_{i_0} x^0, \tilde{\tilde{c}}_{i_0+1} x^1 \geq \tilde{\tilde{c}}_{i_0+1} x^0, \cdots,$$
$$\tilde{\tilde{c}}_m x^1 \geq \tilde{\tilde{c}}_m x^0, \tilde{\tilde{A}}_{m+r} x^1 \geq \tilde{\tilde{A}}_{m+r} x^0) \geq \alpha,$$

where $T_{i_0} = \{(t_1, t_2, \cdots, t_m, t_{m+r}) | t_1 x^1 \geq t_1 x^0, \cdots, t_{i_0-1} x^1 \geq t_{i_0-1} x^0, t_{i_0} x^1 > t_{i_0} x^0, t_{i_0+1} x^1 \geq t_{i_0+1} x^0, \cdots, t_m x^1 \geq t_m x^0, t_{m+r} x^1 \geq t_{m+r} x^0\}, (r = 1, 2, \cdots, p$).

It is contrary that x^0 is not the efficient solution of the model (3.131).

Contrarily, let x^0 is the efficient solution of the model (3.131) and is not the (α, β)-satisfied solution of the model (3.127), then exist $x^2 \in X_\beta$ and

$$s \in \{1, 2, \cdots, m, m+r\}, (r = 1, 2, \cdots, p),$$

$$Pos(\tilde{\tilde{c}}_1 x^2 \geq \tilde{\tilde{c}}_1 x^0, \cdots, \tilde{\tilde{c}}_{s-1} x^2 \geq \tilde{\tilde{c}}_{s-1} x^0, \tilde{\tilde{c}}_s x^2 > \tilde{\tilde{c}}_s x^0,$$
$$\tilde{\tilde{c}}_{s+1} x^2 \geq \tilde{\tilde{c}}_{s+1} x^0, \cdots, \tilde{\tilde{c}}_m x^2 \geq \tilde{\tilde{c}}_m x^0, \tilde{\tilde{A}}_{m+r} x^2 \geq \tilde{\tilde{A}}_{m+r} x^0) \geq \alpha.$$

Following from the Definition 3.16, based on the expand principle, there is a row-vector $p_s \in R^n (s = 1, 2, \cdots, m, \cdots, m+r; r = 1, 2, \cdots, p)$ which satisfy $p_1 x^2 \geq p_1 x^0, \cdots, p_{s-1} x^2 \geq p_{s-1} x^0, p_s x^2 > p_s x^0, p_{s+1} x^2 \geq p_{s+1} x^0, \cdots, p_m x^2 \geq p_m x^0, p_{m+r} x^2 \geq p_{m+r} x^0$ and $\pi_{(\tilde{\tilde{c}}_i, \tilde{\tilde{A}}_{m+r})} \geq \alpha, p_i \in (\tilde{\tilde{c}}_i, \tilde{\tilde{A}}_{m+r})_\alpha$. It is contrary that x^0 is the efficient solution of the model (3.131).

The proof is thus completed. $\square$

Theorem 3.19. *[76] Let $\tilde{\tilde{\xi}}$ be a fuzzy random variable. α_1 is any given possibility level of fuzzy variable, σ is any given probability level of random variable, then the fuzzy random variable can be transformed a (α_1, σ)-level trapezoidal fuzzy variable.*

Proof. Let $\tilde{\tilde{\xi}} = (a_L, \rho, a_R)$, here, ρ be a random variable which has a normal distribution with the probability density function $\varphi(x) = \frac{1}{\sqrt{2\pi}\sigma_0} e^{-\frac{(x-\mu_0)^2}{2\sigma_0^2}}$. Then $\tilde{\tilde{\xi}}$ is a fuzzy random variable. See Figure 3.15. By the concept of random variable σ-level sets, denote the σ-level sets (or σ-cuts) of the random variable ρ as follows

$$\rho_\sigma = [\rho_\sigma^L, \rho_\sigma^R] = \{x \in U | \varphi(x) \geq \sigma\},$$

where $\rho_\sigma^L = \mu_0 - \sqrt{-2\sigma_0^2 ln(\sqrt{2\pi}\sigma\sigma_0)}$ and $\rho_\sigma^R = \mu_0 + \sqrt{-2\sigma_0^2 ln(\sqrt{2\pi}\sigma\sigma_0)}$, here $\mu_0 = \rho_0$. The parameter $\sigma \in [0,1]$ here reflects decision-maker's degree of optimism. These intervals indicate where the group arrival rate and service rate lie at probability level σ. Note that ρ_σ are crisp sets.

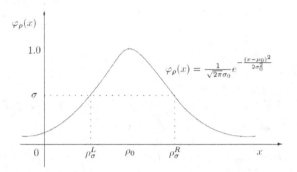

Fig. 3.14 The probability density function of fuzzy random variable ρ

Let $X = \{x_j \in X | \mu(x_j) \geq \sigma, j = 1,2,\cdots,n\}$, so the fuzzy random variable $\tilde{\bar{\xi}} = (a_L, \rho_\sigma, a_R)$ can be denoted as $\tilde{\bar{\xi}} = (a_L, X, a_R)$ or denoted as follows:

$$\tilde{\bar{\xi}} = \begin{cases} \bar{\xi}_1(a_L, x_1, a_R) \\ \bar{\xi}_2(a_L, x_2, a_R) \\ \vdots \\ \bar{\xi}_n(a_L, x_n, a_R), \end{cases}$$

where $x_1 \leq x_2 \leq \cdots \leq x_n$, $X = [x_1, x_n] = [\rho_\sigma^L, \rho_\sigma^R]$, $j = 1,2,\cdots,n$ and $\bar{\xi}_j$ is fuzzy variable. It is easy to reach that $x_1 = \rho_\sigma^L, x_n = \rho_\sigma^R$. In other words, ρ_σ^L is the minimum value that ρ achieves with probability σ, ρ_σ^R is the maximum value that ρ achieves with probability σ. The variable $\tilde{\bar{\xi}}$ can be expressed in another form as $\tilde{\bar{\xi}} = \bar{\xi}_1 \cup \bar{\xi}_2 \cup \cdots \cup \bar{\xi}_n$, here $\bar{\xi}_j$ is fuzzy variable. So the fuzzy variable $\tilde{\bar{\xi}}$ are transformed into some fuzzy variables with membership function $\mu_{\bar{\xi}}(X)$. On the basis of the concept of fuzzy variable α-level sets (or α-cuts), denote the α_1-level sets (or α_1-cuts) of $\bar{\xi}$ as follows

$$\bar{\xi}_{\alpha_1}(X) = [\bar{\xi}_{\alpha_1}^L, \bar{\xi}_{\alpha_1}^R] = \{X \in U | \mu_{\bar{\xi}}(X) \geq \alpha_1\}$$

or

$$\bar{\xi}_{\alpha_1}(X) = \begin{cases} \bar{\xi}_1 = (a_L, \rho_\sigma^L, a_R) \\ \bar{\xi}_2 = (a_L, x_2, a_R) \\ \vdots \\ \bar{\xi}_{n-1} = (a_L, x_{n-1}, a_R) \\ \bar{\xi}_n = (a_L, \rho_\sigma^R, a_R). \end{cases}$$

By the convexity of a fuzzy number, the bounds of these intervals are function of α_1. In order to make the method to be effective, we get $0 \leq \alpha_1 \leq \frac{a_R - a_L}{a_R - a_L + \rho_\sigma^R - \rho_\sigma^L}$ and can be obtained as $\bar{\bar{\xi}}_{\alpha_1} = [\xi_{\alpha_1}^L, \xi_{\alpha_1}^R]$. $\xi_{\alpha_1}^L = \max \xi_{j\alpha_1}^L$, $\xi_{\alpha_1}^R = \min \xi_{j\alpha_1}^R$, and $\xi_{j\alpha_1}^L = \min \mu_{\bar{\xi}_j}^{-1}(\alpha_1)$, $\xi_{j\alpha_1}^R = \max \mu_{\bar{\xi}_j}^{-1}(\alpha_1)$, $j = 1, 2, \cdots, n$, respectively. Obviously,

$$\xi_{\alpha_1}^L = \max \min \mu_{\bar{\xi}_j}^{-1}(\alpha_1) = \min \mu_{\bar{\xi}_n}^{-1}(\alpha_1),$$

$$\xi_{\alpha_1}^R = \min \max \mu_{\bar{\xi}_j}^{-1}(\alpha_1) = \max \mu_{\bar{\xi}_1}^{-1}(\alpha_1).$$

Consequently, we can use its α_1-cuts to construct the corresponding membership function. Let $\xi_{\alpha_1}^L = \underline{a}, \xi_{\alpha_1}^R = \overline{a}$, viz. $\bar{\bar{\xi}} = (a_L, \underline{a}, \overline{a}, a_R)$. Thus, the fuzzy random variable $\bar{\bar{\xi}}$ is transformed into a fuzzy variable which is a similar trapezoidal fuzzy number with the membership function $\mu_{\bar{\xi}}(U)$. The value of $\mu_{\bar{\xi}}(U)$ at $x \in [\underline{a}, \overline{a}]$ is considered subjectively to be 1 approximately. See Figure 3.15.

$$\bar{\bar{\xi}} = \bar{\omega}_{\bar{\xi}} = (a_L, \underline{a}, \overline{a}, a_R). \tag{3.132}$$

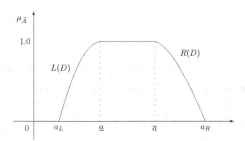

Fig. 3.15 The membership function of $\tilde{\omega}_{\bar{\xi}}$

The Theorem 3.19 is proved.   □

Definition 3.19. (Xu and Liu [76]) Let a domain U. $\bar{\bar{\xi}}$ be a fuzzy random variable defined on U. $\alpha_1 \in [0, 1]$ be any given possibility level of fuzzy variable, $\sigma \in [0, 1]$ be any given probability level of random variable. Basis on the Lemma 3.19. If $\varphi(x) \geq \sigma$ *and* $\mu(x) \geq \alpha_1$ the fuzzy random variable $\bar{\bar{\xi}}$ can be transformed a (α_1, σ)-level trapezoidal fuzzy variable $\tilde{B}$. So B_α consist of all elements whose degrees of membership in $\tilde{B}$ are greater than or equal to α,

$$B_\alpha = \{x \in U | \mu_{\tilde{B}}(x) \geq \alpha, \varphi(x) \geq \sigma, \mu(x) \geq \alpha_1\},$$

then B_α is called the α-level sets of fuzzy random variable $\bar{\bar{\xi}}$ with $\varphi(x) \geq \sigma$ *and* $\mu(x) \geq \alpha_1$, viz. $\bar{\bar{\xi}}_{\alpha_{(\alpha_1, \sigma)}}$.

Definition 3.20. (Xu and Liu [76]) Let $L_\alpha(\tilde{\bar{\mathbf{a}}}, \tilde{\bar{\mathbf{b}}}, \tilde{\bar{\mathbf{c}}})$ be the α-level set of fuzzy random variables $\tilde{\bar{a}}_{rj}, \tilde{\bar{b}}_r, \tilde{\bar{c}}_{ij}$. α be any given possibility level of fuzzy variable, σ be any given probability level of random variable. Then we have

$$L_\alpha(\tilde{\bar{\mathbf{a}}}, \tilde{\bar{\mathbf{b}}}, \tilde{\bar{\mathbf{c}}}) = \{(\mathbf{a}, \mathbf{b}, \mathbf{c}) | \pi_{(\tilde{\bar{a}}_{rj})_{(\alpha_1,\sigma)}}(\tilde{\bar{a}}_{rj})_{(\alpha_1,\sigma)} \geq \alpha, \pi_{(\tilde{\bar{b}}_r)_{(\alpha_1,\sigma)}}(\tilde{\bar{b}}_r)_{(\alpha_1,\sigma)} \geq \alpha,$$
$$\pi_{(\tilde{\bar{c}}_{ij})_{(\alpha_1,\sigma)}}(\tilde{\bar{c}}_{ij})_{(\alpha_1,\sigma)} \geq \alpha, i = 1,2,\cdots,m, j = 1,2,\cdots,n, r = 1,2,\cdots,p\},$$

viz.

$$L_\alpha(\tilde{\bar{\mathbf{a}}}, \tilde{\bar{\mathbf{b}}}, \tilde{\bar{\mathbf{c}}}) = L_\alpha(\tilde{\bar{\mathbf{a}}}_{(\alpha_1,\sigma)}, \tilde{\bar{\mathbf{b}}}_{(\alpha_1,\sigma)}, \tilde{\bar{\mathbf{c}}}_{(\alpha_1,\sigma)}).$$

Theorem 3.20. *[76] Let $\tilde{\bar{a}}, \tilde{\bar{b}}, \tilde{\bar{c}}$ be fuzzy random variables, (α_1, σ) be any given possibility level and probability level respectively, $L_\alpha(\tilde{\bar{a}}, \tilde{\bar{b}}, \tilde{\bar{c}})$ be the α-level set of fuzzy random numbers $(\tilde{\bar{a}}, \tilde{\bar{b}}, \tilde{\bar{c}})$. For different (α_1, σ)-level it is clear that*
(a) if $\sigma^1 \leq \sigma^2, \alpha_1^1 \geq \alpha_1^2$, then $L_\alpha^1(\tilde{\bar{a}}, \tilde{\bar{b}}, \tilde{\bar{c}}) \subseteq L_\alpha^2(\tilde{\bar{a}}, \tilde{\bar{b}}, \tilde{\bar{c}})$,
(b) if $\sigma^1 \geq \sigma^2, \alpha_1^1 \leq \alpha_1^2$, then $L_\alpha^1(\tilde{\bar{a}}, \tilde{\bar{b}}, \tilde{\bar{c}}) \supseteq L_\alpha^2(\tilde{\bar{a}}, \tilde{\bar{b}}, \tilde{\bar{c}})$.

Proof. It follows from the Lemma 3.19 that a fuzzy random variable $\tilde{\bar{\xi}}$ can be transformed into a fuzzy variable, viz. $\tilde{\bar{\xi}} \Rightarrow (a_L, \underline{a}, \bar{a}, a_R)$. Let $\tilde{\bar{\xi}} = (\tilde{\bar{a}}, \tilde{\bar{b}}, \tilde{\bar{c}})$ and depending on the Definition 3.20, we can derive $L_\alpha(\tilde{\bar{\xi}}) = \{a | \pi_{(\tilde{\bar{a}})_{(\alpha_1,\sigma)}}(\tilde{\bar{a}})_{(\alpha_1,\sigma)} \geq \alpha\} = [\xi_{\alpha}^L, \xi_{\alpha}^R]$. $\xi_{\alpha}^L = a_L + (\underline{a} - a_L)\alpha$, $\underline{a} = \xi_{\alpha_1}^L$, $\xi_{\alpha_1}^L = a_L + (\rho_\sigma^R - a_L)\alpha_1$ and $\rho_\sigma^R = \mu_0 + \sqrt{-2\sigma_0^2 ln(\sqrt{2\pi}\sigma\sigma_0)}$.

Finally, we have $\xi_\alpha^L = a_L + \{a_L + [\mu_0 + \sqrt{-2\sigma_0^2 ln(\sqrt{2\pi}\sigma\sigma_0)} - a_L]\alpha_1 - a_L\}\alpha$. Then ξ_α^L is a decreasing function about σ and a increasing function about α_1. For any given $0 \leq \sigma \leq 1, 0 \leq \alpha_1 \leq 1$, if $\sigma^1 \leq \sigma^2$ and $\alpha_1^1 \geq \alpha_1^2$, we have $\xi_\alpha^{L1} \geq \xi_\alpha^{L2}$. Similarly, we may prove $\xi_\alpha^{R1} \leq \xi_\alpha^{R2}$. Thus $[\xi_\alpha^{L1}, \xi_\alpha^{R1}] \subseteq [\xi_\alpha^{L2}, \xi_\alpha^{R2}]$, then we obtain $L_\alpha^1(\tilde{\bar{\xi}}) \subseteq L_\alpha^2(\tilde{\bar{\xi}})$, viz. $L_\alpha^1(\tilde{\bar{a}}, \tilde{\bar{b}}, \tilde{\bar{c}}) \subseteq L_\alpha^2(\tilde{\bar{a}}, \tilde{\bar{b}}, \tilde{\bar{c}})$.

Similarly, we may prove that if $\sigma^1 \geq \sigma^2, \alpha_1^1 \leq \alpha_1^2$, then $L_\alpha^1(\tilde{\bar{a}}, \tilde{\bar{b}}, \tilde{\bar{c}}) \supseteq L_\alpha^2(\tilde{\bar{a}}, \tilde{\bar{b}}, \tilde{\bar{c}})$. The proof is thus completed. $\square$

Thus, the fuzzy random multi-objective model (3.127) is transformed into a fuzzy multi-objective model.

$$\begin{cases} \max \; [\tilde{C}_1 x, \tilde{C}_2 x, \cdots, \tilde{C}_m x] \\ \text{s.t. } x \in X = \{x \in R^n | \tilde{A}'x \leq b, x \geq 0\}, \end{cases} \tag{3.133}$$

where $\tilde{C}_i = \tilde{\omega}_{\tilde{\bar{C}}_{i(\alpha_1,\sigma)}}, (i = 1, 2, \cdots, m)$ and $\tilde{A}' = (\tilde{\omega}_{\tilde{\bar{A}}_1}, \tilde{\omega}_{\tilde{\bar{A}}_2}, \cdots, \tilde{\omega}_{\tilde{\bar{A}}_p})^T$ are fuzzy vectors.

3.7.2 Application to Inventory Problems

For inventory problems, lots of work has been done. Chih Hsun Hsieh [200] gave the optimization of fuzzy production inventory models. In this section, we consider a multi-item inventory system of deteriorating items with an infinite rate of replenishment, no shortages, stock-dependent demand and limited storage space [201]. Here, we consider the selling price, purchase price, holding cost, set-up cost and available total budgetary cost as fuzzy random variables.

3.7.2.1 Denotation and Modelling

In this system, see Figure. 3.16, there are n products in the warehouse. the demand $D_i(t)$ of products of an item are influenced by inventory level $q_i(t)$. The following notations are used in the formulation of the model, For ith ($i = 1,...,n$) item, it is assumed that

n: number of items
A: available floor/storage space
B: available total budgetary cost
s_i: selling price of each product
p_i: purchase price of each product
h_i: holding cost per unit quantity per unit time
u_i: set-up cost per cycle
a_i: constant rate of deterioration, $0 < a_i < 1$
A_i: required storage area per unit quantity
T_i: time period for each cycle
$q_i(t)$: inventory level at time t
Q_i: order quantity
$Z_i(Q_i)$: average profit of the ith item
$D_i(q_i)$: quantity of demand at time t, $D_i(q_i) = b_i + c_i q_i(t)$ (where b_i and c_i being constant, $0 < c_i < 1$)
$TC_i(Q_i)$: Total average cost of the ith item
$PF(Q_i)$: total average profit $PF(Q_i) = \sum_{i=1}^{n} Z_i(Q_i)$
$WC(Q_i)$: total wastage cost of the ith item

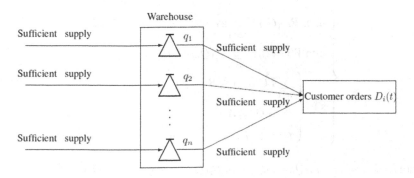

Fig. 3.16 A multi-item inventory system

Here, $D = (D_1, D_2, \ldots, D_n)^T$, $Q = (Q_1, Q_2, \ldots, Q_n)^T$. If $q_i(t)$ is the inventory level at time t of the ith, then $\frac{dq_i}{dt} = -D_i(q_i) - a_i q_i$. So, the length of the cycle of the ith item is

$$T_i = \int_0^{Q_i} \frac{dq_i}{D_i(q_i) + a_i q_i} = \int_0^{Q_i} \frac{dq_i}{b_i + c_i q_i + a_i q_i} = \frac{1}{a_i + c_i} ln\left(\frac{b_i + (a_i + c_i)Q_i}{b_i}\right).$$

The holding cost in each cycle for the ith item is $h_i g_i(Q_i)$, where

$$g_i(Q_i) = q_i T_i = \int_0^{Q_i} \frac{q_i dq_i}{b_i + c_i q_i + a_i q_i} = \frac{Q_i}{(a_i + c_i)} - \frac{b_i}{(a_i + c_i)^2} ln\left(\frac{b_i + (a_i + c_i)Q_i}{b_i}\right).$$

The total number of deteriorating units of ith item is $\theta_i(Q_i) = a_i g_i(Q_i)$. The net revenue of the ith item is $N(Q_i) = (s_i - p_i)Q_i - s_i\theta_i(Q_i)$. Hence, total average profit of the ith item is

$$PF(Q_i) = [N(Q_i) - h_i g_i(Q_i) - u_i]/T_i.$$

Total wastage cost of the ith item

$$WC(Q_i) = \sum_{i=1}^{n} \theta_i(Q_i)p_i/T_i.$$

Total average cost of the ith item is

$$TC_i(Q_i) = [p_i Q_i + h_i g_i(Q_i) + u_i]/T_i.$$

Because the above selling price, purchase price, holding cost, set-up cost and available total budgetary cost are all positive, when they are fuzzy random variables, the random variables part has a log-normal distribution and its probability density function is

$$\varphi(x) = \frac{1}{\sqrt{2\pi}\sigma_0 x} e^{-\frac{(lnx - \mu_0)^2}{2\sigma_0^2}}. \tag{3.134}$$

If the decision maker wants to maximize total average profit and minimize total wastage cost, then the problem can be formulated by the following model:

$$\begin{cases} \max PF(Q_i) = \sum_{i=1}^{n} [\tilde{\bar{N}}(Q_i) - \tilde{\bar{h}}_i g_i(Q_i) - \tilde{\bar{u}}_i]/T_i \\ \min WC(Q_i) = \sum_{i=1}^{n} \theta_i(Q_i)\tilde{\bar{p}}_i/T_i \\ \text{s.t.} \begin{cases} \sum_{i=1}^{n} \tilde{\bar{TC}}_i(Q_i) \leq B \\ \sum_{i=1}^{n} A_i Q_i \leq A \\ Q_i > 0, \ i = 1, 2, 3, \ldots, n, \end{cases} \end{cases} \tag{3.135}$$

where $\tilde{\bar{N}}(Q_i) = (\tilde{\bar{s}}_i - \tilde{\bar{p}}_i)Q_i - \tilde{\bar{s}}_i\theta_i(Q_i)$, $\tilde{\bar{TC}}_i(Q_i) = [\tilde{\bar{p}}_i Q_i + \tilde{\bar{h}}_i g_i(Q_i) + \tilde{\bar{u}}_i]/T_i$.

3.7.2.2 Solving Process of the Proposed Inventory Model

In fuzzy set theory, the coefficients of the fuzzy objectives and fuzzy constraints are taken as fuzzy numbers, namely trapezoidal fuzzy numbers, triangular fuzzy numbers, LR fuzzy numbers, etc. Here, we assume that all coefficients to be fuzzy random variables. To solve the fuzzy multi-objective inventory model described in equations (3.137), we use the (α, β) satisfied method based on Zimmermann[107]. The step-by-step procedure of the (α, β) satisfied approach can be described as follows:

Step 1. Decision-makers give the value of α_1 and σ depending on their experiences or former dates.

Step 2. On the basis of the above discussions about Lemma 4.17, the fuzzy random variable $\tilde{\tilde{\xi}}$ is transformed into a fuzzy variable which is a similar trapezoidal fuzzy number. We consider the above multi-objective inventory model (3.135), where the selling price $\tilde{\tilde{s}}_i$, purchase price $\tilde{\tilde{p}}_i$, holding cost $\tilde{\tilde{h}}_i$ and set-up cost $\tilde{\tilde{u}}_i$ are fuzzy random variables. For each α_1 and σ is given by decision-maker. we can derive that

$$\tilde{\tilde{s}}_i = \tilde{\omega}_{\tilde{s}_i} = (s_L, \underline{s}, \overline{s}, s_R),$$
$$\tilde{\tilde{p}}_i = \tilde{\omega}_{\tilde{p}_i} = (p_L, \underline{p}, \overline{p}, p_R),$$
$$\tilde{\tilde{h}}_i = \tilde{\omega}_{\tilde{h}_i} = (h_L, \underline{h}, \overline{h}, h_R),$$
$$\tilde{\tilde{u}}_i = \tilde{\omega}_{\tilde{u}_i} = (u_L, \underline{u}, \overline{u}, u_R).$$

which can be specified by the $\tilde{\omega}_{\tilde{\xi}} = (a_L, \underline{a}, \overline{a}, a_R)$ with membership function:

$$\mu_{\tilde{\omega}_{\tilde{\xi}}}(t) = \begin{cases} \frac{t-a_L}{\underline{a}-a_L}, & \text{if } a_L \leq t \leq \underline{a} \\ 1, & \text{if } \underline{a} \leq t \leq \overline{a} \\ \frac{a_R-t}{a_R-\overline{a}}, & \text{if } \overline{a} \leq t \leq a_R \\ 0, & \text{if } t < a_L, t > a_R. \end{cases} \tag{3.136}$$

And then, the fuzzy random multi-objective model (3.135) is transformed into a fuzzy multi-objective model.

$$\begin{cases} \max PF(Q_i) = \sum_{i=1}^{n} [\tilde{N}(Q_i) - \tilde{h}_i g_i(Q_i) - \tilde{u}_i]/T_i \\ \min WC(Q_i) = \sum_{i=1}^{n} \theta_i(Q_i)\tilde{p}_i/T_i \\ \text{s.t.} \begin{cases} \sum_{i=1}^{n} \tilde{TC}_i(Q_i) \leq B \\ \sum_{i=1}^{n} A_i Q_i \leq A \\ Q_i > 0, \ i = 1,2,3,\cdots,n, \end{cases} \end{cases} \tag{3.137}$$

where $\tilde{N}(Q_i) = (\tilde{\omega}_{\tilde{s}_i} - \tilde{\omega}_{\tilde{p}_i})Q_i - \tilde{\omega}_{\tilde{s}_i}\theta_i(Q_i)$, $\tilde{TC}_i(Q_i) = [\tilde{\omega}_{\tilde{p}_i}Q_i + \tilde{\omega}_{\tilde{h}_i}g_i(Q_i) + \tilde{\omega}_{\tilde{u}_i}]/T_i$, $\tilde{p}_i = \tilde{\omega}_{\tilde{p}_i}$, $\tilde{h}_i = \tilde{\omega}_{\tilde{h}_i}$, $\tilde{u}_i = \tilde{\omega}_{\tilde{u}_i}$.

Step 3. Give the α-cut of $\widetilde{\omega}$ and be expressed by the following interval

$$\widetilde{\omega}_\alpha = [a_L + (\underline{a} - a_L)\alpha, \underline{a}, \overline{a}, a_R - (a_R - \overline{a})\alpha].$$

Step 4. Based on the (α, β) satisfied approach. Let $(Q_i)_\alpha^\beta$ be a solution of the fuzzy non-linear programming model in equation (3.137), where $\alpha \in [0, 1]$ denotes the level of possibility at which all fuzzy inventory costs and prices are feasible and $\beta \in [0, 1]$ denotes the grade of compromise to which the solution satisfies all of the fuzzy objectives and constraints keeping the coefficients at a feasible level α. Hence, based on the Theorem 3.18, equation (3.137) is reduced to

$$\begin{cases} \max\min\ [\alpha, \beta] \\ \quad \begin{cases} \mu_{\widetilde{PF}}(Q_i, \alpha) \geq \beta \\ \mu_{\widetilde{WC}}(Q_i, \alpha) \geq \beta \\ \mu_{\widetilde{TC}}(Q_i, \alpha) \geq \beta \\ \sum\limits_{i=1}^{n} A_i Q_i \leq A \\ 0 \leq \alpha \leq 1, \beta, \lambda \leq 1. \end{cases} \end{cases} \qquad (3.138)$$

Step 5. Give the concrete forms of the linear membership functions $\mu_{\widetilde{PF}}(Q_i, \alpha)$, $\mu_{\widetilde{WC}}(Q_i, \alpha)$ and $\mu_{\widetilde{TC}}(Q_i, \alpha)$ for two objectives and constraints, respectively, as follows:

$$\mu_{\widetilde{PF}}(Q_i, \alpha) = \begin{cases} 0, & \text{if } PF(Q_i, \alpha) < U_{PF} - P_{PF} \\ 1 + \frac{PF(Q_i, \alpha) - U_{PF}}{P_{PF}}, & \text{if } U_{PF} - P_{PF} \leq PF(Q_i, \alpha) \leq U_{PF} \\ 1, & \text{if } PF(Q_i, \alpha) > U_{PF}, \end{cases} \qquad (3.139)$$

$$\mu_{\widetilde{WC}}(Q_i, \alpha) = \begin{cases} 0, & \text{if } WC(Q_i, \alpha) > L_{WC} + P_{WC} \\ 1 - \frac{WC(Q_i, \alpha) - L_{WC}}{P_{WC}}, & \text{if } L_{WC} \leq WC(Q_i, \alpha) \leq L_{WC} + P_{WC} \\ 1, & \text{if } WC(Q_i, \alpha) < L_{WC}, \end{cases}$$

$$(3.140)$$

$$\mu_{\widetilde{TC}}(Q_i, \alpha) = \begin{cases} 0, & \text{if } \sum\limits_{i=1}^{n} TC_i(Q_i, \alpha) > C + P_{TC} \\ 1 - \frac{\sum\limits_{i=1}^{n} TC_i(Q_i, \alpha) - C}{P_{TC}}, & \text{if } C \leq \sum\limits_{i=1}^{n} TC_i(Q_i, \alpha) \leq C + P_{TC} \\ 1, & \text{if } \sum\limits_{i=1}^{n} TC_i(Q_i, \alpha) < C. \end{cases} \qquad (3.141)$$

Here, P_{PF} is the maximum and P_{WC}, P_{TC} are the minimum acceptable violation of the aspiration levels U_{PF}, L_{WC} and C, respectively.

Step 6. Depending on Definition 3.19 and Theorem 3.20, equation (4.143) can be calculated as

$$
\begin{cases}
\max\min\{\alpha,\beta\} \\
\text{s.t.}
\begin{cases}
1 + \dfrac{PF(Q_i,\alpha)-U_{PF}}{P_{PF}} \geq \beta \\
1 - \dfrac{WC(Q_i,\alpha)-L_{WC}}{P_{WC}} \geq \beta \\
1 - \dfrac{\sum\limits_{i=1}^{n} TC_i(Q_i,\alpha)-C}{P_{TC}} \geq \beta \\
\sum\limits_{i=1}^{n} A_i Q_i \leq A \\
0 \leq \alpha \leq 1, \beta, \lambda \leq 1,
\end{cases}
\end{cases}
\tag{3.142}
$$

where

$$
PF(Q_i,\alpha) = \sum_{i=1}^{n} [\tilde{\omega}^R_{\tilde{s}_{i(\alpha_1,\sigma)}}(\alpha)Q_i - \tilde{\omega}^L_{\tilde{\bar{p}}_{i(\alpha_1,\sigma)}}(\alpha)Q_i - \tilde{\omega}^L_{\tilde{s}_{i(\alpha_1,\sigma)}}(\alpha)\theta_i(Q_i)
$$
$$
- \tilde{\omega}^L_{\tilde{h}_{i(\alpha_1,\sigma)}}(\alpha)g_i(Q_i) - \tilde{\omega}^L_{\tilde{u}_{i(\alpha_1,\sigma)}}(\alpha)]/T_i,
$$
$$
WC(Q_i,\alpha) = \sum_{i=1}^{n} \theta_i(Q_i)\tilde{\omega}^L_{\tilde{\bar{p}}_{i(\alpha_1,\sigma)}}(\alpha)/T_i,
$$
$$
\sum_{i=1}^{n} TC_i(Q_i,\alpha) = \sum_{i=1}^{n} [\tilde{\omega}^L_{\overline{bar}p_{i(\alpha_1,\sigma)}}Q_i + \tilde{\omega}^L_{\tilde{h}_{i(\alpha_1,\sigma)}}g_i(Q_i) + \tilde{\omega}^L_{\tilde{u}_{i(\alpha_1,\sigma)}}]/T_i.
$$

Step 7. Solve and calculate the above certain programming model (3.142).

If the decision maker is satisfied with the current values or results, otherwise, ask the decision maker to the update reference membership levels or the (α_1,σ) values by taking account of the current results and the membership function values, then return to step 1.

Chapter 4
Bifuzzy Multiple Objective Decision Making

When we dig into the uncertainty of a fuzzy set, there are two cases: the memberships are also fuzzy, and the elements are also fuzzy. So there exists a level-2 fuzzy set and type-2 fuzzy set which originally proposed by Zadeh. Then some scholars elaborated these two concepts, respectively.

Based on the type-2 fuzzy set, Liu defined bifuzzy variables and related properties. Similar to the framework of chapter 2, in this chapter, we also discusse the following three models:

(1) Bifuzzy expected value model (Fu-Fu EVM).
(2) Bifuzzy chance constrained model (Fu-Fu EVM).
(3) Bifuzzy dependent chance model (Fu-Fu EVM).

Finally, an application to a purchasing problem under a bifuzzy environment is presented as an illustration.

4.1 Raw Material Purchasing Problem under Bifuzzy Environment

Since the late 1960s, the raw materials purchasing issues have been tackled well by enterprises, and is regarded as a strategic management decision as well as a competitive weapon [345, 346, 347]. Essentially, this is due to the importance of the purchase of raw materials; it is not only the beginning of production operation activities and linked to the connection between production operation activities, but is also the major part of the production costs for an enterprise, accounting for 60%-80% [348]. The question as to how to make effective purchasing policies, i.e. 'optimal purchasing policies' has come to have many new characteristics in the recent 10 years because of that the trend of globalization of market s and development of supply chains. This important research area now is based on mathematical models and is using quantitative methods [349, 350, 351, 352].

Fig. 4.1 is the raw material supply system of a certain large-scale integrated steel plant in China. In this plant, approximately 10 million tons of steel-iron products are produced per year; slabs, hot rolled coils, wire, cold rolled plates, sheets, etc.. Consequently, large quantities of raw materials are required in this plant every year. The bulk of the raw materials has over 100 items, with total quantity reaching 30 million

J. Xu and X. Zhou: Fuzzy-Like Multiple Objective Decision Making, STUDFUZZ 263, pp. 227–294.
springerlink.com   © Springer-Verlag Berlin Heidelberg 2011

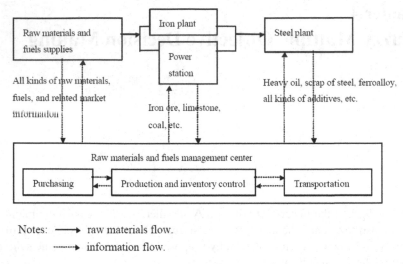

Notes: ⟶ raw materials flow.
........▶ information flow.

Fig. 4.1 Raw materials supply system

tons per year. Its requisite raw materials are mainly used to supply coking, sinter-
ing, power plants, iron production, steel production, and so on. These raw materials
can be divided into four classes: basic raw materials, heavy oil, scrap iron-steel, and
ferroalloy. The basic raw materials include primary raw materials, secondary raw
materials, and fuels. The primary raw materials include iron ores (e.g., crude ore,
size preparation ore, fine ore, ore screenings, pellet ore, sinter ore, etc.) and man-
ganese ore. The secondary raw materials include limestone, light-burned dolomite,
primary lime, etc. Fuels includes coal, coke, etc. Facing such a large requirement
for raw materials in item and quantity, the question as to how to rationally organize
purchasing in order to make it not only meet the production requirements, but also
to obtain maximum benefit from purchasing policies is important to an enterprise.

The main problem the purchasing of raw materials faces is how to make purchasing decisions, in order to obtain required raw materials at a lower price and at the
sametime meet production demand in terms of item, quality, quantity, due date, and
so on. However, in general, this is not easy to achieve, because these criteria are
often in conflict with each other. For example, the better the quality of the product,
the higher the price, while the lower the price, the poorer the quality. Therefore, one
of the main objectives of optimizing purchasing decisions is often a trade-off. These
factors decide the complexity the purchasing decisions. On the other hand, we can
see that the criteria mentioned above are in connection with three kinds of decisions,
i.e., item decision, quantity decision and vendor selection decision. The details are
in the following.

The first is to decide items. In the iron-steel industry, the bulk raw materials are
chiefly are of iron ores, secondary raw materials and fuels. There are approximately
100 items, and the meaning of item is based on the classification to which it belongs.
Iron ores, secondary raw materials, and fuels can be divided further into a number
of classes, and each class can then be subdivided into a number of sub-classes, e.g.

the same kind of iron ore can be divided into a number of grades. Consequently, the number of items is very large. In addition, considering various items combination situations because the steel-iron making process is a very complex metallurgical process, influenced by used raw material items, operation conditions, and working sequence, different items or items combination may produce the same molten iron or molten steel, but different items combination cost varies significantly. Consequently, how to select appropriate items is the key to reduce the production cost.

The second is to decide quantities. The iron-steel production not only requires more items, but also a higher quantity of raw materials compared to other industries. To meet customers requirement and guarantee product quality in the iron-steel production process, different items are required in different quantities and they must satisfy defined proportionality relations. For example, in traditional iron making technology, iron ore, secondary raw materials and fuels are in the proportion 2:0.5:1. Therefore, to make a purchasing decision quantity proportion relations among items needs to be considered. We call such decision the proportionality relation in the process as selecting 'assigning ore and assigning coal schemes'. This decision needs strong professional knowledge of iron-steel metallurgy.

The third phase is to select vendors. To keep the stability and quality of the supply of raw materials, a plant must consider the influence of external suppliers. Actually, this integrated steel plant has about 30 vendors distributed over the world, and every vendor supplies 1-4 kinds of bulk raw materials. Each raw material item differs from vendor to vendor in terms of quality, price, service and so on. Thus selecting-vendor decision will positively affect item, quality, price decisions, etc.. Within the model, we regard the vendor selection as an important consideration. Selecting vendors and deciding the order quantity of a selected vendor is also part of the vendor selection problem. The vendor selection problem has already been discussed in many papers in the past 30 years [353, 354, 355, 356, 357]. We focus on studying a special kind of 'the vendor selection problem' under satisfying 'assigning ore and assigning coal schemes' to the purchase of raw materials of a steel plant.

4.2 Fu-Fu Variable

Generally speaking, a level-2 fuzzy set is a fuzzy set in which the elements are also fuzzy sets, and the Fu-Fu variable is a fuzzy variable with fuzzy parameters.

4.2.1 Level-2 Fuzzy Set, Type-2 Fuzzy Sets and Fu-Fu Variable

Let fuzzy sets, as described above, be called ordinary fuzzy sets. With the appropriate interpretation of their membership function [32], ordinary fuzzy sets can successfully be used to handle imperfect information from one single source, which is either uncertain, imprecise, vague or incomplete. A survey of existing fuzzy modelling approaches is presented in [422]. For handling imperfect information whereby two or more sources of imperfection appear simultaneously, the modelling facilities with ordinary fuzzy sets are limited. Two types of advanced fuzzy sets have been previously discussed by researchers, which are level-2 fuzzy set and type-2 fuzzy set.

Level-2 fuzzy sets were originally presented by Zadeh [11] in 1971 and were further elaborately by Gottwald [21]. Such sets are fuzzy sets whose elements themselves are ordinary fuzzy sets. They are very useful in circumstances where it is difficult to determine some elements for a fuzzy set.

Definition 4.1. (Gottwald [21]) A level-2 fuzzy set $\tilde{\tilde{V}}$ defined over a universal set U is defined by

$$\tilde{\tilde{V}} = \{(\tilde{V}, \mu_{\tilde{\tilde{V}}}(\tilde{V})) | \forall \tilde{V} \in \mathscr{F}(U) : \mu_{\tilde{\tilde{V}}} > 0\}, \tag{4.1}$$

where each ordinary fuzzy set $\tilde{V}$ is defined by

$$\tilde{V} = \{(x, \mu_{\tilde{V}}(x)) | \forall x \in U : \mu_{\tilde{V}} > 0\}. \tag{4.2}$$

For convenience, the membership grades $\mu_{\tilde{\tilde{V}}}(\tilde{V})$ of the fuzzy sets $\tilde{V} \in \mathscr{F}(U)$ are called 'outer-layer' membership grades, whereas the membership grades $\mu_{\tilde{V}}(x)$ of the elements $x \in U$ are called 'inner-layer' membership grades. Since level-2 fuzzy sets are still fuzzy sets, their mathematical behavior is defined by the fuzzy set operators [21].

Type-2 fuzzy sets were introduced by Zadeh [19] in 1975 as another extension of the concept of an ordinary fuzzy set, and it was elaborated by Mendel, Karnik and John [419, 420]. Such sets are fuzzy sets whose membership grades them as ordinary fuzzy sets. They are very useful in circumstances where it is difficult to determine an exact membership function for a fuzzy set.

Definition 4.2. (Mendel [420]) A type-2 fuzzy set, denoted $\tilde{A}$, is characterized by a type-2 membership function $\mu_{\tilde{A}}(x, u)$, where $x \in X$ and $u \in J_x \subseteq [0, 1]$, i.e.,

$$\tilde{A} = \{((x, u), \mu_{\tilde{A}}(x, u)) | \forall x \in X, \forall u \in J_x \subseteq [0, 1]\}, \tag{4.3}$$

in which $0 \leq \mu_{\tilde{A}}(x, u) \leq 1$. $\tilde{A}$ can also be expressed as

$$\tilde{A} = \int_{x \in X} \int_{u \in J_x} \mu_{\tilde{A}}(x, u) / (x, u), J_x \subseteq [0, 1]. \tag{4.4}$$

In the above definition, there are two grades of membership, J_x is the primary membership of x, where $J_x \subseteq [0, 1]$ for $\forall x \in X$. And $\mu_{\tilde{A}}(x, u)(x \in X, u \in J_x)$ is the secondary grade.

Normally speaking, a Fu-Fu variable ξ is a fuzzy variable under fuzzy environment.

Definition 4.3. A Fu-Fu variable ξ is a fuzzy variable with fuzzy parameters.
Example 4.1. Let $\tilde{\rho}_1, \tilde{\rho}_2, \cdots, \tilde{\rho}_n$ be fuzzy numbers and $pos_1, pos_2, \cdots, pos_n$ be real numbers in $[0, 1]$ such that $pos_1 \vee pos_2 \vee \cdots \vee pos_n = 1$. Then

$$\xi = \begin{cases} \tilde{\rho}_1 \text{ with possibility } pos_1 \\ \tilde{\rho}_2 \text{ with possibility } pos_2 \\ \cdots \\ \tilde{\rho}_n \text{ with possibility } pos_n \end{cases}$$

is a Fu-Fu variable.

Example 4.2. $\tilde{\tilde{\xi}} = (s_L, \tilde{\rho}, s_R)$ with $\tilde{\rho} = (\rho_L, \rho_M, \rho_R)$ is called Fu-Fu varaible, see Fig. 4.2, if the outer-layer and inner-layer membership functions are as follows:

$$\mu_{\tilde{\tilde{\xi}}}(x) = \begin{cases} (x - s_L)/(\tilde{\rho} - s_L), & \text{if } s_L \leq x \leq \tilde{\rho} \\ (s_R - x)/(s_R - \tilde{\rho}), & \text{if } \tilde{\rho} \leq x \leq s_R \\ 0, & \text{otherwise} \end{cases}$$

and

$$\mu_{\tilde{\rho}}(x') = \begin{cases} (x' - \rho_L)/(\rho_M - \rho_L), & \text{if } \rho_L \leq x' \leq \rho_M \\ (\rho_R - x')/(\rho_R - \rho_M), & \text{if } \rho_M \leq x' \leq \rho_R \\ 0, & \text{otherwise,} \end{cases}$$

where $\tilde{\rho}$ is the center of $\tilde{\tilde{\xi}}$, which is a triangular fuzzy variable, $s_L \in R$ and $s_R \in R$ are the smallest possible value and the largest possible value of $\tilde{\tilde{\xi}}$. $\rho_L \in R$, $\rho_M \in R$ and $\rho_R \in R$ are the the smallest possible value, the most promising value and the largest possible value of $\tilde{\rho}$, respectively.

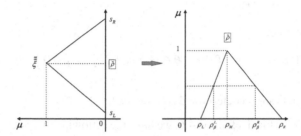

Fig. 4.2 Triangular Fu-Fu variable

4.2.2 Expected Value Operator of Fu-Fu Variables

Definition 4.4. (Liu [220]) The expected value a Fu-Fu variable is defined by

$$E[\xi] = \int_0^{+\infty} Cr\{\theta \in \Theta | E[\xi(\theta)] \geq r\} dr - \int_{-\infty}^0 Cr\{\theta \in \Theta | E[\xi(\theta)] \leq r\} dr$$

provided that at least one of the two integrals is finite.

Theorem 4.1. *Assume that ξ and η are Fu-Fu variables with finite expected values. If (i) for each $\theta \in \Theta$, the fuzzy variables $\xi(\theta)$ and $\eta(\theta)$ are independent, and (ii) $E[\xi(\theta)]$ and $E[\eta(\theta)]$ are independent fuzzy variables, then for any real numbers a and b, we have*

$$E[a\xi + b\eta] = aE[\xi] + bE[\eta]. \tag{4.5}$$

Proof. For any $\theta \in \Theta$ since the fuzzy variables $\xi(\theta)$ and $\eta(\theta)$ are independent, we have $E[a\xi(\theta) + b\eta(\theta)] = aE[\xi(\theta)] + bE[\eta(\theta)]$. In addition, since $E[\xi(\theta)]$ and $E[\eta(\theta)]$ are independent fuzzy variables, we have $E[a\xi + b\eta] = E[a\xi(\theta) + b\eta(\theta)] = aE[\xi(\theta)] + bE[\eta(\theta)] = aE[\xi] + bE[\eta]$. The theorem is proved. □

4.2.3 Chance Operator of Fu-Fu Variables

Definition 4.5 ([220]). Let ξ be a Fu-Fu variable, and B a Borel set of $\mathbf{R}$. Then the primitive chance of Fu-Fu event $\xi \in B$ is a function from $(0,1]$ to $[0,1]$, defined as

$$Ch\{\xi \in B\}(\alpha) = \sup_{Me\{A\} \geq \alpha} \inf_{\theta \in A} Me\{\xi(\theta) \in B\}. \tag{4.6}$$

Usually, we use *Pos* or *Nec* to measure the chance of bifuzzy events. So the following equations will be used frequently.

Definition 4.6. Let $\xi = (\xi_1, \xi_2, \cdots \xi_n)$ be a Fu-Fu vector defined on $(\Theta, P(\Theta), Pos)$, and $f : \mathbf{R}^n \to \mathbf{R}$ is real-valued continuous function. Then the primitive chance of a Fu-Fu event characterized by $f(\xi) \leq 0$ is a function from $(0, 1]$ to $[0, 1]$, defined as the following,
(1) Pos-Pos chance,

$$Ch\{f(\xi) \leq 0\}(\alpha) = \sup_{\alpha \in [0,1]} \{\theta | Pos\{\theta \in \Theta | Pos\{f(\xi(\theta)) \leq 0\} \geq \beta\} \geq \alpha\}. \tag{4.7}$$

(2) Nec-Nec chance,

$$Ch\{f(\xi) \leq 0\}(\alpha) = \sup_{\alpha \in [0,1]} \{\theta | Nec\{\theta \in \Theta | Nec\{f(\xi(\theta)) \leq 0\} \geq \beta\} \geq \alpha\}. \tag{4.8}$$

where $\alpha, \beta \in [0,1]$ are predetermined confidence level.

Remark 4.1. According Definition 4.6, we have the following three equivalent forms respectively

$$Ch\{f(\xi) \leq 0\}(\alpha) \geq \beta$$
$$\Leftrightarrow Pos\{\theta \in \Theta | Pos\{f(\xi(\theta)) \leq 0\} \geq \beta\} \geq \alpha$$
$$\text{or} \Leftrightarrow Nec\{\theta \in \Theta | Nec\{f(\xi(\theta)) \leq 0\} \geq \beta\} \geq \alpha.$$

Remark 4.2. The primitive chance of a Fu-Fu event characterized by $f(\xi) \leq 0$ defined as (4.5) have the equivalent forms respectively.

$$Ch\{f(\xi) \leq 0\}(\alpha) = \sup_{Pos\{A\} \geq \alpha} \inf_{\theta \in A} Pos\{f(\xi(\theta)) \leq 0\}, \tag{4.9}$$

$$Ch\{f(\xi) \leq 0\}(\alpha) = \sup_{Nec\{A\} \geq \alpha} \inf_{\theta \in A} Nec\{f(\xi(\theta)) \leq 0\}. \tag{4.10}$$

Remark 4.3. The primitive chance represents that the Fu-Fu event holds with possibility $Ch\{f(\xi) \leq 0\}(\alpha)$ at possibility α.

Remark 4.4. It is obvious that $Ch\{f(\xi) \leq 0\}(\alpha)$ is a decreasing function of α.

Remark 4.5. If the Fu-Fu vector ξ becomes a fuzzy vector, then the chance $Ch\{f_j(\xi) \leq 0, j = 1, 2, \cdots, m\}(\alpha)$ (with $(\alpha) > 0$) is exactly the possibility or the necessity of the event.

4.3 Fu-Fu EVM

In order to obtain the decision with optimizing the expected objective values subject to expected constraints, we may employ the following Fu-Fu EVM.

4.3.1 General Model for Fu-Fu EVM

First we give the general model of Fu-Fu multi-objective decision making model as follows,

$$\begin{cases} \max \ [f_1(x,\xi), f_2(x,\xi), \cdots, f_m(x,\xi)] \\ \text{s.t.} \begin{cases} g_r(x,\xi) \leq 0, \quad r = 1,2,\cdots,p \\ x \in X. \end{cases} \end{cases}$$

If ξ is a Fu-Fu vector, $x = (x_1, x_2, \cdots, x_n)$ is decision vector, then the objective function $f_i(x,\xi)$ and constraint functions $g_r(x,\xi)$ are also Fu-Fu variables, $i = 1,2,\cdots,m, r = 1,2,\cdots,p$. In order to rank Fu-Fu objective $f_i(x,\xi)$, we may employ the expected value operator to deal with the objective functions and constraints, and we can get the following model (4.11). For the expected value of the objective $E[f_i(x,\xi)], i = 1,2,\cdots,m$, it means that the larger the expected returns $E[f_i(x,\xi)]$, the better the decision x. The first type of Fu-Fu decision-making model is expected value multi-objective decision-making model in which the underlying philosophy is based on selecting the decision with maximum expected objective values.

$$\begin{cases} \max \ [E[f_1(x,\xi)], E[f_2(x,\xi)], \cdots, E[f_m(x,\xi)]] \\ \text{s.t.} \begin{cases} E[g_r(x,\xi)] \leq 0, \quad r = 1,2,\cdots,p \\ x \in X. \end{cases} \end{cases} \quad (4.11)$$

Definition 4.7. A solution is called a Fu-Fu expected feasible solution of model (4.11) if it satisfies

$$E[g_r(x,\xi)] \leq 0,$$

for $r = 1,2,\cdots,p$.

Definition 4.8. A feasible point, x^*, is said to be a Fu-Fu expected efficient solution(nondominated solution, Pareto solution) for Problem (4.11) such that $E[f_i(x,\xi)] \geq E[f_i(x^*,\xi)]$ for $i = 1,2,\cdots,m$ with strict inequality holding for at least one i.

We can also formulate a Fu-Fu decision system as an expected value goal programming (EVGP) model according to the priority structure and target levels set by the decision-maker:

$$\begin{cases} \min \ \sum_{j=1}^{l} P_j \sum_{i=1}^{m} (u_{ij}d_i^+ + v_{ij}d_i^-) \\ \text{s.t.} \begin{cases} E[f_i(x,\xi)] + d_i^- - d_i^+ = b_i, \ i = 1,2,\cdots,m \\ E[g_r(x,\xi)] \leq 0, \ r = 1,2,\cdots,p \\ d_i^-, d_i^+ \geq 0, \ i = 1,2,\cdots,m \\ x \in X. \end{cases} \end{cases} \quad (4.12)$$

where P_j is the preemptive priority factor which expresses the relative importance of various goals, $P_j >> P_{j+1}$, for all j, u_{ij} is the weighting factor corresponding to positive deviation for goal i with priority j assigned, v_{ij} is the weighting factor corresponding to negative deviation for goal i with priority j assigned, d_i^+ is the positive deviation from the target of goal i, defined as

$$d_i^+ = [E[f_i(x, \xi)] - b_i] \vee 0,$$

d_i^- is the negative deviation from the target of goal i, defined as

$$d_i^- = [b_i - E[f_i(x, \xi)]] \vee 0,$$

f_i is a function in goal constraints, g_r is a function in real constraints, b_i is the target value according to goal i, l is the number of priorities, m is the number of goal constraints, and p is the number of real constraints.

4.3.2   Linear Fu-Fu EVM and Step Method

For the regular linear multi-objective decision making problem with Fu-Fu coefficients, we can give the Fu-Fu EVM (4.13),

$$\begin{cases} \max E[\sum_{j=1}^{n} \tilde{\tilde{c}}_{ij} x_j] \\ \text{s.t.} \begin{cases} E[\tilde{\tilde{a}}_{rj} x_j] \geq E[\tilde{\tilde{b}}_r], \ r = 1, 2, \cdots, p \\ x_j \geq 0, \ j = 1, 2, \cdots, n. \end{cases} \end{cases} \quad (4.13)$$

Because of the introduction of the expected value operator, the model (4.13) is crisp, then we can use the section 4.4.1 to obtain the expected value.

4.3.2.1   Crisp Equivalent

In order to solve the multi-objective decision making problem (4.13), we must compute the crisp expected value of ξ. However, as we know, this process is usually a hard work at most of time. In this section, we will consider some special

$$\begin{cases} \max E[\tilde{\tilde{c}}_1^T x, \tilde{\tilde{c}}_2^T x, \cdots, \tilde{\tilde{c}}_m^T x] \\ \text{s.t.} \begin{cases} E[\tilde{\tilde{a}}_r^T x] \leq E[\tilde{\tilde{b}}_r], r = 1, 2, \cdots, p \\ x_j \geq 0, j = 1, 2, \cdots, n, \end{cases} \end{cases} \quad (4.14)$$

where $\tilde{\tilde{c}}_i = (\tilde{\tilde{c}}_{i1}, \tilde{\tilde{c}}_{i1}, \cdots, \tilde{\tilde{c}}_{in})^T, \tilde{\tilde{a}}_r = (\tilde{\tilde{a}}_{r1}, \tilde{\tilde{a}}_{r1}, \cdots, \tilde{\tilde{a}}_{rn})^T$ are Fu-Fu vectors, $\tilde{\tilde{b}}_r$ are Fu-Fu variables, $i = 1, 2, \cdots, m, r = 1, 2, \cdots, p$. If these Fu-Fu vectors have special forms, we have the following theorems. In this section, we use the Cr to compute the expected value.

Theorem 4.2. *If Fu-Fu variable $\tilde{\tilde{c}}_{ij}$ is characterized as follows:*

$$\tilde{\tilde{c}}_{ij}(\theta) = (\tilde{c}_{ij1}(\theta), \tilde{c}_{ij2}(\theta), \tilde{c}_{ij3}(\theta), \tilde{c}_{ij4}(\theta)),$$
$$\text{with } \mu_{\tilde{c}_{ijt}(\theta)}(x) = \begin{cases} 1, & \text{if } x \in [c_{ijt1}, c_{ijt2}] \\ 0, & \text{otherwise,} \end{cases}$$

for $i = 1, 2, \cdots, m, j = 1, 2, \cdots, n, t = 1, 2, 3, 4,$ then

$$E[\tilde{\tilde{c}}_1^T x], E[\tilde{\tilde{c}}_2^T x], \cdots, E[\tilde{\tilde{c}}_m^T x]$$

is equivalent to

$$\frac{1}{8} \sum_{j=1}^{n} \sum_{t=1}^{4} \sum_{k=1}^{2} c_{1jtk} x_j, \frac{1}{8} \sum_{j=1}^{n} \sum_{t=1}^{4} \sum_{k=1}^{2} c_{2jtk} x_j, \cdots, \frac{1}{8} \sum_{j=1}^{n} \sum_{t=1}^{4} \sum_{k=1}^{2} c_{mjtk} x_j.$$

Proof. For any $i \in 1, 2, \cdots, m, \theta \in \Theta, \tilde{\tilde{c}}_{ij}(\theta) = (\tilde{c}_{ij1}(\theta), \tilde{c}_{ij2}(\theta), \tilde{c}_{ij3}(\theta), \tilde{c}_{ij4}(\theta))$ is a trapezoidal fuzzy variable. It follows from Theorem 2.4 we have

$$E[\sum_{j=1}^{n} \tilde{\tilde{c}}_{ij}(\theta) x_j] = \frac{1}{4} \sum_{j=1}^{n} (\tilde{c}_{ij1}(\theta) + \tilde{c}_{ij2}(\theta) + \tilde{c}_{ij3}(\theta) + \tilde{c}_{ij4}(\theta)) x_j.$$

It follows from Definition 4.4 and Theorem 4.1 that

$$\begin{aligned} E[\tilde{\tilde{c}}_i^T x] &= E[E[\tilde{\tilde{c}}^T(\theta)_i x]] \\ &= E[\frac{1}{4} \sum_{j=1}^{n} (\tilde{c}_{ij1}(\theta) + \tilde{c}_{ij2}(\theta) + \tilde{c}_{ij3}(\theta) + \tilde{c}_{ij4}(\theta)) x_j] \\ &= \frac{1}{4}(\sum_{j=1}^{n} E[\tilde{c}_{ij1}(\theta)] x_j + \sum_{j=1}^{n} E[\tilde{c}_{ij2}(\theta)] x_j + \sum_{j=1}^{n} E\tilde{c}_{ij3}(\theta)] x_j + \sum_{j=1}^{n} E[\tilde{c}_{ij4}(\theta)] x_j) \\ &= \frac{1}{4}(\sum_{j=1}^{n} \frac{1}{2}(c_{ij11} + c_{ij12}) x_j + \sum_{j=1}^{n} \frac{1}{2}(c_{ij21} + c_{ij22}) x_j \\ &\quad + \sum_{j=1}^{n} \frac{1}{2}(c_{ij31} + c_{ij32}) x_j + \sum_{j=1}^{n} \frac{1}{2}(c_{ij41} + c_{ij42}) x_j) \\ &= \frac{1}{8} \sum_{j=1}^{n} \sum_{t=1}^{4} \sum_{k=1}^{2} c_{ijtk} x_j, \end{aligned}$$

for any $i \in 1, 2, \cdots, m$.

Thus this theorem is proved. $\square$

Theorem 4.3. *If Fu-Fu variables $\tilde{\tilde{a}}_{rj}, \tilde{\tilde{b}}_r$ are characterized as follows:*

$$\tilde{\tilde{a}}_{rj}(\theta) = (\tilde{a}_{rj1}(\theta), \tilde{a}_{rj2}(\theta), \tilde{a}_{rj3}(\theta), \tilde{a}_{rj4}(\theta)),$$
$$\text{with } \mu_{\tilde{a}_{rjt}(\theta)}(x) = \begin{cases} 1, & \text{if } x \in [a_{rjt1}, a_{rjt2}] \\ 0, & \text{otherwise,} \end{cases}$$

$$\tilde{\tilde{b}}_r(\theta) = (\tilde{a}_{r1}(\theta), \tilde{a}_{r2}(\theta), \tilde{a}_{r3}(\theta), \tilde{a}_{r4}(\theta)),$$
$$\text{with } \mu_{\tilde{a}_{rt}(\theta)}(x) = \begin{cases} 1, & \text{if } x \in [b_{rt1}, b_{rt2}] \\ 0, & \text{otherwise,} \end{cases}$$

for $i = 1, 2, \cdots, m, r = 1, 2, \cdots, p, j = 1, 2, \cdots, n, t = 1, 2, 3, 4$, then

$$E[\tilde{\bar{a}}_r^T x] \leq E[\tilde{\bar{b}}_r], r = 1, 2, \cdots, p$$

is equivalent to

$$\sum_{j=1}^{n} \sum_{t=1}^{4} \sum_{k=1}^{2} a_{rjtk} x_j \leq \sum_{t=1}^{4} \sum_{k=1}^{2} h_{rtk}, r = 1, 2, \cdots, p.$$

Proof. Similar to the proof of the Theorem 4.2, we have

$$E[\tilde{\bar{a}}_r^T x] = \frac{1}{8} \sum_{j=1}^{n} \sum_{t=1}^{4} \sum_{k=1}^{2} a_{rjtk} x_j$$

and

$$E[\tilde{\bar{b}}_r^T] = \frac{1}{8} \sum_{t=1}^{4} \sum_{k=1}^{2} b_{rtk},$$

for any $r \in 1, 2, \cdots, p$.

The theorem is proved.   □

According to the above Theorem 4.2-4.3, the linear Fu-Fu multi-objective model (4.14) with the coefficients describe in Theorem 4.2 and 4.3 can be transformed into the following crisp equivalent model:

$$\begin{cases} \max \; [\; \frac{1}{8} \sum_{j=1}^{n} \sum_{t=1}^{4} \sum_{k=1}^{2} c_{1jtk} x_j, \frac{1}{8} \sum_{j=1}^{n} \sum_{t=1}^{4} \sum_{k=1}^{2} c_{2jtk} x_j, \cdots, \\ \qquad \frac{1}{8} \sum_{j=1}^{n} \sum_{t=1}^{4} \sum_{k=1}^{2} c_{mjtk} x_j \;] \\ \text{s.t.} \begin{cases} \sum_{j=1}^{n} \sum_{t=1}^{4} \sum_{k=1}^{2} a_{rjtk} x_j \leq \sum_{t=1}^{4} \sum_{k=1}^{2} b_{rtk}, r = 1, 2, \cdots, p \\ x_j \geq 0, j = 1, 2, \cdots, n. \end{cases} \end{cases} \tag{4.15}$$

Theorem 4.4. *If Fu-Fu variables $\tilde{\bar{c}}_{ij}, \tilde{\bar{a}}_{rj}, \tilde{\bar{b}}_r$ are characterized as follows:*

$$\tilde{\bar{c}}_{ij}(\theta) = (\tilde{c}_{ij1}(\theta), \tilde{c}_{ij2}(\theta), \tilde{c}_{ij3}(\theta), \tilde{c}_{ij4}(\theta)), \quad \tilde{c}_{ijt}(\theta) = (c_{ijt1}, c_{ijt2}, c_{ijt3}, c_{ijt4});$$
$$\tilde{\bar{a}}_{rj}(\theta) = (\tilde{a}_{rj1}(\theta), \tilde{a}_{rj2}(\theta), \tilde{a}_{rj3}(\theta), \tilde{a}_{rj4}(\theta)), \quad \tilde{a}_{rjt}(\theta) = (a_{ijt1}, a_{ijt2}, a_{ijt3}, a_{ijt4});$$
$$\tilde{\bar{b}}_r(\theta) = (\tilde{b}_{r1}(\theta), \tilde{b}_{r2}(\theta), \tilde{b}_{r3}(\theta), \tilde{b}_{r4}(\theta)), \quad \tilde{b}_{rt}(\theta) = (b_{rt1}, b_{rt2}, b_{rt3}, b_{rt4}),$$

for $i = 1, 2, \cdots, m, r = 1, 2, \cdots, p, j = 1, 2, \cdots, n, t = 1, 2, 3, 4$, then problem (4.14) is equivalent to the conventional multi-objective linear programming

$$
\begin{cases}
\max \ [\ \frac{1}{16} \sum\limits_{j=1}^{n} \sum\limits_{t=1}^{4} \sum\limits_{k=1}^{4} c_{1jtk}x_j, \ \frac{1}{16} \sum\limits_{j=1}^{n} \sum\limits_{t=1}^{4} \sum\limits_{k=1}^{4} c_{2jtk}x_j, \cdots, \\
\qquad \frac{1}{16} \sum\limits_{j=1}^{n} \sum\limits_{t=1}^{4} \sum\limits_{k=1}^{4} c_{mjtk}x_j\ ] \\
s.t. \begin{cases}
\sum\limits_{j=1}^{n} \sum\limits_{t=1}^{4} \sum\limits_{k=1}^{4} a_{rjtk}x_j \le \sum\limits_{t=1}^{4} \sum\limits_{k=1}^{4} b_{rtk}, r = 1,2,\cdots,p \\
x_j \ge 0, j = 1,2,\cdots,n.
\end{cases}
\end{cases}
\tag{4.16}
$$

Proof. For any $i \in 1,2,\cdots,m, \theta \in \Theta, \tilde{\bar{c}}_{ij}(\theta) = (\tilde{c}_{ij1}(\theta), \tilde{c}_{ij2}(\theta), \tilde{c}_{ij3}(\theta), \tilde{c}_{ij4}(\theta))$ is a fuzzy variable. According to Proposition 2.2 and Theorem 2.4, we have

$$
E[\sum_{j=1}^{n} \tilde{\bar{c}}_{ij}(\theta)x_j] = \frac{1}{4} \sum_{j=1}^{n} (\tilde{c}_{ij1}(\theta) + \tilde{c}_{ij2}(\theta) + \tilde{c}_{ij3}(\theta) + \tilde{c}_{ij4}(\theta))x_j.
$$

It follows from Definition 4.4 and Theorem 4.1 that

$$
\begin{aligned}
E[\tilde{\bar{c}}_i^T x] &= E[E[\tilde{\bar{c}}^T(\theta)_i x]] \\
&= E[\frac{1}{4} \sum_{j=1}^{n} (\tilde{c}_{ij1}(\theta) + \bar{c}_{ij2}(\theta) + \tilde{c}_{ij3}(\theta) + \tilde{c}_{ij4}(\theta))x_j] \\
&= \frac{1}{4} (\sum_{j=1}^{n} E[\tilde{c}_{ij1}(\theta)]x_j + \sum_{j=1}^{n} E[\tilde{c}_{ij2}(\theta)]x_j \\
&\quad + \sum_{j=1}^{n} E[\tilde{c}_{ij3}(\theta)]x_j + \sum_{j=1}^{n} E[\tilde{c}_{ij4}(\theta)]x_j) \\
&= \frac{1}{4}(\sum_{j=1}^{n}(\frac{1}{4} \sum_{j=k}^{4} c_{ij1k})x_j + \sum_{j=1}^{n}(\sum_{j=1}^{n}(\frac{1}{4} \sum_{k=1}^{4} c_{ij2k})x_j \\
&\quad + \sum_{j=1}^{n}(\sum_{j=1}^{n}(\frac{1}{4} \sum_{k=1}^{4} c_{ij3k})x_j + \sum_{j=1}^{n}(\sum_{j=1}^{n}(\frac{1}{4} \sum_{k=1}^{4} c_{ij4k})x_j) \\
&= \frac{1}{16} \sum_{t=1}^{4} \sum_{j=1}^{n} \sum_{k=1}^{4} \sum c_{ijtk}x_j.
\end{aligned}
$$

Similarly, we have

$$
E[\tilde{\bar{a}}_r^T x] = \frac{1}{16} \sum_{t=1}^{4} \sum_{j=1}^{n} \sum_{k=1}^{4} a_{rjtk}x_j
$$

and

$$
E[\tilde{\bar{b}}_r^T] = \frac{1}{16} \sum_{t=1}^{4} \sum_{k=1}^{4} b_{rtk},
$$

for any $r \in 1,2,\cdots,p$. Then this theorem is proved. $\qquad\square$

Remark 4.6. If the fuzzy variables in Theorem 4.4 are specified as triangular fuzzy variables, then the result (4.17) of Theorem 4.4 can be rewritten as

$$
\begin{cases}
\max \ [\ \frac{1}{16} \sum\limits_{j=1}^{n} \sum\limits_{t=1}^{4} (c_{1jt1} + 2c_{1jt2} + c_{1jt3})x_j, \ \frac{1}{16} \sum\limits_{j=1}^{n} \sum\limits_{t=1}^{4} (c_{2jt1} + 2c_{2jt2} + c_{2jt3})x_j, \cdots, \\
\quad \frac{1}{16} \sum\limits_{j=1}^{n} \sum\limits_{t=1}^{4} (c_{mjt1} + 2c_{mjt2} + c_{mjt3})x_j\ ] \\
\text{s.t.} \begin{cases} \sum\limits_{j=1}^{n} \sum\limits_{t=1}^{4} (a_{1jt1} + 2a_{1jt2} + a_{1jt3})x_j \le \sum\limits_{t=1}^{4} (b_{rt1} + 2b_{rt2} + b_{rt3}), r = 1,2,\cdots,p \\ x_j \ge 0, j-1,2, \quad ,n. \end{cases}
\end{cases}
$$

$$(4.17)$$

Theorem 4.5. *If Fu-Fu variables $\tilde{\tilde{c}}_{ij}, \tilde{\tilde{a}}_{rj}, \tilde{\tilde{b}}_r$ are characterized as follows:*

$$
\tilde{\tilde{c}}_{ij}(\theta) = (\tilde{c}_{ij1}(\theta), \tilde{c}_{ij2}(\theta), \tilde{c}_{ij3}(\theta), \tilde{c}_{ij4}(\theta)),
$$

$$
\mu_{\tilde{c}_{ijt}(\theta)}(x) = \begin{cases} \mu_{ijt1}, & if \quad x = c_{ijt1} \\ \mu_{ijt2}, & if \quad x = c_{ijt2} \\ \cdots & \cdots \\ \mu_{ijts}, & if \quad x = c_{ijts}, \end{cases}
$$

$$
\tilde{\tilde{a}}_{rj}(\theta) = (\tilde{a}_{rj1}(\theta), \tilde{a}_{rj2}(\theta), \tilde{a}_{rj3}(\theta), \tilde{a}_{rj4}(\theta)),
$$

$$
\mu_{\tilde{a}_{rjt}(\theta)} = \begin{cases} \mu_{rjt1}, & if \quad x = a_{rjt1} \\ \mu_{rjt2}, & if \quad x = a_{rjt2} \\ \cdots & \cdots \\ \mu_{rjts}, & if \quad x = a_{rjts}, \end{cases}
$$

$$
\tilde{\tilde{b}}_r(\theta) = (\tilde{b}_{r1}(\theta), \tilde{b}_{r2}(\theta), \tilde{b}_{r3}(\theta), \tilde{b}_{r4}(\theta)),
$$

$$
\mu_{\tilde{b}_{rt}(\theta)} = \begin{cases} \mu_{rt1}, & if \quad x = b_{rt1} \\ \mu_{rt2}, & if \quad x = b_{rt2} \\ \cdots & \cdots \\ \mu_{rts}, & if \quad x = b_{rts}, \end{cases}
$$

for $i = 1,2,\cdots,m, r = 1,2,\cdots,p, j = 1,2,\cdots,n, t = 1,2,3,4, s \in \mathbf{N}$, then problem (4.14) is equivalent to

$$
\begin{cases}
\max \ [\ \frac{1}{4} \sum\limits_{j=1}^{n} \sum\limits_{t=1}^{4} \sum\limits_{k=1}^{s} w_{1jtk} c_{1jtk} x_j, \ \frac{1}{4} \sum\limits_{j=1}^{n} \sum\limits_{t=1}^{4} \sum\limits_{k=1}^{s} w_{2jtk} c_{2jtk} x_j, \cdots, \\
\quad \frac{1}{8} \sum\limits_{j=1}^{n} \sum\limits_{t=1}^{4} \sum\limits_{k=1}^{s} w_{mjtk} c_{mjtk} x_j\ ] \\
\text{s.t.} \begin{cases} \sum\limits_{j=1}^{n} \sum\limits_{t=1}^{4} \sum\limits_{k=1}^{s} w_{rjtk} a_{rjtk} x_j \le \sum\limits_{t=1}^{4} \sum\limits_{k=1}^{s} w_{rtk} b_{rtk}, r = 1,2,\cdots,p \\ x_j \ge 0, j = 1,2,\cdots,n, \end{cases}
\end{cases}
$$

$$(4.18)$$

where the weights $w_{ijtk}, w_{rjtk}, w_{rtk}, i = 1,2,\cdots,m, j = 1,2,\cdots,n, r = 1,2,\cdots,p, t = 1,2,3,4, k = 1,2,\cdots,s$ are given by

$$w_{ijt1} = \tfrac{1}{2}(\mu_{ijt1} + \max_{1\le l\le s}\mu_{ijtl} - \max_{1<l\le s}\mu_{ijtl}),$$

$$w_{ijtk} = \tfrac{1}{2}(\max_{1\le l\le k}\mu_{ijtl} - \max_{1\le l<s}\mu_{ijtl} + \max_{k\le l\le s}\mu_{ijtl} - \max_{k<l\le s}\mu_{ijtl}),\ 2\le k\le s-1,$$

$$w_{ijts} = \tfrac{1}{2}(\max_{1\le l\le s}\mu_{ijtl} - \max_{1\le l<s}\mu_{ijtl} + \mu_{ijts}),$$

$$w_{rjt1} = \tfrac{1}{2}(\mu_{rjt1} + \max_{1\le l\le s}\mu_{rjtl} - \max_{1<l\le s}\mu_{rjtl}),$$

$$w_{rjtk} = \tfrac{1}{2}(\max_{1\le l\le k}\mu_{rjtl} - \max_{1\le l<s}\mu_{rjtl} + \max_{k\le l\le s}\mu_{rjtl} - \max_{k<l\le s}\mu_{rjtl}),\ 2\le k\le s-1,\quad (4.19)$$

$$w_{rjts} = \tfrac{1}{2}(\max_{1\le l\le s}\mu_{rjtl} - \max_{1\le l<s}\mu_{rjtl} + \mu_{rjts}),$$

$$w_{rt1} = \tfrac{1}{2}(\mu_{rt1} + \max_{1\le l\le s}\mu_{rtl} - \max_{1<l\le s}\mu_{rtl}),$$

$$w_{rtk} = \tfrac{1}{2}(\max_{1\le l\le k}\mu_{rtl} - \max_{1\le l<s}\mu_{rtl} + \max_{k\le l\le s}\mu_{rtl} - \max_{k<l\le s}\mu_{rtl}),\ 2\le k\le s-1,$$

$$w_{rts} = \tfrac{1}{2}(\max_{1\le l\le s}\mu_{rtl} - \max_{1\le l<s}\mu_{rtl} + \mu_{rts}).$$

Proof. For any $i \in 1,2,\cdots,m$, $\theta \in \Theta$, $\tilde{\tilde{c}}_{ij}(\theta) = (\tilde{c}_{ij1}(\theta),\tilde{c}_{ij2}(\theta),\tilde{c}_{ij3}(\theta),\tilde{c}_{ij4}(\theta))$ is a fuzzy variable. It follows from Proposition 2.2 and Theorem 2.4 we have

$$E[\sum_{j=1}^{n}\tilde{\tilde{c}}_{ij}(\theta)x_j] = \frac{1}{4}\sum_{j=1}^{n}(\tilde{c}_{ij1}(\theta) + \tilde{c}_{ij2}(\theta) + \tilde{c}_{ij3}(\theta) + \tilde{c}_{ij4}(\theta))x_j.$$

It follows from Definition 4.4 and Theorem 4.1 that

$$
\begin{aligned}
E[\tilde{\tilde{c}}_i^T x] &= E[E[\tilde{\tilde{c}}^T(\theta)_i x]]\\
&= E[\tfrac{1}{4}\sum_{j=1}^{n}(\tilde{c}_{ij1}(\theta) + \tilde{c}_{ij2}(\theta) + \tilde{c}_{ij3}(\theta) + \tilde{c}_{ij4}(\theta))x_j]\\
&= \tfrac{1}{4}(\sum_{j=1}^{n}E[\tilde{c}_{ij1}(\theta)]x_j + \sum_{j=1}^{n}E[\tilde{c}_{ij2}(\theta)]x_j + \sum_{j=1}^{n}E[\tilde{c}_{ij3}(\theta)]x_j + \sum_{j=1}^{n}E[\tilde{c}_{ij4}(\theta)]x_j)\\
&= \tfrac{1}{4}(\sum_{j=1}^{n}\sum_{k=1}^{s}w_{ij1k}c_{ij1s}x_j + \sum_{j=1}^{n}\sum_{k=1}^{s}w_{ij2k}c_{ij2s}x_j\\
&\quad + \sum_{j=1}^{n}\sum_{k=1}^{s}w_{ij3k}c_{ij3s}x_j + \sum_{j=1}^{n}\sum_{k=1}^{s}w_{ij4k}c_{ij4s}x_j)\\
&= \tfrac{1}{4}\sum_{j=1}^{n}\sum_{t=1}^{4}\sum_{k=1}^{s}w_{ijtk}c_{ijtk}x_j,
\end{aligned}
$$

for any $i \in 1,2,\cdots,m$.

Similarly, we have

$$E[\tilde{\tilde{a}}_r^T x] = \sum_{j=1}^{n}\sum_{t=1}^{4}\sum_{k=1}^{s}w_{rjtk}a_{rjtk}x_j$$

and

$$E[\tilde{\tilde{b}}_r^T] = \sum_{t=1}^{4}\sum_{k=1}^{s}w_{rtk}b_{rtk},$$

for any $r \in 1, 2, \cdots, p$, where the weights $w_{ijtk}, w_{rjtk}, w_{rtk}, i = 1, 2, \cdots, m, j = 1, 2, \cdots, n, r = 1, 2, \cdots, p, t = 1, 2, 3, 4, k = 1, 2, \cdots, s$ are given by (4.19). Then this theorem is proved.   □

4.3.2.2 Step Method

In this section, we use the step method, which is also called STEM method to deal with the crisp linear multi-objective programming problem [367].

The STEM method is based on the norm ideal point method and its resolving process includes the analysis and decision stage. In the analysis stage, an analyzer resolves the problem by the norm ideal point method and provides the decision makers with the solutions and the related objective values and the ideal objective values. In the decision stage, the decision maker gives the tolerance level of the satisfied object to the dissatisfied object to make its objective value better after comparing the objective values obtained in the analysis stage with the ideal point, then provides the analyzer with the information to go on resolving. Done repeatedly and a decision maker will get a final satisfactory solution.

Shimizu once extent the STEM method to deal with a general nonlinear multi-objective programming problem. Interested readers can refer to the literature [368, 369] and others [370, 371, 372] regarding further development.

Consider the following multi-objective programming problem,

$$\begin{cases} \min f(x) = (f_1(x), f_2(x), \cdots, f_m(x)) \\ \text{s.t. } x \in X, \end{cases} \tag{4.20}$$

where $x = (x_1, x_2, \cdots, x_n)$ and $X = \{x \in \mathbf{R}^n | Ax = b, x \geq 0\}$. Let x^i be the optimal solution of the problem $\min_{x \in X} f_i(x)$ and compute each objective function $f_i(x)$ at x^k, then we get m^2 objective function value,

$$f_{ik} = f_i(x^k), \ i, k = 1, 2, \cdots, m.$$

Denote $f_i^* = f_i(x^i)$, $f^* = (f_1^*, f_2^*, \cdots, f_m^*)^T$ and f_i^* is a ideal point of the problem (4.20). Compute the maximum value of the objective function $f_i(x)$ at every minimum point x^k

$$f_i^{\max} = \max_{1 \leq k \leq m} f_{ik}, i = 1, 2, \cdots, m.$$

To make it more clearly, we list it in Table 4.1.

According to Table 4.1, we only look for the solution x such that the distance between $f(x)$ and f^* is minimum, that is, the solution such that each objective is close to the ideal point. Consider the following problem,

$$\min_{x \in X} \max_{1 \leq i \leq m} w_i |f_i(x) - f_i^*| = \min_{x \in X} \max_{1 \leq i \leq m} w_i |\sum_{j=1}^n c_{ij} x_j - f_i^*|, \tag{4.21}$$

Table 4.1 Payoff table

f	x^1	$\cdots$	x^i	$\cdots$	x^m	max
f_1	$f_{11} = f_1^*$	$\cdots$	f_{1i}	$\cdots$	f_{1m}	$f_1^{\max}$
$\vdots$	$\vdots$	$\vdots$	$\vdots$	$\vdots$	$\vdots$	
f_i	f_{i1}	$\cdots$	$f_{ii} = f_i^*$	$\cdots$	f_{im}	$f_i^{\max}$
$\vdots$	$\vdots$	$\vdots$	$\vdots$	$\vdots$	$\vdots$	
f_m	f_{m1}	$\cdots$	f_{mi}	$\cdots$	$f_{mm} = f_m^*$	$f_m^{\max}$

where $w = (w_1, w_2, \cdots, w_m)^T$ is the weight vector and w_i is the ith weight which can be decided as follows,

$$\alpha_i = \begin{cases} \frac{f_i^{\max} - f_i^*}{f_i^{\max}} \frac{1}{||c_i||}, & f_i^{\max} > 0 \\ \frac{f_i^* - f_i^{\max}}{f_i^{\max}} \frac{1}{||c_i||}, & f_i^{\max} \leq 0 \end{cases} \quad i = 1, 2, \cdots, m, \tag{4.22}$$

$$w_i = \alpha_i / \sum_{i=1}^m \alpha_i, \ i = 1, 2, \cdots, m, \tag{4.23}$$

where $||c_i|| = \sqrt{\sum_{j=1}^n c_{ij}^2}$. Then the problem (4.20) is equivalent to

$$\begin{cases} \min \lambda \\ \text{s.t.} \begin{cases} w_i \left(\sum_{j=1}^n c_{ij}x_j - f_i^* \right) \leq \lambda, \ i = 1, 2, \cdots, m \\ \lambda \geq 0, x \in X. \end{cases} \end{cases} \tag{4.24}$$

Assume that the optimal solution of the problem (4.24) is $(\tilde{x}, \tilde{\lambda})^T$. It is obvious that $(\tilde{x}, \tilde{\lambda})^T$ is a weak efficient solution of the problem (4.1). In order to check if $\tilde{x}$ is satisfied, the decision maker needs to compare $f_i(\tilde{x})$ with the ideal objective value $f_i^*, i = 1, 2, \cdots, m$. If the decision maker has been satisfied with $f_s(\tilde{x})$, but dissatisfied with $f_t(\tilde{x})$, we add the following constraint in the next step in order to improve the objective value f_t,

$$f_t(x) \leq f_t(\tilde{x}).$$

For the satisfied object f_s, we add one tolerance level δ_s,

$$f_s(x) \leq f_s(\tilde{x}) + \delta_s.$$

Thus, in the problem (4.24), we replace X with the following constraint set,

$$X^1 = \{x \in X | f_s(x) \leq f_s(\tilde{x}) + \delta_s, f_t(x) \leq f_t(\tilde{x})\},$$

and delete the objective f_s (do it by letting $w_s = 0$), then resolve the new problem to get better solutions.

In a word, the STEM method can be summarized as follows:

Step 1. Compute every single objective programming problem,

$$f_i(x^i) = \min_{x \in X} f_i(x), \ i = 1, 2, \cdots, m.$$

If $x^1 = \cdots = x^m$, we obtain the optimal solution $x^* = x^1 = \cdots = x^m$ and stop.

Step 2. Compute the objective value of $f_i(x)$ at every minimum point x^k, then get m^2 objective values $f_{ik} = f_i(x^k)(i, k = 1, 2, \cdots, m)$. List Table 4.1 and we have

$$f_i^* = f_{ii}, \ f_i^{\max} = \max_{1 \le k \le m} f_{ik}, \ i = 1, 2, \cdots, m.$$

Step 3. Give the initial constraint set and let $X^1 = X$.

Step 4. Compute the weight coefficients $w_1, w_2, \cdots, w_m$ by equation (4.22) and (4.23).

Step 5. Solve the auxiliary problem,

$$\begin{cases} \min \lambda \\ \text{s.t.} \begin{cases} w_i \left(\sum_{j=1}^n c_{ij} x_j - f_i^* \right) \le \lambda, \ i = 1, 2, \cdots, m \\ \lambda \ge 0, x \in X^k. \end{cases} \end{cases} \qquad (4.25)$$

Let the optimal of problem (4.25) be $(x^k, \lambda^k)^T$.

Step 6. The decision maker compare the reference value $f_i(x^k)(i = 1, 2, \cdots, m)$ with the ideal objective value f_i^*. (1) If the decision maker is satisfied with all objective values, output $\tilde{x} = x^k$. (2) If the decision maker is dissatisfied with all objective values, there doesn't exists any satisfied solutions and stop the process. (3) If the decision maker is satisfied with the object $f_{s_k} (1 \le s_k \le m, k < m)$, turn to *Step 7*.

Step 7. The decision maker gives the tolerance level $\delta_{s_k} > 0$ to the object f_{s_k} and construct the new constraint set as follows,

$$X^{k+1} = \{x \in X^k | f_{s_k}(x) \le f_{s_k}(x^k) + \delta_{s_k}, f_i(x) \le f_i(x^k), i \ne s_k\}.$$

Let $\delta_{s_k} = 0, k = k+1$ and turn to *Step 4*.

4.3.2.3 Numerical Example

Example 4.3. Let us consider the following example with Fu-Fu variables,

$$\begin{cases} \min \ \xi_1 x_1 + \xi_2 x_2 \\ \min \ -3x_1 - x_2 \\ \text{s.t.} \begin{cases} x_1 + x_2 \le 6 \\ \xi_3 x_1 + \xi_4 x_2 \le \xi_5 \\ 0 \le x_1 \le 4 \\ 0 \le x_2 \le 5, \end{cases} \end{cases} \tag{4.26}$$

where the coefficients are as follows,

$$\begin{aligned} \xi_1 &= (\mu_1, 0.2, 0.2)_{LR}, \text{with } \mu_1 = (-1.1, -1, -0.9), \\ \xi_2 &= (\mu_2, 0.2, 0.2)_{LR}, \text{with } \mu_2 = (-2.5, -2, -1.5), \\ \xi_3 &= (\mu_3, 0.2, 0.2)_{LR}, \text{with } \mu_3 = (1.5, 2, 2.5), \\ \xi_4 &= (\mu_4, 0.2, 0.2)_{LR}, \text{with } \mu_4 = (0.5, 1, 1.5), \\ \xi_5 &= (\mu_5, 0.2, 0.2)_{LR}, \text{with } \mu_5 = (8.5, 9, 9.5). \end{aligned}$$

In order to solve it, we use the expected operator to deal with bifuzzy objectives and constraints, then we can obtain the model,

$$\begin{cases} \min \ E[\xi_1 x_1 + \xi_2 x_2] \\ \min \ -3x_1 - x_2 \\ \text{s.t.} \begin{cases} x_1 + x_2 \le 6 \\ E[\xi_3 x_1 + \xi_4 x_2] \le E[\xi_5] \\ 0 \le x_1 \le 4 \\ 0 \le x_2 \le 5. \end{cases} \end{cases} \tag{4.27}$$

By theorem 4.4, we know that the problem (4.10) is equivalent to model (4.28),

$$\begin{cases} \min \ F_1 = -x_1 - 2x_2 \\ \min \ F_2 = -3x_1 - x_2 \\ \text{s.t.} \begin{cases} x_1 + x_2 \le 6 \\ 2x_1 + x_2 \le 9 \\ 0 \le x_1 \le 4 \\ 0 \le x_2 \le 5. \end{cases} \end{cases} \tag{4.28}$$

Since the model (4.28) is crisp multi-objective model, so we can use the step method to handle it. According to the step method, we need to compute the two single objective models (4.29) and (4.30),

$$\begin{cases} \min \ F_1 = -x_1 - 2x_2 \\ \text{s.t.} \begin{cases} x_1 + x_2 \le 6 \\ 2x_1 + x_2 \le 9 \\ 0 \le x_1 \le 4 \\ 0 \le x_2 \le 5 \end{cases} \end{cases} \tag{4.29}$$

and

$$\begin{cases} \min F_2 = -3x_1 - x_2 \\ \text{s.t.} \begin{cases} x_1 + x_2 \leq 6 \\ 2x_1 + x_2 \leq 9 \\ 0 \leq x_1 \leq 4 \\ 0 \leq x_2 \leq 5. \end{cases} \end{cases} \tag{4.30}$$

First we let $y - 1$ and can get the solution for these two single objective model as follows,

$$(x_1, x_2) = (1, 5), F_1^* = -11, F_{12} = -8,$$
$$(x_1, x_2) = (4, 1), F_1^* = -13, F_{21} = -6.$$

Then according to Equations (4.22) and (4.23), we can calculate the weights as follows,

$$\alpha_1 = 3/11\sqrt{5} = 0.122, \alpha_2 = 7/13\sqrt{10} = 0.17, w_1 = 0.418, w_2 = 0.582;$$

We let $X^1 = \{x \in R^2 | x_1 + x_2 \leq 6, 2x_1 + x_2 \leq 9, 0 \leq x_1 \leq 4, 0 \leq x_2 \leq 5\}$, and we can construct the following model (4.31) and get the solution as follows,

$$\begin{cases} \min \lambda \\ \text{s.t.} \begin{cases} \lambda \geq -0.418x_1 - 0.836x_2 - 4.598 \\ \lambda \geq -1.746x_1 - 0.582x_2 - 7.566 \\ x \in X^1. \end{cases} \end{cases} \tag{4.31}$$

We obtain that $x^1 = (2.8, 3.2)$, $f_1(x^1) = -9.2$, $f_2(x^1) = -11.6$.

We can provide this result to the decision maker, he will compare this result $(-9.2, -11.6)$ with the ideal point $(-11, -13)$, and decide that which objective value is too high and which is too bad.

Here we suppose that the decision maker consider to improve F_1, and reduce F_2 for one unit, then we let $X^2 = \{x \in R^2 | -3x_1 - x_2 \leq -10.6, -x_1 - 2x_2 \leq -9.2, \text{ and } x_1 + x_2 \leq 6, 2x_1 + x_2 \leq 9, 0 \leq x_1 \leq 4, 0 \leq x_2 \leq 5\}$ and $w_1 = 1, w_2 = 0, q = 2$. We construct the model (4.32) for the second iteration as follows,

$$\begin{cases} \min \lambda \\ \text{s.t.} \begin{cases} \lambda \geq -x_1 - 2x_2 + 11 \\ x \in X^2. \end{cases} \end{cases} \tag{4.32}$$

And we obtain $x^2 = (2.3, 2.7)$, $F_1(x^2) = -9.7$, $F_2(x^2) = -10.6$.

Then we provide this solution to the decision maker, if he is satisfied with it, then the solution x^2 is the best compromise solution. However, if the decision maker is not satisfied, then we need to adjust the value of the objective function to compute another iteration.

4.3.3 Nonlinear Fu-Fu EVM and Fu-Fu Simulation Based SA

For the Fu-Fu EVM, we use the Fu-Fu simulation based simulated annealing algorithm (SA, for short) to solve.

4.3.3.1   Fu-Fu Simulation 1 for Expected Value

First we present the simulation for the expected value of Fu-Fu variables as follows:

Assume that ξ is an n-dimensional Fu-Fu vector on the possibility space $(\Theta, P(\Theta), Pos)$. Then $f(\xi)$ is a Fu-Fu variable whose expected value $E[f(\xi)]$ is

$$E[\xi] = \int_0^{+\infty} Cr\{\theta \in \Theta | E[\xi(\theta)] \geq r\}dr - \int_{-\infty}^0 Cr\{\theta \in \Theta | E[\xi(\theta)] \leq r\}dr.$$

A Fu-Fu simulation will be introduced to compute the expected value $E[f(\xi)]$. We randomly sample θ_k from Θ such that $Pos\{\theta_k\} \geq \varepsilon$, and denote $v_k = Pos\{\theta_k\}, k = 1, 2, \cdots N$, respectively, where ε is a sufficiently small number. Then for any number $r \geq 0$, the credibility $Cr\{\theta \in \Theta | E[\xi(\theta)] \geq r\}$ can be estimated by

$$\frac{1}{2}\left(\max_{1 \leq k \leq N}\{v_k | E[f(\xi(\theta_k))] \geq r\} + \min_{1 \leq k \leq N}\{1 - v_k | E[f(\xi(\theta_k))] < r\}\right). \quad (4.33)$$

And for any number $r < 0$, the credibility $Cr\{\theta \in \Theta | E[\xi(\theta)] \leq r\}$ can be estimated by

$$\frac{1}{2}\left(\max_{1 \leq k \leq N}\{v_k | E[f(\xi(\theta_k))] \leq r\} + \min_{1 \leq k \leq N}\{1 - v_k | E[f(\xi(\theta_k))] > r\}\right). \quad (4.34)$$

provided that N is sufficiently large, where $E[f(\xi(\theta_k))], k = 1, 2, \cdots, N$ may be estimated by the fuzzy simulation.

Step 1. Set $e = 0$.

Step 2. Randomly sample θ_k from Θ such that $Pos\{\theta_k\} \geq \varepsilon$ for $k = 1, 2, \cdots, N$, where ε is a sufficiently small number.

Step 3. Let $a = \min_{1 \leq k \leq N} E[f(\xi(\theta_k))]$ and $b = \max_{1 \leq k \leq N} E[f(\xi(\theta_k))]$.

Step 4. Randomly generate r from [a, b].

Step 5. If $r \geq 0$, then $e \leftarrow e + Cr\{\theta \in \Theta | E[f(\xi(\theta_k))] \geq r\}$.

Step 6. If $r < 0$, then $e \leftarrow e - Cr\{\theta \in \Theta | E[f(\xi(\theta_k))] \leq r\}$.

Step 7. Repeat the fourth to sixth steps for N times.

Step 8. $E[f(\xi)] = a \vee 0 + b \wedge 0 + e(b - a)/N$.

Example 4.4. Suppose that the fuzzy variables $\xi_1, \xi_2, \xi_3, \xi_4$ are defined as

$$\begin{aligned}
\xi_1 &= (\rho_1 - 1, \rho_1, \rho_1 + 1), \text{ with } \rho_1 = (1, 2, 3), \\
\xi_2 &= (\rho_2 - 1, \rho_2, \rho_2 + 1), \text{ with } \rho_2 = (2, 3, 4), \\
\xi_3 &= (\rho_3 - 1, \rho_3, \rho_3 + 1), \text{ with } \rho_3 = (3, 4, 5), \\
\xi_4 &= (\rho_4 - 1, \rho_4, \rho_4 + 1), \text{ with } \rho_4 = (4, 5, 6).
\end{aligned}$$

After a run of Fu-Fu simulation 1 with 1000 cycles to get the expected value

$$E[\sqrt{\xi_1 + \xi_2 + \xi_3 + \xi_4}] = 3.7.$$

4.3.3.2 SA

Simulated annealing algorithm (SA) is the method proposed for the problem of finding, numerically, a point of the global minimum of a function defined on a subset of a k-dimensional Euclidean space. The motivation of the methods lies in the physical process of annealing, in which a solid is heated to a liquid state and, when cooled sufficiently slowly, takes up the configuration with minimal inner energy. Metropolis, Rosenbluth, and Teller [358] described this process mathematically. Simulating annealing uses this mathematical description for the minimization of other functions than the energy. The first results have been published by Černý [359], Kirpatrick, Gelatt Jr., and Vecchi [360], and Geman and Geman [361].

SA is a step-by-step method which could be considered an improvement of the local optimization algorithm. The local optimization algorithm proceeds by generating, at each iteration, a solution in the neighbourhood of the previous one. If the value of the criterion corresponding to the new solution is better than the previous one, the new solution is selected, otherwise it is rejected. In both cases, we restart the process by choosing a solution in the neighbourhood of the solution at hand. The algorithm stops either when it is no longer possible to improve the solution or the maximal number of trials decided by the user has been reached. The drawback of the local optimization algorithm is that it terminates at a local minimum which depends on the initial solution and which may be far from a global minimum.

The SA algorithm avoids entrapment in a local optimum. The difference with the local optimization algorithm is that a solution S_j derived from a solution S_0 is not only accepted if S_j is better than S_0 according to the criterion, but it may also be accepted with a given probability if B is worse than S_0. This probability is equal to $\exp(-df/T)$, where T is a given parameter called temperature which decreases with the number of trials, and $df = f(S_j) - f(S_0) > 0$, where $f(\cdot)$ is the criterion. This is called the Metropolis acceptance rule. This acceptance rule implies that: (i) the smaller the increase of the criterion value, the more likely the new solution is selected, and (ii) the lower the value of T (i.e. the greater the number of trials), the less likely the new solution is selected.

The SA algorithm starts with an initial solution and a relatively high value of T to avoid being prematurely entrapped in a local optimum. At each iteration the algorithm generates several solutions in the neighbourhood of the previous one and decreases the temperature. A new solution is chosen at random among the ones which have been previously generated. This new solution is selected or not according to the Metropolis acceptance rule. The process restarts with the new solution (if accepted) or with the previous one (if the new solution has been rejected). Several tests can be applied to stop the process, for instance: number of trials, minimal value of T, minimal mean value of the improvement of the criterion during the last n trials.

In the general SA algorithm illustrated hereafter we have two nested loops: the outer loop controls the temperature decreasing process and the inner loop controls the number of feasible solutions generated at a given temperature, called the epoch length. The general steps of simulated annealing algorithm are presented in Fig. 4.3.

Give initial temperature To, randomly generate inital state So, and let k=0;
Repeat
    Reapt
        Generate a new state Sj;
        If min{1,exp[-(f(Sj)-f(So))/Tk]}>=randrom[0,1]   So=Sj;
    Until the sampling stability rule is satisfied;
    Lower the temperature Tk+1=update(Tk), and let k=k+1
Until the stopping criterion is satisfied;
Output the results.

Fig. 4.3 Basic steps for simulated annealing

In the simulated annealing algorithm, weather a new solution S_j is be accepted is the key, so we elaborate this by the following Fig. 4.4.

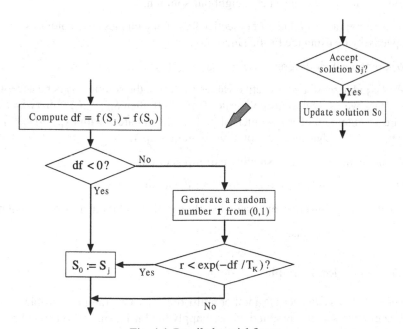

Fig. 4.4 Detailed partial figure

In the temperature update process of the simulated annealing algorithm, there are usually two kinds of formulations to lower the temperature:

(1) $T_{k+1} = \alpha T_k, k \geq 0, 0 < \alpha < 1$, where α is the temperature decrement parameter.
(2) $T_{k+1} = \frac{K-k}{K} T_0$, where T_0 is the initial temperature, K is the total times that the temperature decreases in the algorithm.

The stopping criterion of the simulated annealing algorithm is as follows:

(1) Set the threshold of termination temperature.
(2) Set the time of iteration cycle.
(3) The optimal value that the algorithm searched do not change in several continuous times.

According to the above statement, we conclude the Fu-Fu simulation based simulated annealing algorithm as the following steps:

Step 1. Initialize the parameters of simulated annealing algorithm, including the initial temperature T, the temperature decrement parameter α.

Step 2. Generate a solution S_0 randomly in the range.

Step 3. Checking every chance constraints by employing the Fu-Fu simulation, if the the solution S_i satisfy all of the constraints, return 1, the solution S_0 turn into the initial solution and continue step 4, otherwise return 0 and go back to step 2.

Step 4. Generate solution S_j in the neighbourhood of S_0. We can use $S_j = S +$ RandomDisturb to generate a new neighbour solution.

Step 5. We compute the value F of objective function with these two solutions S_0 and S_j respectively by using the Fu-Fu simulation.

Step 6. Compute their difference $df = f(S_j) - f(S_0)$.

Step 7. We judge these two objective values to decide the solution S_j is be accepted or not. Here we use the Metropolis rule: If $df < 0$, then $S_0 := S_j$; Otherwise, we generate a random number r in interval $[0, 1]$ and compare it with $\exp(-df/T_k)$, if $r < \exp(-df/T_k)$, then $S_0 := S_j$, otherwise, go to step 4.

Step 8. Judge the temperature should be dropped or not.

Step 9. Use the formula $T_{k+1} = \alpha T_k$ to lower the temperature.

Step 10. If the stopping criterion is satisfied, then go to step 11; otherwise, go to step 4.

Step 11. Output the results.

4.3.3.3   Numerical Example

Example 4.5. Let's consider the another multi-objective programming problem with Fu-Fu coefficients as follows, and we will apply the Fu-Fu simulation-based SA to resolve it.

$$\begin{cases} \max\ f_1(x,\xi) = 3\tilde{\tilde{\xi}}_1^2 x_1 - 2\tilde{\tilde{\xi}}_1\tilde{\tilde{\xi}}_2 x_2^2 + 1.3\tilde{\tilde{\xi}}_2^2 x_3 \\ \max\ f_2(x,\xi) = 2.5\tilde{\tilde{\xi}}_3^2 x_1 + 3\tilde{\tilde{\xi}}_3\tilde{\tilde{\xi}}_4 x_2^2 + 5\tilde{\tilde{\xi}}_4^2 x_3 \\ \text{s.t.} \begin{cases} x_1 + x_2 + x_3 \leq 10 \\ 3x_1 + 5x_2 + 3x_3 \geq 4 \\ x_1, x_2, x_3 \geq 0, \end{cases} \end{cases} \tag{4.35}$$

where $\xi_i (i = 1, \cdots, 4)$ are all independently triangular LR Fu-Fu variables as follows,

$$\begin{aligned} \tilde{\tilde{\xi}}_1 &= (\tilde{\mu}_1, 1, 1), \text{ with } \tilde{\mu}_1 = (5, 6, 7), \\ \tilde{\tilde{\xi}}_2 &= (\tilde{\mu}_2, 0.5, 0.5), \text{ with } \tilde{\mu}_2 = (6.5, 8, 10), \\ \tilde{\tilde{\xi}}_3 &= (\tilde{\mu}_3, 0.7, 0.7), \text{ with } \tilde{\mu}_3 = (4, 5, 6), \\ \tilde{\tilde{\xi}}_4 &= (\tilde{\mu}_4, 1, 1), \text{ with } \tilde{\mu}_4 = (5, 7, 8), \end{aligned}$$

where $\tilde{\mu}_i$ are all triangular fuzzy numbers, $i = 1, \cdots, 4$. By the expected value operator of Fu-Fu variables, we have the following expected model of problem (4.35),

$$\begin{cases} \max H_1(x) = E[3\tilde{\tilde{\xi}}_1^2 x_1 - 2\tilde{\tilde{\xi}}_1\tilde{\tilde{\xi}}_2 x_2^2 + 1.3\tilde{\tilde{\xi}}_2^2 x_3] \\ \max H_2(x) = E[2.5\tilde{\tilde{\xi}}_3^2 x_1 + 3\tilde{\tilde{\xi}}_3\tilde{\tilde{\xi}}_4 x_2^2 + 5\tilde{\tilde{\xi}}_4^2 x_3] \\ \text{s.t.} \begin{cases} x_1 + x_2 + x_3 \leq 10 \\ 3x_1 + 5x_2 + 3x_3 \geq 4 \\ x_1, x_2, x_3 \geq 0. \end{cases} \end{cases} \tag{4.36}$$

Then we use the Fu-Fu simulation 1 based SA to solve this problem, we set the weights for each objective as $w_1 = w_2 = 0.5$, and we can get the solutions as follows:

$$x_1 = 0.014, x_2 = 0.032, x_3 = 9.954.$$

And when we set the weights as $w_1 = 0.8, w_2 = 0.2$, we can get the following solutions:

$$x_1 = 0.01, x_2 = 0, x_3 = 9.99.$$

4.4 Fu-Fu CCM

In order to obtain the decision with optimize the critical values subject to chance constraints, we may employ the following Fu-Fu CCM.

4.4.1 General Model for Fu-Fu CCM

Let's introduce the general Fu-Fu CCM as follows.

$$\begin{cases} \max\ [f_1, f_2, \cdots, f_m] \\ \text{s.t.} \begin{cases} Ch\{f_i(x,\xi) \geq \bar{f}_i\}(\gamma_i) \geq \delta_i, \ i = 1, 2, \cdots, m \\ Ch\{g_r(x,\xi) \leq 0\}(\eta_r) \geq \theta_r, \ r = 1, 2, \cdots, p \\ x \in X, \end{cases} \end{cases}$$

where Ch is the chace measure of the Fu-Fu events, $\gamma_i, \delta_i, \eta_r, \theta_r$ are the predetermined confidence level, f_i and x_i are the decision variables, $i = 1, 2, \cdots, m$.

According to the Definition 4.5, we have

$$Ch\{\tilde{\tilde{e}}_r^T x \leq \tilde{\tilde{b}}_r\}(\eta_r) \geq \theta_r \Leftrightarrow Me\{\theta | Me\{\tilde{\tilde{e}}_r(\theta)^T x \leq \tilde{\tilde{b}}_r(\theta)\} \geq \theta_r\} \geq \eta_r, \quad (4.37)$$

$$Ch\{\tilde{\tilde{c}}_i^T x \geq f_i\}(\gamma_i) \geq \delta_i \Leftrightarrow Me\{\theta | Me\{\tilde{\tilde{c}}_i(\theta)^T x \geq f_i\} \geq \delta_i\} \geq \gamma_i. \quad (4.38)$$

Theorem 4.6. *Let ξ be a Fu-Fu variable, and B a Borel set of $\mathbf{R}$. For any given $\alpha^* > 0.5$, we write $\beta^* = Ch\{\xi \in B\}(\alpha^*)$. Then we have*

$$Cr\{\theta \in \Theta | Cr\{\xi(\theta) \in B\} \geq \beta^*\} \geq \alpha^*. \quad (4.39)$$

Proof. Since β^* is the supremum of β satisfying

$$Cr\{\theta \in \Theta | Cr\{\xi(\theta) \in B\} \geq \beta^*\} \geq \alpha^*,$$

there exists an increasing sequence $\{\beta_i\}$ such that

$$Cr\{\theta \in \Theta | Cr\{\xi(\theta) \in B\} \geq \beta_i\} \geq \alpha^* > 0.5. \quad (4.40)$$

and $\beta_i \uparrow \beta^*$ as $i \to \infty$. It is also easy to verify that

$$\{\theta \in \Theta | Cr\{\xi(\theta) \in B\} \geq \beta_i\} \downarrow \{\theta \in \Theta | Cr\{\xi(\theta) \in B\} \geq \beta^*\}$$

as $i \to \infty$. It follows from definition (4.6) and the credibility semicontinuity law that

$$\begin{aligned} &Cr\{\theta \in \Theta | Cr\{\xi(\theta) \in B\} \geq \beta^*\} \\ &= \lim_{i \to \infty} Cr\{\theta \in \Theta | Cr\{\xi(\theta) \in B\} \geq \beta_i\} \\ &\geq \alpha^*. \end{aligned}$$

The proof is complete.   □

Example 4.6. When $\alpha^* \leq 0.5$, generally speaking, the inequality

$$Cr\{\theta \in \Theta | Cr\{\xi(\theta) \in B\} \geq \beta^*\} \geq \alpha^*$$

does not hold. For example, let $\Theta = \{\theta_1, \theta_2, \cdots\}$ and $Pos\{\theta_i\} = 1$ for $i = 1, 2 \cdots$. A Fu-Fu variable is defined on $(\Theta, P(\Theta), Pos)$ as

$$\xi(\theta_i) = \begin{cases} 1 & \text{with possibility } 1 \\ 0 & \text{with possibility } (i-1)/i, \end{cases}$$

for $i = 1, 2, \cdots$ Then we have

$$\beta^* = Ch\{\xi \leq 0\}(0.5) = \sup_{1 \leq i \leq \infty} \frac{i-1}{2i} = \frac{1}{2}.$$

However,

$$Cr\{\theta \in \Theta | Cr\{\xi(\theta) \in B\} \geq \beta^*\} = Cr\{\emptyset\} = 0 < 0.5.$$

4.4.2 Linear Fu-Fu CCM and Lexicographic Method

Let's consider the multi-objective linear programming problem with Fu-Fu coefficients:

$$\begin{cases} \max \; [\tilde{\tilde{c}}_1^T x, \tilde{\tilde{c}}_2^T x, \cdots, \tilde{\tilde{c}}_m^T x] \\ \text{s.t.} \begin{cases} \tilde{\tilde{e}}_r^T x \leq \tilde{\tilde{b}}_r, \quad r = 1, 2, \cdots, p \\ x \geq 0. \end{cases} \end{cases} \tag{4.41}$$

So we can get the Fu-Fu multi-objective chance-constrained linear decision making model,

$$\begin{cases} \max \; [f_1, f_2, \cdots, f_m] \\ \text{s.t.} \begin{cases} Cr\{\theta | Cr\{\tilde{\tilde{c}}_i(\theta)^T x \geq f_i\} \geq \delta_i\} \geq \gamma_i, \quad i = 1, 2, \cdots, m \\ Cr\{\theta | Cr\{\tilde{\tilde{e}}_r(\theta)^T x \leq \tilde{\tilde{b}}_r(\theta)\} \geq \theta_r\} \geq \eta_r, \quad r = 1, 2, \cdots, p \\ x \geq 0, \end{cases} \end{cases} \tag{4.42}$$

where $\delta_i, \gamma_i, \theta_r, \eta_r \in [0,1]$ are the predetermined confidence level, $Cr\{\cdot\}$ denotes the credibility of the fuzzy events in $\{\cdot\}$.

For simpleness, the parameters $\delta, \gamma, \theta, \eta$ can be the same confidence level, i.e. $\delta_i = \delta, \gamma_i = \gamma, \theta_r = \theta, \eta_r = \eta, i = 1, 2, \cdots, m, r = 1, 2, \cdots, p$.

Remark 4.7. If the Fu-Fu vector $\tilde{\tilde{c}}_i$ delegates to fuzzy vector $\tilde{c}_i$, then $\tilde{c}_i^T x \geq f_i$ is a fuzzy event. For $\theta \in \Theta$, $Pos\{\tilde{c}_i(\theta)^T x \geq f_i\} \geq \delta_i$ means $\tilde{c}_i(\theta)^T x \geq f_i$. So,

$$Cr\{\theta | Cr\{\tilde{c}_i(\theta)^T x \geq f_i\} \geq \delta_i\} \geq \gamma_i$$

is equivalent to $Cr\{\theta | \tilde{c}_i(\theta)^T x \geq f_i\} \geq \gamma_i, i = 1, 2, \cdots, m$.

And similarly, If the Fu-Fu vector $\tilde{\tilde{e}}_r$ and $\tilde{\tilde{b}}_r$ delegate to fuzzy vector $\tilde{e}_r$ and $\tilde{b}_r$ respectively, then the constraint

$$Pos\{\theta | Cr\{\tilde{e}_r(\theta)^T x \leq \tilde{b}_r(\theta)\} \geq \theta_r\} \geq \eta_r$$

is equivalent to $Cr\{\theta | \tilde{e}_r(\theta)^T x \leq \tilde{b}_r(\theta)\} \geq \eta_r, r = 1, 2, \cdots, p$. So, the model (4.43) can be rewritten as

$$\begin{cases} \max \; [f_1, f_2, \cdots, f_m] \\ \text{s.t.} \begin{cases} Cr\{\theta | \tilde{c}_i(\theta)^T x \geq f_i\} \geq \gamma_i, \quad i = 1, 2, \cdots, m \\ Cr\{\theta | \tilde{e}_r(\theta)^T x \leq \tilde{b}_r(\theta)\} \geq \eta_r, \quad r = 1, 2, \cdots, p \\ x \geq 0. \end{cases} \end{cases}$$

This is coincident to the fuzzy chance-constrained multi-objective programming in the chapter 2 of the book.

4.4.2.1   Crisp Equivalent Model

In this book, we concentrate on the problem (4.41). One way of solving the problem (4.41) is to convert it into its crisp equivalent. As we know the process is usually hard work and only successful in some cases, so let us consider the following situations.

(1). Pos-Pos constrained multi-objective linearity model

Assume that $\tilde{\tilde{c}}_{ij}$, $\tilde{\tilde{e}}_{rj}$ and $\tilde{\tilde{b}}_r$ are Fu-Fu variables, we give the following two theorems to transform the chance-constrained model (4.43) into its crisp model based on $Pos - Pos$ measure.

$$
\begin{cases}
\max\ [f_1, f_2, \cdots, f_m] \\
\text{s.t.} \begin{cases}
Pos\{\theta | Pos\{\tilde{\tilde{c}}_i(\theta)^\mathrm{T} x \geq f_i\} \geq \delta_i\} \geq \gamma_i, & i = 1, 2, \cdots, m \\
Pos\{\theta | Pos\{\tilde{\tilde{e}}_r(\theta)^\mathrm{T} x \leq \tilde{\tilde{b}}_r(\theta)\} \geq \theta_r\} \geq \eta_r, & r = 1, 2, \cdots, p \\
x \geq 0.
\end{cases}
\end{cases}
\tag{4.43}
$$

Theorem 4.7. *Assume that $\tilde{\tilde{c}}_{ij}$ is LR Fu-Fu variable with the membership function as follows, for any $\theta \in \Theta$,*

$$
\mu_{\tilde{\tilde{c}}_{ij}(\theta)}(t) = \begin{cases}
L(\frac{c_{ij}(\theta) - t}{\alpha^c_{ij1}}), & t \leq c_{ij}(\theta), \alpha^c_{ij1} > 0 \\
R(\frac{t - c_{ij}(w)}{\beta^c_{ij1}}), & t \geq c_{ij}(w), \beta^c_{ij1} > 0
\end{cases} \qquad \theta \in \Theta,
\tag{4.44}
$$

where the fuzzy vector $(c_i(\theta))_{n \times 1} = (c_{i1}(\theta), c_{i2}(\theta), \cdots, c_{in}(\theta))^\mathrm{T}$ is also LR fuzzy variable with the membership functions as follows,

$$
\mu_{c_{ij}(\theta)}(t) = \begin{cases}
L(\frac{c_{ij} - t}{\alpha^c_{ij2}}), & t \leq c_{ij}, \alpha^c_{ij2} > 0 \\
R(\frac{t - c_{ij}}{\beta^c_{ij2}}), & t \geq c_{ij}, \beta^c_{ij2} > 0
\end{cases} \qquad \theta \in \Theta,
\tag{4.45}
$$

And α^c_{ij1}, α^c_{ij2}, β^c_{ij1} and β^c_{ij2} are the left and right spread of $\tilde{\tilde{c}}_{ij}(\theta)$ and $c_{ij}(\theta)$, $i = 1, 2, \cdots, m, j = 1, 2, \cdots, n$, the basis function $L, R : [0, 1] \to [0, 1]$ satisfies that $L(1) = R(1) = 0, L(0) = R(0) = 1$, and it is monotone function. Then $Pos\{\theta | Pos\{\tilde{\tilde{c}}_i(\theta)^\mathrm{T} x \geq f_i\} \geq \delta_i\} \geq \gamma_i$ is equivalent to

$$
c_i^\mathrm{T} x + R^{-1}(\delta_i)\beta_{i1}^{c\mathrm{T}} x + R^{-1}(\gamma_i)\beta_{i2}^{c\mathrm{T}} x \geq f_i, \quad i = 1, 2, \cdots, m,
\tag{4.46}
$$

Where $\delta_i, \gamma_i \in [0, 1]$ are predetermined confidence level.

Proof. For certain $\theta \in \Theta$, $\tilde{c}_{ij}(\theta)$ are fuzzy number, its membership function is $\mu_{\tilde{c}_{ij}(\theta)}(t)$. By extension principle[22], the membership function of fuzzy number $\tilde{c}_i(\theta)^\mathrm{T} x$ is

$$
\mu_{\tilde{c}_i(\theta)^\mathrm{T} x}(r) = \begin{cases}
L(\frac{c_i(\theta)^\mathrm{T} x - r}{\alpha_{i1}^{c\mathrm{T}} x}), & r \leq c_i(\theta)^\mathrm{T} x \\
R(\frac{r - c_i(\theta)^\mathrm{T} x}{\beta_{i1}^{c\mathrm{T}} x}), & r \geq c_i(\theta)^\mathrm{T} x
\end{cases} \qquad i = 1, 2, \cdots, m.
\tag{4.47}
$$

For convenience, denote $\tilde{\tilde{c}}_{ij}(\theta) = (c_{ij}(\theta), \alpha^c_{ij1}, \beta^c_{ij1})_{LR}$, $\tilde{\tilde{c}}_i(\theta)^T x = (c_i(\theta)^T x, \alpha^{cT}_{i1} x, \beta^{cT}_{i1} x)_{LR}$.

Since $(c_{ij}(\theta))_{n \times 1}$ is also a LR fuzzy vector with the left and right spread α^c_{ij2} and β^c_{ij2}, so $c_i(\theta)^T x = (c_i^T x, \alpha^{cT}_{i2} x, \beta^{cT}_{i2} x)_{LR}$.

According to Lemma 2.2 we can get

$$Pos\{\tilde{\tilde{c}}_i(\theta)^T x \geq f_i\} \geq \delta_i \Leftrightarrow c_i(\theta)^T x + R^{-1}(\delta_i)\beta^{cT}_{i1} x \geq f_i, \quad i = 1, 2, \cdots, m.$$

So for predetermined level $\delta_i, \gamma_i \in [0, 1]$,

$$Pos\{\theta | Pos\{\tilde{\tilde{c}}_i(\theta)^T x > f_i\} \geq \delta_{i1}\} \geq \gamma_i$$
$$\Leftrightarrow Pos\{\theta | c_i(\theta)^T x \geq f_i - R^{-1}(\delta_i)\beta^{cT}_{i1} x\} \geq \gamma_i$$
$$\Leftrightarrow c_i^T x + R^{-1}(\delta_i)\beta^{cT}_{i1} x + R^{-1}(\gamma_i)\beta^{cT}_{i2} x \geq f_i.$$

The proof is completed. $\square$

Remark 4.8. Especially, when the basis function of the variables $\tilde{\tilde{c}}_{ij}(\theta)$ and $c_{ij}(\theta)$ are both $L(x) = R(x) = 1 - x$ $(x \in [0, 1])$, then the LR fuzzy variable is specified as the triangular fuzzy variable, and $R^{-1}(\delta_i) = 1 - \delta_i$, and $R^{-1}(\gamma_i) = 1 - \gamma_i$, so we have the following equivalent expressions:

$$Pos\{\theta | Pos\{\tilde{\tilde{c}}_i(\theta)^T x \geq f_i\} \geq \delta_i\} \geq \gamma_i$$
$$\Leftrightarrow c_i^T x + (1 - \delta_i)\beta^{cT}_{i1} x + (1 - \gamma_i)\beta^{cT}_{i2} x \geq f_i, \quad i = 1, 2, \cdots, m.$$

Similarly, the chance-constraint $Pos\{\theta | Pos\{\tilde{\tilde{e}}_r(\theta)^T x \leq \tilde{\tilde{b}}_r(\theta)\} \geq \theta_r\} \geq \eta_r$ can also be transformed into crisp equivalent constraint by Theorem 4.9.

Before the Theorem 4.9, we give Theorem 4.8 which will be used in Theorem 4.9 first.

Theorem 4.8. *Let* $M = (m, \alpha, \beta)_{LR}$, $N = (n, \gamma, \delta)_{LR}$, *then we have the following conclusions:*

$$M(+)N = (m + n, \alpha + \gamma, \beta + \delta)_{LR}, \quad M(-)N = (m - n, \alpha + \delta, \beta + \gamma)_{LR}.$$

Proof. Let $\omega \in [0, 1]$, Let $L(\frac{m-x}{\alpha}) = \omega = L(\frac{n-y}{\gamma})$, then

$$x = m - \alpha L^{-1}(\omega), \quad y = n - \gamma L^{-1}(\omega).$$

It follows that, $z = x + y = m + n - (\alpha + \gamma)L^{-1}(\omega)$. So we have $L(\frac{m+n-z}{\alpha+\gamma}) = \omega$.
Similarly, we can prove $R(\frac{z-(m+n)}{\beta+\delta}) = \omega$.

The other conclusion can be proved in the similar way. $\square$

Theorem 4.9. *Assume that* $\tilde{\tilde{e}}_{rj}$ *and* $\tilde{\tilde{b}}_r$ *are LR Fu-Fu variables, for* $w \in \Omega$, *the membership function of* $\tilde{\tilde{e}}_{rj}(w)$ *are* $\tilde{\tilde{b}}_r(w)$ *are*

$$\mu_{\tilde{e}_{rj}(\theta)}(t) = \begin{cases} L(\frac{e_{rj}(\theta)-t}{\alpha^e_{rj1}}), & t \le e_{rj}(\theta), \alpha^e_{rj1} > 0 \\ R(\frac{t-e_{rj}(\theta)}{\beta^e_{rj1}}), & t \ge e_{rj}(\theta), \beta^e_{rj1} > 0, \end{cases} \tag{4.48}$$

$$\mu_{\tilde{b}_r(\theta)}(t) = \begin{cases} L(\frac{b_r(\theta)-t}{\alpha^b_{r1}}), & t \le b_r(\theta), \alpha^b_{r1} > 0 \\ R(\frac{t-b_r(\theta)}{\beta^b_{r1}}), & t \ge b_r(\theta), \beta^b_{r1} > 0, \end{cases} \tag{4.49}$$

where $(e_{rj}(\theta))_{n \times 1} = (e_{r1}(\theta), e_{r2}(\theta), \cdots, e_{rn}(\theta))^T$, $b_r(w)$ are also fuzzy variables which the membership functions are

$$\mu_{e_{rj}(\theta)}(t) = \begin{cases} L(\frac{e_{rj}-t}{\alpha^e_{rj2}}), & t \le e_{rj}, \alpha^e_{rj2} > 0 \\ R(\frac{t-e_{rj}}{\beta^e_{rj2}}), & t \ge e_{rj}, \beta^e_{rj2} > 0, \end{cases} \tag{4.50}$$

$$\mu_{b_r(\theta)}(t) = \begin{cases} L(\frac{b_r-t}{\alpha^b_{r2}}), & t \le b_r, \alpha^b_{r2} > 0 \\ R(\frac{t-b_r}{\beta^b_{r2}}), & t \ge b_r, \beta^b_{r2} > 0, \end{cases} \tag{4.51}$$

and α^e_{rj1}, α^e_{rj2}, β^e_{rj1}, and β^e_{rj2} are left and right spreads of $\tilde{e}_{rj}(\theta)$ and $e_{rj}(\theta)$, α^b_{r1}, α^b_{r2}, β^b_{r1} and β^b_{r2} are the left and right spread of $\tilde{b}_r(\theta)$ and $b_r(\theta)$, $r = 1, 2, \cdots, p$, $j = 1, 2, \cdots, n$, the basis function $L, R : [0,1] \to [0,1]$ are monotone decreasing continuous function, and it satisfies $L(1) = R(1) = 0, L(0) = R(0) = 1$. For any $j = 1, 2, \cdots, n$, If $e_{rj}(\theta)$ and $b_r(\theta)$ are independent fuzzy variables. Then

$$Pos\{\theta | Pos\{\tilde{e}_r(\theta)^T x \le \tilde{b}_r(\theta)\} \ge \theta_r\} \ge \eta_r$$

is equivalent to

$$R^{-1}(\theta_r)\beta^b_{r1} + L^{-1}(\theta_r)\alpha^{e T}_{r1} x - e^T_r x + b_r + L^{-1}(\eta_r)(\alpha^{e}_{r2}{}^T x + \beta^b_{r2}) \ge 0.$$

Proof. By extension principle, the membership function of fuzzy number $\tilde{e}_r(\theta)^T x$ is

$$\mu_{\tilde{e}_r(\theta)^T x}(r) = \begin{cases} L(\frac{e_r(\theta)^T x - t}{\alpha^{e T}_{r1} x}), & r \le e_r(\theta)^T x, \alpha^e_{rj1} > 0 \\ R(\frac{r - e_{rj}(\theta)^T x}{\beta^{e T}_{r1} x}), & r \ge e_r(\theta)^T x, \beta^e_{r1} > 0 \end{cases} \quad i = 1, 2, \cdots, m. \tag{4.52}$$

For convenience, denote $\tilde{e}_r(\theta)^T x = (e_r(\theta)^T x, \alpha^{e T}_{r1} x, \beta^{e T}_{r1} x)_{LR}$.

According to Lemma 2.2, we have

$$Pos\{\tilde{e}_r(\theta)^T x \le \tilde{b}_r(\theta)\} \ge \theta_r \Leftrightarrow b_r(\theta) + R^{-1}(\theta_r)\beta^b_{1r} \ge e_r(\theta)^T x - L^{-1}(\theta_r)\alpha^{e T}_{1r} x.$$

Since

$$(e_{rj}(\theta)) = (e_{rj}, \alpha_{rj2}^e, \beta_{rj2}^e)_{LR},$$
$$(e_r(\theta)^T x) = (e_r^T x, \alpha_{r2}^{eT} x, \beta_{r2}^{eT} x)_{LR},$$
$$(b_r(\theta)) = (b_r, \alpha_{r2}^b, \beta_{r2}^b)_{LR}.$$

so

$$Pos\{\theta | Pos\{\tilde{\tilde{e}}_r(\theta)^T x \leq \tilde{\tilde{b}}_r(\theta)\} \geq \theta_r\} \geq \eta_r$$
$$\Leftrightarrow Pos\{\theta | e_r(\theta)^T x - b_r(\theta) \leq R^{-1}(\theta_r)\beta_{r1}^b + L^{-1}(\theta_r)\alpha_{r1}^{eT} x\} \geq \eta_r.$$

By Theorem 4.8, we can derive that

$$e_r(\theta)^T x - b_r(\theta) = (e_r^T x - b_r, \alpha_{r2}^{eT} x + \beta_{r2}^b, \beta_{r2}^{eT} x + \alpha_{r2}^b).$$

It follows that

$$Pos\{\theta | e_r(\theta)^T x - b_r(\theta) \leq R^{-1}(\theta_r)\beta_{r1}^b + L^{-1}(\theta_r)\alpha_{r1}^{eT} x\} \geq \eta_r$$
$$\Leftrightarrow e_r^T x - b_r - L^{-1}(\eta_r)(\alpha_{r2}^{eT} x + \beta_{r2}^b) \leq R^{-1}(\theta_r)\beta_{r1}^b + L^{-1}(\theta_r)\alpha_{r1}^{eT} x$$
$$\Leftrightarrow g_r(x) \geq 0$$

where

$$g_r(x) = R^{-1}(\theta_r)\beta_{r1}^b + L^{-1}(\theta_r)\alpha_{r1}^{eT} x - e_r^T x + b_r + L^{-1}(\eta_r)(\alpha_{r2}^{eT} x + \beta_{r2}^b).$$

The proof is completed.   □

We denote $X := \{x \in R^n | R^{-1}(\theta_r)\beta_{r1}^b + L^{-1}(\theta_r)\alpha_{r1}^{eT} x - e_r^T x + b_r + L^{-1}(\eta_r)(\alpha_{r2}^{eT} x + \beta_{r2}^b) \geq 0, r = 1, 2, \cdots, p; x \geq 0\}$.

By Theorems 4.7 and 4.9, the model (4.43) is equivalent to the following multi-objective model,

$$\begin{cases} \max \ [f_1, f_2, \cdots, f_m] \\ \text{s.t.} \begin{cases} f_i \leq c_i^T x + R^{-1}(\delta_i)\beta_{i1}^{cT} x + R^{-1}(\gamma_i)\beta_{i2}^{cT} x, & i = 1, 2, \cdots, m \\ x \in X \end{cases} \end{cases} \quad (4.53)$$

or

$$\begin{cases} \max \ [H_1(x), H_2(x), \cdots, H_m(x)] \\ \text{s.t.} \ x \in X, \end{cases} \quad (4.54)$$

where $H_i(x) := c_i^T x + R^{-1}(\delta_i)\beta_{i1}^{cT} x + R^{-1}(\gamma_i)\beta_{i2}^{cT} x, i = 1, 2, \cdots, m$.

(2). Nec-Nec constrained multi-objective linearity model
Similar to the $Pos - Pos$ constrained multi-objective linearity model, we assume that $\tilde{\tilde{c}}_{ij}, \tilde{\tilde{e}}_{rj}$ and $\tilde{\tilde{b}}_r$ are Fu-Fu variables, we give the following two theorems to transform the chance-constrained model (4.55) into its crisp model based on $Nec - Nec$ if the decision maker is comparatively pessimistic.

$$\begin{cases} \max \ [f_1, f_2, \cdots, f_m] \\ \text{s.t.} \begin{cases} Nec\{\theta | Nec\{\tilde{\tilde{c}}_i(\theta)^T x \geq f_i\} \geq \delta_i\} \geq \gamma_i, & i = 1, 2, \cdots, m \\ Nec\{\theta | Nec\{\tilde{\tilde{e}}_r(\theta)^T x \leq \tilde{\tilde{b}}_r(\theta)\} \geq \theta_r\} \geq \eta_r, & r = 1, 2, \cdots, p \\ x \geq 0. \end{cases} \end{cases} \quad (4.55)$$

Theorem 4.10. *Assume that the Fu-Fu variable $\tilde{\tilde{c}}_{ij}$ is as same as the assumption in Theorem 4.7, $i = 1, 2, \cdots, m$, $j = 1, 2, \cdots, n$. For confidence level $\delta_i, \gamma_i \in [0, 1]$, $i = 1, 2, \cdots, m$, then we have*

$$Nec\{\theta | Nec\{\tilde{\tilde{c}}_i(\theta)^{\mathrm{T}} x \geq f_i\} \geq \delta_i\} \geq \gamma_i$$

is equivalent to

$$c_i^{\mathrm{T}} x + L^{-1}(1 - \delta_i)\alpha_{i1}^{c\mathrm{T}} x + L^{-1}(1 - \gamma_i)\alpha_{i2}^{c\mathrm{T}} x \geq f_i, \quad i = 1, 2, \cdots, m, \qquad (4.56)$$

where $\delta_i, \gamma_i \in [0, 1]$ are predetermined confidence level.

Proof. For certain $\theta \in \Theta$, $\tilde{\tilde{c}}_{ij}(\theta)$ is a fuzzy number, its membership function is $\mu_{\tilde{\tilde{c}}_{ij}(\theta)}(t)$. By extension principle[22], the membership function of fuzzy number $\tilde{\tilde{c}}_i(\theta)^{\mathrm{T}} x$ is

$$\mu_{\tilde{\tilde{c}}_i(\theta)^{\mathrm{T}} x}(r) = \begin{cases} L\left(\frac{c_i(\theta)^{\mathrm{T}} x - r}{\alpha_i^{c\mathrm{T}} x}\right), & r \leq c_i(\theta)^{\mathrm{T}} x \\ R\left(\frac{r - c_i(\theta)^{\mathrm{T}} x}{\beta_i^{c\mathrm{T}} x}\right), & r \geq c_i(\theta)^{\mathrm{T}} x \end{cases} \quad i = 1, 2, \cdots, m. \qquad (4.57)$$

For convenience, denote $\tilde{\tilde{c}}_{ij}(\theta) = (c_{ij}(\theta), \alpha_{ij1}^c, \beta_{ij2}^c)_{\mathrm{LR}}$, $\tilde{\tilde{c}}_i(w)^{\mathrm{T}} x = (c_i(\theta)^{\mathrm{T}} x, \alpha_{i1}^{c\mathrm{T}} x, \beta_{i1}^{c\mathrm{T}} x)_{\mathrm{LR}}$.

Since $(c_{ij}(\theta))_{n \times 1}$ is also a LR fuzzy vector with the left and right spread α_{ij2}^c and β_{ij2}^c, so $c_i(\theta)^{\mathrm{T}} x = (c_i^{\mathrm{T}} x, \alpha_{i2}^{c\mathrm{T}} x, \beta_{i2}^{c\mathrm{T}} x)_{\mathrm{LR}}$.

According to Lemma 2.2 we can get

$$Nec\{\tilde{\tilde{c}}_i(\theta)^{\mathrm{T}} x \geq f_i\} \geq \delta_i \Leftrightarrow c_i(\theta)^{\mathrm{T}} x + L^{-1}(1 - \delta_i)\alpha_{i1}^{c\mathrm{T}} x \geq f_i, \quad i = 1, 2, \cdots, m.$$

So for predetermined level $\delta_i, \gamma_i \in [0, 1]$,

$$\begin{aligned} &Nec\{\theta | Nec\{\tilde{\tilde{c}}_i(\theta)^{\mathrm{T}} x \geq f_i\} \geq \delta_{i1}\} \geq \gamma_i \\ \Leftrightarrow &Nec\{\theta | c_i(\theta)^{\mathrm{T}} x \geq f_i + L^{-1}(1 - \delta_i)\alpha_{i1}^{c\mathrm{T}} x\} \geq \gamma_i \\ \Leftrightarrow &c_i^{\mathrm{T}} x - L^{-1}(1 - \delta_i)\alpha_{i1}^{c\mathrm{T}} x - L^{-1}(1 - \gamma_i)\alpha_{i2}^{c\mathrm{T}} x \geq f_i. \end{aligned}$$

The proof is completed.   □

Remark 4.9. Especially, when the basis function of the variables $\tilde{\tilde{c}}_{ij}(\theta)$ and $c_{ij}(\theta)$ are both $L(x) = R(x) = 1 - x$ ($x \in [0, 1]$), then the LR fuzzy variable is specified as the triangular fuzzy variable, and $L^{-1}(1 - \delta_i) = \delta_i$, and $L^{-1}(1 - \gamma_i) = \gamma_i$, so we have the following equivalent expressions:

$$\begin{aligned} &Nec\{\theta | Nec\{\tilde{\tilde{c}}_i(\theta)^{\mathrm{T}} x \geq f_i\} \geq \delta_i\} \geq \gamma_i \\ \Leftrightarrow &c_i^{\mathrm{T}} x - \delta_i \alpha_{i1}^{c\mathrm{T}} x - \gamma_i \alpha_{i2}^{c\mathrm{T}} x \geq f_i, \quad i = 1, 2, \cdots, m. \end{aligned}$$

Similarly, the chance constraints $Nec\{\theta | Nec\{\tilde{\tilde{e}}_r(\theta)^{\mathrm{T}} x \leq \tilde{\tilde{b}}_r(\theta)\} \geq \theta_r\} \geq \eta_r$ can also be transformed into crisp equivalent constraint.

Theorem 4.11. *Assume that the Fu-Fu variable $\tilde{\tilde{c}}_{ij}$ is as same as the assumption in Theorem 4.9, $i = 1, 2, \cdots, m$, $j = 1, 2, \cdots, n$. For confidence level $\delta_i, \gamma_i \in [0, 1]$, $i = 1, 2, \cdots, m$, then we have $\mathrm{Nec}\{\theta | \mathrm{Nec}\{\tilde{\tilde{e}}_r(\theta)^{\mathrm{T}} x \le \tilde{\tilde{b}}_r(\theta)\} \ge \theta_r\} \ge \eta_r$ is equivalent*

$$b_r - e_r^{\mathrm{T}} x - L^{-1}(1 - \eta_r)(\alpha_{r2}^b + \beta_{r2}^e{}^{\mathrm{T}} x) - L^{-1}(1 - \theta_r)\alpha_{1r}^b - R^{-1}(\theta_r)\beta_{r1}^e{}^{\mathrm{T}} x \ge 0.$$

Proof. By Theorem 4.9 we know that $\tilde{\tilde{e}}_r(\theta)^{\mathrm{T}} x = (e_r(\theta)^{\mathrm{T}} x, \alpha_{r1}^e{}^{\mathrm{T}} x, \beta_{r1}^e{}^{\mathrm{T}} x)_{\mathrm{LR}}$.

According to Lemma 2.2, we have

$$Nec\{\tilde{\tilde{e}}_r(\theta)^{\mathrm{T}} x \le \tilde{\tilde{b}}_r(\theta)\} \ge \theta_r \Leftrightarrow b_r(\theta) - L^{-1}(1 - \theta_r)\alpha_{1r}^b \ge e_r(\theta)^{\mathrm{T}} x + R^{-1}(\theta_r)\beta_{1r}^e{}^{\mathrm{T}} x.$$

Since

$$(e_{rj}(\theta)) = (e_{rj}, \alpha_{rj2}^e, \beta_{rj2}^e)_{\mathrm{LR}},$$
$$(e_r(\theta)^{\mathrm{T}} x) = (e_r^{\mathrm{T}} x, \alpha_{r2}^e{}^{\mathrm{T}} x, \beta_{r2}^e{}^{\mathrm{T}} x)_{\mathrm{LR}},$$
$$(b_r(\theta)) = (b_r, \alpha_{r2}^b, \beta_{r2}^b)_{\mathrm{LR}}.$$

so

$$Nec\{\theta | Nec\{\tilde{\tilde{e}}_r(\theta)^{\mathrm{T}} x \le \tilde{\tilde{b}}_r(\theta)\} \ge \theta_r\} \ge \eta_r$$
$$\Leftrightarrow Nec\{\theta | b_r(\theta) - e_r(\theta)^{\mathrm{T}} x \ge L^{-1}(1 - \theta_r)\alpha_{1r}^b + R^{-1}(\theta_r)\beta_{r1}^e{}^{\mathrm{T}} x\} \ge \eta_r.$$

By Theorem 4.8, we can derive that

$$b_r(\theta) - e_r(\theta)^{\mathrm{T}} x = (b_r - e_r^{\mathrm{T}} x, \alpha_{r2}^b + \beta_{r2}^e{}^{\mathrm{T}} x, \beta_{r2}^b + \alpha_{r2}^e{}^{\mathrm{T}} x)_{\mathrm{LR}}.$$

It follows that

$$Nec\{\theta | b_r(\theta) - e_r(\theta)^{\mathrm{T}} x \ge L^{-1}(1 - \theta_r)\alpha_{1r}^b + R^{-1}(\theta_r)\beta_{r1}^e{}^{\mathrm{T}} x\} \ge \eta_r$$
$$\Leftrightarrow b_r - e_r^{\mathrm{T}} x - L^{-1}(1 - \eta_r)(\alpha_{r2}^b + \beta_{r2}^e{}^{\mathrm{T}} x) \ge L^{-1}(1 - \theta_r)\alpha_{1r}^b + R^{-1}(\theta_r)\beta_{r1}^e{}^{\mathrm{T}} x$$
$$\Leftrightarrow g_r(x) \ge 0,$$

where

$$g_r(x) = b_r - e_r^{\mathrm{T}} x - L^{-1}(1 - \eta_r)(\alpha_{r2}^b + \beta_{r2}^e{}^{\mathrm{T}} x) - L^{-1}(1 - \theta_r)\alpha_{1r}^b - R^{-1}(\theta_r)\beta_{r1}^e{}^{\mathrm{T}} x.$$

The proof is completed. $\square$

We denote $X' := \{x \in R^n | b_r - e_r^{\mathrm{T}} x - L^{-1}(1 - \eta_r)(\alpha_{r2}^b + \beta_{r2}^e{}^{\mathrm{T}} x) - L^{-1}(1 - \theta_r)\alpha_{1r}^b - R^{-1}(\theta_r)\beta_{r1}^e{}^{\mathrm{T}} x \ge 0, r = 1, 2, \cdots, p; x \ge 0\}$.

By Theorems 4.10 and 4.11, the model (4.55) is equivalent to the following multi-objective problem,

$$\begin{cases} \max \ [f_1, f_2, \cdots, f_m] \\ \text{s.t.} \begin{cases} f_i \le c_i^{\mathrm{T}} x + L^{-1}(1 - \delta_i)\alpha_{i1}^{c\mathrm{T}} x + L^{-1}(1 - \gamma_i)\alpha_{i2}^{c\mathrm{T}} x, & i = 1, 2, \cdots, m \\ x \in X' \end{cases} \end{cases} \quad (4.58)$$

or

$$\begin{cases} \max \ [G_1(x), G_2(x), \cdots, G_m(x)] \\ \text{s.t. } x \in X', \end{cases} \quad (4.59)$$

where $G_i(x) := c_i^{\mathrm{T}} x + L^{-1}(1 - \delta_i)\alpha_{i1}^{c\mathrm{T}} x + L^{-1}(1 - \gamma_i)\alpha_{i2}^{c\mathrm{T}} x$, $i = 1, 2, \cdots, m$.

Theorem 4.12. *Assume that the Fu-Fu variable $\tilde{\bar{c}}_{ij}$ is as the same as the assumption in Theorem 4.9, $i = 1, 2, \cdots, m$, $j = 1, 2, \cdots, n$. For confidence levels $\delta_i, \gamma_i \in [0, 1]$, $i = 1, 2, \cdots, m$, then we have $\mathrm{Nec}\{\theta | \mathrm{Nec}\{\tilde{\bar{e}}_r(\theta)^T x \geq \tilde{\bar{b}}_r(\theta)\} \geq \theta_r\} \geq \eta_r$ is equivalent to*

$$e_r^T x - b_r - L^{-1}(1 - \eta_r)(\alpha_{r2}^e{}^T x + \alpha_{r2}^b) - L^{-1}(1 - \theta_r)\alpha_{r1}^e{}^T x - R^{-1}(\theta_r)\beta_{1r}^b \geq 0.$$

4.4.2.2 Lexicographic Method

The basic idea of lexicographic method is to rank the objective function by its importance to decision makers and then resolve the next objective function after resolving the above one. We take the solution of the last programming problem as the final solution.

Consider the following multi-objective programming problem,

$$\begin{cases} \min [f_1(x), f_2(x), \cdots, f_m(x)] \\ \text{s.t. } x \in X. \end{cases} \tag{4.60}$$

Without loss of generality, assume the rank as $f_1(x), f_2(x), \cdots, f_m(x)$ according to different importance. Solve the following single objective problem in turn,

$$\min_{x \in X} f_1(x) \tag{4.61}$$

$$\begin{cases} \min f_i(x) \\ \text{s.t.} \begin{cases} f_k(x) = f_k(x^k), k = 1, 2, \cdots, i-1 \\ x \in X, \end{cases} \end{cases} \tag{4.62}$$

where $i = 1, 2, \cdots, m$, X is the feasible area and denote the feasible area of problem (4.62) as X^i.

Theorem 4.13. *Let $X \subset \mathbf{R}^n$, $f : X \to \mathbf{R}^m$. If x^m be the optimal solution by the lexicographic method, then x^m is an efficient solution of problem (4.60).*

Proof. If x^m is not an efficient solution of problem (4.60), there exists $\bar{x} \in X$ such that $f(\bar{x}) \leq f(x^m)$. Since $f_1(x^m) = f_1^* = f_1(x^1)$, $f_1(\bar{x}) < f_1(x^m)$ cannot hold. It necessarily follows that $f_1(\bar{x}) = f_1(x^m)$.

If we have proved $f_k(\bar{x}) = f_k(x^m)(k = 1, 2, \cdots, i-1)$, but $f_i(\bar{x}) < f_i(x^m)$. It follows that $\bar{x}$ is a feasible solution of problem (4.62). Since $f_i(\bar{x}) < f_i(x^m) = f_i(x^i)$, this results in the conflict with that x^i the the optimal solution of problem (4.62). Thus, $f_k(\bar{x}) = f_k(x^m)(k = 1, 2, \cdots, i)$ necessarily holds. Then we can prove $f_k(\bar{x}) = f_k(x^m)(k = 1, 2, \cdots, m)$ by the mathematical induction. This conflicts with $f(\bar{x}) \leq f(x^m)$. This completes the proof. $\square$

4.4.2.3 Numerical Example

Example 4.7. We use the following example to illustrate this interactive fuzzy satisfied method. Consider the following *Pos − Pos* constrained multi-objective linear model.

Let us consider the following example with fuzzy variables,

$$\begin{cases} \min \ \xi_1 x_1 + \xi_2 x_2 \\ \min \ -3x_1 - x_2 \\ \text{s.t.} \begin{cases} x_1 + x_2 \leq 6 \\ \xi_3 x_1 + \xi_4 x_2 \leq \xi_5 \\ 0 \leq x_1 \leq 4 \\ 0 \leq x_2 \leq 5, \end{cases} \end{cases} \tag{4.63}$$

where the coefficients are triangular LR Fu-Fu variables,

$$\begin{aligned} \xi_1 &= (\theta_1, 0.2, 0.2)_{LR}, \text{ with } \theta_1 = (-1.1, -1, -0.9), \\ \xi_2 &= (\theta_2, 0.2, 0.2)_{LR}, \text{ with } \theta_2 = (-2.1, -2, -1.9), \\ \xi_3 &= (\theta_3, 0.2, 0.2)_{LR}, \text{ with } \theta_3 = (1.9, 2, 2.1), \\ \xi_4 &= (\theta_4, 0.2, 0.2)_{LR}, \text{ with } \theta_4 = (0.9, 1, 1.1), \\ \xi_5 &= (\theta_5, 0.4, 0.4)_{LR}, \text{ with } \theta_5 = (8.5, 9, 9.5). \end{aligned}$$

In order to solve it, we use the expected operator to deal with fuzzy objectives and fuzzy constraints, then we can obtain the model,

$$\begin{cases} \min \ [F_1, F_2] \\ \text{s.t.} \begin{cases} Pos\{\theta | Pos\{\xi_1 x_1 + \xi_2 x_2 \leq F_1\} \geq \delta\} \geq \gamma \\ -3x_1 - x_2 \leq F_2 \\ x_1 + x_2 \leq 6 \\ Pos\{\theta | Pos\{\xi_3 x_1 + \xi_4 x_2 \leq \xi_5\} \geq \theta\} \geq \eta \\ 0 \leq x_1 \leq 4 \\ 0 \leq x_2 \leq 5. \end{cases} \end{cases} \tag{4.64}$$

Then we use the lexicographic method to solve it. Here we suppose that the objective $H_1(x)$ is more important that the second. So wo solve the first important single objective first.

$$\begin{cases} \min \ [F_1, F_2] \\ \text{s.t.} \begin{cases} -x_1 - 2x_2 + (1-\delta)(0.2x_1 + 0.2x_2) + (1-\gamma)(0.1x_1 + 0.1x_2) \leq F_1 \\ -3x_1 - x_2 \leq F_2 \\ x_1 + x_2 \leq 6 \\ (1-\theta)(0.2x_1 + 0.2x_2 + 0.4) - 2x_1 - x_2 + 9 \\ \qquad\qquad + (1-\eta)(0.1x_1 + 0.1x_2 + 0.5) \geq 0 \\ 0 \leq x_1 \leq 4 \\ 0 \leq x_2 \leq 5. \end{cases} \end{cases}$$

$$\tag{4.65}$$

We set $\delta = \gamma = \theta = \eta = 0.95$, and we get

$$\begin{cases} \min \ [F_1, F_2] \\ \text{s.t.} \begin{cases} -0.85x_1 - 1.85x_2 \leq F_1 \\ -3x_1 - x_2 \leq F_2 \\ x_1 + x_2 \leq 6 \\ 1.85x_1 + 0.85x_2 \leq 9.45 \\ 0 \leq x_1 \leq 4 \\ 0 \leq x_2 \leq 5 \end{cases} \end{cases} \tag{4.66}$$

or

$$\begin{cases} \min \ H_1(\mathbf{x}^*) = -0.85x_1 - 1.85x_2 \\ \min \ H_2(\mathbf{x}^*) = -3x_1 - x_2 \\ \text{s.t.} \begin{cases} x_1 + x_2 \leq 6 \\ 1.85x_1 + 0.85x_2 \leq 9.45 \\ 0 \leq x_1 \leq 4 \\ 0 \leq x_2 \leq 5. \end{cases} \end{cases} \tag{4.67}$$

Then we get the optimal soliton $\mathbf{x}^* = (1, 5)$ and the corresponding objective value $H_1(\mathbf{x}^*) = -10.1$.

According to the lexicographic method, we construct the second model (4.68),

$$\begin{cases} \min \ H_2(\mathbf{x}^*) = -3x_1 - x_2 \\ \text{s.t.} \begin{cases} -0.85x_1 - 1.85x_2 = -10.1 \\ x_1 + x_2 \leq 6 \\ 1.85x_1 + 0.85x_2 \leq 9.45 \\ 0 \leq x_1 \leq 4 \\ 0 \leq x_2 \leq 5. \end{cases} \end{cases} \tag{4.68}$$

Finally we get the final optimal solution $\mathbf{x}^* = (1, 5)$ and the corresponding objective values $H_1(\mathbf{x}^*) = -10.1, H_2(\mathbf{x}^*) = -8$.

4.4.3 Nonlinear Fu-Fu CCM and Fu-Fu Simulation Based Parallel SA

For the Fu-Fu CCM, we use the Fu-Fu simulation 2 based parallel SA to solve.

4.4.3.1 Fu-Fu Simulation 2 for Critical Value

We introduce the simulation for critical value of Fu-Fu variables as follows:

Assume that ξ is an n-dimensional Fu-Fu vector on the possibility space $(\Theta, P(\Theta), Pos)$, and $f : \mathbf{R}^n \to \mathbf{R}$ is a measurable function. For any given confidence levels α and β the problem is to find the maximal value $\bar{f}$ such that

$$Ch\{f(\xi) \geq \bar{f}\}(\alpha) \geq \beta$$

holds. That is, we should compute the maximal value $\bar{f}$ such that

$$Cr\{\theta \in \Theta | Cr\{f(\xi(\omega)) \geq \bar{f}\} \geq \beta\} \geq \alpha.$$

We randomly generate θ_k from Θ such that $Pos\{\theta_k\} \geq \varepsilon$, and write $v_k Pos\{\theta_k\}, k = 1, 2, \cdots N$, respectively, where ε is a sufficiently small number. For any number θ_k, we search for the maximal value $\bar{f}(\theta_k)$ such that

$$Cr\{f(\xi(\theta_k)) \geq \bar{f}(\theta_k)\} \geq \beta. \tag{4.69}$$

For any number r, we have

$$L(r) = \frac{1}{2} \left(\max_{1 \leq k \leq N} \{v_k | \bar{f}(\theta_k) \geq r\} + \min_{1 \leq k \leq N} \{1 - v_k | \bar{f}(\theta_k) < r\} \right). \tag{4.70}$$

It follows from monotonicity that we may employ bisection search to find the maximal value r such that

$$L(r) \geq \alpha. \tag{4.71}$$

This value is an estimation of L. We summarize this process as follows.

Step 1. Generate θ_k from Θ such that $Pos\{\theta_k\} \geq \varepsilon$ for $k = 1, 2, \cdots, N$, where ε is a sufficiently small number.

Step 2. Find the maximal value r such that $L(r) \geq \alpha$ holds.

Step 3. Return r.

Example 4.8. In order to find the maximal value $\bar{f}$ such that $Ch\{\xi_1^2 + \xi_2^2 + \xi_3^2 \geq \bar{f}\}(0.9) \geq 0.8$, where ξ_1, ξ_2, ξ_3 are defined as

$$\mu_{\xi_1}(x) = exp(-|x - \rho_1|), \text{ with } \mu_{\rho_1}(x) = [1 - (x-1)^2] \vee 0,$$
$$\mu_{\xi_2}(x) = exp(-|x - \rho_2|), \text{ with } \mu_{\rho_2}(x) = [1 - (x-2)^2] \vee 0,$$
$$\mu_{\xi_3}(x) = exp(-|x - \rho_3|), \text{ with } \mu_{\rho_3}(x) = [1 - (x-3)^2] \vee 0.$$

We perform the Fu-Fu simulation 2 with 10000 cycles and obtain that $\bar{f} = 1.89$.

4.4.3.2 Parallel SA

Some Parallel SA schemes, which share the feature that Multiple PEs follow a Single Markov Chain, have an inherent drawback. A very high overhead is to be paid due to too frequent communication and too much idle time of PEs. Only one search path during the whole calculation process leads to that only one move can be accept/reject in the same step. In other words, the scheme allows only one PE to operate a move while other PEs are idle. Even though some schemes allow more than one PE to operate a perturbation at the same time and choose a best move for next step, the frequent communication between PEs still causes efficiency problems.

In order to improve the performance of SMC parallel SA, aMMC Parallel SA-based layout algorithm is proposed, being able to solve multiple objective and complex constraints in 3D engineering layout designs. The perturbation (move),

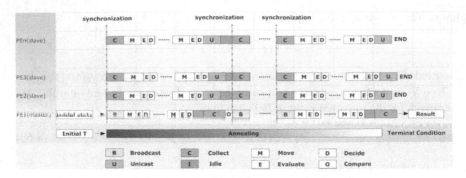

Fig. 4.5 Time line of the MMC parallel SA

evaluation and decision are performed by each PE individually, and all PEs trace multiple search paths at the same time. As a timing diagram, Fig. 4.5 shows how PEs work independently and communicate with each other according to the time line. The PEs are organized in a master-slaves structure, including one master PE and some slave PEs. The duties of the master PE are not only to start and initialize the parallel SA, but also to collect, choose and send intermediate solutions during the synchronization process. As shown in Fig. 4.6, during the whole solution process, the search path of a PE is not completely independent, and PEs exchange local information including the intermediate solutions within a fixed period. The periodic exchange scheme, which introduces flexibility to our MMC parallel SA, becomes the key to solving the 3D engineering layout problem efficiently.

Figure 4.6 illustrates the detailed flow of our parallel SA and shows the interaction between the master PE and slave PEs. According to this flow chart, the synchronization condition plays an important role in communication, and the communication overhead depends on a suitable synchronization condition. So as the synchronization condition may influence the performance of MMCPSA algorithm, it should be selected carefully as a part of the annealing schedule.

Due to the different computing abilities of the different PEs and the stochastic nature of SA, idle time almost cannot be avoided in a synchronous parallel computing process, as shown in Figure 4.5. Therefore it is necessary to carefully choose the control parameters in a parallel cooling schedule to balance the idle time and communication overhead.

Early in the annealing process, nearly all design state perturbations are accepted as new design states. A long independent computing time should be set on each PE to encourage search the path to escape from the local optima. But later in the process, a low acceptance probability causes a mass of inferior solutions to be rejected, and frequent synchronization will lead the search path rapidly to the optimum points.

Furthermore, the number of slave PEs is still an important measure which influences the efficiency of our algorithm. Unfortunately, because of the heavy communication and idle overhead, the increase in processors does not guarantee a better speedup.

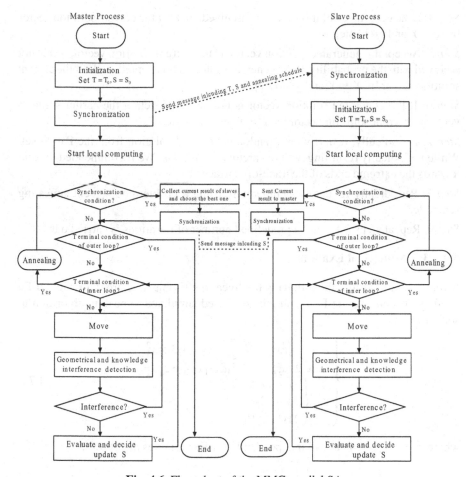

Fig. 4.6 Flow chart of the MMC parallel SA

So making reasonable choices of the control parameters in a cooling schedule, especially for terminal conditions and the number of PEs, is a big challenge for algorithm designers.

A general procedure for parallel SA for MODM is stated as follows.

Step 1. Start with a randomly generated initial solution vector, X, (a $k1$ vector whose elements are decision variables) and evaluate all objective functions and put them into a Pareto set of solutions.

Step 2. Give a random perturbation and generate a new solution vector, Y, in the neighborhood of the current solution vector, X, revaluate the objective functions and apply a penalty function approach to the corresponding objective functions, if necessary.

Step 3. Compare the generated solution vector with all solutions in the Pareto set and update the Pareto set, if necessary.

Step 4. If the generated solution vector is archived, make it the current solution vector by $X = Y$ and go to step 7.

Step 5. Accept the generated solution vector as the current solution vector, if it is not archived with the probability. If the generated solution is accepted, make it the current solution vector by $X = Y$ and go to step 7.

Step 6. If the generated solution vector is not accepted, retain the earlier solution vector as current solution vector by $X = X$ and go to step 7.

Step 7. Periodically, restart with a randomly selected solution from the Pareto set. While periodically restarting with the archived solutions, We recommended biasing towards the extreme ends of the trade-off surface.

Step 8. Periodically reduce the temperature using a problem dependent annealing schedule.

Step 9. Repeat steps 2-8, until a predefined number of iterations is carried out.

4.4.3.3 Numerical Example

Example 4.9. Consider the following nonlinear multi-objective problem with Fu-Fu coefficients and use the Fu-Fu simulation-based simulated annealing algorithm to solve it.

$$\begin{cases} \max 3\bar{\bar{\xi}}_1^2 x_1^2 - 2\bar{\bar{\xi}}_1\bar{\bar{\xi}}_2 x_1 x_2 + 1.3\bar{\bar{\xi}}_2^2 x_2^2 \\ \max 2.5\bar{\bar{\xi}}_3^2 x_1^2 + 3\bar{\bar{\xi}}_3\bar{\bar{\xi}}_4 x_1 x_2 + 5\bar{\bar{\xi}}_4^2 x_2^2 \\ \text{s.t.} \begin{cases} x_1 + x_2 \le 10 \\ 5x_1 - 2x_2 \ge 2 \\ x_1, x_2 \ge 0, \end{cases} \end{cases} \qquad (4.72)$$

where $\xi_i (i = 1, \cdots, 4)$ are all independently Fu-Fu variables as follows,

$$\begin{aligned} \bar{\bar{\xi}}_1 &\sim (\tilde{\mu}_1, 1, 1), \text{ with } \tilde{\mu}_1 = (5, 6, 7), \\ \bar{\bar{\xi}}_2 &\sim (\tilde{\mu}_2, 0.5, 0.5), \text{ with } \tilde{\mu}_2 = (6.5, 8, 10), \\ \bar{\bar{\xi}}_3 &\sim (\tilde{\mu}_3, 0.7, 0.7), \text{ with } \tilde{\mu}_3 = (4, 5, 6), \\ \bar{\bar{\xi}}_4 &\sim (\tilde{\mu}_4, 1, 1), \text{ with } \tilde{\mu}_4 = (5, 7, 8), \end{aligned}$$

where $\tilde{\mu}_i$ are all triangular fuzzy numbers, $i = 1, \cdots, 4$.

We use the Fu-Fu CCM to deal with the above multi-objective model, and we can get the following model:

$$\begin{cases} \max[\bar{f}_1, \bar{f}_2] \\ \begin{cases} Pos\{\theta | Pos\{3\bar{\bar{\xi}}_1^2 x_1^2 - 2\bar{\bar{\xi}}_1\bar{\bar{\xi}}_2 x_1 x_2 + 1.3\bar{\bar{\xi}}_2^2 x_2^2 \ge \bar{f}_1\} \ge 0.9\} \ge 0.8 \\ Pos\{\theta | Pos\{2.5\bar{\bar{\xi}}_3^2 x_1^2 + 3\bar{\bar{\xi}}_3\bar{\bar{\xi}}_4 x_1 x_2 + 5\bar{\bar{\xi}}_4^2 x_2^2 \ge \bar{f}_2\} \ge 0.9\} \ge 0.8 \\ \text{s.t.} \quad x_1 + x_2 \le 10 \\ \quad 5x_1 - 2x_2 \ge 2 \\ \quad x_1, x_2 \ge 0. \end{cases} \end{cases} \qquad (4.73)$$

We employ the Fu-Fu simulation based SA and we can get the following solutions:

$$x_1 = 3.151, x_2 = 6.849;$$
$$\bar{f}_1 = 2903.04, \bar{f}_2 = 14378.24.$$

4.5 Fu-Fu DCM

This section provides Fu-Fu DCM in which the underling philosophy is based on selecting the decision with the maximum chance of meeting the event.

4.5.1 General Model for Fu-Fu DCM

A generally Fu-Fu DCM has the following form,

$$\begin{cases} \max\ Ch\{h_i(x,\xi) \le 0,\ i = 1,2,\cdots,m\} \\ \text{s.t.}\ \begin{cases} g_r(x,\xi) \le 0,\ r = 1,2,\cdots,p \\ x \in X, \end{cases} \end{cases} \tag{4.74}$$

where x is an n-dimensional decision vector, ξ is a Fu-Fu vector, the event ξ is characterized by $h_i(x,\xi) \le 0, i = 1,2,\ldots m$, and the Fu-Fu environment is described by the uncertain constraints $g_r(x,\xi) \le 0, r = 1,2,\ldots p$.

If there are multiple events in the Fu-Fu environment, a typical formulation of Fu-Fu DCM ois given as follows,

$$\begin{cases} \max\ \begin{bmatrix} Ch\{h_{1k}(x,\xi) \le 0,\ k = 1,2,\cdots,q_1\} \\ Ch\{h_{2k}(x,\xi) \le 0,\ k = 1,2,\cdots,q_2\} \\ \cdots \\ Ch\{h_{mk}(x,\xi) \le 0,\ k = 1,2,\cdots,q_m\} \end{bmatrix} \\ \text{s.t.}\ \begin{cases} g_r(x,\xi) \le 0,\ r = 1,2,\cdots,p \\ x \in X, \end{cases} \end{cases} \tag{4.75}$$

where $h_{ik}(x,\xi) \le 0,\ k = 1,2,\cdots,q_i$ represent events ε_i for $i = 1,2,\cdots,m$, respectively.

Fu-Fu dependent-chance goal programming is employed to formulate Fu-Fu decision systems according to the priority structure and target levels set by the decision-maker,

$$\begin{cases} \max\ \sum_{j=1}^{l} P_j \sum_{i=1}^{m} (u_{ij} d_i^+ \vee 0 + v_{ij} d_i^- \vee 0) \\ \text{s.t.}\ \begin{cases} Ch\{h_{1k}(x,\xi) \le 0,\ k = 1,2,\cdots,q_i\} - b_i = d_i^+,\ i = 1,2,\cdots,m \\ b_i - Ch\{h_{1k}(x,\xi) \le 0,\ k = 1,2,\cdots,q_i\} = d_i^-,\ i = 1,2,\cdots,m \\ g_r(x,\xi) \le 0,\ r = 1,2,\cdots,p, \\ x \in X, \end{cases} \end{cases} \tag{4.76}$$

where P_j is the preemptive priority factor which expresses the relative importance of various goals, $P_j \gg P_{j+1}$, for all j, u_{ij} is the weighting factor corresponding to positive deviation for goal i with priority j assigned, v_{ij} is the weighting factor corresponding to negative deviation for goal i with priority j assigned, $d_i^+ \vee 0$ is the positive deviation from the target of goal i, $d_i^- \vee 0$ is the negative deviation from the target of goal i, g_j is a function in system constraints, b_i is the target value according to goal i, l is the number of priorities, m is the number of goal constraints, and p is the number of system constraints.

4.5.2   Linear Fu-Fu DCM and Weight Sum Method

We consider when the objective function and constraints function are linear, let $f_i(x, \xi) = \tilde{\tilde{c}}_i^T x$, $g_r(x, \xi) = \tilde{\tilde{e}}_r^T x - \tilde{\tilde{b}}_r$, then the model (4.74) can be written as,

$$\begin{cases} \max \; [Ch\{\tilde{\tilde{c}}_1^T x \geq \bar{f}_1\}(\gamma_1), \cdots, Ch\{\tilde{\tilde{c}}_m^T x \geq \bar{f}_m\}(\gamma_m)] \\ \text{s.t.} \begin{cases} Ch\{\tilde{\tilde{e}}_r^T x \leq \tilde{\tilde{b}}_r\}(\eta_r) \geq \theta_r, \quad r = 1, 2, \cdots, p \\ x \geq 0. \end{cases} \end{cases} \tag{4.77}$$

If we introduce the new variables δ_i, $i = 1, 2, \cdots, m$, then the model (4.77) can be written as the following equivalent form,

$$\begin{cases} \max \; [\delta_1, \delta_2, \cdots, \delta_m] \\ \text{s.t.} \begin{cases} Ch\{\tilde{\tilde{c}}_i^T x \geq \bar{f}_i\}(\gamma_i) \geq \delta_i, \quad i = 1, 2, \cdots, m \\ Ch\{\tilde{\tilde{e}}_r^T x \leq \tilde{\tilde{b}}_r\}(\eta_r) \geq \theta_r, \quad r = 1, 2, \cdots, p \\ x \geq 0. \end{cases} \end{cases} \tag{4.78}$$

According to the definition of the chance measure, model (4.78) is also can be written as:

$$\begin{cases} \max \; [\delta_1, \delta_2, \cdots, \delta_m] \\ \text{s.t.} \begin{cases} Pos\{\theta | Pos\{\tilde{c}_i(\theta)^T x \geq f_i\} \geq \delta_i\} \geq \gamma_i, \quad i = 1, 2, \cdots, m \\ Pos\{\theta | Pos\{\tilde{e}_r(\theta)^T x \leq \tilde{b}_r(\theta)\} \geq \theta_r\} \geq \eta_r, \quad r = 1, 2, \cdots, p \\ x \geq 0. \end{cases} \end{cases} \tag{4.79}$$

Let $x \in R^n$, for determined confidence level η_r, θ_r, if $x \geq 0$ and $Pos\{\theta | Pos\{\tilde{e}_r(\theta)^T x \leq \tilde{b}_r(\theta)\} \geq \theta_r\} \geq \eta_r$ comes into existence, then x is a feasible solution of model (4.78) or (4.79). And let X be these set of the whole feasible solution of model (4.77) or (4.79).

Remark 4.10. If the Fu-Fu vectors $\tilde{\tilde{c}}_i$, $\tilde{\tilde{e}}_r$ and $\tilde{\tilde{b}}_r$ delegate to fuzzy vectors $\tilde{c}_i$, $\tilde{e}_r$ and $\tilde{b}_r$, then we have

$$\begin{aligned} Pos\{\theta | Pos\{\tilde{c}_i^T x \geq \bar{f}_i\} \geq \delta_i\} \geq \gamma_i &\Leftrightarrow Pos\{\tilde{c}_i^T x \geq \bar{f}_i\} \geq \delta_i, \quad i = 1, 2, \cdots, m, \\ Pos\{\theta | Pos\{\tilde{e}_r^T x \leq \tilde{b}_r\} \geq \theta_r\} \geq \eta_r &\Leftrightarrow Pos\{\tilde{e}_r^T x \leq \tilde{b}_r\} \geq \theta_r, \quad r = 1, 2, \cdots, p. \end{aligned}$$

So the model(4.77)is equivalent to

$$
\begin{cases}
\max \ [Pos\{\tilde{\tilde{c}}_i^\mathrm{T} x \geq \bar{f}_i\}, i = 1, 2, \cdots, m] \\
\text{s.t.} \begin{cases} Pos\{\tilde{\tilde{e}}_r^\mathrm{T} x \leq \tilde{\tilde{b}}_r\} \geq \theta_r, \quad r = 1, 2, \cdots, p \\ x \geq 0. \end{cases}
\end{cases}
\tag{4.80}
$$

The model (4.80) is the extension of the Modality model proposed by Inuiguchi[188], If there is no possibility constraint in (4.80), it is coincident to the Modality model.

In the models (4.77)or (4.79), We can also use the chance measure based on the necessary measure, then we have the following model

$$
\begin{cases}
\max \ [\delta_1, \delta_2, \cdots, \delta_m] \\
\text{s.t.} \begin{cases} Nec\{\theta | Nec\{\tilde{\tilde{c}}_i(\theta)^\mathrm{T} x \geq f_i\} \geq \delta_i\} \geq \gamma_i, \quad i = 1, 2, \cdots, m \\ Nec\{\theta | Nec\{\tilde{\tilde{e}}_r(\theta)^\mathrm{T} x \leq \tilde{\tilde{b}}_r(\theta)\} \geq \theta_r\} \geq \eta_r, \quad r = 1, 2, \cdots, p \\ x \geq 0. \end{cases}
\end{cases}
\tag{4.81}
$$

Denote the feasible region of (4.81) as,

$$
X^N := \{x \in \mathbf{R}^n | Nec\{\theta | Nec\{\tilde{\tilde{e}}_r(\theta)^\mathrm{T} x \leq \tilde{\tilde{b}}_r(\theta)\} \geq \theta_r\} \geq \eta_r, r = 1, 2, \cdots, p; x \geq 0\}.
\tag{4.82}
$$

4.5.2.1 Crisp Equivalent Model

In the following content we discuss how to transform the models (4.79) and (4.81) to their crisp equivalent models when the Fu-Fu coefficients are some special Fu-Fu variables, then we give the method to obtain the weak efficient solution.

(1) Crisp equivalent model based on Pos-Pos
First we assume that for $x \in X, \forall i = 1, 2, \cdots, m, \beta_i^{c\mathrm{T}} x > 0$ holds. Here X denote the feasible region of problem (4.77) and (4.79).

Theorem 4.14. *Assume that $\tilde{\tilde{c}}_{ij}$ is LR Fu-Fu variable with the membership function as follows, for any $\theta \in \Theta$,*

$$
\mu_{\tilde{\tilde{c}}_{ij}(\theta)}(t) = \begin{cases} L(\frac{c_{ij}(\theta)-t}{\alpha_{ij1}^c}), & t \leq c_{ij}(\theta), \alpha_{ij1}^c > 0 \\ R(\frac{t-c_{ij}(\theta)}{\beta_{ij1}^c}), & t \geq c_{ij}(\theta), \beta_{ij1}^c > 0 \end{cases} \quad \theta \in \Theta,
\tag{4.83}
$$

where the fuzzy vector $(c_i(\theta))_{n \times 1} = (c_{i1}(\theta), c_{i2}(\theta), \cdots, c_{in}(\theta))^\mathrm{T}$ is also LR fuzzy variable with the membership functions as follows,

$$
\mu_{c_{ij}(\theta)}(t) = \begin{cases} L(\frac{c_{ij}-t}{\alpha_{ij2}^c}), & t \leq c_{ij}, \alpha_{ij2}^c > 0 \\ R(\frac{t-c_{ij}}{\beta_{ij2}^c}), & t \geq c_{ij}, \beta_{ij2}^c > 0 \end{cases} \quad \theta \in \Theta,
\tag{4.84}
$$

and α_{ij1}^c, α_{ij2}^c, β_{ij1}^c and β_{ij2}^c are the left and right spread of $\tilde{\tilde{c}}_{ij}(\theta)$ and $c_{ij}(\theta)$, $i = 1, 2, \cdots, m$, $j = 1, 2, \cdots, n$, the reference function $L, R : [0, 1] \rightarrow [0, 1]$ satisfies that

$L(1) = R(1) = 0, L(0) = R(0) = 1$, and it is decreasing, monotone function. Then $Pos\{\theta|Pos\{\tilde{\tilde{c}}_i(\theta)^\mathrm{T}x \geq f_i\} \geq \delta_i\} \geq \gamma_i$ is equivalent to

$$R^{-1}(\delta_i) \geq \frac{f_i - c_i^\mathrm{T}x - R^{-1}(\gamma_i)\beta_{i2}^{c\mathrm{T}}x}{\beta_{i1}^{c\mathrm{T}}x}, \quad i = 1, 2, \cdots, m, \tag{4.85}$$

where $\delta_i, \gamma_i \in [0, 1]$ are predetermined confidence level.

Proof. We can easily prove this by Theorem 4.7.

Because the reference function $R(\cdot)$ is decreasing monotone function, so max δ_i is equivalent to $\min R^{-1}(\delta_i)$, by Theorem 4.14 and Theorem 4.9, model 4.79 is equivalent to the following model 4.86,

$$\begin{cases} \max \left[\frac{f_i - c_i^\mathrm{T}x - R^{-1}(\gamma_i)\beta_{i2}^{c\mathrm{T}}x}{\beta_{i1}^{c\mathrm{T}}x}, \ i = 1, 2, \cdots, m \right] \\ \text{s.t.} \begin{cases} R^{-1}(\theta_r)\beta_{r1}^b + L^{-1}(\theta_r)\alpha_{r1}^{e\mathrm{T}}x - e_r^\mathrm{T}x + b_r + L^{-1}(\eta_r)(\alpha_{r2}^{e\mathrm{T}}x + \beta_{r2}^b) \geq 0, \\ \qquad\qquad r = 1, 2, \cdots, p \\ x \geq 0. \end{cases} \end{cases} \tag{4.86}$$

(2) Crisp equivalent model based on Nec-Nec

Theorem 4.15. *Assume that* $\tilde{\tilde{c}}_{ij}$ *is LR Fu-Fu variable with the membership function as follows, for any* $\theta \in \Theta$,

$$\mu_{\tilde{c}_{ij}(\theta)}(t) = \begin{cases} L(\frac{c_{ij}(\theta) - t}{\alpha_{ij1}^c}), & t \leq c_{ij}(\theta), \alpha_{ij1}^c > 0 \\ R(\frac{t - c_{ij}(\theta)}{\beta_{ij1}^c}), & t \geq c_{ij}(\theta), \beta_{ij1}^c > 0 \end{cases} \quad \theta \in \Theta, \tag{4.87}$$

where the fuzzy vector $(c_i(\theta))_{n \times 1} = (c_{i1}(\theta), c_{i2}(\theta), \cdots, c_{in}(\theta))^\mathrm{T}$ is also L-R fuzzy variable with the membership functions as follows,

$$\mu_{c_{ij}(\theta)}(t) = \begin{cases} L(\frac{c_{ij} - t}{\alpha_{ij2}^c}), & t \leq c_{ij}, \alpha_{ij2}^c > 0 \\ R(\frac{t - c_{ij}}{\beta_{ij2}^c}), & t \geq c_{ij}, \beta_{ij2}^c > 0 \end{cases} \quad \theta \in \Theta, \tag{4.88}$$

and $\alpha_{ij1}^c, \alpha_{ij2}^c, \beta_{ij1}^c$ and β_{ij2}^c are the left and right spread of $\tilde{\tilde{c}}_{ij}(\theta)$ and $c_{ij}(\theta)$, $i = 1, 2, \cdots, m$, $j = 1, 2, \cdots, n$, the reference function $L, R : [0, 1] \to [0, 1]$ satisfies that $L(1) = R(1) = 0, L(0) = R(0) = 1$, and it is decreasing, monotone function. Then $Nec\{\theta|Nec\{\tilde{\tilde{c}}_i(\theta)^\mathrm{T}x \geq f_i\} \geq \delta_i\} \geq \gamma_i$ is equivalent to

$$L^{-1}(1 - \delta_i) \leq \frac{c_i^\mathrm{T}x - L^{-1}(1 - \gamma_i)\alpha_{i2}^{c\mathrm{T}}x - f_i}{\alpha_{i1}^{c\mathrm{T}}x} \quad i = 1, 2, \cdots, m, \tag{4.89}$$

where $\delta_i, \gamma_i \in [0, 1]$ are predetermined confidence levels.

Proof. We can easily prove this by Theorem 4.10.

Because the reference function $L(\cdot)$ is decreasing monotone function, so max δ_i is equivalent to $\max L^{-1}(1 - \delta_i)$, by Theorem 4.15 and Theorem 4.11, model 4.79 is equivalent to the following model 4.90,

$$
\begin{cases}
\max \left[\dfrac{c_i^T x - L^{-1}(1-\gamma_i)\alpha_{i2}^{cT}x - f_i}{\alpha_{i1}^{cT}x} \; i = 1,2,\cdots,m \right] \\
\text{s.t.} \begin{cases} b_r - e_r^T x - L^{-1}(1 - \eta_r)(\alpha_{r2}^b + \beta^{e_{r2}^T}x) - L^{-1}(1 - \theta_r)\alpha_{1r}^b - R^{-1}(\theta_r)\beta_{1r}^{eT}x \geq 0, \\ \qquad\qquad r = 1,2,\cdots,p \\ x \geq 0. \end{cases}
\end{cases}
$$
(4.90)

4.5.2.2 Weight Sum Method

The weight sum method is one of the techniques which are broadly applied to solve the multi-objective programming problem (4.91).

$$
\begin{cases}
\max\{H_1(x),\cdots,H_i(x),\cdots,H_m(x)\} \\
\text{s.t. } x \in X.
\end{cases}
$$
(4.91)

Assume that the related weight of the objective function $H_i(x)$ is w_i such that $\sum_{i=1}^{m} w_i = 1$ and $w_i \geq 0$. Construct the evaluation function as follows,

$$
u(\mathbf{H}(x)) = \sum_{i=1}^{m} w_i H_i(x) = \mathbf{w}^T \mathbf{H}(x),
$$

where w_i expresses the importance of the object $H_i(x)$ for DM. Then we get the following weight problem,

$$
\max_{x \in X} u(\mathbf{H}(x)) = \max_{x \in X} \sum_{i=1}^{m} w_i H_i(x) = \max_{x \in X} \mathbf{w}^T \mathbf{H}(x).
$$
(4.92)

Let $\bar{x}$ be an optimal solution of the problem (4.92), we can easily deduce that if $w > 0$, $\bar{x}$ is an efficient solution of the problem (4.91). By changing w, we can obtain a set composed of the efficient solutions of the problem (4.91) by solving the problem (4.92).

4.5.2.3 Numerical Example

Example 4.10. We use this example to explain the linear Fu-Fu DCM and the the most easy weight sum method.

Let us consider the following example with fuzzy variables,

$$
\begin{cases}
\max \ \xi_1 x_1 + \xi_2 x_2 \\
\max \ \xi_3 x_1 + \xi_4 x_2 \\
\text{s.t.} \begin{cases}
x_1 + x_2 \leq 6 \\
2x_1 + x_2 \leq 9 \\
0 \leq x_1 \leq 4 \\
0 \leq x_2 \leq 5,
\end{cases}
\end{cases}
\tag{4.93}
$$

where the coefficients are triangular LR Fu-Fu variables,

$$
\begin{aligned}
\xi_1 &= (\theta_1, 0.2, 0.2)_{LR}, \text{with } \theta_1 = (0.9, 1, 1.1), \\
\xi_2 &= (\theta_2, 0.2, 0.2)_{LR}, \text{with } \theta_2 = (1.9, 2, 2.1), \\
\xi_3 &= (\theta_3, 0.2, 0.2)_{LR}, \text{with } \theta_3 = (2.0, 3, 3.1), \\
\xi_4 &= (\theta_4, 0.2, 0.2)_{LR}, \text{with } \theta_4 = (0.9, 1, 1.1).
\end{aligned}
$$

In order to solve it, we use the expected operator to deal with fuzzy objectives and fuzzy constraints, then we can obtain the model,

$$
\begin{cases}
\max \ Ch\{\xi_1 x_1 + \xi_2 x_2 \geq F_1\} \\
\max \ Ch\{\xi_3 x_1 + \xi_4 x_2 \geq F_2\} \\
\text{s.t.} \begin{cases}
x_1 + x_2 \leq 6 \\
2x_1 + x_2 \leq 9 \\
0 \leq x_1 \leq 4 \\
0 \leq x_2 \leq 5.
\end{cases}
\end{cases}
\tag{4.94}
$$

If we adopted the Nec-Nec DCM, and we get

$$
\begin{cases}
\max \ Nec\{\theta | Nec\{\xi_1 x_1 + \xi_2 x_2 \geq F_1\} \geq \delta\} \\
\max \ Nec\{\theta | Nec\{\xi_3 x_1 + \xi_4 x_2 \geq F_2\} \geq \delta\} \\
\text{s.t.} \begin{cases}
x_1 + x_2 \leq 6 \\
2x_1 + x_2 \leq 9 \\
0 \leq x_1 \leq 4 \\
0 \leq x_2 \leq 5.
\end{cases}
\end{cases}
\tag{4.95}
$$

We use the crisp equivalent to transform the model (4.95) into the following model,

$$
\begin{cases}
\max \ Nec\{\theta | Nec\{\xi_1 x_1 + \xi_2 x_2 \geq F_1\} \geq \delta\} \\
\max \ Nec\{\theta | Nec\{\xi_3 x_1 + \xi_4 x_2 \geq F_2\} \geq \delta\} \\
\text{s.t.} \begin{cases}
x_1 + x_2 \leq 6 \\
2x_1 + x_2 \leq 9 \\
0 \leq x_1 \leq 4 \\
0 \leq x_2 \leq 5.
\end{cases}
\end{cases}
\tag{4.96}
$$

Then we use the lexicographic method to solve it. Here we suppose that the objective $H_1(x)$ is more important that the second. So wo solve the first important single objective first.

$$
\begin{cases}
\max \ G_1(x) = \frac{x_1 + 2x_2 - 0.9(0.1x_1 + 0.1x_2) - F_1}{0.2x_1 + 0.2x_2} \\
\max \ G_2(x) = \frac{3x_1 + x_2 - 0.9(0.1x_1 + 0.1x_2) - F_2}{0.2x_1 + 0.2x_2} \\
\text{s.t.} \begin{cases} x_1 + x_2 \le 6 \\ 2x_1 + x_2 \le 9 \\ 0 \le x_1 \le 4 \\ 0 \le x_2 \le 5. \end{cases}
\end{cases} \tag{4.97}
$$

We set $F_1 = 8, F_2 = 5$ and compute the two single objective model respectively,

$$
G_1(x^*) = 2.05, x^* = (1,5),
$$
$$
G_2(x^*) = 8.3, x^* = (4,0).
$$

Then we set $w_1 = 0.3, w_2 = 0.7$, and we construct the following model,

$$
\begin{cases}
\max \ w_1 \frac{x_1 + 2x_2 - 0.9(0.1x_1 + 0.1x_2) - F_1}{0.2x_1 + 0.2x_2} + w_2 \frac{3x_1 + x_2 - 0.9(0.1x_1 + 0.1x_2) - F_2}{0.2x_1 + 0.2x_2} \\
\text{s.t.} \begin{cases} x_1 + x_2 \le 6 \\ 2x_1 + x_2 \le 9 \\ 0 \le x_1 \le 4 \\ 0 \le x_2 \le 5. \end{cases}
\end{cases} \tag{4.98}
$$

After computing, we can get the following solution,

$$
(x_1, x_2) = (4, 1.45).
$$

4.5.3 Nonlinear Fu-Fu DCM and Fu-Fu Simulation Based Adaptive SA

Consider the following model,

$$
\begin{cases}
\max \ [Pos\{\theta | Pos\{f_i(x,\xi) \ge \bar{f}_i\} \ge \gamma_i\}, i = 1, 2, \cdots, m] \\
\text{s.t.} \begin{cases} Pos\{\theta | Pos\{g_r(x,\xi) \le 0\} \ge \theta_r\} \ge \eta_r, r = 1, 2, \cdots, p \\ x \in X, \end{cases}
\end{cases}
$$

where γ_i, η_r and θ_r are predetermined confidence levels, $i = 1, 2, \cdots, m, r = 1, 2, \cdots, p$. If f_i or g_r or even both of them are nonlinear functions, we cannot directly convert it into crisp model, then another method is introduced to solve it.

For the nonlinear Fu-Fu DCM, we use the Fu-Fu simulation based adaptive simulated annealing algorithm (ASA, for short) to solve.

4.5.3.1 Fu-Fu Simulation 3 for Chance

Assume that ξ is an n-dimensional Fu-Fu vector defined on the possibility space $(\Theta, P(\Theta), Pos)$, and $f : \mathbf{R}^n \to \mathbf{R}$ is a function. For any confidence level α. we design a Fu-Fu simulation to compute the α-chance $Ch\{f(\xi) \le 0\}(\alpha)$. Equivalently, we should find the supremum $\bar{\beta}$ such that

$$Cr\{\theta \in \Theta | Cr\{f(\xi(\theta)) \leq 0\} \geq \bar{\beta}\} \geq \alpha. \qquad (4.99)$$

We randomly generate θ_k from Θ such that $Pos\{\theta_k\} \geq \varepsilon$, and write $v_k Pos\{\theta_k\}, k = 1, 2, \cdots N$, respectively, where ε is a sufficiently small number. For any number θ_k, by using fuzzy simulation, we can estimate the credibility

$$g(\theta_k) = Cr\{f(\xi(\theta_k)) \leq 0\}. \qquad (4.100)$$

For any number r, we have

$$L(r) = \frac{1}{2} \left(\max_{1 \leq k \leq N} \{v_k | g(\theta_k) \geq r\} + \min_{1 \leq k \leq N} \{1 - v_k | g(\theta_k) < r\} \right). \qquad (4.101)$$

It follows from monotonicity that we may employ bisection search to find the maximal value r such that $L(r) \geq \alpha$. This value is an estimation of L. We summarize this process as follows.

Step 1. Generate θ_k from Θ such that $Pos\{\theta_k\} \geq \varepsilon$ for $k = 1, 2, \cdots, N$, where ε is a sufficiently small number.

Step 2. Find the maximal value r such that $L(r) \geq \alpha$ holds.

Step 3. Return r.

Example 4.11. Suppose that the Fu-Fu variables ξ_1, ξ_2, ξ_3 are defined as

$$\xi_1 = (\rho_1 - 1, \rho_1, \rho_1 + 1), \text{ with } \rho_1 = (0, 1, 2),$$
$$\xi_2 = (\rho_2 - 1, \rho_2, \rho_2 + 1), \text{ with } \rho_2 = (1, 2, 3),$$
$$\xi_3 = (\rho_3 - 1, \rho_3, \rho_3 + 1), \text{ with } \rho_3 = (2, 3, 4).$$

After a run of Fu-Fu simulation 3 with 10000 cycles, we get that

$$Ch\{\xi_1 + \xi_2 + \xi_3 \geq 2\}(0.9) = 0.61.$$

4.5.3.2 Adaptive SA

The adaptive SA, also known as the very fast simulated reannealing, is a very efficient version of SA. Detailed analysis of the algorithm can be found in[365, 362, 363, 364]. Many signal processing applications pose the following general optimization problem:

$$\min_{w \in W} J(w), \qquad (4.102)$$

where $w = [w_1, \cdots, w_n]^T$ is the n-dimensional parameter vector to be optimized,

$$W := \{w : (L_i \leq w_i \leq U_i, 1 \leq i \leq n) \cap (\alpha_j \leq g_j(w) \leq \beta_j, 1 \leq j \leq m)\} \qquad (4.103)$$

is the feasible set of w, L_i and U_i are the lower and upper bounds of w_i, respectively, and $\alpha_j \leq g_j(w) \leq \beta_j, 1 \leq j \leq m$, are the m inequality constraints. The cost function $J(w)$

can be multi-modal and non-smooth. Adaptive SA is a global optimization scheme for solving this kind of constrained optimization problems.

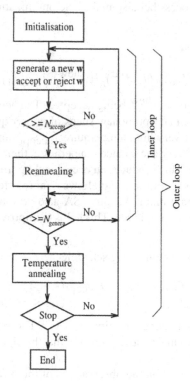

Fig. 4.7 Flow chart of the adaptive simulated annealing.

Let us briefly introduce the search guiding mechanisms. Adaptive SA evolves from a single point w in the parameter or state space W. The seemingly random search is guided by certain underlying probability distributions. An elegant discussion on how the general SA algorithm works can be found in [366]. Specifically, the general SA algorithm is described by three functions.

(1). Generating probability density function

$$G(w_i^{old}, w_i^{new}, T_{i,gen}; 1 \leq i \leq n). \tag{4.104}$$

This determines how a new state w^{new} is created, and from what neighbourhood and probability distributions it is generated, given the current state w^{old}. The generating "temperatures" a $1i$, $T_{i,gen}$ describe the widths or scales of the generating distribution along each dimension w_i of the state space.

Often a cost function has different sensitivities along different dimensions of the state space. Ideally, the generating distribution used to search a steeper and more sensitive dimension should have a narrower width than that of the distribution used in searching a dimension less sensitive to change. Adaptive SA adopts a so-called re-annealing scheme to periodically re-scale $T_{i,gen}$, so that they optimally adapt to the current status of the cost function. This is an important mechanism, which not only speeds up the search process but also makes the optimization process robust to different problems.

(2). Acceptance function

$$P_{accept}(J(w^{old}), J(w^{new}), T_{accept}).$$ (4.105)

This gives the probability of w^{new} being accepted. The acceptance temperature determines the frequency of accepting new states of poorer quality. Probability of acceptance is very high at very high temperature T_{accept}, and it becomes smaller as T_{accept} is reduced. At every acceptance temperature, there is a finite probability of accepting the new state. This produces an occasional uphill move, enables the algorithm to escape from local minima, and allows a more effective search of the state space to find a global minimum. Adaptive SA also periodically adapts T_{accept} to best suit the status of the cost function. This helps to improve convergence speed and robustness.

(3). Reduce temperatures or annealing schedule

$$T_{accept}(k_a) \to T_{accept}(k_a + 1),$$
$$T_{i,gen}(k_i) \to T_{i,gen}(k_i + 1), 1 \le i \le n,$$ (4.106)

where k_a and k_i are some annealing time indexes. The reduction of temperatures should be sufficiently gradual in order to ensure that the algorithm finds a global minimum.

This mechanism is based on the observations of the physical annealing process. When the metal is cooled from a high temperature, if the cooling is sufficiently slow, the atoms line themselves up and form a crystal, which is the state of minimum energy in the system. The slow convergence of many SA algorithms is rooted in this slow annealing process. Adaptive SA, however, can employ a very fast annealing schedule, as it has self adaptation ability to re-scale temperatures.

We state adaptive SA implementation. Although there are many possible realizations of adaptive SA, an implementation is illustrated in Figure 1, and this algorithm is detailed here. How adaptive SA realizes the above three functions will also become clear during the description.

Step 1. In the initialization, an initial $w \in W$ is randomly generated, the initial temperature of the acceptance probability function, $T_{accept}(0)$, is set to $J(w)$, and the initial temperatures of the parameter generating probability functions, $T_{i,gen}(0), 1 \le i \le n$, are set to 1.0. A user-defined control parameter c in annealing is given, and the annealing times, k_i for $1 \le i \le n$ and k_a, are all set to 0.

Step 2. The algorithm generates a new point in the parameter space with

$$w_i^{new} = w_i^{old} + q_i(U_i - L_i), 1 \le i \le n \tag{4.107}$$

and

$$w^{new} \in W, \tag{4.108}$$

where q_i is calculated as

$$q_i = sgn(v_i - \frac{1}{2})T_{i,gen}(k_i) \times \left(\left(1 + \frac{1}{T_{i,gen}(k_i)}\right)^{|2v_i - 1|} - 1 \right), \tag{4.109}$$

and v_i a uniformly distributed random variable in $[0,1]$. Notice that if a generated w^{new} is not in W, it is simply discarded and a new point is tried again until $w^{new} \in W$.

The value of the cost function $J(w^{new})$ is then evaluated and the acceptance probability function of w^{new} is given by

$$P_{accept} = \frac{1}{1 + exp((J(w^{new}) - J(w^{old}))/T_{accept}(K_a))}. \tag{4.110}$$

A uniform random variable P_{unif} is generated in $[0,1]$. If $P_{unif} \le P_{accept}$, w^{new} is accepted; otherwise it is rejected.

Step 3. After every N_{accept} acceptance points, reannealing takes place by first calculating the sensitivies

$$s_i = \left| \frac{J(w^{best} + e_i\delta) - J(w^{best})}{\delta} \right|, 1 \le i \le n, \tag{4.111}$$

where w^{best} is the best point found so far, δ is a small step size, the n-dimensional vector e_i has unit ith element and the rest of elements of e_i are all zeros. Let $s_{max} = max\{s_i, 1 \le i \le n\}$. Each parameter generating temperature $T_{i,gen}$ is scaled by a factor s_{max}/s_i and the annealing time k_i is reset,

$$T_{i,gen}(k_i) = \frac{s_{max}}{s_i}T_{i,gen}(k_i), \; k_i = \left(-\frac{1}{c}log\left(\frac{T_{i,gen}(k_i)}{T_{i,gen}(0)} \right) \right)^n. \tag{4.112}$$

Similarly, $T_{accept}(0)$ is reset to the value of the last accepted cost function, $T_{accept}(k_a)$ is reset to $J(w^{best})$ and the annealing time k_a is rescaled accordingly,

$$k_a = \left(-\frac{1}{c}log\left(\frac{T_{i,gen}(k_a)}{T_{i,gen}(0)} \right) \right)^n. \tag{4.113}$$

Step 4. After every N_{genera} generated points, annealing takes place with

$$k_i = k_i + 1,$$
$$T_{i,gen}(k_i) = T_{i,gen}(0)exp(-ck_i^{1/n}), 1 \le i \le n; \tag{4.114}$$

and

$$k_a = k_a + 1,$$
$$T_{accept}(k_a) = T_{accept}(0)exp(-ck_a^{1/n}); \tag{4.115}$$

otherwise, go to step (2).

Step 5. The algorithm is terminated if the parameters have remained unchanged for a few successive reannealings or a preset maximum number of cost function evaluations has been reached; otherwise, go to step (2).

As in a standard SA algorithm, this adaptive SA contains two loops. The inner loop ensures that the parameter space is searched sufficiently at a given temperature, which is necessary to guarantee that the algorithm finds a global optimum. Adaptive SA also uses only the value of the cost function in the optimization process and is very simple to program.

Last, we discuss algorithm parameter tuning. For the above adaptive SA, most of the algorithm parameters are automatically set and "tuneda", and the user only needs to assign a control parameter c and set two values N_{accept} and N_{gen}. Obviously, the optimal values of N_{accept} and N_{gen} are problem dependent, but our experience suggests that an adequate choice for N_{accept} is in the range of tens to hundreds and an appropriate value for N_{gen} is in the range of hundreds to thousands. The annealing rate control parameter c can be determined from the chosen initial temperature, final temperature and the predetermined number of annealing steps. We have found out that a choice of c in the range 1.0 to 10.0 is often adequate.

It should be emphasized that, as adaptive SA has excellent self adaptation abilities, the performance of the algorithm is not critically influenced by the specific chosen values of c, N_{accept} and N_{gen}.

4.5.3.3 Numerical Example

Example 4.12. Let's consider the following non-linear multi-objective programming with Fu-Fu coefficients,

$$\begin{cases} \max \ f_1 = 3\tilde{\tilde{\xi}}_1^2 x_1^2 - 2\tilde{\tilde{\xi}}_1\tilde{\tilde{\xi}}_2 x_1 x_2 + 1.3\tilde{\tilde{\xi}}_2^2 x_2^2 \\ \max f_2 = 2.5\tilde{\tilde{\xi}}_3^2 x_1^2 + 3\tilde{\tilde{\xi}}_3\tilde{\tilde{\xi}}_4 x_1 x_2 + 5\tilde{\tilde{\xi}}_4^2 x_2^2 \\ s.t. \begin{cases} x_1 + x_2 \le 10 \\ 5x_1 - 2x_2 \ge 2 \\ x_1, x_2 \ge 0, \end{cases} \end{cases} \tag{4.116}$$

where $\xi_i (i = 1, \cdots, 4)$ are all independently Fu-Fu variables as follows,

$$\tilde{\tilde{\xi}}_1 \sim (\tilde{\mu}_1, 1, 1), \text{with } \tilde{\mu}_1 = (5, 6, 7),$$
$$\tilde{\tilde{\xi}}_2 \sim (\tilde{\mu}_2, 0.5, 0.5), \text{with } \tilde{\mu}_2 = (6.5, 8, 10),$$
$$\tilde{\tilde{\xi}}_3 \sim (\tilde{\mu}_3, 0.7, 0.7), \text{with } \tilde{\mu}_3 = (4, 5, 6),$$
$$\tilde{\tilde{\xi}}_4 \sim (\tilde{\mu}_4, 1, 1), \text{with } \tilde{\mu}_4 = (5, 7, 8),$$

where $\tilde{\mu}_i$ are all triangular fuzzy numbers, $i = 1, \cdots, 4$.

When we set $\bar{f}_1 = 2900$, $\bar{f}_2 = 14000$ and $\delta_1 = \delta_2 = 0.9$, and use the Fu-Fu DCM to deal with the above multi-objective model under bifuzzy environment.

$$
\begin{cases}
\max f_1(x, \xi) = Ch\left\{3\tilde{\tilde{\xi}}_1^2 x_1^2 - 2\tilde{\tilde{\xi}}_1\tilde{\tilde{\xi}}_2 x_1 x_2 + 1.3\tilde{\tilde{\xi}}_2^2 x_2^2 \geq 5\right\}(0.9) \\
\max f_2(x, \xi) = Ch\left\{2.5\tilde{\tilde{\xi}}_3^2 x_1^2 + 3\tilde{\tilde{\xi}}_3\tilde{\tilde{\xi}}_4 x_1 x_2 + 5\tilde{\tilde{\xi}}_4^2 x_2^2 \geq 12\right\}(0.9) \\
\text{s.t.} \begin{cases} x_1 + x_2 \leq 10 \\ 5x_1 - 2x_2 \geq 2 \\ x_1, x_2 \geq 0, \end{cases}
\end{cases} \tag{4.117}
$$

which is equivalent to

$$
\begin{cases}
\max f_1(x, \xi) = Pos\{\theta | Pos\{3\tilde{\tilde{\xi}}_1^2 x_1^2 - 2\tilde{\tilde{\xi}}_1\tilde{\tilde{\xi}}_2 x_1 x_2 + 1.3\tilde{\tilde{\xi}}_2^2 x_2^2 \geq 5\} \geq 0.9\} \\
\max f_2(x, \xi) = Pos\{\theta | Pos\{2.5\tilde{\tilde{\xi}}_3^2 x_1^2 + 3\tilde{\tilde{\xi}}_3\tilde{\tilde{\xi}}_4 x_1 x_2 + 5\tilde{\tilde{\xi}}_4^2 x_2^2 \geq 12\} \geq 0.9\} \\
\text{s.t.} \begin{cases} x_1 + x_2 \leq 10 \\ 5x_1 - 2x_2 \geq 2 \\ x_1, x_2 \geq 0. \end{cases}
\end{cases} \tag{4.118}
$$

We use the Fu-Fu simulation 3 based SA to solve this problem, and we can get the following solutions:

$$x_1 = 3.143, x_2 = 6.857;$$
$$f_1 = 0.81, f_2 = 0.86.$$

4.6 Application to Purchasing Problem in a Large-Scale Integrated Steel Plant

In this section, for the problem proposed in section 4.1, we propose a model which is a single time phase with the unit of time measure being 1 month. The demand and inventory level are given for the required raw materials.

Usually, there are many objectives for the purchasing decision, but the most important three factors for the decision maker are quality, price, and due date[353]. The unit of measure of quality is the scrap ratio of raw materials, and the unit of measurement for the due date is a tardy-delivery fraction. Therefore, we proposed a multi-objective linear programming model with three objectives.

4.6.1 Modelling

Our aims of proposing the model is to decide which raw material items should be purchased and how big a quantity order ought to be placed with each selected vendor. Therefore, we use the two-dimensional vector x_{ij} denoting the order quantity of the jth item of raw materials from the ith vendor.

4.6.1.1 Objective Functions

We selected the minimization objectives for all factors because of to its ready-to-use linear programming solving techniques. In addition, a multi-objective approach has several advantages over a single-objective; firstly, it allows various criteria to be evaluated in their natural units of measurement and, therefore, eliminates the necessity of transforming them into a common unit of measurement such as dollars; secondly, multi-objective techniques permit the decision maker to incorporate personal experience and insight in the decision process; thirdly, the multi-objective approach provides the decision maker with a method to systematically analyze the effects of policy decisions on the relevant criteria space when making decisions. This very important feature makes the decision maker see the potential effects of different decisions. The formulations are as follows:

The first objective is minimum cost,

$$\min \left\{ F_1 = \sum_{i=1}^{m} \sum_{j=1}^{n} \tilde{\tilde{c}}_{ij} x_{ij} \right\}, \tag{4.119}$$

the second is lowest scrap ratio,

$$\min \left\{ F_2 = \sum_{i=1}^{m} \sum_{j=1}^{n} \tilde{\tilde{r}}_{ij} x_{ij} \right\}, \tag{4.120}$$

the last is least tardy-delivery fraction

$$\min \left\{ F_3 = \sum_{i=1}^{m} \sum_{j=1}^{n} \tilde{\tilde{s}}_{ij} x_{ij} \right\}, \tag{4.121}$$

where $\tilde{\tilde{c}}_{ij}$ is the one unit purchasing cost of the jth item from the ith vendor, $\tilde{\tilde{r}}_{ij}$ the tardy-delivery ratio of the jth item from the ith vendor, $\tilde{\tilde{s}}_{ij}$ the scrap fraction of the jth item from the ith vendor, m the number of vendors, and n the number of items.

4.6.1.2 Constraints

There are several constraints we need to consider.

Purchasing budget

The purchasing capital budget depends on purchase supply planning and price fluctuation situation if considerable. Raw materials are classified into a number of classes, such as iron-ores, secondary raw materials, fuels, etc. They ought to satisfy their own the purchasing budget and at the same time, satisfy total quantity control. Thus, the following relations exist:

$$\sum_{i=1}^{m} \tilde{\tilde{c}}_{ij} x_{ij} \leq c_j, \ j = 1, 2, \cdots, n, \tag{4.122}$$

where m is the number of vendors, n the number of items, $\tilde{\tilde{c}}_{ij}$ the one unit purchasing cost of the jth item of raw materials from the ith vendor, c_j the purchasing budget of the jth item of raw materials, and c the total budget for purchasing of raw materials.

Production demand

The raw materials purchased must satisfy the demands in terms of items, quality, and quantities in a given time period. Therefore, we have

$$\sum_{i=1}^{m} \tilde{\tilde{d}}_{ij} x_{ij} \leq \tilde{\tilde{d}}_j, \ j = 1, 2, \cdots, n, \tag{4.123}$$

where $\tilde{\tilde{d}}_{ij}$ is the unit converting rate to the requisite from the jth item of raw materials from the ith vendor, and $\tilde{\tilde{d}}_j$ the total demand for the jth item of raw materials.

Inventory capacity constraints

Holding safe-stock at a certain level is necessary. In general, safe-stock held in the integrated steel plant is between 15 and 50 days quantity required by the production and market supply situation, and changes as this is usually small. Here, the inventory considers quantity as well as volume. So, the relation is as follows:

$$\sum_{i=1}^{m} x_{ij} \geq 1.5 \tilde{\tilde{d}}_j - I_{j0}, \ j = 1, 2, \cdots, n, \tag{4.124}$$

$$\sum_{i=1}^{m} x_{ij} \leq 2.5 \tilde{\tilde{d}}_j - I_{j0}, \ j = 1, 2, \cdots, n, \tag{4.125}$$

where $\tilde{\tilde{d}}_j$ is the demand of the jth item of raw materials (1 month), and I_{j0} the initial stock for the jth item of raw materials.

The formulations of (4.124), (4.125) are given by the following deductive relation: Since, $I_j = I_{j0} + P_j - \tilde{\tilde{d}}_j$; but exists: $0.5 d_j \leq I_j \leq 1.5 \tilde{\tilde{d}}_j$ (by given), (4.124), (4.125) are rational, where I_j is the denoting the end inventory for the jth item of raw materials, P_j denotes the purchasing amount for the jth item of raw materials.

Technology constraints

In the purchasing process, we need to consider proportionality relations among a variety of items. For example, in iron-making technology, making 1 unit molten-iron needs 2 units of iron ore, 0.5 unit lime and 1 unit coal, etc. So, the following relations hold:

$$\sum_{i=1}^{m} x_{ik} = a_{kl} \sum_{i=1}^{m} x_{il}, \ k,l \in \{1,2,\cdots,n\}, k \neq l, \quad (4.126)$$

where a_{kl} are proportional constants.

Vendor resource constraints

Due to different geographical places, resource situations, transportation conditions, etc., the plant can get the ordering quantity from each vendor in different ways. So, we have

$$x_{il} \leq U_{il}, \ l \in L, \quad (4.127)$$

where U_{il} is the maximum ordering quantity for the lth item of raw materials from the ith vendor, and L the given special set.

4.6.1.3 Multi-objective Linear Programming Model

We list the multi-objective model for the purchase of bulk raw materials of a large-scale integrated steel plant as follows,

$$
\begin{cases}
\min F_1 = \sum_{i=1}^{m} \sum_{j=1}^{n} \tilde{\tilde{c}}_{ij} x_{ij} \\
\min F_2 = \sum_{i=1}^{m} \sum_{j=1}^{n} \tilde{\tilde{r}}_{ij} x_{ij} \\
\min F_3 = \sum_{i=1}^{m} \sum_{j=1}^{n} \tilde{\tilde{s}}_{ij} x_{ij} \\
\text{s.t.}
\begin{cases}
\sum_{i=1}^{m} \tilde{\tilde{c}}_{ij} x_{ij} \leq c_j, \ j=1,2,\cdots,n \\
\sum_{i=1}^{m} \tilde{\tilde{d}}_{ij} x_{ij} \leq \tilde{\tilde{d}}_j, \ j=1,2,\cdots,n \\
\sum_{i=1}^{m} x_{ij} \geq 1.5\tilde{\tilde{d}}_j - I_{j0}, \ j=1,2,\cdots,n \\
\sum_{i=1}^{m} x_{ij} \leq 2.5\tilde{\tilde{d}}_j - I_{j0}, \ j=1,2,\cdots,n \\
\sum_{i=1}^{m} x_{ik} = a_{kl} \sum_{i=1}^{m} x_{il}, \ k,l \in \{1,2,\cdots,n\}, k \neq l \\
x_{il} \leq U_{il}, \ l \in L.
\end{cases}
\end{cases}
\quad (4.128)
$$

Then according to the Fu-Fu EVM and Fu-Fu CCM, we get the expectation model with chance constraints (ECM, for short) as follows:

$$\begin{cases} \min F_1 = E\left[\sum\limits_{i=1}^{m} \sum\limits_{j=1}^{n} \tilde{\tilde{c}}_{ij} x_{ij}\right] \\ \min F_2 = E\left[\sum\limits_{i=1}^{m} \sum\limits_{j=1}^{n} \tilde{\tilde{r}}_{ij} x_{ij}\right] \\ \min F_3 = E\left[\sum\limits_{i=1}^{m} \sum\limits_{j=1}^{n} \tilde{\tilde{s}}_{ij} x_{ij}\right] \\ \text{s.t.} \begin{cases} Ch\{\sum\limits_{i=1}^{m} \tilde{\tilde{c}}_{ij} x_{ij} \le c_j\}(\alpha_1) \ge \beta_1, \ j = 1,2,\cdots,n \\ Ch\{\sum\limits_{i=1}^{m} \tilde{\tilde{d}}_{ij} x_{ij} \le \tilde{\tilde{d}}_j\}(\alpha_1) \ge \beta_1, \ j = 1,2,\cdots,n \\ \sum\limits_{i=1}^{m} x_{ij} \ge 1.5 \tilde{\tilde{d}}_j - I_{j0}, \ j = 1,2,\cdots,n \\ \sum\limits_{i=1}^{m} x_{ij} \le 2.5 \tilde{\tilde{d}}_j - I_{j0}, \ j = 1,2,\cdots,n \\ \sum\limits_{i=1}^{m} x_{ik} = a_{kl} \sum\limits_{i=1}^{m} x_{il}, \ k,l \in \{1,2,\cdots,n\}, k \neq l \\ x_{il} \le U_{il}, \ l \in L. \end{cases} \end{cases}$$ (4.129)

4.6.2 Data Collection

The data resource for this concrete model comes from a certain large-scale steel plant. This plant has 30 vendors and 100 items. We selected the 3 months data from real purchasing business and shrunk this real problem model to have only 7 vendors and 13 items that belong to four kinds of large bulk raw materials F, L, P and C, respectively. F, L, P, and C denote fine ore, lump ore, pellet, and coal, respectively. In addition, due to the point estimate weighted-sums method to our problem, we often have need to scale the objective functions for normalization; we used this kind of scaling method on our model. The coordinated model is as follows:

These four kinds of materials need to satisfy the production demand constraints, and the data are in Table 4.5.

4.6.3 Model Processing Specification

When we use the expected value of Fu-Fu variable to substitute the Fu-Fu variable, and we get the specific model as follows,

Table 4.2 Unit purchasing cost of the jth item from the ith vendor

$\tilde{\tilde{c}}_{ij}$	1	2
1	$(0.01, c_{11}, 0.01)$ with $c_{11}=(0.002, 0.112, 0.002)$	$(0.002, c_{12}, 0.002)$ with $c_{12}=(0.004, 0.0654, 0.004)$
2	-	$(0.002, \mu_{22}, 0.002)$ with $c_{22}=(0.0003, 0.0621, 0.0003)$
3	$(0.02, c_{31}, 0.01)$ with $c_{31}=(0.002, 0.127, 0.002)$	$(0.002, c_{32}, 0.002)$ with $c_{32}=(0.0003, 0.0586, 0.0003)$
4	$(0.01, c_{41}, 0.01)$ with $c_{41}=(0.002, 0.122, 0.002)$	-
5	$(0.01, c_{51}, 0.01)$ with $c_{51}=(0.002, 0.115, 0.002)$	-
6	-	$(0.005, c_{62}, 0.005)$ with $c_{62}=(0.001, 0.0602, 0.001)$
7	$(0.01, c_{71}, 0.01)$ with $c_{71}=(0.001, 0.119, 0.001)$	-

$\tilde{\tilde{c}}_{ij}$	3	4
1	-	$(0.004, c_{14}, 0.004)$ with $c_{14}=(0.0001, 0.09521, 0.0001)$
2	-	-
3	$(0.02, c_{33}, 0.01)$ with $c_{33}=(0.002, 0.195, 0.002)$	$(0.003, c_{34}, 0.003)$ with $c_{34}=(0.0002, 0.0975, 0.0002)$
4	-	-
5	$(0.02, c_{53}, 0.02)$ with $c_{53}=(0.003, 0.185, 0.003)$	-
6	-	-
7	-	-

$$
\begin{cases}
\min F_1 = \tilde{\tilde{c}}_{11}x_{11} + \tilde{\tilde{c}}_{31}x_{31} + \tilde{\tilde{c}}_{41}x_{41} + \tilde{\tilde{c}}_{51}x_{51} + \tilde{\tilde{c}}_{71}x_{71} \\
\qquad + \tilde{\tilde{c}}_{12}x_{12} + \tilde{\tilde{c}}_{22}x_{22} + \tilde{\tilde{c}}_{32}x_{32} + \tilde{\tilde{c}}_{62}x_{62} \\
\qquad + \tilde{\tilde{c}}_{33}x_{33} + \tilde{\tilde{c}}_{53}x_{53} + \tilde{\tilde{c}}_{14}x_{14} + \tilde{\tilde{c}}_{34}x_{34} \\
\min F_2 = \tilde{\tilde{r}}_{11}x_{11} + \tilde{\tilde{r}}_{31}x_{31} + \tilde{\tilde{r}}_{41}x_{41} + \tilde{\tilde{r}}_{51}x_{51} + \tilde{\tilde{r}}_{71}x_{71} + \tilde{\tilde{r}}_{12}x_{12} + \tilde{\tilde{r}}_{22}x_{22} \\
\qquad + \tilde{\tilde{r}}_{32}x_{32} + \tilde{\tilde{r}}_{62}x_{62} + \tilde{\tilde{r}}_{33}x_{33} + \tilde{\tilde{r}}_{53}x_{53} + \tilde{\tilde{r}}_{14}x_{14} + \tilde{\tilde{r}}_{34}x_{34} \\
\min F_3 = \tilde{\tilde{s}}_{11}x_{11} + \tilde{\tilde{s}}_{31}x_{31} + \tilde{\tilde{s}}_{41}x_{41} + \tilde{\tilde{s}}_{51}x_{51} + \tilde{\tilde{s}}_{71}x_{71} + \tilde{\tilde{s}}_{12}x_{12} + \tilde{\tilde{s}}_{22}x_{22} \\
\qquad + \tilde{\tilde{s}}_{32}x_{32} + \tilde{\tilde{s}}_{62}x_{62} + \tilde{\tilde{s}}_{33}x_{33} + \tilde{\tilde{s}}_{53}x_{53} + \tilde{\tilde{s}}_{14}x_{14} + \tilde{\tilde{s}}_{34}x_{34} \\
\text{s.t.} \begin{cases}
\tilde{\tilde{c}}_{11}x_{11} + \tilde{\tilde{c}}_{31}x_{31} + \tilde{\tilde{c}}_{41}x_{41} + \tilde{\tilde{c}}_{51}x_{51} + \tilde{\tilde{c}}_{71}x_{71} \\
\quad + \tilde{\tilde{c}}_{12}x_{12} + \tilde{\tilde{c}}_{22}x_{22} + \tilde{\tilde{c}}_{32}x_{32} + \tilde{\tilde{c}}_{62}x_{62} \\
\quad + \tilde{\tilde{c}}_{33}x_{33} + \tilde{\tilde{c}}_{53}x_{53} + \tilde{\tilde{c}}_{14}x_{14} + \tilde{\tilde{c}}_{34}x_{34} \leq 16.373 \\
\tilde{\tilde{d}}_{11}x_{11} + \tilde{\tilde{d}}_{31}x_{31} + \tilde{\tilde{d}}_{41}x_{41} + \tilde{\tilde{d}}_{51}x_{51} + \tilde{\tilde{d}}_{71}x_{71} \geq \tilde{\tilde{d}}_1 \\
\tilde{\tilde{d}}_{12}x_{12} + \tilde{\tilde{d}}_{32}x_{22} + \tilde{\tilde{d}}_{32}x_{32} + \tilde{\tilde{d}}_{62}x_{62} \geq \tilde{\tilde{d}}_2 \\
\tilde{\tilde{d}}_{33}x_{33} + \tilde{\tilde{d}}_{53}x_{53} \geq \tilde{\tilde{d}}_3 \\
\tilde{\tilde{d}}_{14}x_{14} + \tilde{\tilde{d}}_{34}x_{34} \geq \tilde{\tilde{d}}_4 \\
2x_{12} + 2x_{22} + 2x_{32} + 2x_{62} + 3x_{33} + 3x_{53} = x_{11} + x_{31} + x_{41} + x_{51} \\
\quad + x_{71} + x_{14} + x_{34} \\
x_{ij} \geq 0, \ i = 1, 2, \cdots, 7; \ j = 1, 2, 3, 4.
\end{cases}
\end{cases}
\tag{4.130}
$$

Table 4.3 Tardy-delivery ratio of the jth item from the ith vendor

$\tilde{\tilde{r}}_{ij}$	1	2
1	$(0.02, r_{11}, 0.02)$ with $r_{11}=(0.01, 0.1, 0.01)$	$(0.02, r_{12}, 0.02)$ with $r_{12}=(0.01, 0.1, 0.01)$
2	-	$(0.03, r_{22}, 0.03)$ with $r_{22}=(0.02, 0.25, 0.02)$
3	$(0.02, r_{31}, 0.01)$ with $r_{31}=(0.003, 0.155, 0.003)$	$(0.02, r_{32}, 0.02)$ with $r_{32}=(0.003, 0.15, 0.003)$
4	$(0.01, r_{41}, 0.01)$ with $r_{41}=(0.003, 0.17, 0.003)$	-
5	$(0.01, r_{51}, 0.01)$ with $r_{51}=(0.003, 0.12, 0.003)$	-
6	-	$(0.05, r_{62}, 0.05)$ with $r_{62}=(0.003, 0.3, 0.003)$
7	$(0.01, r_{71}, 0.01)$ with $r_{71}=(0.01, 0.2, 0.01)$	-

$\tilde{\tilde{r}}_{ij}$	3	4
1	-	$(0.02, r_{14}, 0.02)$ with $r_{14}=(0.01, 0.1, 0.01)$
2	-	-
3	$(0.02, r_{33}, 0.01)$ with $r_{33}=(0.002, 0.155, 0.002)$	$(0.02, r_{34}, 0.02)$ with $r_{34}=(0.002, 0.15, 0.002)$
4	-	-
5	$(0.01, r_{53}, 0.01)$ with $r_{53}=(0.002, 0.12, 0.002)$	-
6	-	-
7	-	-

Since there are Fu-Fu variables in Model (4.130), so we need to employ the expected value operator or the chance constrained operator to handle the above Fu-Fu multi-objective model. In this section, we use the expected value operator and the chance constrained operator to deal with the objective functions and the constraints, respectively.

Table 4.4 Scrap fraction of the jth item from the ith vendor

$\bar{\tilde{s}}_{ij}$	1	2
1	(0.02,s_{11},0.02) with s_{11}=(0.01,0.2,0.01)	(0.02,s_{12},0.02) with s_{12}=(0.01,0.2,0.01)
2	-	(0.01,s_{22},0.01) with s_{22}=(0.01,0.1,0.01)
3	(0.02,s_{31},0.01) with μ_{31}=(0.01,0.1,0.01)	(0.02,s_{32},0.02) with s_{32}=(0.005,0.15,0.005)
4	(0.01,s_{41},0.01) with s_{41}=(0.003,0.15,0.003)	-
5	(0.01,s_{51},0.01) with s_{51}=(0.003,0.17,0.003)	-
6	-	(0.05,s_{62},0.05) with s_{62}=(0.003,0.22,0.003)
7	(0.01,s_{71},0.01) with s_{71}=(0.003,0.13,0.003)	-

$\bar{\tilde{s}}_{ij}$	3	4
1	-	(0.02,s_{14},0.02) with s_{14}=(0.01,0.2,0.01)
2	-	-
3	(0.02,s_{33},0.01) with s_{33}=(0.001,0.15,0.001)	(0.02,s_{34},0.02) with s_{34}=(0.002,0.15,0.002)
4	-	-
5	(0.01,s_{53},0.01) with s_{53}=(0.003,0.17,0.002)	-
6	-	-
7	-	-

$$
\left\{
\begin{aligned}
&\min F_1 = E[\bar{\tilde{c}}_{11}]x_{11}+E[\bar{\tilde{c}}_{31}]x_{31}+E[\bar{\tilde{c}}_{41}]x_{41}+E[\bar{\tilde{c}}_{51}]x_{51}+E[\bar{\tilde{c}}_{71}]x_{71} \\
&\qquad +E[\bar{\tilde{c}}_{12}]x_{12}+E[\bar{\tilde{c}}_{22}]x_{22}+E[\bar{\tilde{c}}_{32}]x_{32}+E[\bar{\tilde{c}}_{62}]x_{62} \\
&\qquad +E[\bar{\tilde{c}}_{33}]x_{33}+E[\bar{\tilde{c}}_{53}]x_{53}+E[\bar{\tilde{c}}_{14}]x_{14}+E[\bar{\tilde{c}}_{34}]x_{34} \\
&\min F_2 = E[\bar{\tilde{r}}_{11}]x_{11}+E[\bar{\tilde{r}}_{31}]x_{31}+E[\bar{\tilde{r}}_{41}]x_{41}+E[\bar{\tilde{r}}_{51}]x_{51}+E[\bar{\tilde{r}}_{71}]x_{71}+E[\bar{\tilde{r}}_{12}]x_{12} \\
&\qquad +E[\bar{\tilde{r}}_{22}]x_{22}+E[\bar{\tilde{r}}_{32}]x_{32}+E[\bar{\tilde{r}}_{62}]x_{62}+E[\bar{\tilde{r}}_{33}]x_{33}+E[\bar{\tilde{r}}_{53}]x_{53}+E[\bar{\tilde{r}}_{14}]x_{14} \\
&\qquad +E[\bar{\tilde{r}}_{34}]x_{34} \\
&\min F_3 = E[\bar{\tilde{s}}_{11}]x_{11}+E[\bar{\tilde{s}}_{31}]x_{31}+E[\bar{\tilde{s}}_{41}]x_{41}+E[\bar{\tilde{s}}_{51}]x_{51}+E[\bar{\tilde{s}}_{71}]x_{71}+E[\bar{\tilde{s}}_{12}]x_{12} \\
&\qquad +E[\bar{\tilde{s}}_{22}]x_{22}+E[\bar{\tilde{s}}_{32}]x_{32}+E[\bar{\tilde{s}}_{62}]x_{62}+E[\bar{\tilde{s}}_{33}]x_{33}+E[\bar{\tilde{s}}_{53}]x_{53}+E[\bar{\tilde{s}}_{14}]x_{14} \\
&\qquad +E[\bar{\tilde{s}}_{34}]x_{34} \\
&\text{s.t.}
\left\{
\begin{aligned}
&E[\bar{\tilde{c}}_{11}x_{11}+\bar{\tilde{c}}_{31}x_{31}+\bar{\tilde{c}}_{41}x_{41}+\bar{\tilde{c}}_{51}x_{51}+\bar{\tilde{c}}_{71}x_{71} \\
&\qquad +\bar{\tilde{c}}_{12}x_{12}+\bar{\tilde{c}}_{22}x_{22}+\bar{\tilde{c}}_{32}x_{32}+\bar{\tilde{c}}_{62}x_{62} \\
&\qquad +\bar{\tilde{c}}_{33}x_{33}+\bar{\tilde{c}}_{53}x_{53}+\bar{\tilde{c}}_{14}x_{14}+\bar{\tilde{c}}_{34}x_{34}] \leq 16.373 \\
&Ch\{\bar{\tilde{d}}_{11}x_{11}+\bar{\tilde{d}}_{31}x_{31}+\bar{\tilde{d}}_{41}x_{41}+\bar{\tilde{d}}_{51}x_{51}+\bar{\tilde{d}}_{71}x_{71} \geq \bar{\tilde{d}}_1\}(\alpha_1) \geq \beta_1 \\
&Ch\{\bar{\tilde{d}}_{12}x_{12}+\bar{\tilde{d}}_{32}x_{22}+\bar{\tilde{d}}_{32}x_{32}+\bar{\tilde{d}}_{62}x_{62} \geq \bar{\tilde{d}}_2\}(\alpha_2) \geq \beta_2 \\
&Ch\{\bar{\tilde{d}}_{33}x_{33}+\bar{\tilde{d}}_{53}x_{53} \geq \bar{\tilde{d}}_3\}(\alpha_3) \geq \beta_3 \\
&Ch\{\bar{\tilde{d}}_{14}x_{14}+\bar{\tilde{d}}_{34}x_{34} \geq \bar{\tilde{d}}_4\}(\alpha_4) \geq \beta_5 \\
&2x_{12}+2x_{22}+2x_{32}+2x_{62}+3x_{33}+3x_{53}=x_{11}+x_{31}+x_{41}+x_{51} \\
&\qquad +x_{71}+x_{14}+x_{34} \\
&x_{ij} \geq 0,\ i=1,2,\cdots,7;\ j=1,2,3,4.
\end{aligned}
\right.
\end{aligned}
\right.
$$

$$\text{(4.131)}$$

Table 4.5 Scrap fraction of the jth item from the ith vendor

$\tilde{\tilde{d}}_{ij}$	1	2
1	(0.02,d_{11},0.02) with d_{11}=(0.005,1.2,0.005)	(0.02,d_{12},0.02) with d_{12}=(0.005,1.25,0.005)
2	-	(0.01,d_{22},0.01) with μ_{22}=(0.005,0.95,0.005)
3	(0.02,d_{31},0.01) with d_{31}=(0.005,0.9,0.005)	(0.02,d_{32},0.02) with d_{32}=(0.005,1.15,0.005)
4	(0.01,d_{41},0.01) with μ_{41}=(0.02,1,0.02)	-
5	(0.01,d_{51},0.01) with μ_{51}=(0.02,1.1,0.02)	-
6	-	(0.05,d_{62},0.05) with d_{62}=(0.02,1.05,0.02)
7	(0.01,d_{71},0.01) with μ_{71}=(0.02,0.95,0.02)	-
$\tilde{\tilde{d}}_j$	(3,d_1,3) with d_1=(1,60,1)	(2,d_2,2) with d_2=(0.5,30,0.5)

$\tilde{\tilde{d}}_{ij}$	3	4
1	-	(0.02,d_{14},0.02) with d_{14}=(0.02,1.12,0.02)
2	-	-
3	(0.02,d_{33},0.01) with d_{33}=(0.02,1.3,0.02)	(0.02,d_{34},0.02) with d_{34}=(0.02,1.24,0.02)
4	-	-
5	(0.05,d_{53},0.05) with d_{53}=(0.02,1.1,0.02)	-
6	-	-
7	-	-
$\tilde{\tilde{d}}_j$	(1,d_3,1) with d_3=(0.5,10,0.5)	(3,d_4,3) with d_4=(2,70,2)

Following Theorem 4.4, the expected value for the Fu-Fu variable $\tilde{\tilde{c}} = (l_1,\tilde{c},r_1)$ with $\tilde{c} = (l_2,c,r_2)$ we obtain that

$$E[\tilde{\tilde{c}}] = c + \frac{(r_1+r_2)-(l_1+l_2)}{4}.$$

By Theorem 4.12, we use the expected value operator and $Nec - Nec$ chance definition, so we get the following crisp specific model.

$$
\begin{cases}
\min F_1 = 0.112x_{11} + 0.127x_{31} + 0.122x_{41} + 0.115x_{51} + 0.119x_{71} \\
\quad\quad +0.0654x_{12} + 0.0621x_{22} + 0.0586x_{32} + 0.0602x_{62} \\
\quad\quad +0.195x_{33} + 0.185x_{53} + 0.09521x_{14} + 0.0975x_{34} \\
\min F_2 = 0.1x_{11} + 0.155x_{31} + 0.17x_{41} + 0.12x_{51} + 0.2x_{71} + 0.1x_{12} + 0.25x_{22} \\
\quad\quad +0.15x_{32} + 0.3x_{62} + 0.15x_{33} + 0.12x_{53} + 0.1x_{14} + 0.15x_{34} \\
\min F_3 = 0.2x_{11} + 0.1x_{31} + 0.15x_{41} + 0.17x_{51} + 0.13x_{71} + 0.2x_{12} + 0.1x_{22} \\
\quad\quad +0.15x_{32} + 0.22x_{62} + 0.15x_{33} + 0.17x_{53} + 0.2x_{14} + 0.15x_{34} \\
\text{s.t.}
\begin{cases}
0.112x_{11} + 0.127x_{31} + 0.122x_{41} + 0.115x_{51} + 0.119x_{71} + 0.0654x_{12} \\
\quad +0.0621x_{22} + 0.0586x_{32} + 0.0602x_{62} + 0.195x_{33} + 0.185x_{53} \\
\quad +0.09521x_{14} + 0.0975x_{34} \leq 16.373 \\
1.2x_{11} + 0.9x_{31} + x_{41} + 1.1x_{51} + 0.95x_{71} \\
\quad -\beta_1(0.005x_{11} + 0.005x_{31} + 0.02x_{41} + 0.02x_{51} + 0.02x_{71}) \\
\quad -\alpha_1(0.02x_{11} + 0.02x_{31} + 0.01x_{41} + 0.02x_{51} + 0.01x_{71}) \\
\quad \geq 60 + \beta_1 + 3(1 - \alpha_1) \\
1.25x_{12} + 0.95x_{22} + 1.15x_{32} + 1.05x_{62} \\
\quad -\beta_2(0.005x_{12} + 0.005x_{22} + 0.005x_{32} + 0.02x_{62}) \\
\quad -\alpha_2(0.02x_{12} + 0.01x_{22} + 0.02x_{32} + 0.05x_{62}) \geq 30 + 0.5\beta_2 + 2(1 - \alpha_2) \\
1.3x_{33} + 1.1x_{53} - \beta_3(0.02x_{33} + 0.02x_{53}) - \alpha_3(0.02x_{33} + 0.05x_{53}) \\
\quad \geq 10 + 0.5\beta_3 + (1 - \alpha_3) \\
1.12x_{14} + 1.24x_{34} - \beta_4(0.02x_{14} + 0.02x_{34}) - \alpha_4(0.02x_{14} + 0.02x_{34}) \\
\quad \geq 70 + 2\beta_4 + 3(1 - \alpha_4) \\
2x_{12} + 2x_{22} + 2x_{32} + 2x_{62} + 3x_{33} + 3x_{53} = x_{11} + x_{31} + x_{41} + x_{51} \\
\quad +x_{71} + x_{14} + x_{34} \\
x_{ij} \geq 0, \ i = 1,2,\cdots,7; \ j = 1,2,3,4.
\end{cases}
\end{cases}
$$

$$(4.132)$$

We set the $\alpha_1 = \alpha_2 = \alpha_3 = \alpha_4 = 0.8$, $\beta_1 = \beta_2 = \beta_3 = \beta_4 = 0.7$. We employ the SA algorithm to solve the above model and get the following results.

$$x_{11} = 53.33, x_{22} = 31.58, x_{33} = 5.93, x_{53} = 9.62, x_{34} = 56.45.$$

The values of the three objective functions are

$$F_1 = 1.6373, F_2 = 2.3739, F_3 = 2.4816.$$

The purchasing of raw materials is a very important problem in today's companies. This application is not only valuable to the steel industry, but also valuable to other production and manufacturing industries. Using a mathematical model to study purchasing issues is very necessary, as it supplies helpful information to the purchasing decision maker.

4.7 Another Way to Deal with Fu-Fu Multi-objective Decision Making Model

The multi-objective model with Fu-Fu variables:

$$\begin{cases} \min \ [f_1(x,\tilde{\tilde{a}}_1), f_2(x,\tilde{\tilde{a}}_2), \cdots, f_m(x,\tilde{\tilde{a}}_m)] \\ \text{s.t.} \begin{cases} g_i(x,\tilde{\tilde{b}}_i) \leq 0, i = 1,2,\cdots,p \\ h_r(x,\tilde{\tilde{c}}_r) = 0, r = 1,2,\cdots,q, \end{cases} \end{cases} \quad (4.133)$$

where $\tilde{\tilde{a}}_k(k = 1,2,\cdots,m), \tilde{\tilde{b}}_i(i = 1,2,\cdots,p)$ and $\tilde{\tilde{c}}_r(r = 1,2,\cdots,q)$ are Fu-Fu variables.

In order to study conveniently, we consider the multi-objective model with Fu-Fu coefficients as follow:

$$\begin{cases} \max \ [\tilde{\tilde{c}}_1 x, \tilde{\tilde{c}}_2 x, \cdots, \tilde{\tilde{c}}_K x] \\ \text{s.t.} \ x \in X = \{x \subset R^n | \tilde{\tilde{A}} x \leq b, x \geq 0\}, \end{cases} \quad (4.134)$$

where $\tilde{\tilde{c}}_k = (\tilde{\tilde{c}}_{k1}, \tilde{\tilde{c}}_{k2}, \cdots, \tilde{\tilde{c}}_{kn}) \ (k = 1,2,\cdots,K), \tilde{\tilde{A}} = (\tilde{\tilde{A}}_1, \tilde{\tilde{A}}_2, \cdots, \tilde{\tilde{A}}_m)^T, \tilde{\tilde{A}}_i = (\tilde{\tilde{a}}_{i1}, \tilde{\tilde{a}}_{i2}, \cdots, \tilde{\tilde{a}}_{in})$ are Fu-Fu vectors, and $b = (b_1, b_2, \cdots, b_m)^T$.

Definition 4.9. Let $\beta = (\beta_1, \beta_2, \cdots, \beta_m)^T$ be possibility level vector, $\beta_i \in [0,1], x \in \mathfrak{R}^n$, and if

$$pos(\tilde{\tilde{A}}_i x \leq b_i, \tilde{\tilde{c}}_k x) \geq \beta_i, \quad i = 1,2,\cdots,m, k = 1,2,\cdots,K,$$

then x is called β-possible feasible solution to (4.134). All β-possible feasible solutions are called β-possible feasible set X_β of the model (4.134).

Consider the form of a multi-objective problem written as follows:

$$\max_{x \in X}\{\tilde{\tilde{c}}_1 x, \tilde{\tilde{c}}_2 x, \cdots, \tilde{\tilde{c}}_K x, \tilde{\tilde{A}}_i x\}, \quad i = 1,2,\cdots,m. \quad (4.135)$$

Definition 4.10. Let α be a possibility level, $\alpha \in [0,1], D \in \mathbf{R}^n$ and $x_0 \in D$. if do not exist $x \in D$ and $k \in \{1,2,\cdots,K\}, x$ satisfy

$$pos(\tilde{\tilde{c}}_1 x \geq \tilde{\tilde{c}}_1 x_0, \cdots, \tilde{\tilde{c}}_{k-1} x \geq \tilde{\tilde{c}}_{k-1} x_0, \tilde{\tilde{c}}_k x > \tilde{\tilde{c}}_k x_0, \tilde{\tilde{c}}_{k+1} x \geq \tilde{\tilde{c}}_{k+1} x_0, \cdots, \tilde{\tilde{c}}_K x \geq \tilde{\tilde{c}}_K x_0, \tilde{\tilde{A}}_{K+i} x \geq \tilde{\tilde{A}}_{K+i} x_0) \geq \alpha, \quad i = 1,2,\cdots,m, k = 1,2,\cdots,K,$$

then x_0 is called α-possible efficient solution of the model (4.135).

Definition 4.11. Let $x^0 \in X$, if x^0 is the

$$\begin{cases} \max \ [\tilde{\tilde{c}}_1 x, \tilde{\tilde{c}}_2 x, \cdots, \tilde{\tilde{c}}_K x, \tilde{\tilde{A}}_i x] \\ \text{s.t.} \ x \in X_\beta, \end{cases} \quad (4.136)$$

α-possible efficient solution of Problem (4.135), and then x^0 is called (α, β)-satisfied solution of the model (4.134).

In fact, to solve the model (4.134) and find its (α, β)-satisfied solution, we may consider the multi-objective problem as follows:

$$\begin{cases} \max \ [(\tilde{\tilde{c}}_1)_\alpha x, (\tilde{\tilde{c}}_2)_\alpha x, \cdots, (\tilde{\tilde{c}}_K)_\alpha x, (\tilde{\tilde{A}}_i)_\alpha x] \\ \text{s.t.} \ x \in X_\beta, \end{cases} \quad (4.137)$$

where $(\tilde{\bar{c}}_k)_\alpha, (\tilde{\bar{A}}_i)_\alpha$ is α-level set of Fu-Fu variables $\tilde{\bar{c}}_k, \tilde{\bar{A}}_i$ ($i = 1, 2, \cdots, m, k = 1, 2, \cdots, K$) respectively.

Theorem 4.16. x^0 *is the (α, β)-satisfied solution of the model (4.134) if and only if x^0 is the efficient solution of the model (4.137).*

Proof. Let x^0 be the (α, β)-satisfied solution of the model (4.134). Following from the Definition 4.11, x^0 is the β-possible feasible solution and α-possible efficient solution to the model (4.136). If x^0 is not the efficient solution of the model (4.137), then exist $x^1 \in X_\beta$ and row-vector $q_k, q_k \in (\tilde{\bar{c}}_k, \tilde{\bar{A}}_{K+i})_\alpha (k = 1, 2, \cdots, K, i = 1, 2, \cdots, m)$, and $k_0 \in \{1, 2, \cdots, K, K+i\}$, we can derive that

$$q_k x^1 \ge q_k x^0, \text{and } q_{k_0} x^1 > q_{k_0} x^0, \text{when } \forall k \in \{1, 2, \cdots, K, K+i\} \backslash \{k_0\}.$$

Due to $q_k \in (\tilde{\bar{c}}_k, \tilde{\bar{A}}_{K+i})_\alpha$, then $\min\{\pi_{\tilde{\bar{c}}_1}(q_1), \pi_{\tilde{\bar{c}}_2}(q_2), \cdots, \pi_{\tilde{\bar{c}}_K}(q_K), \pi_{\tilde{\bar{A}}_{K+i}}(q_{K+i})\} \ge \alpha, (i = 1, 2, \cdots, m)$. Following from the Definition 4.9, basis on the expand principle, we have

$$\sup_{(t_1, t_2, \cdots, t_K, t_{K+i}) \in T_{k_0}} \min\{\pi_{\tilde{\bar{c}}_1}(t_1), \cdots, \pi_{\tilde{\bar{c}}_{k_0-1}}(t_{k_0-1}), \pi_{\tilde{\bar{c}}_{k_0}}(t_{k_0}),$$
$$\pi_{\tilde{\bar{c}}_{k_0+1}}(t_{k_0+1}), \cdots, \pi_{\tilde{\bar{c}}_K}(t_K), \pi_{\tilde{\bar{A}}_{K+i}}(t_{K+i})\}$$
$$= pos(\tilde{\bar{c}}_1 x^1 \ge \tilde{\bar{c}}_1 x^0, \cdots, \tilde{\bar{c}}_{k_0-1} x^1 \ge \tilde{\bar{c}}_{k_0-1} x^0, \tilde{\bar{c}}_{k_0} x^1 > \tilde{\bar{c}}_{k_0} x^0, \tilde{\bar{c}}_{k_0+1} x^1 \ge \tilde{\bar{c}}_{k_0+1} x^0,$$
$$\cdots, \tilde{\bar{c}}_K x^1 \ge \tilde{\bar{c}}_K x^0, \tilde{\bar{A}}_{K+i} x^1 \ge \tilde{\bar{A}}_{K+i} x^0) \ge \alpha,$$

where

$$T_{k_0} = \{(t_1, t_2, \cdots, t_K, t_{K+i}) | t_1 x^1 \ge t_1 x^0, \cdots, t_{k_0-1} x^1 \ge t_{k_0-1} x^0, t_{k_0} x^1 > t_{k_0} x^0,$$
$$t_{k_0+1} x^1 \ge t_{k_0+1} x^0, \cdots, t_K x^1 \ge t_K x^0, t_{K+i} x^1 \ge t_{K+i} x^0\}, (i = 1, 2, \cdots, m).$$

It is contrary that x^0 is not the efficient solution of the model (4.137).

Contrarily, let x^0 is the efficient solution of the model (4.137) and is not the (α, β)-satisfied solution of the model (4.134), then exist $x^2 \in X_\beta$ and $s \in \{1, 2, \cdots, K, K+i\}, i = (1, 2, \cdots, m)$,

$$Pos(\tilde{\bar{c}}_1 x^2 \ge \tilde{\bar{c}}_1 x^0, \cdots, \tilde{\bar{c}}_{s-1} x^2 \ge \tilde{\bar{c}}_{s-1} x^0, \tilde{\bar{c}}_s x^2 > \tilde{\bar{c}}_s x^0, \tilde{\bar{c}}_{s+1} x^2 \ge \tilde{\bar{c}}_{s+1} x^0,$$
$$\cdots, \tilde{\bar{c}}_K x^2 \ge \tilde{\bar{c}}_K x^0, \tilde{\bar{A}}_{K+i} x^2 \ge \tilde{\bar{A}}_{K+i} x^0) \ge \alpha.$$

Following from the Definition 4.9, basis on the expand principle, there is a row-vector $p_k \in \mathbf{R}^n (k = 1, 2, \cdots, K, \cdots, K + i; i = 1, 2, \cdots, m)$ which satisfy $p_1 x^2 \ge p_1 x^0, \cdots, p_{s-1} x^2 \ge p_{s-1} x^0, p_s x^2 > p_s x^0, p_{s+1} x^2 \ge p_{s+1} x^0, \cdots, p_K x^2 \ge p_K x^0, p_{K+i} x^2 \ge p_{K+i} x^0$ and $\pi_{(\tilde{\bar{c}}_k, \tilde{\bar{A}}_{K+i})} \ge \alpha, p_k \in (\tilde{\bar{c}}_k, \tilde{\bar{A}}_{K+i})_\alpha$. It is contrary that x^0 is the efficient solution of the model 4.137. The proof is thus completed. $\square$

In the following, we introduce how to transform the Fu-Fu variables into their corresponding fuzzy variables.

Theorem 4.17. *(Zhou and Xu [402]) Let $\tilde{\tilde{\xi}}$ be a triangular LR Fu-Fu variable, that is, $\tilde{\tilde{\xi}} = (a_L, \tilde{\rho}, a_R)_{LR}$ with $\tilde{\rho} = (\rho_L, \rho_M, \rho_R)$, α_1, α_2 are any given possibility of the outer-layer and the inner-layer fuzzy variables, respectively. Then the Fu-Fu variable can be transformed into a (α_1, α_2)-level trapezoidal fuzzy variable.*

Proof. Since $\tilde{\rho} = (\rho_L, \rho_M, \rho_R)$ is the inner-layer fuzzy variable, based on Definition 1.2, we get the α_2-cut of $\tilde{\rho}$ as follows:

$$\tilde{\rho}_\beta = [\rho^L_{\alpha_2}, \rho^R_{\alpha_2}] = \{x \in U | \mu_{\tilde{\rho}}(x) \geq \alpha_2\},$$

where $\rho^L_{\alpha_2} = \rho_L + \alpha_2(\rho_M - \rho_L)$, $\rho^R_{\alpha_2} = \rho_R - \alpha_2(\rho_R - \rho_M)$, and U is the universe.

Note that $\tilde{\rho}_{\alpha_2}$ are crisp sets. The parameter $\beta \in (0,1)$ reflects the decision maker's degree of optimism. These intervals indicate the range at the possibility level α_2. In other words, $\rho^L_{\alpha_2}$ is the minimum value that $\tilde{\rho}$ achieves with probability α_2, and $\rho^R_{\alpha_2}$ is the maximum value that $\tilde{\rho}$ achieves with probability α_2.

So the α_2-level Fu-Fu variable $\tilde{\tilde{\xi}}_{(\alpha_2)}$ can be defined as $\tilde{\tilde{\xi}}_{(\alpha_2)} = (a_L, \tilde{\rho}_{\alpha_2}, a_R)$. Let $X = \{x_j \in U | \mu_{\tilde{\rho}}(x_j) \geq \alpha_2, j = 1, 2, \cdots, n\}$, so the Fu-Fu variable $\tilde{\tilde{\xi}}_{(\alpha_2)}$ can be also denoted as $\tilde{\tilde{\xi}}_{(\alpha_2)} = (a_L, X, a_R)$ or denoted as follows:

$$\tilde{\tilde{\xi}}_{(\alpha_2)} = \begin{cases} \tilde{\xi}^1_{(\alpha_2)} = (a_L, x_1, a_R) \\ \tilde{\xi}^2_{(\alpha_2)} = (a_L, x_2, a_R) \\ \vdots \\ \tilde{\xi}^n_{(\alpha_2)} = (a_L, x_n, a_R), \end{cases}$$

where $x_1 \leq x_2 \leq \cdots \leq x_n$, $X = [x_1, x_n] = [\rho^L_{\alpha_2}, \rho^R_{\alpha_2}]$ and $\tilde{\xi}^j_{(\alpha_2)}, j = 1, 2, \cdots, n$ is a fuzzy variable.

Thus the Fu-Fu variable $\tilde{\tilde{\xi}}_{(\alpha_2)}$ is transformed into a group of fuzzy variables $\tilde{\xi}^j_{(\alpha_2)}, j = 1, 2, \cdots, n$, see Figure 4.8. We have the following two special extreme points as follows:

$$\tilde{\xi}^1_{(\alpha_2)} = (a_L, \rho^R_{\alpha_2}, a_R) = (a_L, \rho_R - \alpha_2(\rho_R - \rho_M), a_R),$$
$$\tilde{\xi}^n_{(\alpha_2)} = (a_L, \rho^L_{\alpha_2}, a_R) = (a_L, \rho_L + \alpha_2(\rho_M - \rho_L), a_R).$$

Again, on the basis of the concept of α_1-cut, denote the α_1-cut of $\tilde{\tilde{\xi}}_{(\alpha_2)}$ as follows $\tilde{\tilde{\xi}}_{(\alpha_1, \alpha_2)} = [\tilde{\xi}_{(\alpha_2)}{}^L_{\alpha_1}, \tilde{\xi}_{(\alpha_2)}{}^R_{\alpha_1}]$, where

$$\xi_{(\alpha_2)}{}^L_{\alpha_1} = \max \xi^j_{(\alpha_2)}{}^L_{\alpha_1}, \text{ and } \xi^j_{(\alpha_2)}{}^L_{\alpha_1} = \min \mu^{-1}_{\xi^j_{(\alpha_2)}}(\alpha_1),$$

$$\xi_{(\alpha_2)}{}^R_{\alpha_1} = \min \xi^j_{(\alpha_2)}{}^R_{\alpha_1}, \text{ and } \xi^j_{(\alpha_2)}{}^R_{\alpha_1} = \max \mu^{-1}_{\xi^j_{(\alpha_2)}}(\alpha_1).$$

Obviously, we have that

$$\xi_{(\alpha_2)}{}^L_{\alpha_1} = \max\min_{\xi_{(\alpha_2)}}\mu^{-1}_{\xi_i}(\alpha_1) = \min_{\xi_{(\alpha_2)}}\mu^{-1}_{\xi_n}(\alpha_1),$$

$$\xi_{(\alpha_2)}{}^R_{\alpha_1} = \min\max_{\xi_{(\alpha_2)}}\mu^{-1}_{\xi_i}(\alpha_1) = \min_{\xi_{(\alpha_2)}}\mu^{-1}_{\xi_1}(\alpha_1).$$

By the convexity of a fuzzy variable, the bounds of these intervals are function of α_1. In order to make the method to be effective, the level α_1 should satisfy that

$$0 \le \alpha_1 \le \frac{a_R - a_L}{(a_R - a_L) + (\rho^R_{\alpha_2} - \rho^L_{\alpha_2})} = \frac{a_R - u_L}{a_R - a_L + (1 - \alpha_2)(\rho_R - \rho_L)}.$$

Consequently, we can use its α_1 cut to construct the corresponding membership function. Let $\xi^L_{\alpha_1} = \underline{a}, \xi^R_{\alpha_1} = \overline{a}$, then the Fu-Fu variable $\tilde{\tilde{\xi}} = (a_L, \tilde{\rho}, a_R)$ can be transformed in the (α, α_2)-level trapezoidal fuzzy variable $\tilde{\xi}_{(\alpha_1,\alpha_2)}$ as shown in Figure 3.10, that is,

$$\tilde{\tilde{\xi}} \to \tilde{\xi}_{(\alpha_1,\alpha_2)} = (a_L, \underline{a}, \overline{a}, a_R),$$

where the parameter α and β both reflect the decision-maker's degree of optimism. And the values of $\underline{a}$ and $\overline{a}$ can be obtained by the follow equation (4.138).

$$\underline{a} = a_L + \alpha\{[\rho_R - \alpha_2(\rho_R - \rho_M)] - a_L\},$$
$$\overline{a} = a_R - \alpha\{a_R - [\rho_L + \alpha_2(\rho_M - \rho_L)]\}. \qquad (4.138)$$

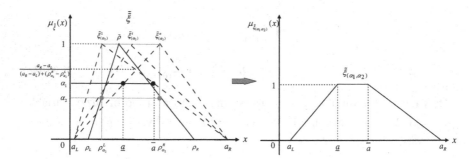

Fig. 4.8 Transformation process from the Fu-Fu variable to the (α_1, α_2)-level fuzzy variable

The value of $\mu_{\tilde{\xi}_{(\alpha_1,\alpha_2)}}(x)$ at $x \in [\underline{a}, \overline{a}]$ is considered subjectively to be 1 approximately, and thus the membership function of the fuzzy variable $\tilde{\xi}_{(\alpha_1,\alpha_2)}$ is

$$\mu_{\tilde{\xi}_{(\alpha_1,\alpha_2)}}(x) = \begin{cases} 0, & \text{if } x < a_L, x > a_R \\ \frac{x - a_L}{\underline{s} - a_L}, & \text{if } a_L \le x < \underline{a} \\ 1, & \text{if } \underline{a} \le x \le \overline{a} \\ \frac{a_R - x}{a_R - \overline{a}}, & \text{if } \overline{s} \le x < a_R. \end{cases}$$

Thus Theorem 4.17 is proved. $\square$

Thus, base on the Theorem 4.17, the Fu-Fu multi-objective model (4.133) is transformed into a fuzzy multi-objective model as follow:

$$
\begin{cases}
\min \ [f_1(x,\widetilde{A}_1), f_2(x,\widetilde{A}_2), \cdots, f_k(x,\widetilde{A}_m)] \\
\text{s.t.} \begin{cases} g_i(x,\widetilde{B}_i) \leq 0, i = 1,2,\cdots,p \\ h_r(x,\widetilde{C}_r) = 0, r = 1,2,\cdots,q, \end{cases}
\end{cases}
\tag{4.139}
$$

where $\widetilde{A}_k = \widetilde{\omega}_{\widetilde{a}_{k(\alpha_1,\alpha_2)}}$ $(k = 1,2,\cdots,m), \widetilde{B}_i = \widetilde{\omega}_{\widetilde{b}_{i(\alpha_1,\alpha_2)}}$ $(i = 1,2,\cdots,p)$ and $\widetilde{C}_r = \widetilde{\omega}_{\widetilde{c}_{r(\alpha_1,\alpha_2)}}$ $(r = 1,2,\cdots,q)$ are fuzzy trapezoidal numbers.

Definition 4.12. Let a domain U, $\widetilde{\widetilde{\xi}}$ be a Fu-Fu variable defined on U. $\alpha_1, \alpha_2 \in [0,1]$ be any given possibility level of fuzzy variable. Basis on the Theorem 4.17. If $\mu_1(x) \geq \alpha_1$ and $\mu_2(x) \geq \alpha_2$ the Fu-Fu variable $\widetilde{\widetilde{\xi}}$ can be transformed a (α_1,α_2)-level trapezoidal fuzzy variable $\widetilde{B}$. So B_α consist of all elements whose degrees of membership in $\widetilde{B}$ are greater than or equal to α,

$$
B_\alpha = \{x \in U \, | \, \mu_{\widetilde{B}}(x) \geq \alpha, \mu_1(x) \geq \alpha_1, \mu_2(x) \geq \alpha_2\},
$$

then B_α is called the α-level sets of Fu-Fu variable $\widetilde{\widetilde{\xi}}$ with $\mu_1(x) \geq \alpha_1$ and $\mu_2(x) \geq \alpha_2$, viz. $\widetilde{\widetilde{\xi}}_{\alpha_{(\alpha_1,\alpha_2)}}$.

Theorem 4.18. *Let $\widetilde{a}, \widetilde{b}, \widetilde{c}$ be Fu-Fu variables on the possibility space $(\Theta, P(\Theta), Pos)$, (α_1, α_2) are any given possibility level respectively, $L_\alpha(\widetilde{a}, \widetilde{b}, \widetilde{c})$ is the α-level set of Fu-Fu numbers $(\widetilde{a}, \widetilde{b}, \widetilde{c})$. For different (α_1, α_2)-level it is clear that*

(a) if $\alpha_1^1 \leq \alpha_1^2, \alpha_2^1 \geq \alpha_2^2$, then $L_\alpha^1(\widetilde{a}, \widetilde{b}, \widetilde{c}) \subseteq L_\alpha^2(\widetilde{a}, \widetilde{b}, \widetilde{c})$,
(b) if $\alpha_1^1 \geq \alpha_1^2, \alpha_2^1 \leq \alpha_2^2$, then $L_\alpha^1(\widetilde{a}, \widetilde{b}, \widetilde{c}) \supseteq L_\alpha^2(\widetilde{a}, \widetilde{b}, \widetilde{c})$.

Proof. It follows from the theorem 4.17 that a Fu-Fu variable $\widetilde{\widetilde{\xi}}$ can be transformed into a fuzzy variable, viz. $\widetilde{\widetilde{\xi}} \Rightarrow (a_L, \underline{a}, \overline{a}, a_R)$. Let $\widetilde{\widetilde{\xi}} = (\widetilde{a}, \widetilde{b}, \widetilde{c})$ and depending on the Definition 4.14, we can derive $L_\alpha(\widetilde{\widetilde{\xi}}) = \{a \, | \, \pi_{\widetilde{(a)}_{(\alpha_1,\alpha_2)}} (\widetilde{a})_{(\alpha_1,\alpha_2)} \geq \alpha\} = [\xi_\alpha^L, \xi_\alpha^R]. \, \xi_\alpha^L = a_L + (\underline{a} - a_L)\alpha, \underline{a} = \xi_{\alpha_2}^L, \, \xi_{\alpha_2}^L = a_L + (\rho_{\alpha_1}^R - a_L)\alpha_2$ and $\rho_{\alpha_1}^R = \rho_R - (\rho_R - \rho_0)\alpha_1$.

Finally we have $\xi_\alpha^L = a_L + \{a_L + [\rho_R - (\rho_R - \rho_0)\alpha_1 - a_L]\alpha_2 - a_L\}\alpha$. Then ξ_α^L is a decreasing function about α_1 and a increasing function about α_2. For any given $0 \leq \alpha_1 \leq 1, 0 \leq \alpha_2 \leq 1$, if $\alpha_1^1 \leq \alpha_1^2$ and $\alpha_2^1 \geq \alpha_2^2$, we have $\xi_\alpha^{L1} \geq \xi_\alpha^{L2}$.

Similarly, we may prove $\xi_\alpha^{R1} \leq \xi_\alpha^{R2}$. Thus $[\xi_\alpha^{L1}, \xi_\alpha^{R1}] \subseteq [\xi_\alpha^{L2}, \xi_\alpha^{R2}]$, then we obtain $L_\alpha^1(\widetilde{\widetilde{\xi}}) \subseteq L_\alpha^2(\widetilde{\widetilde{\xi}})$, viz. $L_\alpha^1(\widetilde{a}, \widetilde{b}, \widetilde{c}) \subseteq L_\alpha^2(\widetilde{a}, \widetilde{b}, \widetilde{c})$. Similarly, we may prove that if $\alpha_1^1 \geq \alpha_1^2, \alpha_2^1 \leq \alpha_2^2$, then $L_\alpha^1(\widetilde{a}, \widetilde{b}, \widetilde{c}) \supseteq L_\alpha^2(\widetilde{a}, \widetilde{b}, \widetilde{c})$.

The proof is thus completed. $\qquad\square$

Definition 4.13. [57] Let $L_\alpha(\widetilde{\mathbf{a}}, \widetilde{\mathbf{b}}, \widetilde{\mathbf{c}})$ be the α-level set of fuzzy numbers $\widetilde{a}_{ij}, \widetilde{b}_i, \widetilde{c}_{ki}$. Then we have $L_\alpha(\widetilde{\mathbf{a}}, \widetilde{\mathbf{b}}, \widetilde{\mathbf{c}}) = \{(\mathbf{a}, \mathbf{b}, \mathbf{c}) | \pi_{\widetilde{a}_{ij}}(\widetilde{a}_{ij}) \geq \alpha, \pi_{\widetilde{b}_i}(\widetilde{b}_i) \geq \alpha, \pi_{\widetilde{c}_{kj}}(\widetilde{c}_{kj}) \geq \alpha, i = 1, 2, \cdots, m, j = 1, 2, \cdots, n, k = 1, 2, \cdots, K\}$.

Definition 4.14. Let $L_\alpha(\widetilde{\widetilde{\mathbf{a}}}, \widetilde{\widetilde{\mathbf{b}}}, \widetilde{\widetilde{\mathbf{c}}})$ be the α-level set of Fu-Fu numbers $\widetilde{\widetilde{a}}_{ij}, \widetilde{\widetilde{b}}_i, \widetilde{\widetilde{c}}_{ki}$. (α_1, α_2) are any given possibility level of Fu-Fu numbers $\widetilde{\widetilde{a}}_{ij}, \widetilde{\widetilde{b}}_i, \widetilde{\widetilde{c}}_{ki}$ respectively. Then we have

$$L_\alpha(\widetilde{\widetilde{\mathbf{a}}}, \widetilde{\widetilde{\mathbf{b}}}, \widetilde{\widetilde{\mathbf{c}}}) = \{(\mathbf{a}, \mathbf{b}, \mathbf{c}) | \pi_{(\widetilde{\widetilde{a}}_{ij})_{(\alpha_1, \alpha_2)}}(\widetilde{\widetilde{a}}_{ij})_{(\alpha_1, \alpha_2)} \geq \alpha,$$

$$\pi_{(\widetilde{\widetilde{b}}_i)_{(\alpha_1, \alpha_2)}}(\widetilde{\widetilde{b}}_i)_{(\alpha_1, \alpha_2)} \geq \alpha,$$

$$\pi_{(\widetilde{\widetilde{c}}_{kj})_{(\alpha_1, \alpha_2)}}(\widetilde{\widetilde{c}}_{kj})_{(\alpha_1, \alpha_2)} \geq \alpha, i = 1, 2, \cdots, m, j = 1, 2, \cdots, n, k = 1, 2, \cdots, K\}.$$

Thus, base on the Theorem 4.17, the Fu-Fu multi-objective model (4.133) is transformed into a fuzzy multi-objective model as follow:

$$\begin{cases} \min \ [f_1(x, \widetilde{A}_1), f_2(x, \widetilde{A}_2), \cdots, f_k(x, \widetilde{A}_m)] \\ \text{s.t.} \begin{cases} g_i(x, \widetilde{B}_i) \leq 0, i = 1, 2, \cdots, p \\ h_r(x, \widetilde{C}_r) = 0, r = 1, 2, \cdots, q, \end{cases} \end{cases} \quad (4.140)$$

where $\widetilde{A}_k = \widetilde{\omega}_{\widetilde{a}_{k(\alpha_1, \alpha_2)}} \ (k = 1, 2, \cdots, m), \widetilde{B}_i = \widetilde{\omega}_{\widetilde{b}_{i(\alpha_1, \alpha_2)}} \ (i = 1, 2, \cdots, p)$ and $\widetilde{C}_r = \widetilde{\omega}_{\widetilde{c}_{r(\alpha_1, \alpha_2)}} \ (r = 1, 2, \cdots, q)$ are fuzzy trapezoidal numbers.

Solving procedure

In order to solve the above Fu-Fu multi-objective decision making models, we may integrate Fu-Fu simulation, NN and (α, β)-satisfied method[107] to produce a hybrid intelligent algorithm for solving Fu-Fu programming models. And then analyze the sensitivity of the Fu-Fu model about possibility level α. By transforming Fu-Fu variables into trapezoidal fuzzy numbers, we can derive the fuzzy model (4.140). The step-by-step procedure of the hybrid intelligent algorithm can be described as follows:

Step 1. Generate training input-output coefficients for uncertain functions α_1 and α_2 by the fuzzy simulation.

$$U_1 : \alpha_1 \longrightarrow \phi(x),$$

$$U_2 : \alpha_2 \longrightarrow \varphi(x).$$

Step 2. Train a neural network to approximate the uncertain functions according to the generated training input-output coefficients α_1 and α_2.

Step 3. On the basis of the above discussions about Theorem 4.17, the Fu-Fu variables are transformed into fuzzy variables which are similar trapezoidal fuzzy number. We consider the above multi-objective model (4.133) where $\widetilde{\bar{a}}_k(k = 1, 2, \cdots, m), \widetilde{\bar{b}}_i(i = 1, 2, \cdots, p)$ and $\widetilde{\bar{c}}_r(r = 1, 2, \cdots, q)$ are Fu-Fu variables. For each α_1 and α_2 are given by decision-maker. we can derive that

$$\widetilde{\bar{a}}_k = \widetilde{\bar{\omega}}_{\bar{a}_{k(\alpha_1,\alpha_2)}} = (a_L, \underline{a}, \overline{a}, a_R) \quad (k = 1, 2, \cdots, m),$$

$$\widetilde{\bar{b}}_i = \widetilde{\bar{\omega}}_{\bar{b}_{i(\alpha_1,\alpha_2)}} = (b_L, \underline{b}, \overline{b}, b_R) \quad (i = 1, 2, \cdots, p),$$

$$\widetilde{\bar{c}}_r = \widetilde{\bar{\omega}}_{\bar{c}_{r(\alpha_1,\alpha_2)}} = (c_L, \underline{c}, \overline{c}, c_R) \quad (r = 1, 2, \cdots, q),$$

which can be specified by the $\widetilde{\bar{\omega}}_{\bar{\xi}} = (a_L, \underline{a}, \overline{a}, a_R)$ with membership function:

$$\mu_{\widetilde{\bar{\omega}}_{\bar{\xi}}}(t) = \begin{cases} \frac{t - a_L}{\underline{a} - a_L}, & \text{if } a_L \leq t \leq \underline{a} \\ 1, & \text{if } \underline{a} \leq t \leq \overline{a} \\ \frac{a_R - t}{a_R - \overline{a}}, & \text{if } \overline{a} \leq t \leq a_R \\ 0, & \text{if } t < a_L, t > a_R. \end{cases} \quad (4.141)$$

And then, the Fu-Fu multi-objective model (4.133) is transformed into a fuzzy multi-objective model (4.140), viz.

$$\begin{cases} \min \{f_1(x, \widetilde{A}_1), f_2(x, \widetilde{A}_2), \cdots, f_k(x, \widetilde{A}_m)\} \\ \text{s.t.} \begin{cases} g_i(x, \widetilde{B}_i) \leq 0, i = 1, 2, \cdots, p \\ h_r(x, \widetilde{C}_r) = 0, r = 1, 2, \cdots, q, \end{cases} \end{cases} \quad (4.142)$$

where $\widetilde{A}_k = \widetilde{\bar{\omega}}_{\bar{a}_{k(\alpha_1,\alpha_2)}} (k = 1, 2, \cdots, m), \widetilde{B}_i = \widetilde{\bar{\omega}}_{\bar{b}_{i(\alpha_1,\alpha_2)}} (i = 1, 2, \cdots, p)$ and $\widetilde{C}_r = \widetilde{\bar{\omega}}_{\bar{c}_{r(\alpha_1,\alpha_2)}} (r = 1, 2, \cdots, q)$ are fuzzy trapezoidal numbers.

Step 4. Give the α-cut of $\widetilde{\omega}$ and be expressed by the following interval

$$\widetilde{\omega}_\alpha = [a_L + (\underline{a} - a_L)\alpha, \underline{a}, \overline{a}, a_R - (a_R - \overline{a})\alpha].$$

Step 5. Based on the (α, β)-satisfied approach. Let $(Q_i)_\alpha^\beta$ be a solution of the fuzzy non-linear programming model in equation (4.142), where $\alpha \in [0, 1]$ denotes the level of possibility at which all trapezoidal fuzzy numbers $\widetilde{A}_k, \widetilde{B}_i, \widetilde{C}_r$ and $\beta \in [0, 1]$ denotes the grade of compromise to which the solution satisfies all of the fuzzy objectives and constraints keeping the coefficients at a feasible level α. Hence, based on the Theorem 4.16, equation (4.142) is reduced to

$$\begin{cases} \max \min \{\alpha, \beta\} \\ \text{s.t.} \begin{cases} \mu_{\widetilde{f}_k}(x, \alpha) \geq \beta & (k = 1, 2, \cdots, m) \\ \mu_{\widetilde{g}_i}(x, \alpha) \geq \beta & (i = 1, 2, \cdots, p) \\ h_r(x, \widetilde{C}_r) = 0 & (r = 1, 2, \cdots, q) \\ 0 \leq \alpha, \beta \leq 1. \end{cases} \end{cases} \quad (4.143)$$

Step 6. Give the concrete forms of the linear membership functions $\mu_{\widetilde{f}_k}(x, \alpha), \mu_{\widetilde{g}_i}(x, \alpha)$ for objectives and constraints, respectively, as follows:

$$\mu_{\widetilde{f}_k}(x, \alpha) = \begin{cases} 0, & \text{if } f_k(x, \alpha) < f_{0_k} - \Delta f_k \\ 1 + \frac{f_k(x,\alpha) - f_{0_k}}{\Delta f_k}, & \text{if } f_{0_k} - \Delta f_k \leq f_k(x, \alpha) \leq f_{0_k} \\ 1, & \text{if } f_k(x, \alpha) > f_{0_k}, \end{cases} \tag{4.144}$$

$$\mu_{\widetilde{g}_i}(x, \alpha) = \begin{cases} 0, & \text{if } g_i(x, \alpha) > g_{0_i} + \Delta g_i \\ 1 - \frac{g_i(x,\alpha) - g_{0_i}}{\Delta g_i}, & \text{if } g_{0_i} \leq g_i(x, \alpha) \leq g_{0_i} + \Delta g_i \\ 1, & \text{if } g_i(x, \alpha) < g_{0_i}. \end{cases} \tag{4.145}$$

Here, Δf_k is the maximum and Δg_i is the minimum acceptable violation of the aspiration levels f_{0_k}, g_{0_i} respectively.

Step 7. Depending on the Definition 4.12 and the Theorem 4.18, equation (4.143) can be calculated as

$$\begin{cases} \max \min \{\alpha, \beta\} \\ \text{s.t.} \begin{cases} 1 + \frac{f_k(x,\alpha) - f_{0_k}}{\Delta f_k} \geq \beta \\ 1 - \frac{g_i(x,\alpha) - g_{0_i}}{\Delta g_i} \geq \beta \\ h_r(x, \widetilde{C}_r) = 0 \\ 0 \leq \alpha, \beta \leq 1, \end{cases} \end{cases}$$

where $(k = 1, 2, \cdots, m), (i = 1, 2, \cdots, p)$ and $(r = 1, 2, \cdots, q)$

Step 8. Calculate the above certain programming model (4.146) and solve the (α, β)-satisfied solution x^*. Finally, we analyze the sensitivity of Fu-Fu model about the coefficients α_1, α_2, respectively. By comparing the ratio Δh, viz.

$$\Delta h = \frac{\frac{\alpha_1 - \alpha_2}{\alpha_1}}{\frac{x_1^* - x_2^*}{x_1^*}} = \frac{\Delta \alpha_1 x_1^*}{\Delta x^* \alpha_1}.$$

We can derive the sensitivity of Fu-Fu model about α_1, α_2.

If the decision maker is not satisfied with the current values or results, ask the decision maker to update reference membership levels or the (α_1, α_2) values by taking account of the current results and the values of membership functions, then repeat the first to eighth steps.

Chapter 5
Fuzzy Rough Multiple Objective Decision Making

The concept of a rough set was first raised by Pawlak [340]. Then Liu [226] proposed the fuzzy rough (Fu-Ro) variable by combining the fuzzy variable and rough variable. Xu and Zhao [343] discussed the properties of Fu-Ro variable, and introduced the Fu-Ro multi-objective decision making models and the ways to deal with them, some crisp equivalent models are given and relative algorithms are proposed to solve the problems.

In this chapter, we first introduce the Fu-Ro variable, then the arithmetic and the properties of the Fu-Ro variable. Based on the expected value operator and chance operator of the Fu-Ro variable, three parts are presented respectively:

(1) Fuzzy rough expected value decision-making model(Fu-Ro EVM).

(2) Fuzzy rough chance constraint decision-making model(Fu-Ro CCM).

(3) Fuzzy rough dependent chance decision-making model(Fu-Ro DCM).

Finally, an application to reuse an integrated logistics network design problem under fuzzy rough environment is presented to show the effectiveness of the above three models.

5.1 Integrated Logistics Problem under Fuzzy Rough Environment

In recent years, the logistics system has been gaining importance due to the increasing market globalization competitiveness. At first, most of the scholars just researched forward logistics, and there are many papers on this subject. Then scholars realized a growth of interest in transporting items that could be recycled or reused, and the possible commercial returns and saw that reverse logistics is an important problem requiring careful consideration.

Reverse logistics was first mentioned in the early 1990s. Two papers about reverse logistics from the American GLM (Council of Logistics Management) mark the start of research into reverse logistics [81, 82]. The first paper represents the research results of Stock (1992), who proposed the relevance among the fields of reverse

J. Xu and X. Zhou: Fuzzy-Like Multiple Objective Decision Making, STUDFUZZ 263, pp. 295–374.
springerlink.com

logistics, business and social. A year later, Kopicki and other scholars researched the actual operation and rules of reverse logistics, including the factors of re-use and recycling. Kostecki (1998) discussed reverse logistics as the way to extend the life cycle of products. The same year, Stock reported in detail how to set up a reverse logistics and implementation plan. Roger (1999) and Tibber-Lembke collected extensive data on reverse logistics business operation examples, especially in the United States, where the two researchers wrote a lot of logistics optimization management articles. After that, many problems regarding reverse logistics were discussed. Some papers under assumed a crisp environment, like [83, 84, 85, 86, 87], but many uncertainties were found to exist in a reverse logistics system, so the research of uncertain reverse logistics began, as in [88, 89].

In more recent years, forward and reverse logistics were integrated to build the integrated logistics system. In 1997, Fleischmann and Jacqueline first did some research on integrated logistics[90], wherein they integrated forward logistics and reverse logistics to construct a close-loop integrated logistics system. Hyun Jeung Ko[91] presented a mixed integer nonlinear programming model from the perspective of the third party for the design of a dynamic integrated distribution network to account for the integrated goals of simultaneously optimizing the forward and return network. There was very little literature about integrated logistics. The integrated logistics system is often applied in practice, but the theoretical study of integrated logistics has lagged behind. It is therefore necessary for scholars to research and develop this field.

The integrated logistics system is composed of the forward logistics system and the reverse logistics system. Forward logistics systems are usually the same, they all deliver the new product from the first producers to the last customers. However, there are different reverse logistics networks structures according to different kinds of reverse goods, such as reuse, remanufacturing, recycling and commercial return. If the purpose of the reverse logistics system is for reusing items, then the integrated logistics can be called a reuse integrated logistics system.

In every day life, re-usable packages and containers such as glass bottles, plastic bottles, cans, boxes and pallets are widely used in the food and chemical industries. As more and more people realize the importance of environmental protection, more and more producers want to reuse the recycled items to reduce resource waste. So in this paper, we concentrate on the reuse reverse logistics system.

The process of the reuse integrated logistics network is that the re-usable packages are gathered by collectors and processed by recyclers to then be sent to the producers to reuse. New products which use the recycled packages are produced and again get into the forward logistics and are finally consumed by the customers. After that, the same process recurs. Thus, the whole integrated logistics system operates in cycles like this and forms a closed loop system.

There are three main establishments in an integrated logistics network:
(1) Collectors: have the responsibility of collecting reusable packages that are scattered around.

(2) Recyclers or expanded distributors: receive the items from collectors, and their work concentrates on detecting, cleaning and processing the used items to such a state that they are undifferentiated from new items, and then these items will be delivered to enterprises again.

(3) Disposal places: process the waste items that can no longer be used. The integrated logistics system also can be described as in the following Figure. 5.1.

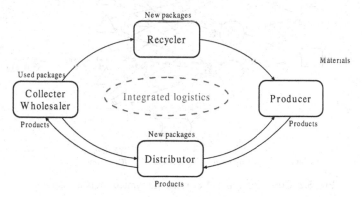

Fig. 5.1 Integrated logistics system

The integrated logistics network also includes the forward logistics network and the reverse logistics network. In forward logistics, there are producers, distributors and main wholesalers, and in reuse reverse logistics network there are collectors, recyclers/ expanded distributors, final disposal places and producers, and the associated transport routes.

Unfortunately, the integrated logistics network design problem is subject to many sources of uncertainty. In a practical decision-making process, we often face a hybrid uncertain environment. To deal with this twofold uncertainty, the concept of the fuzzy rough variable was proposed to depict the phenomena in which fuzziness and roughness appear simultaneously.

In this next section, we consider the reuse integrated logistics system, items which can be re-used after simple treatment, mainly package containers and auxiliary materials such as trays. In the integrated logistics network problem, it is hard to describe these problem parameters as known variables. For instance, since people usually drink more beer in summer and autumn, and less beer in winter and spring, that is, the demand of beer is seasonal. When we forecast the demand in a period, we may use the fuzzy variable to estimate, for example, we give a middle value μ, two spread α and β. Further more, the middle value μ is usually not a certain number, because when we design the network of the network of a reuse integrated logistics network, the period we consider will definitely cover the whole season, so the it is appropriate to use a rough variable to describe the middle value μ. So until now, in this situation, we can use fuzzy rough variables to describe the demand of the beer. Because the amount of the used packages is relevant to the consumption of the product, so it is natural to consider that the quantity of used packages is also a fuzzy rough variable, just as is that of the demand for the products.

The following Figure. 5.2 describes this kind of reuse integrated logistics network.

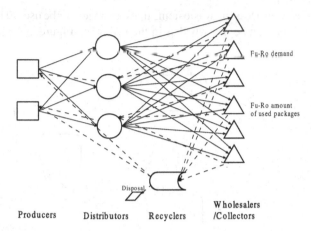

Fig. 5.2 Conceptual model of reuse integrated logistics network

However, there is no attempt to research another mixed environment, where fuzziness and roughness both appear simultaneously. For some seasonal items (Ice cream, Christmas trees, woolen materials), the demand may vary year to year. According to the historical data or abundance of information or the experiences of experts, we can know the demand in one year is a certain fuzzy variable. However, the middle value of the fuzzy variable is vague and varies year to year. The result is that decision makers are unable to achieve a better decision. Hence, we have to consider it as an uncertain variable. A rough variable can be applied to depict it well if the average sold amount is clear according to the statistical data of every year. Thus, the demand of some seasonal items can be described as a fuzzy rough variable to help decision makers develop better strategies.

5.2   Fu-Ro Variable

Let's introduce the basic knowledge of Fu-Ro variables, which include the definition, the chance measure, the expected value, and the optimistic and pessimistic value.

5.2.1   Definition of Fu-Ro Variable

Before the introduction of the concept of fuzzy rough variables, let's recall some definitions and properties of rough sets.

The rough sets theory introduced by Pawlak [340, 377] has often proved to be an excellent mathematical tool for the analysis of a vague description of objects (called actions in decision problems). The adjective vague, referring to the quality of information, means inconsistency or ambiguity which follows from information granulation. The rough sets philosophy is based on the assumption that with every object of the universe there is associated a certain amount of information (data, knowledge), expressed by means of some attributes used for object description. Objects having the same description are indiscernible (similar) with respect to the available information. The indiscernibility relation thus generated constitutes a mathematical basis for the rough sets theory; it induces a partition of the universe into blocks of indiscernible objects, called elementary sets, that can be used to build knowledge about a real or abstract world. The use of the indiscernibility relation results in information granulation. The rough sets theory, dealing with the representation and processing of vague information, presents a series of intersections and complements with respect to many other theories and mathematical techniques handling imperfect information, like probability theory, evidence theory of DempsterShafer, fuzzy sets theory, discriminant analysis and mereology [373, 374, 375, 376, 341, 378, 379, 380].

For algorithmic reasons, the information regarding the objects is supplied in the form of a data table, whose separate rows refer to distinct objects (actions), and whose columns refer to di.erent attributes considered. Each cell of this table indicates an evaluation (quantitative or qualitative) of the object placed in that row by means of the attribute in the corresponding column.

Formally, a data table is the 4-tuple $\mathbf{S} = (U, Q, V, f)$, where U is a finite set of objects (universe), $Q = q_1, q_2, \cdots, q_n$ is a finite set of attributes, V_q is the domain of the attribute, $V = \bigcup_{q \in Q} V_q$ and $f : U \times Q \rightarrow V$ is a total function such that $f(x, q) \in V_q$ for each $x \in U, q \in Q$, called *information function*.

Therefore, each object x of U is described by a vector (string) $Des_Q(x) = (f(x, q_1), f(x, q_2), \cdots, f(x, q_m))$, called description of x in terms of the evaluations of the attributes from Q; it represents the available information about x.

To every (non-empty) subset of attributes P is associated an indiscernibility relation on U, denoted by I_P:

$$I_p = \{(x, y) | \in U \times U : f(x, q) = f(y, q) \forall q \P\}.$$

If $(x, y) \in I_p$, it is said that the objects x and y are P-indiscernible. Clearly, the indiscernibility relation thus de.ned is an equivalence relation (reflexive, symmetric and transitive). The family of all the equivalence classes of the relation I_P is denoted by $U|I_P$ and the equivalence class containing an element $x \in U$ by $I_p(x)$. The equivalence classes of the relation I_P are called *P-elementary sets*. If $P = Q$, the Q-elementary sets are called *atoms*.

Let $\mathbf{S}$ be a data table, X a non-empty subset of U and $\Phi \neq P \subseteq Q$. The *P-lower* approximation and the *P-upper* approximation of X in $\mathbf{S}$ are defined, respectively, by:

$$\underline{P}(X) = \{x \in U : I_p(x) \subseteq X\},$$

$$\bar{P}(X) = \bigcup_{x \in X} I_P(X).$$

The elements of $\underline{P}(X)$ are all and only those objects $x \in U$ which belong to the equivalence classes generated by the indiscernibility relation I_P, contained in X; the elements of $\bar{P}(X)$ are all and only those objects $x \in U$ which belong to the equivalence classes generated by the indiscernibility relation I_P, containing at least one object x belonging to X. In other words, $\underline{P}X$ is the largest union of the P-elementary sets included in X, while $\bar{P}(X)$ is the smallest union of the P-elementary sets containing X.

(1) The *P-boundary* of X in S, denoted by $Bn_P(X)$, is $Bn_P(X) = \bar{P}(X) - \underline{P}(X)$.

(2) The following relation holds: $\underline{P}(X) \subseteq X \subseteq \bar{P}(X)$.

Therefore, if an object x belongs to $\underline{P}(X)$, it is certainly also an element of X, while if x belongs to $\bar{P}(X)$, it may belong to the set X. $Bn_P(X)$ constitutes the "doubtful region" of X: nothing can be said with certainty about the belonging of its elements to the set X.

The following relation, called *complementarity property*, is satisfied: $\underline{P}(X) = U - \bar{P}(U - X)$.

If the P-boundary of X is empty, $Bn_P(X) = \Phi$, then the set X is an ordinary (exact) set with respect to P, that is, it may be expressed as the union of a certain number of P-elementary sets; otherwise, if $Bn_P(X) \neq \Phi$, the set X is an approximate (rough) set with respect to P and may be characterized by means of the approximations $\underline{P}(X)$ and $\bar{P}(X)$. The family of all the sets $X \subseteq U$ having the same P-lower and P-upper approximations is called a *rough set*.

The following ratio defines an accuracy of the approximation of X, $X \neq \Phi$ by means of the attributes from P:

$$\alpha_P(X) = \frac{|\underline{P}(X)|}{|\bar{P}(X)|},$$

where $|Y|$ indicates the cardinality of a (finite) set Y. Obviously, $0 \leq \alpha_P(X) \leq 1$; if $\alpha_P(X) = 1$, X is an ordinary (exact) set with respect to P; if $\alpha_P(X) = 1$, X is a rough (vague) set with respect to P.

Another ratio defines a quality of the approximation of X by means of the attributes from P:

$$\gamma_P(X) = \frac{|\underline{P}(X)|}{|X|}.$$

The quality $\gamma_P(X)$ represents the relative frequency of the objects correctly classified by means of the attributes from P. Moreover, $0 \leq \alpha_P(X) \leq \gamma_P(X) \leq 1$, and $\gamma_P(X) = 0$ iff $\alpha_P(X) = 0$, while $\gamma_P(X) = 1$ iff $\alpha_P(X) = 1$.

The definition of approximations of a subset $X \subseteq U$ can be extended to a classi.cation, i.e. a partition $Y = \{Y_1, Y_2, \cdots, Y_n\}$ of U. Subsets $Y_i, i = 1, 2, \cdots, n$ are disjunctive classes of Y. By P-lower (P-upper) approximation of Y in S, we mean sets $\underline{P}(Y) = \{\underline{P}(Y_1), \underline{P}(Y_2), \cdots, \underline{P}(Y_n)\}$ and $\bar{P}(Y) = \{\bar{P}(Y_1), \bar{P}(Y_2), \cdots, \bar{P}(Y_n)\}$, respectively. The coefficient

$$\gamma_P(X) = \frac{|\sum\limits_{1=1}^{n} \underline{P}(X)|}{|U|}$$

is called quality of the approximation of classication Y by set of attributes P, or in short, quality of classification. It expresses the ratio of all P-correctly classified objects to all objects in the system.

The main preoccupation of the rough sets theory is approximation of subsets or partitions of U, representing a knowledge about U, with other sets or partitions built up using available information about U. From the viewpoint of a particular object $x \in U$, it may be interesting, however, to use the available information to assess the degree of its membership to a subset X of U. The subset X can be identified with a concept of knowledge to be approximated. Using the rough set approach one can calculate the membership function $\mu_X^P(x)$ (rough membership function) as

$$\mu_X^P(x) = \frac{X \cap I_p(x)}{I_p(x)}.$$

The value of $\mu_X^P(x)$ may be interpreted analogously to conditional probability and may be understood as the degree of certainty (credibility) to which x belongs to X. Observe that the value of the membership function is calculated from the available data, and not subjectively assumed, as it is the case of membership functions of fuzzy sets.

Between the rough membership function and the approximations of X the following relationships hold (Pawlak [340]):

$$\underline{P}(X) = \{x \in U : \mu_X^P(x) = 1\}, \bar{P}(X) = \{x \in U : \mu_X^P(x) > 0\},$$

$$Bn_P(X) = \{x \in U : 0 < \mu_X^P(x) < 1\}, \underline{P}(U - X) = \{x \in U : \mu_X^P(x) = 0\}.$$

In the rough sets theory there is, therefore, a close link between vagueness (granularity) connected with rough approximation of sets and uncertainty connected with rough membership of objects to sets.

Trust theory [145] is the branch of mathematics that studies the behavior of rough events. It is the foundation for rough programming as the probability theory for stochastic programming as well as the possibility theory for fuzzy programming. Liu [145] also combined trust measure with probability measure and possibility measure to describe the two-fold uncertain events, such as random rough variable, fuzzy rough variable, rough random variable and rough fuzzy variable. In this section, we will define the fuzzy rough variable from another perspective, i.e. the rough approximation.

After the rough set was initialized by Pawlak [340], it has been applied to many fields to deal with the vague description of objectives. He asserted that any vague information can be approximated by other crisp information. In this section, we will recall these fundamental concepts and introduce its application to the statistical field and programming problem.

Definition 5.1. (Slowinski and Vanderpooten [381]) Let U be a universe, and X a set representing a concept. Then its lower approximation is defined by

$$\underline{X} = \{x \in U | R^{-1}(x) \subset X\}, \tag{5.1}$$

and the upper approximation is defined by

$$\overline{X} = \bigcup_{x \in X} R(x), \tag{5.2}$$

where R is the similarity relationship on U. Obviously, we have $\underline{X} \subseteq X \subseteq \overline{X}$.

Definition 5.2. (Pawlak [340]) The collection of all sets having the same lower and upper approximations is called a rough set, denoted by $(\underline{X}, \overline{X})$. Its boundary is defined as follows,

$$Bn_R(X) = \overline{X} - \underline{X}. \tag{5.3}$$

In order to know the degree of the upper and lower approximation describing the set X, the concept of the *accuracy* of approximation is proposed by Greco et al. [382],

$$\alpha_R(X) = \frac{|\underline{X}|}{|\overline{X}|}, \tag{5.4}$$

where $X \neq \Phi$, $|X|$ expresses the cardinal number of the set X when X is a finite set, otherwise it expresses the Lebesgue measure.

Another ratio defines a *quality* of the approximation of X by means of the attributes from R according to Greco et al. [382],

$$\gamma_R(X) = \frac{|\underline{X}|}{|X|}. \tag{5.5}$$

The quality $\gamma_R(X)$ represents the relative frequency of the objects correctly classified by means of the attributes from R.

Remark 5.1. For any set A we can represents its frequency of the objects correctly approximated by $(\underline{X}, \overline{X})$ as follows,

$$\beta_R(A) = \frac{|\underline{X} \cap A|}{|\overline{X} \cap A|}.$$

If $\underline{X} \subseteq A \subseteq \overline{X}$, namely, A has the upper approximation $\overline{X}$ and the lower approximation $\underline{X}$, we have that $\beta_R(A)$ degenerates to the *quality* $\gamma_R(A)$ of the approximation.

As we know, the *quality* $\gamma_R(A)$ of the approximation describes the frequency of A, and when $\gamma_R(A) = 1$, we only have $|A| = |\underline{X}|$, namely, the set A is well approximated by the lower approximation. If we we want to make A be a definable set, there must be $\gamma_R(A) = 1$ and $\alpha_R(X) = 1$ both holds. Then we could make use of the following definition to combine them into together.

Definition 5.3. Let $(\underline{X}, \overline{X})$ be a rough set under the similarity relationship R and A be any set satisfying $\underline{X} \subseteq A \subseteq \overline{X}$. Then we define the approximation function as follows expressing the relative frequency of the objects of A correctly classified into $(\underline{X}, \overline{X})$,

$$Appr_R(A) = 1 - \eta(1 - \frac{|A|}{|\overline{X}|}), \tag{5.6}$$

where η is a predetermined by the decision maker's preference.

From Definition 5.3, we know that $\frac{|A|}{|\overline{X}|}$ which keeps accord with $\gamma_R(A)$ describes the relative frequency of the objects correctly classified by R from the view of the upper approximation $\overline{X}$. Obviously, $Appr_R(A)$ is a number between 0 and 1, and is increasing along with the increase of $|A|$. The extreme case $Appr_R(A) = 1$ means that $|A| = |\overline{X}|$, namely, A is completely described by $\overline{X}$.

Lemma 5.1. *Let $(\underline{X}, \overline{X})$ be a rough set under the similarity relationship R and A be any set satisfying $\underline{X} \subseteq A \subseteq \overline{X}$. Then we have*

$$Appr_R(A) = \frac{\eta \alpha_R(A) + (1 - \eta)\gamma_R(A)}{\gamma_R(A)}.$$

Proof. Since $\underline{X} \subseteq A \subseteq \overline{X}$, it means that A has the lower approximation $\underline{X}$ and the upper approximation $\overline{X}$, and it follows from Greco et al. [382] that

$$\alpha_R(A) = \frac{|\underline{X}|}{|\overline{X}|}, \quad \gamma_R(A) = \frac{|\underline{X}|}{|A|}.$$

Thus,

$$\frac{|A|}{|\overline{X}|} = \frac{\alpha_R(A)}{\gamma_R(A)}.$$

It follows that

$$\begin{aligned}
Appr_R(A) &= 1 - \eta(1 - \frac{|A|}{|\overline{X}|}) \\
&= 1 - \eta(1 - \frac{\alpha_R(A)}{\gamma_R(A)}) \\
&= \frac{\eta \alpha_R(A) + (1 - \eta)\gamma_R(A)}{\gamma_R(A)}.
\end{aligned}$$

This completes the proof. □

Lemma 5.2. *Let $(\underline{X}, \overline{X})$ be a rough set on the finite universe under the equivalence relationship R, A be any set satisfying $\underline{X} \subseteq A \subseteq \overline{X}$ and $\eta \in (0,1)$. Then $Appr_R(A) = 1$ holds if and only if $\underline{X} = A = \overline{X}$.*

Proof. If $\underline{X} = A = \overline{X}$ holds, it is obvious that $Appr_R(A) = 1$ according to Definition 5.4. Let's proved the necessity of the condition.

If $Appr_R(A) = 1$ holds for any A satisfying $\underline{X} \subseteq A \subseteq \overline{X}$, it follows from Lemma 5.1 that, for $0 < \eta \le 1$,

$$\frac{\eta \alpha_R(A) + (1 - \eta)\gamma_R(A)}{\gamma_R(A)} = 1 \Rightarrow \alpha_R(A) = \gamma_R(A) \Rightarrow |\overline{X}| = |A|.$$

Since $A \subseteq \overline{X}$ and the universe is finite, we have that $A = \overline{X}$. Because A is any set satisfying $\underline{X} \subseteq A \subseteq \overline{X}$, let $A = X$, then we have $X = \overline{X}$. It follows from the property proposed by Pawlak [340] that $\underline{X} = X = \overline{X}$. Thus, we have $\underline{X} = A = \overline{X}$.   □

Lemma 5.1 shows that the approximation function Appr inherits the accuracy and quality of the approximation, and extends it to the relationship between any set A and the rough set $(\underline{X}, \overline{X})$. Lemma 5.2 shows that the approximation function is complete and well describes the property in traditional rough set theory, and describe the property only by one index.

Lemma 5.3. *Let $(\underline{X}, \overline{X})$ be a rough set on the infinite universe under the similarity relationship R, A be any set satisfying $\underline{X} \subseteq A \subseteq \overline{X}$ and $\eta \in (0,1)$. If $Appr_R(A) = 1$ holds, then there exist the similarity relationship R^* such that $|\underline{X}| = |A| = |\overline{X}|$, where $|\cdot|$ expresses the Lebesgue measure.*

Proof. According to Lemma 5.2, we know that $|A| = |\overline{X}|$ must hold. Let $\underline{X} = \overline{X}/\partial\overline{X}$ under the similarity relationship R^*, where $\partial\overline{X}$ is composed by all the elements such that $|\partial\overline{X}| = 0$, namely, the measure of $\partial\overline{X}$ is 0. Next, we will prove that $\overline{X}/\partial\overline{X} \subseteq A$.

(1) If $|\overline{X}| = 0$, then $\overline{X}/\partial\overline{X} = \Phi$. Thus, $|\underline{X}| = |A| = |\overline{X}| = 0$.

(2) If $|\overline{X}| \neq 0$, we only need to prove that for any $x^0 \in \overline{X}/\partial\overline{X}, x^0 \in A$. In fact, when $x^0 \in \overline{X}/\partial\overline{X}$, then $x^0 \in int(\overline{X})$ holds, where $int(\overline{X})$ is the internal part of $\overline{X}$. It follows that there exists $r > 0$ such that $N(x^0, r) \subset int(\overline{X})$ and $|N(x^0, r)| > 0$. There exist four cases describing the relationship between A and $N(x^0, r)$.

Case 1. $A \cap N(x^0, r) = \Phi$ (see Figure 5.3) . Since $N(x^0, r) \subset int(\overline{X}) \subset \overline{X}$ and $A \subseteq \overline{X}$, we have that

$$|\overline{X}| \geq |N(x^0, r) \cup A| = |N(x^0, r)| + |A|.$$

This conflicts with $|A| = |\overline{X}|$.

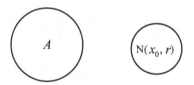

Fig. 5.3 Apartment

Case 2. $A \cap N(x^0, r) = P$, where the set P includes countable points (see Figure 5.4). Obviously, we have $|P| = 0$, thus $|N(x_0, r)/P| = |N(x_0, r)| > 0$. Then we have

$$|\overline{X}| \geq |N(x^0, r) \cup A| = |N(x^0, r)/P| + |A|.$$

This also conflicts with $|A| = |\overline{X}|$.

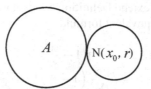

Fig. 5.4 Tangent

Case 3. $A \cap N(x^0, r) = P'$, where $P' \subset N(x^0, r)/\{x_0\}$. As Figure 5.5 shows, we can divide it into three parts, namely, $(N(x^0, r)/P) = P' \cup \{x_0\} \cup T$, where P', T and $\{x_0\}$ don't have the same element with each other. Then $|T| > 0$, it follows that

$$|\overline{X}| \geq |N(x^0, r) \cup A| = |T| + |A|.$$

This also conflicts with $|A| = |\overline{X}|$.

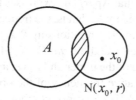

Fig. 5.5 Intersection

Case 4. $A \supset (N(x^0, r)/x_0)$ (see Figure 5.6). This means that for any $x_0 \in int(A), x_0 \notin A$. It follows that $A \cap int(A) = \Phi$, then we have

$$|\overline{X}| \geq |int(A) \cup A| = |int(A)| + |A|.$$

This also conflicts with $|A| = |\overline{X}|$. In above, we can get $\overline{X}/\partial \overline{X} \subseteq A$. Thus, there exists the lower approximation $\underline{X} = \overline{X}/\partial \overline{X}$ such that $\underline{X} \subseteq A \subseteq \overline{X}$ under the similarity relationship $R*$. $\qquad \square$

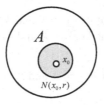

Fig. 5.6 Inclusion

Remark 5.2. In fact, we can extend Definition 5.3 to more general set. when $\underline{X} \subseteq A \subseteq \overline{X}$, we have the following equivalent formula,

$$
\begin{aligned}
\mathrm{Appr}_R(A) &= 1 - \eta(1 - \tfrac{|A|}{|\overline{X}|}) \\
&= \tfrac{|A \cap \underline{X}|}{|\underline{X}|} \left(1 - \eta\left(1 - \tfrac{|A \cap \overline{X}|}{|\overline{X}|}\right)\right) \\
&= \tfrac{|A \cap \underline{X}|}{|\underline{X}|} + \eta\left(\tfrac{|A \cap \overline{X}|}{|\overline{X}|} - \tfrac{|A \cap \underline{X}|}{|\underline{X}|}\right).
\end{aligned}
$$

Furthermore, we get the definition of the approximation function for any set A.

Definition 5.4. Let $(\underline{X}, \overline{X})$ be the rough set generated by X under the similarity relationship R, for any set A, the approximation function of event A by $(\underline{X}, \overline{X})$ is defined as follows

$$
Appr_R(A) = \frac{|A \cap \underline{X}|}{|\underline{X}|} + \eta\left(\frac{|A \cap \overline{X}|}{|\overline{X}|} - \frac{|A \cap \underline{X}|}{|\underline{X}|}\right),
$$

where η is a given parameter predetermined by the decision maker's preference.

From Definition 5.4, we know that $\mathrm{Appr}_R(A)$ expresses the relationship between the set A and the set $(\underline{X}, \overline{X})$ generated by X, that is, the frequency of A correctly classified into $(\underline{X}, \overline{X})$ according to the similarity relationship R. It has the internal link with the *accuracy* α_R of the approximation and the *quality* γ_R of the approximation in some extent. α_R expresses the degree of the upper and lower approximation describing the set X. $\gamma_R(X)$ represents the relative frequency of the objects correctly classified by means of the attributes from R. Then Appr_R combines both of them together and considers the level which A has the attributes correctly classified by $(\underline{X}, \overline{X})$ for any A.

Lemma 5.4. *Let $(\underline{X}, \overline{X})$ be a rough set, for any set A, we have the following conclusion,*

$$
Appr_R = \begin{cases}
1, & \text{if } A \supseteq \overline{X} \\
1 - \eta(1 - \tfrac{\alpha_R(A)}{\gamma_R(A)}), & \text{if } \underline{X} \subset A \subset \overline{X} \\
\tfrac{1 - \eta(1 - \alpha_R(A))}{\gamma_R(A)}, & \text{if } A \subseteq \underline{X} \\
0, & \text{if } A \cap \overline{X} = \Phi \\
\tfrac{|A \cap \overline{X}|}{|\overline{X}|}\left(\tfrac{\beta_R(A)}{\alpha_R(A)} + \eta(1 - \tfrac{\beta_R(A)}{\alpha_R(A)})\right), & \text{otherwise.}
\end{cases}
\tag{5.7}
$$

Proof. (1) If $A \supseteq \overline{X}$, we have that $A \cap \underline{X} = \underline{X}$ and $A \cap \overline{X} = \overline{X}$. Then $\mathrm{Appr}_R = 1$.

(2) If $\underline{X} \subseteq A \subseteq \overline{X}$, we have that $A \cap \underline{X} = \underline{X}$ and $A \cap \overline{X} = A$. It follows that $\mathrm{Appr}_R = 1 - \eta(1 - \tfrac{|A|}{|\overline{X}|})$.

(3) If $A \subset \underline{X}$, we have that $A \cap \underline{X} = A$ and $A \cap \overline{X} = A$. It follows that $\mathrm{Appr}_R = \tfrac{1 - \eta(1 - \alpha_R(A))}{\gamma_R(A)}$.

(4) If $A \cap \overline{X} = \Phi$, we have that $A \cap \underline{X} = \Phi$ and $A \cap \overline{X} = \Phi$. It follows that $Appr_R = 0$.

(5) For the others, we have

$$Appr_R(A) = \frac{|A \cap \underline{X}|}{|X|} + \eta \left(\frac{|A \cap \overline{X}|}{|X|} - \frac{|A \cap \underline{X}|}{|X|} \right)$$

$$= \frac{|A \cap \overline{X}|}{|X|} \left(\frac{|A \cap \underline{X}|}{|X|} \cdot \frac{|X|}{|A \cap \overline{X}|} + \eta \left(1 - \frac{|A \cap \underline{X}|}{|X|} \cdot \frac{|X|}{|A \cap \overline{X}|} \right) \right)$$

$$= \frac{|A \cap \overline{X}|}{|X|} \left(\frac{\beta_R(A)}{\alpha_R(A)} + \eta \left(1 - \frac{\beta_R(A)}{\alpha_R(A)} \right) \right).$$

This completes the proof.   □

For the different purposes, we can respectively discuss the extreme case as follows.

Remark 5.3. When $\eta = 1$, we have $Appr_R(A) = \frac{|A \cap \overline{X}|}{|X|}$. It means that the decision maker only consider the level that A includes the frequency of A correctly classified into $\underline{X}$ according to the similarity relationship R.

Remark 5.4. When $\eta = 0$, we have $Appr_R(A) = \frac{|A \cap \overline{X}|}{|X|}$. It means that the decision maker only consider the level that A includes the frequency of A correctly classified into $\overline{X}$ according to the similarity relationship R.

In fact, the rough set theory is increasingly developed by many scholars and applied to many fields, for example, data mining, decision reduction, system analysis and so on. Figure 5.7 shows that the rough approximation. The curves including the internal points is X. The two thick curves including their internal points are the upper and lower approximation.

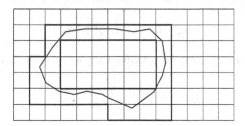

Fig. 5.7 Rough approximation

Let's focus on the continuous set in the one dimension real space **R**. There are still some vague sets which cannot be directly fixed and need to be described by the rough approximation. For example, set **R** be the universe, a similarity relation $\simeq$ is defined as $a \simeq b$ if and only if $|a - b| \leq 10$. We have that for the set $[20, 50]$, its lower approximation $\underline{[20, 50]} = [30, 40]$ and its upper approximation $\overline{[20, 50]} = [10, 60]$. Then the upper and lower approximation of the set $[20, 50]$ make up a rough set $([30, 40], [10, 60])$ which is the collection of all sets having the same lower approximation $[30, 40]$ and upper approximation $[10, 60]$.

Definition 5.5. A fuzzy rough variable ξ is a fuzzy variable with uncertain parameter $\rho \in X$, where X is approximated by $(\underline{X}, \overline{X})$ according to the similarity relation R, namely, $\underline{X} \subseteq X \subseteq \overline{X}$.

For convenience, we usually denote $\rho \vdash (\underline{X}, \overline{X})_R$ expressing that ρ is in some set A which is approximated by $(\underline{X}, \overline{X})$ according to the similarity relation R, namely, $\underline{X} \subseteq A \subseteq \overline{X}$.

Example 5.1. Let's consider the *LR* fuzzy variable ξ with the following membership function,

$$\mu_{\xi}(x) = \begin{cases} L\left(\frac{\rho - x}{\alpha}\right), & \text{if } \rho - \alpha < x < \rho \\ 1, & \text{if } x = \rho \\ R\left(\frac{x - \rho}{\beta}\right), & \text{if } \rho < x_j < \rho + \beta, \end{cases} \tag{5.8}$$

where $\rho \vdash ([1,2],[0,3])_R$. Then ξ is a fuzzy rough variable.

5.2.2 Expected Value Operator of Fu-Ro Variables

Definition 5.6. Let ξ be a Fu-Ro variable with the uncertain parameter λ, where $\lambda \vdash (\underline{X}, \overline{X})_R$, then its expected value is defined by

$$E[\xi] = \int_0^{\infty} \text{Appr}\{E[\xi(\lambda)] \geq r\}dr - \int_{-\infty}^0 \text{Appr}\{E[\xi(\lambda)] \leq r\}dr \tag{5.9}$$

Lemma 5.5 ([220]). *Assume that ξ and η are the introduction of variables with finite expected values. Then for any real numbers a and b, we have*

$$E[a\xi + b\eta] = aE[\xi] + bE[\eta]. \tag{5.10}$$

Proposition 5.1. *Let ξ be a Fu-Ro variable with the membership function*

$$\mu_{\xi}(x) = \begin{cases} 1, & \text{if } x \in [\bar{a}, \bar{b}] \\ 0, & \text{otherwise}, \end{cases}$$

where $\bar{a}, \bar{b}$ are rough variables defined on $(\Lambda, \Theta, \mathscr{A}, \pi)$, and $\bar{a} = ([m_2, m_3], [m_1, m_4])$, $0 < m_1 \leq m_2 < m_3 \leq m_4$, $\bar{b} = ([n_2, n_3], [n_1, n_4]), 0 < n_1 \leq n_2 < n_3 \leq n_4$.
Then the expected value of ξ is

$$E[\xi] = \frac{1}{8} \sum_{i=1}^4 (m_i + n_i).$$

Proof. Since $\xi(\lambda) = [\bar{a}, \bar{b}]$, and according to proposition 2.1, we have

$$E^{Cr}[\xi(\lambda)] = \frac{1}{2}(\bar{a} + \bar{b}).$$

By the rough arithmetic operators, it follows that,

$$\frac{\bar{a}+\bar{b}}{2} = \frac{1}{2}\{([m_2,m_3],[m_1,m_4]) + ([n_2,n_3],[n_1,n_4])\}$$

$$= \left([\frac{m_2+n_2}{2}, \frac{m_3+n_3}{2}],[\frac{m_1+n_1}{2}, \frac{m_4+n_4}{2}]\right). \tag{5.11}$$

From the definition of trust measure, we have

$$Appr\{\xi \geq r\} = \begin{cases} 0, & \text{if } \frac{m_4+n_4}{2} \leq r \\ \frac{\frac{m_4+n_4}{2}-r}{m_4+n_4-m_1-n_1}, & \text{if } \frac{m_3+n_3}{2} \leq r \leq \frac{m_4+n_4}{2} \\ \frac{1}{2}(\frac{\frac{m_4+n_4}{2}-r}{m_4+n_4-m_1-n_1} + \frac{\frac{n_3+m_3}{2}-r}{m_3+n_3-m_2-n_2}), & \text{if } \frac{m_2+n_2}{2} \leq r \leq \frac{m_3+n_3}{2} \\ \frac{\frac{m_4+n_4}{2}-r}{m_4+n_4-m_1-n_1}+\frac{1}{2}, & \text{if } \frac{m_1+n_1}{2} \leq r \leq \frac{m_2+n_2}{2} \\ 1, & \text{if } r \leq \frac{m_1+n_1}{2}. \end{cases}$$

It follows that,

$$E[\xi] = \int_0^{+\infty} Appr\{\frac{\bar{a}+\bar{b}}{2} \geq r\}dr - \int_{-\infty}^0 Appr\{\frac{\bar{a}+\bar{b}}{2} \leq r\}dr$$

$$= \int_0^{\frac{m_1+n_1}{2}} 1 dr + \int_{\frac{m_1+n_1}{2}}^{\frac{m_2+n_2}{2}} (\frac{\frac{m_4+n_4}{2}-r}{m_4+n_4-m_1-n_1} + \frac{1}{2})dr$$

$$+ \int_{\frac{m_2+n_2}{2}}^{\frac{m_3+n_3}{2}} \frac{1}{2}(\frac{\frac{m_4+n_4}{2}-r}{m_4+t_4-m_1-n_1} + \frac{\frac{m_3+n_3}{2}-r}{m_3+n_3-m_2-n_2})dr + \int_{\frac{m_3+n_3}{2}}^{\frac{m_4+n_4}{2}} \frac{\frac{m_4+n_4}{2}-r}{m_4+n_4-m_1-n_1}dr$$

$$= \frac{1}{8} \sum_{i=1}^{4} (m_i + n_i).$$

The proof is complete. □

Proposition 5.2. *Let ξ be a trapezoidal Fu-Ro variable $\xi = (\bar{r}_1, \bar{r}_2, \bar{r}_3, \bar{r}_4)$, where $\bar{r}_1, \bar{r}_2, \bar{r}_3, \bar{r}_4$ are rough variables defined on $(\Lambda, \Theta, \mathscr{A}, \pi)$, and*

$$\bar{r}_1 = ([m_2,m_3],[m_1,m_4]), 0 < m_1 \leq m_2 < m_3 \leq m_4,$$
$$\bar{r}_2 = ([n_2,n_3],[n_1,n_4]), 0 < n_1 \leq n_2 < n_3 \leq n_4,$$
$$\bar{r}_3 = ([s_2,s_3],[s_1,s_4]), 0 < s_1 \leq s_2 < s_3 \leq s_4,$$
$$\bar{r}_4 = ([t_2,t_3],[t_1,t_4]), 0 < t_1 \leq t_2 < t_3 \leq t_4.$$

Then the expected value of ξ is

$$E[\xi] = \frac{1}{16} \sum_{i=1}^{4} (m_i + n_i + s_i + t_i).$$

Proof. Since $\xi(\lambda) = [\bar{r}_1, \bar{r}_2, \bar{r}_3, \bar{r}_4]$, and according to proposition 2.2, we have

$$E^{Cr}[\xi(\lambda)] = \frac{1}{4}(\bar{r}_1 + \bar{r}_2 + \bar{r}_3 + \bar{r}_4).$$

By the rough arithmetic operators, it follows that,

$$\frac{\tilde{r}_1+\tilde{r}_2+\tilde{r}_3+\tilde{r}_4}{4}$$

$$= \frac{1}{4}\{([m_2,m_3],[m_1,m_4]) + ([n_2,n_3],[n_1,n_4]) + ([s_2,s_3],[s_1,s_4]) + ([t_2,t_3],[t_1,t_4])\}$$

$$= ([\frac{m_2+n_2+s_2+t_2}{4}, \frac{m_3+n_3+s_3+t_3}{4}],[\frac{m_1+n_1+s_1+t_1}{4}, \frac{m_4+n_4+s_4+t_4}{4}]).$$

From the definition of trust measure, we have

$$Appr\{\xi \geq r\} =$$

$$\begin{cases}
0, & \text{if } \frac{m_4+n_4+s_4+t_4}{4} \leq r \\[2mm]
\frac{\frac{m_4+n_4+s_4+t_4}{2}-2r}{m_4+n_4+s_4+t_4-(m_1+n_1+s_1+t_1)}, & \text{if } \frac{m_3+n_3+s_3+t_3}{4} \leq r \leq \frac{m_4+n_4+s_4+t_4}{4} \\[2mm]
\frac{1}{2}\left(\frac{\frac{m_4+n_4+s_4+t_4}{2}-2r}{m_4+n_4+s_4+t_4-(m_1+n_1+s_1+t_1)}\right. \\
\left. +\frac{\frac{m_3+n_3+s_3+t_3}{2}-2r}{m_3+n_3+s_3+t_3-(m_2+n_2+s_2+t_2)}\right), \\
& \text{if } \frac{m_2+n_2+s_2+t_2}{4} \leq r \leq \frac{m_3+n_3+s_3+t_3}{4} \\[2mm]
\frac{\frac{m_4+n_4+s_4+t_4}{2}-2r}{m_4+n_4+s_4+t_4-(m_1+n_1+s_1+t_1)} + \frac{1}{2}, & \text{if } \frac{m_1+n_1+s_1+t_1}{4} \leq r \leq \frac{m_2+n_2+s_2+t_2}{4} \\[2mm]
1, & \text{if } r \leq \frac{m_1+n_1+s_1+t_1}{4}.
\end{cases}$$

It follows that,

$$E[\xi] = \int_0^{+\infty} Appr\{\frac{\tilde{r}_1+\tilde{r}_2+\tilde{r}_3+\tilde{r}_4}{4} \geq r\}dr - \int_{-\infty}^0 Appr\{\frac{\tilde{r}_1+\tilde{r}_2+\tilde{r}_3+\tilde{r}_4}{4} \leq r\}dr$$

$$= \int_0^{\frac{m_1+n_1+s_1+t_1}{4}} 1dr + \int_{\frac{m_1+n_1+s_1+t_1}{4}}^{\frac{m_2+n_2+s_2+t_2}{4}} (\frac{\frac{m_4+n_4+s_4+t_4}{2}-2r}{m_4+n_4+s_4+t_4-(m_1+n_1+s_1+t_1)} + \frac{1}{2})dr$$

$$+ \int_{\frac{m_2+n_2+s_2+t_2}{2}}^{\frac{m_3+n_3+s_3+t_3}{2}} \frac{1}{2}(\frac{\frac{m_4+n_4+s_4+t_4}{2}-2r}{m_4+t_4+s_4+t_4-(m_1+n_1+s_1+t_1)} + \frac{\frac{m_3+n_3+s_3+t_3}{2}-2r}{m_3+n_3+s_3+t_3-(m_2+n_2+s_2+t_2)})dr$$

$$+ \int_{\frac{m_3+n_3+s_3+t_3}{2}}^{\frac{m_4+n_4+s_4+t_4}{2}} \frac{\frac{m_4+n_4+s_4+t_4}{2}-2r}{m_4+n_4+s_4+t_4-(m_1+n_1+s_1+t_1)}dr$$

$$= \frac{1}{16}\sum_{i=1}^{4}(m_i + n_i + s_i + t_i).$$

The proof is complete. □

Proposition 5.3. *Let ξ be a LR Fu-Ro variable with the membership function of fuzzy variable ξ has the following form*

$$\mu_\xi(x) = \begin{cases} L(\frac{\bar{z}-x}{\alpha}), & \bar{z}-\alpha < x \leq \bar{z} \\ 1, & x = \bar{z} \\ R(\frac{x-\bar{z}}{\beta}), & \bar{z} < x < \bar{z}+\beta, \end{cases} \tag{5.12}$$

where $\bar{z}$ is a rough variable and $\bar{z} = ([z_2,z_3],[z_1,z_4]), \alpha < z_1 < z_2 < z_3 < z_4$. And here we just consider the situation when the reference function $L(x) = R(x) = 1-x$, then this LR fuzzy rough variable is triangular type, and the left and right spread $\alpha, \beta > 0$.

Then the expected value of ξ is

$$E[\xi] = \frac{1}{4}(z_1 + z_2 + z_3 + z_4 + \alpha + \beta).$$

Proof. Since $\xi(\lambda) = (\bar{z}, \alpha, \beta)_{LR}$ is triangular LR Fu-Ro variable, which means the reference functions $L(x) = R(x) = 1 - x$, according to remark 2.11, we have

$$E^{Cr}[\xi(\lambda)] = \bar{z} + \frac{1}{4}(\alpha + \beta).$$

Then we have

$$E[\xi] = E[\bar{z} + \tfrac{1}{4}(\alpha + \beta)]$$
$$= E[\bar{z}] + E[\tfrac{1}{4}(\alpha + \beta)].$$

It follows that

$$E[\xi] = \tfrac{1}{4}(z_1 + z_2 + z_3 + z_4) + \tfrac{1}{4}(\alpha + \beta)$$
$$= \tfrac{z_1 + z_2 + z_3 + z_4 + \alpha + \beta}{4}.$$

The proof is complete.   □

5.2.3 Chance Operator of Fu-Ro Variables

To begin with, we give the three types of primitive chance of Fu-Ro event as follows.

Definition 5.7. Let $\xi = (\xi_1, \xi_2, \cdots \xi_n)$ be a Fu-Ro vector defined on $(\Lambda, \Delta, \mathscr{A}, \pi)$, and $f : \mathbf{R}^n \to \mathbf{R}$ is Borel measurable function. Then the primitive chance of a Fu-Ro event characterized by $f(\xi) \leq 0$ is a function from $(0, 1]$ to $[0, 1]$, defined as

(1). $Appr - Pos$ chance,

$$Ch\{f(\xi) \leq 0\}(\alpha) = \sup_{\alpha \in [0,1]} \{\beta | Appr\{\lambda \in \Lambda | Pos\{f(\xi(\lambda)) \leq 0\} \geq \beta\} \geq \alpha\}.$$
(5.13)

(2). $Appr - Nec$ Chance,

$$Ch\{f(\xi) \leq 0\}(\alpha) = \sup_{\alpha \in [0,1]} \{\beta | Appr\{\lambda \in \Lambda | Nec\{f(\xi(\lambda)) \leq 0\} \geq \beta\} \geq \alpha\}.$$
(5.14)

(3). $Appr - Cr$ Chance,

$$Ch\{f(\xi) \leq 0\}(\alpha) = \sup_{\alpha \in [0,1]} \{\beta | Appr\{\lambda \in \Lambda | Cr\{f(\xi(\lambda)) \leq 0\} \geq \beta\} \geq \alpha\}.$$
(5.15)

Remark 5.5. The primitive chance of a Fu-Ro event characterized by $f(\xi) \leq 0$ defined as (5.13), (5.14), (5.15) have the equivalent forms respectively.

$$Ch\{f(\xi) \leq 0\}(\alpha) = \sup_{Appr\{A\} \geq \alpha} \inf_{\lambda \in A} Pos\{f(\xi(\lambda)) \leq 0\},$$
(5.16)

$$Ch\{f(\xi) \leq 0\}(\alpha) = \sup_{Appr\{A\} \geq \alpha} \inf_{\lambda \in A} Nec\{f(\xi(\lambda)) \leq 0\},$$
(5.17)

$$Ch\{f(\xi) \leq 0\}(\alpha) = \sup_{Appr\{A\} \geq \alpha} \inf_{\lambda \in A} Cr\{f(\xi(\lambda)) \leq 0\}. \tag{5.18}$$

Lemma 5.6. *For any confidence levels* α, β.
(1). Appr-Pos Chance $Ch\{f(\xi) \leq 0\}(\alpha) \geq \beta$ *holds if and only if*

$$Appr\{\lambda \in \Lambda | Pos\{f(\xi(\lambda)) \leq 0\} \geq \beta\} \geq \alpha.$$

(2). Appr-Nec Chance $Ch\{f(\xi) \leq 0\}(\alpha) \geq \beta$ *holds if and only if*

$$Appr\{\lambda \in \Lambda | Nec\{f(\xi(\lambda)) \leq 0\} \geq \beta\} \geq \alpha.$$

(3). Appr-Cr Chance $Ch\{f(\xi) \leq 0\}(\alpha) \geq \beta$ *holds if and only if*

$$Appr\{\lambda \in \Lambda | Cr\{f(\xi(\lambda)) \leq 0\} \geq \beta\} \geq \alpha.$$

5.3 Fu-Ro EVM

For the multi-objective model (5.19) with Fu-Ro parameters, we cannot deal with it directly, we should use some tools to make it have mathematical meaning, we then can solve it. In this section, we employ the expected value operator to transform the fuzzy rough model into Fu-Ro EVM. Consider the following multi-objective decision making model (5.19) with fuzzy rough coefficients:

$$\begin{cases} \max \{f_1(x,\xi), f_2(x,\xi), \cdots, f_m(x,\xi)\} \\ \text{s.t.} \begin{cases} g_r(x,\xi) \leq 0, r = 1, 2, \cdots, p \\ x \in X, \end{cases} \end{cases} \tag{5.19}$$

where x is a n-dimensional decision vector, $\xi = (\xi_1, \xi_2, \cdots, \xi_n)$ is a Fu-Ro vector, $f_i(x,\xi)$ are objective functions, $i = 1, 2, \cdots, m$. Because of the existence of Fu-Ro vector ξ, problem (5.19) is not well-defined. That is, the meaning of maximizing $f_i(x,\xi), i = 1, 2, \cdots, m$ is not clear and constraints $g_r(x,\xi) \leq 0, r = 1, 2, \cdots, p$ do not define a deterministic feasible set. In the following, we use Fu-Ro EVM to deal with the meaningless model.

5.3.1 General Model for Fu-Ro EVM

Based on the definition of the expected value of fuzzy rough events f_i and g_r, the general model for Fu-Ro EVM is proposed as follows,

$$\begin{cases} \max E[f_1(x,\xi), f_2(x,\xi), \cdots, f_m(x,\xi)] \\ \text{s.t.} \begin{cases} E[g_r(x,\xi)] \leq 0, r = 1, 2, \cdots, p \\ x \in X, \end{cases} \end{cases} \tag{5.20}$$

where x is n-dimensional decision vector and ξ is n-dimensional fuzzy rough variable.

Definition 5.8. If x^* is an efficient solution of problem (5.20), we call it as a fuzzy rough expected efficient solution.

Clearly, the problem (5.20) is a multi-objective with crisp parameters. Then we can convert it into a single-objective programming by traditional method of weight sum.

$$\begin{cases} \max \sum_{i=1}^{m} w_i E[f_i(x,\xi)] \\ \text{s.t.} \begin{cases} E[g_r(x,\xi)] \le 0, \ r = 1,2,\cdots,p \\ x \in X \\ w_1 + w_2 + \cdots + w_m = 1. \end{cases} \end{cases} \tag{5.21}$$

Theorem 5.1. *Problem (5.21) is equivalent to problem (5.20), i.e., the efficient solution of problem (5.20) is the optimal solution of problem (5.21) and the optimal solution of problem (5.21) is the efficient solution of problem (5.20).*

Proof. The proof is the same as the proof of Theorem 3.1.

Theorem 5.2. *Let $\xi = (\xi_1,\xi_2,\cdots,\xi_n)$ be a fuzzy rough vector on the rough space $(\Lambda,\Delta,\mathscr{A},\pi)$, and f_i and $g_r : \mathscr{A}^n \to \mathscr{A}$ be convex continuous functions with respect to x, $i = 1,2,\cdots,m; r = 1,2,\cdots,p$. Then the expected value programming problem (5.21) is a convex programming.*

Proof. It is similar to the proof of Theorem 3.2, and thus omit. □

We can also formulate a fuzzy rough decision system as an expected value goal programming (EVGP) model according to the priority structure and target levels set by the decision-maker:

$$\begin{cases} \min \sum_{j=1}^{l} P_j \sum_{i=1}^{m} (u_{ij}d_i^+ + v_{ij}d_i^-) \\ \text{s.t.} \begin{cases} E[f_i(x,\xi)] + d_i^- - d_i^+ = b_i, \ i = 1,2,\cdots,m \\ E[g_r(x,\xi)] \le 0, \ r = 1,2,\cdots,p \\ d_i^-, d_i^+ \ge 0, \ i = 1,2,\cdots,m \\ x \in X, \end{cases} \end{cases} \tag{5.22}$$

where P_j is the preemptive priority factor which expresses the relative importance of various goals, $P_j \gg P_{j+1}$, for all j, u_{ij} is the weighting factor corresponding to positive deviation for goal i with priority j assigned, v_{ij} is the weighting factor corresponding to negative deviation for goal i with priority j assigned, d_i^+ is the positive deviation from the target of goal i, defined as

$$d_i^+ = [E[f_i(x,\xi)] - b_i] \vee 0,$$

d_i^- is the negative deviation from the target of goal i, defined as

$$d_i^- = [b_i - E[f_i(x,\xi)]] \vee 0,$$

f_i is a function in goal constraints, g_j is a function in real constraints, b_i is the target value according to goal i, l is the number of priorities, m is the number of goal constraints, and p is the number of real constraints.

5.3.2 Linear Fu-Ro EVM and Minimax Point Method

For the regular fuzzy rough linear programming problem, we can use the expected value operator to handle it,

$$\begin{cases} \max E[\sum\limits_{j=1}^{n} \tilde{\bar{c}}_{ij}x_j, \ i = 1,2,\cdots,m] \\ \text{s.t.} \begin{cases} E[\tilde{\bar{a}}_{rj}x_j] \geq E[\tilde{\bar{b}}_r], \ r = 1,2,\dots,p \\ x_j \geq 0, \ j = 1,2,\dots,n, \end{cases} \end{cases} \tag{5.23}$$

where $\tilde{\bar{c}}, \tilde{\bar{a}}, \tilde{\bar{b}}$ are fuzzy rough variables.

5.3.2.1 Crisp Equivalent Model

In order to solve the model (5.23), we must compute the crisp expected value of ξ. However, as we know, this process is usually a hard work at most of time. In this section, we will consider a special cases and present their results.

$$\begin{cases} \max \ [E[\tilde{\bar{c}}_1^T x], E[\tilde{\bar{c}}_2^T x], \cdots, E[\tilde{\bar{c}}_m^T x]] \\ \text{s.t.} \begin{cases} E[\tilde{\bar{a}}_r^T x] \leq E[\tilde{\bar{b}}_r], r = 1,2,\cdots,p \\ x \geq 0, \end{cases} \end{cases} \tag{5.24}$$

where $\tilde{\bar{c}}_i = (\tilde{\bar{c}}_{i1}, \tilde{\bar{c}}_{i1}, \cdots, \tilde{\bar{c}}_{in})^T, \tilde{\bar{a}}_r = (\tilde{\bar{a}}_{r1}, \tilde{\bar{a}}_{r1}, \cdots, \tilde{\bar{a}}_{rn})^T$ are fuzzy rough vectors, $\tilde{\bar{b}}_r$ are fuzzy rough variables, $i = 1,2,\cdots,m, r = 1,2,\cdots,p$. If these fuzzy vectors, as well as rough variables have special forms, we have the following theorem.

Theorem 5.3. *If fuzzy rough variables $\tilde{\bar{c}}_{ij}$ are defined as*

$$\tilde{\bar{c}}_{ij}(\lambda) = (\bar{c}_{ij1}, \bar{c}_{ij2}, \bar{c}_{ij3}, \bar{c}_{ij4}), \quad with \quad \bar{c}_{ijt} \vdash ([c_{ijt1}, c_{ijt2}], [c_{ijt3}, c_{ijt4}])$$

for $i = 1,2,\cdots,m, j = 1,2,\cdots,n, t = 1,2,3,4$, then

$$E[\tilde{\bar{c}}_1^T x], E[\tilde{\bar{c}}_2^T x], \cdots, E[\tilde{\bar{c}}_m^T x]$$

is equivalent to

$$\frac{1}{16}\sum_{j=1}^{n}\sum_{t=1}^{4}\sum_{k=1}^{4} c_{1jtk}x_j, \frac{1}{16}\sum_{j=1}^{n}\sum_{t=1}^{4}\sum_{k=1}^{4} c_{2jtk}x_j, \cdots, \frac{1}{16}\sum_{j=1}^{n}\sum_{t=1}^{4}\sum_{k=1}^{4} c_{mjtk}x_j.$$

Proof. First, we verify that $E[\tilde{\bar{c}}_{ij}] = \frac{1}{16}\sum\limits_{t=1}^{4}\sum\limits_{k=1}^{4} c_{ijtk}, i = 1,2,\cdots,m$. In fact $\forall \lambda \in \Lambda$,

$$E[\tilde{\bar{c}}_{ij}(\lambda)] = \frac{1}{4}(\bar{c}_{ij1} + \bar{c}_{ij2} + \bar{c}_{ij3} + \bar{c}_{ij4})$$

$$= ([\frac{1}{4}\sum_{t=1}^{4} c_{ijt1}, \frac{1}{4}\sum_{t=1}^{4} c_{ijt2}], [\frac{1}{4}\sum_{t=1}^{4} c_{ijt3}, \frac{1}{4}\sum_{t=1}^{4} c_{ijt4}]).$$

Suppose $A = \frac{1}{4}\sum\limits_{t=1}^{4} c_{ijt1}, B = \frac{1}{4}\sum\limits_{t=1}^{4} c_{ijt2}, C = \frac{1}{4}\sum\limits_{t=1}^{4} c_{ijt3}, D = \sum\limits_{t=1}^{4} c_{ijt4}$, then we have

$$Appr\{E[\tilde{\bar{c}}_{ij}(\lambda)] \geq r\} = \begin{cases} 0, & if \quad D \leq r \\ \frac{D-r}{2(D-C)}, & if \quad B \leq r \leq D \\ \frac{1}{2}(\frac{D-r}{D-C} + \frac{B-r}{B-A}), & if \quad A \leq r \leq B \\ \frac{1}{2}(\frac{D-r}{D-C} + 1), & if \quad C \leq r \leq A \\ 1, & if \quad r \leq C \end{cases}$$

and

$$Appr\{E[\tilde{\bar{c}}_{ij}(\lambda)] \leq r\} = \begin{cases} 0, & if \quad r \leq C \\ \frac{r-C}{2(D-C)}, & if \quad C \leq r \leq A \\ \frac{1}{2}(\frac{r-C}{D-C} + \frac{r-A}{B-A}), & if \quad A \leq r \leq B \\ \frac{1}{2}(\frac{r-C}{D-C} + 1), & if \quad B \leq r \leq D \\ 1, & if \quad D \leq r. \end{cases}$$

There are five cases when we compute the expected value of ξ. Let's discuss every case in turn.

Case 1: $0 \leq C \leq A \leq B \leq D$.

$$\begin{aligned} E[\tilde{\bar{c}}_{ij}] &= \int_0^{+\infty} Appr\{\lambda \in \Lambda | E[\tilde{\bar{c}}_{ij}(\lambda)] \geq r\} dr - \int_{-\infty}^0 Appr\{\lambda \in \Lambda | E[\tilde{\bar{c}}_{ij}(\lambda)] \leq r\} dr \\ &= \int_0^C 1 dr + \int_C^A \frac{1}{2}(\frac{D-r}{D-C} + 1) dr \\ &\quad + \int_A^B \frac{1}{2}(\frac{D-r}{D-C} + \frac{B-r}{B-A}) dr + \int_B^D \frac{D-r}{2(D-C)} dr \\ &= \frac{1}{4}(A+B+C+D). \end{aligned}$$

Case 2: $C \leq 0 \leq A \leq B \leq D$.

$$\begin{aligned} E[\tilde{\bar{c}}_{ij}] &= \int_0^{+\infty} Appr\{\lambda \in \Lambda | E[\tilde{\bar{c}}_{ij}(\lambda)] \geq r\} dr - \int_{-\infty}^0 Appr\{\lambda \in \Lambda | E[\tilde{\bar{c}}_{ij}(\lambda)] \leq r\} dr \\ &= \int_0^A \frac{1}{2}(\frac{D-r}{D-C} + 1) dr + \int_A^B \frac{1}{2}(\frac{D-r}{D-C} + \frac{B-r}{B-A}) dr \\ &\quad + \int_B^D \frac{D-r}{2(D-C)} dr - \int_C^0 \frac{r-C}{2(D-C)} dr \\ &= \frac{1}{4}(A+B+C+D). \end{aligned}$$

Case 3: $C \leq A \leq 0 \leq B \leq D$.

$$\begin{aligned} E[\tilde{\bar{c}}_{ij}] &= \int_0^{+\infty} Appr\{\lambda \in \Lambda | E[\tilde{\bar{c}}_{ij}(\lambda)] \geq r\} dr - \int_{-\infty}^0 Appr\{\lambda \in \Lambda | E[\tilde{\bar{c}}_{ij}(\lambda)] \leq r\} dr \\ &= \int_0^B \frac{1}{2}(\frac{D-r}{D-C} + \frac{B-r}{B-A}) dr + \int_B^D \frac{D-r}{2(D-C)} dr \\ &\quad - \int_C^A \frac{r-C}{2(D-C)} dr - \int_A^0 \frac{1}{2}(\frac{r-C}{D-C} + \frac{r-A}{B-A}) dr \\ &= \frac{1}{4}(A+B+C+D). \end{aligned}$$

Case 4: $C \leq A \leq B \leq 0 \leq D$.

$$\begin{aligned} E[\tilde{\bar{c}}_{ij}] &= \int_0^{+\infty} Appr\{\lambda \in \Lambda | E[\tilde{\bar{c}}_{ij}(\lambda)] \geq r\} dr - \int_{-\infty}^0 Appr\{\lambda \in \Lambda | E[\tilde{\bar{c}}_{ij}(\lambda)] \leq r\} dr \\ &= \int_0^D \frac{D-r}{2(D-C)} dr - \int_C^A \frac{r-C}{2(D-C)} dr \\ &\quad - \int_A^B \frac{1}{2}(\frac{r-C}{D-C} + \frac{r-A}{B-A}) dr - \int_B^0 \frac{1}{2}(\frac{r-C}{D-C} + 1) dr \\ &= \frac{1}{4}(A+B+C+D). \end{aligned}$$

Case 5: $C \leq A \leq B \leq D \leq 0$.

$$
\begin{aligned}
E[\tilde{\bar{c}}_{ij}] &= \int_0^{+\infty} Appr\{\lambda \in \Lambda | E[\tilde{\bar{c}}_{ij}(\lambda)] \geq r\} dr - \int_{-\infty}^0 Appr\{\lambda \in \Lambda | E[\tilde{\bar{c}}_{ij}(\lambda)] \leq r\} dr \\
&= -\int_C^A \frac{r-C}{2(D-C)} dr - \int_A^B \frac{1}{2}(\frac{r-C}{D-C} + \frac{r-A}{B-A}) dr \\
&\quad - \int_B^D \frac{1}{2}(\frac{r-C}{D-C} + 1) - dr \int_D^0 1 dr \\
&= \frac{1}{4}(A+B+C+D).
\end{aligned}
$$

So we always have $E[\tilde{\bar{c}}_{ij}] = \frac{1}{16} \sum_{t=1}^4 \sum_{k=1}^4 c_{ijtk}, i = 1,2,\cdots,m$.

It follows from the nonnegativity of $x_j (j = 1,2,\cdots,n)$ and linearity of expected value operator that

$$
\begin{aligned}
E[\tilde{\bar{c}}_i^T x] &= E[\sum_{j=1}^n \tilde{\bar{c}}_{ij} x_j] \\
&= \sum_{j=1}^n E[\tilde{\bar{c}}_{ij}] x_j \\
&= \frac{1}{16} \sum_{j=1}^n \sum_{t=1}^4 \sum_{k=1}^4 c_{rjtk} x_j.
\end{aligned}
$$

Thus the theorem is proved.   □

Theorem 5.4. *If fuzzy rough variables $\tilde{\bar{a}}_{rj}, \tilde{\bar{b}}_r$ are defined as follows,*

$$
\begin{aligned}
\tilde{\bar{a}}_{rj}(\lambda) &= (\bar{a}_{rj1}, \bar{c}_{rj2}, \bar{a}_{rj3}, \bar{a}_{rj4}), \quad \text{with} \quad \bar{a}_{rjt} \vdash ([a_{rjt1}, a_{rjt2}], [a_{rjt3}, a_{rjt4}]), \\
\tilde{\bar{b}}_r(\lambda) &= (\bar{b}_{r1}, \bar{b}_{r2}, \bar{b}_{r3}, \bar{c}_{r4}), \quad \text{with} \quad \bar{b}_{rt} \vdash ([b_{rt1}, b_{rt2}], [b_{rt3}, b_{rt4}]),
\end{aligned}
$$

for $r = 1,2,\cdots,p, j = 1,2,\cdots,n, t = 1,2,3,4$, then

$$
E[\tilde{\bar{a}}_r^T x] \leq E[\tilde{\bar{b}}_r], r = 1,2,\cdots,p
$$

is equivalent to

$$
\sum_{j=1}^n \sum_{t=1}^4 \sum_{k=1}^4 a_{rjtk} x_j \leq \sum_{t=1}^4 \sum_{k=1}^4 b_{rtk}, r = 1,2,\cdots,p.
$$

Proof. Similar to Theorem 5.4, we have

$$
E[\tilde{\bar{a}}_i^T x] = \frac{1}{16} \sum_{j=1}^n \sum_{t=1}^4 \sum_{k=1}^4 a_{rjtk} x_j
$$

and

$$
E[\tilde{\bar{b}}_r] = \frac{1}{16} \sum_{t=1}^4 \sum_{k=1}^4 b_{rtk},
$$

for $i = 1,2,\cdots,m, r = 1,2,\cdots,p$.

Thus the theorem holds.   □

According to Theorems 5.3-5.4, Model (5.24) with the Fu-Ro coefficients described as Theorem 5.3 and Theorem 5.4 is equivalent to the conventional multi-objective linear programming

$$
\begin{cases}
\max \left[\dfrac{1}{16} \sum\limits_{j=1}^{n} \sum\limits_{t=1}^{4} \sum\limits_{k=1}^{4} c_{1jtk} x_j, \dfrac{1}{16} \sum\limits_{j=1}^{n} \sum\limits_{t=1}^{4} \sum\limits_{k=1}^{4} c_{2jtk} x_j, \cdots, \dfrac{1}{16} \sum\limits_{j=1}^{n} \sum\limits_{t=1}^{4} \sum\limits_{k=1}^{4} c_{mjtk} x_j \right] \\
\text{s.t.} \begin{cases} \sum\limits_{j=1}^{n} \sum\limits_{t=1}^{4} \sum\limits_{k=1}^{4} a_{rjtk} x_j \leq \sum\limits_{t=1}^{4} \sum\limits_{k=1}^{4} b_{rtk}, r = 1, 2, \cdots, p \\ x_j \geq 0, j = 1, 2, \cdots, n. \end{cases}
\end{cases}
$$

$$(5.25)$$

5.3.2.2 Minimax Point Method

In this section, we use the minimax point method proposed in [92] to deal with the crisp multiobjective problem (5.26).

$$
\begin{cases}
\max \ [H_1(x), H_2(x), \cdots, H_m(x)] \\
\text{s.t.} \ x \in X.
\end{cases}
$$

$$(5.26)$$

To maximize the objectives, the minimax point method firstly constructing an evaluation function by seeking the minimal objective value after respectively computing all objective functions, that is, $u(\mathbf{H}(x)) = \min_{1 \leq i \leq m} H_i(x)$, where $\mathbf{H}(x) = (H_1(x), H_2(x), \cdots, H_m(x))^T$. Then the objective function of problem (5.26) is came down to solve the maximization problem as follows,

$$\max_{x \in X'} u(\mathbf{H}(x)) = \max_{x \in X'} \min_{1 \leq i \leq m} H_i(x).$$

$$(5.27)$$

Sometimes, decision makers need considering the relative importance of various goals, then the weight can be combined into the evaluation function as follows,

$$\max_{x \in X'} u(\mathbf{H}(x)) = \max_{x \in X'} \min_{1 \leq i \leq m} \{\omega_i H_i(x)\},$$

$$(5.28)$$

where the weight $\sum_{i=1}^{m} \omega_i = 1 (\omega_i > 0)$ and is predetermined by decision makers.

Theorem 5.5. *The optimal solution x^* of problem (5.28) is the weak efficient solution of problem (5.26).*

Proof. Assume that $x^* \in X'$ is the optimal solution of the problem (5.28). If there exists an x such that $H_i(x) \geq H_i(x^*) (i = 1, 2, \cdots, m)$, we have

$$\min_{1 \leq i \leq m} \{\omega_i H_i(x^*)\} \leq \omega_i H_i(x^*) \leq \omega_i H_i(x), \ 0 < \omega_i < 1.$$

Denote $\delta = \min_{1 \leq i \leq m} \{\omega_i H_i(x)\}$, then $\delta \geq \min_{1 \leq i \leq m} \{\omega_i H_i(x^*)\}$. This means that x^* isn't the optimal solution of the problem (5.28). This conflict with the condition.

Thus, there doesn't exist $x \in X'$ such that $H_i(x) \geq H_i(x^*)$, namely, x^* is a weak efficient solution of the problem (5.26).   □

By introducing an auxiliary variable, the minimax problem (5.28) can be converted into a single objective problem. Let

$$\lambda = \min_{1 \leq i \leq m} \{\omega_i H_i(x)\},$$

then the problem (5.28) is converted into

$$\begin{cases} \max \lambda \\ \text{s.t.} \begin{cases} \omega_i H_i(x) \geq \lambda, i = 1, 2, \cdots, m \\ x \in X'. \end{cases} \end{cases} \tag{5.29}$$

Theorem 5.6. *The problem (5.28) is equivalent to the problem (5.29).*

Proof. Assume that $x^* \in X'$ is the optimal solution of the problem (5.28) and let $\lambda^* = \min_{1 \leq i \leq m} \{\omega_i H_i(x^*)\}$, then it is apparent that $H_i(x^*) \geq \lambda^*$. This means that (x^*, λ^*) is a feasible solution of the problem (5.29). Assume that (x, λ) is any feasible solution of the problem (5.29). Since x^* is the optimal solution of the problem (5.28), wa have

$$\lambda^* = \min_{1 \leq i \leq m} \{\omega_i H_i(x^*)\} \geq \min_{1 \leq i \leq m} \{\omega_i H_i(x)\} \geq \lambda,$$

namely, (x^*, λ^*) is the optimal solution of the problem (5.29).

On the contrary, assume that (x^*, λ^*) is an optimal solution of the problem (5.29). Then $\omega_i H_i(x^*) \geq \lambda^*$ holds for any i, this means $\min_{1 \leq i \leq m} \{\omega_i H_i(x^*)\} \geq \lambda^*$. It follows that for any any feasible $x \in X'$,

$$\min_{1 \leq i \leq m} \{\omega_i H_i(x)\} = \lambda \leq \lambda^* \leq \min_{1 \leq i \leq m} \{\omega_i H_i(x^*)\}$$

holds, namely, x^* is the optimal solution of the problem (5.28).   □

In a word, the minimax point method can be summarized as follows:
Step 1. Compute the weight for each objective function by solving the two problems, $\max_{x \in X'} H_i(x)$ and $\omega_i = H_i(x^*) / \sum_{i=1}^{m} H_i(x^*)$.
Step 2. Construct the auxiliary problem as follows,

$$\begin{cases} \max \lambda \\ \text{s.t.} \begin{cases} \omega_i H_i(x) \geq \lambda, i = 1, 2, \cdots, m \\ x \in X'. \end{cases} \end{cases}$$

Step 3. Solve the above problem to obtain the optimal solution.

5.3.2.3 Numerical Example

Example 5.2. Let us consider the following problem.

$$\begin{cases} \max\ 0.375x_1 + 0.625x_2 + 0.875x_3 \\ \max\ E[c_1\xi_1 x_1 + c_2\xi_2 x_2 + c_3\xi_3 x_3] \\ \text{s.t.} \begin{cases} x_1 + x_2 + x_3 \le 250 \\ x_1 + x_2 + x_3 \ge 200 \\ \xi_4 x_1 + \xi_5 x_2 + \xi_6 x_3 \le 600 \\ x_1 \ge 20, x_2 \ge 20, x_3 \ge 20. \end{cases} \end{cases} \qquad (5.30)$$

The following is the relevant data, ξ is fuzzy rough variable,

$(c_1, c_2, c_3) = (1.2, 0.8, 1.5),$
$\xi_1 = (\rho_1 - 2, \rho_1 - 1, \rho_1 + 1, \rho_1 + 2),$ with $\rho_1 \vdash ([0,1], [0,3]),$
$\xi_2 = (\rho_2 - 2, \rho_2 - 1, \rho_2 + 1, \rho_2 + 2),$ with $\rho_2 \vdash ([1,2], [0,3]),$
$\xi_3 = (\rho_3 - 2, \rho_3 - 1, \rho_3 + 1, \rho_3 + 2),$ with $\rho_3 \vdash ([2,3], [0,3]).$
$\xi_4 = (\rho_4 - 2, \rho_4 - 1, \rho_4 + 1, \rho_4 + 2),$ with $\rho_1 \vdash ([2,5], [0,9]),$
$\xi_5 = (\rho_5 - 2, \rho_5 - 1, \rho_5 + 1, \rho_5 + 2),$ with $\rho_2 \vdash ([10,20], [4,30]),$
$\xi_6 = (\rho_6 - 2, \rho_6 - 1, \rho_6 + 1, \rho_6 + 2),$ with $\rho_3 \vdash ([6,10], [4,12]).$

It follows from Proposition 5.4 that problem (5.30) is equivalent to

$$\begin{cases} \max\ F_1(x) = 0.375x_1 + 0.625x_2 + 0.875x_3 \\ \max\ F_2(x) = 0.3x_1 + 0.3x_2 + 0.75x_3 \\ \text{s.t.} \begin{cases} x_1 + x_2 + x_3 \le 250 \\ x_1 + x_2 + x_3 \ge 200 \\ x_1 + 2x_2 + 4x_3 \le 600 \\ x_1 \ge 20, x_2 \ge 20, x_3 \ge 20. \end{cases} \end{cases} \qquad (5.31)$$

According to the minimax point method, first we compute the weight by solving the two single objective models,

$$w_1 = F_1^*/(F_1^* + F_2^*) = 166.25/(166.25 + 124.5) = 0.572,$$
$$w_2 = F_2^*/(F_1^* + F_2^*) = 0.428.$$

Then according to Equation (5.29) we construct the following mode (5.32),

$$\begin{cases} \max\ \lambda \\ \text{s.t.} \begin{cases} w_1 * (0.375x_1 + 0.625x_2 + 0.875x_3) + w_2 * (0.3x_1 + 0.3x_2 + 0.75x_3) \ge \lambda \\ x_1 + x_2 + x_3 \le 250 \\ x_1 + x_2 + x_3 \ge 200 \\ x_1 + 2x_2 + 4x_3 \le 600 \\ x_1 \ge 20, x_2 \ge 20, x_3 \ge 20. \end{cases} \end{cases}$$

$$(5.32)$$

After solving the model (5.32), we can get a efficient solution as follows,

$$(x_1, x_2, x_3) = (120, 20, 110).$$

5.3.3 Non-linear Fu-Ro EVM and Fu-Ro Simulation-Based TS

For the non-linear Fu-Ro EVM, we use the Fu-Ro simulation 1 based TS to solve.

5.3.3.1 Fu-Ro Simulation 1 for Expected Valie

First, we introduce the procedure to simulate the expected value of a Fu-Ro variable.

Assume that ζ is an n-dimensional Fu-Ro vector defined on the rough space $(\Lambda, \Delta, \mathscr{A}, \pi)$, and $f : \mathbf{R}^n \to \mathbf{R}^m$ is a measurable function. In order to calculate the expected value $E[f(\xi)]$, we sample $\underline{\lambda}_1, \underline{\lambda}_2, \cdots, \underline{\lambda}_N$ from Δ and $\overline{\lambda}_1, \overline{\lambda}_2, \cdots, \overline{\lambda}_N$ from Λ. For each $\underline{\lambda}_n$ and $\overline{\lambda}_n$, $n = 1, 2, \cdots, N$, $\xi(\underline{\lambda}_n)$ and $\xi(\overline{\lambda}_n)$ are both fuzzy variables, and $f(\xi(\underline{\lambda}_n))$ and $F(\xi(\overline{\lambda}_n))$ are both fuzzy variables. Then we can apply the fuzzy simulation 1 to get their expected values $E[f(\xi(\underline{\lambda}_n))]$ and $E[f(\xi(\overline{\lambda}_n))]$.

Since $E[f(\xi)]$ is essentially the expected value of rough variable $E[f(\xi(\lambda))]$, and the following (5.33) will be used to get the expected value of the rough variables.

$$E[f(\xi)] = \frac{\sum_{n=1}^{N}(\eta E[f(\xi(\overline{\lambda}_n))] + (1 - \eta)E[f(\xi(\underline{\lambda}_n))])}{2N}. \tag{5.33}$$

So we may combine rough simulation and fuzzy simulation to produce a fuzzy rough simulation as follows.

Step 1. Set $L = 0$.
Step 2. Generate $\underline{\underline{\lambda}}$ from Δ according to the measure.
Step 3. Generate $\overline{\lambda}$ from Λ according to the measure π.
Step 4. $L \leftarrow L + E[f(\xi(\underline{\lambda}))] + E[f(\xi(\overline{\lambda}))]$.
Step 5. Repeat the second to fourth steps N times.
Step 6. Return $L/(2N)$.

Example 5.3. We employ the Fu-Ro simulation 1 to calculate the expected value of $\xi_1 \xi_2$, where ξ_1 and ξ_2 are Fu-Ro variables defined as

$$\xi_1 = (\rho_1, \rho_1 + 1, \rho_1 + 2), \text{ with } \rho_1 = ([1,2],[0,3]),$$
$$\xi_2 = (\rho_2, \rho_2 + 1, \rho_2 + 2), \text{ with } \rho_1 = ([2,3],[1,4]).$$

After a run of Fu-Ro simulation 1 with 5000 cycles, and we get $E[\xi_1 \xi_2] = 8.93$.

5.3.3.2 TS

In the following, let's introduce the Tabu search algorithm (TS) algorithm.

Local search employs the idea that a given solution x may be improved by making small changes. Those solutions obtained by modifying solution x are called neighbors of x. The local search algorithm starts with some initial solution and moves from neighbor to neighbor as long as possible while decreasing the objective function value. The main problem with this strategy is to escape from local minima where the search cannot find any further neighborhood solution that decreases the objective function value. Different strategies have been proposed to solve this problem. One of the most efficient strategies is tabu search. Tabu search allows the search to explore solutions that do not decrease the objective function value only in those cases where these solutions are not forbidden. This is usually obtained by keeping track of the last solutions in term of the action used to transform one solution to the next. When an action is performed it is considered tabu for the next T iterations, where T is the tabu status length. A solution is forbidden if it is obtained by applying a tabu action to the current solution. The Tabu Search metaheuristic has been defined by Fred Glover [385]. The basic ideas of TS have also been sketched by P. Hansen [386]. After that, TS has achieved widespread success in solving practical optimization problems in different domains(such as resource management, process design, logistics and telecommunications).

A *tabu list* is a set of solutions determined by historical information from the last t iterations of the algorithm, where t is fixed or is a variable that depends on the state of the search, or a particular problem. At each iteration, given the current solution x and its corresponding neighborhood $N(x)$, the procedure moves to the solution in the neighborhood $N(x)$ that most improves the objective function. However, moves that lead to solutions on the tabu list are forbidden, or are tabu . If there are no improving moves, TS chooses the move which least changes the objective function value. The tabu list avoids returning to the local optimum from which the procedure has recently escaped. A basic element of tabu search is the *aspiration criterion*, which determines when a move is admissible despite being on the tabu list. One *termination* criterion for the tabu procedure is a limit in the number of consecutive moves for which no improvement occurs. Given an objective function $f(x)$ over a feasible domain D, a generic tabu search for finding an approximation of the global minimum of $f(x)$ is given in Figure 5.8.

We introduce the detailed steps on how to apply a special TS algorithm–Enhanced Continuous Tabu Search(ECTS) proposed by R. Chelouah and P. Siarry [323] based on fuzzy rough simulation to solve a multi-objective expected value model with fuzzy rough parameters.

Setting of parameters. Two of the parameters must be set before any execution of ECTS:

(1) initialization,

(2) control parameters.

For each of these categories, some parameter values must be chosen by the user and some parameter values must be calculated. These four subsets of parameters are listed in Table 5.1.

Initialization. In this stage, we will list the representation of the solution. We have resumed and adapted the method described in detail in [324]. Randomly generate a solution x and check its feasibility by the fuzzy rough simulation such that $E[g_r(x,\xi)] \leq 0 (r = 1,2,\cdots,p)$. Then generate its neighborhood by the concept of 'ball' defined in [324]. A ball $B(x,r)$ is centered on x with radius r, which contains all points x' such that $||x' - x|| \leq 4$(the symbol $||\cdot||$ denotes the Euclidean norm). To obtain a homogeneous exploration of the space, we consider a set of balls centered on the current solution x, with $h_0, h_1, \cdots, h_\eta$. Hence the space is partitioned into concentric 'crowns' $C_i(x, h_{i-1}, h_i)$, such that

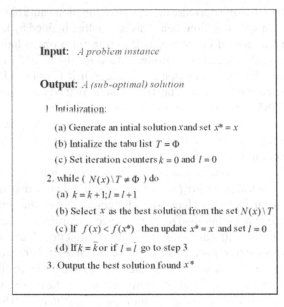

Input: *A problem instance*

Output: *A (sub-optimal) solution*

1 Intialization:

    (a) Generate an intial solution x and set $x^* = x$

    (b) Intialize the tabu list $T = \Phi$

    (c) Set iteration counters $k = 0$ and $l = 0$

2. while ($N(x) \backslash T \neq \Phi$) do

    (a) $k = k+1; l = l+1$

    (b) Select x as the best solution from the set $N(x) \backslash T$

    (c) If $f(x) < f(x^*)$ then update $x^* = x$ and set $l = 0$

    (d) If $k = \bar{k}$ or if $l = \bar{l}$ go to step 3

3. Output the best solution found x^*

Fig. 5.8 Layout of tabu search

$$C_i(x, h_{i-1}, h_i) = \{x' | h_{i-1} \leq ||x' - x|| \leq h_i\}.$$

The η neighbors of s are obtained by random selection of one point inside each crown C_i, for i varying from 1 to η. Finally, we select the best neighbor of x among these η neighbors, even if it is worse than x. In ECTS, we replace the balls by hyperrectangles for the partition of the current solution neighborhood (see Figure 5.9), and we gener-

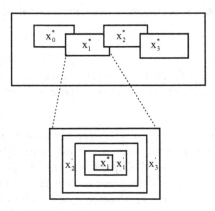

Fig. 5.9 Partition of current solution neighborhood

Table 5.1 Listing of the ECTS parameters

A. *Initialization parameters chosen by the user*
Search domain of each function variable
Starting point
Content of the tabu list
Content of the promising list

B. *Initialization parameters calculated*
Length δ of the smallest edge of the initial hyperrectangular search domain
Initial threshold for the acceptance of a promising area
Initial best point
Number η of neighbors of the current solution investigated at each iteration
Maximum number of successive iterations without any detection of a promising area
Maximum number of successive iterations without any improvement of the objective
function value
Maximum number of successive reductions of the hyperrectangular neighborhood and of the
radius of tabu balls with out any improvement
Maximum number of iterations

C. *Control parameters chosen by the user*
Length N_t of the tabu list
Length N_p of the promising list
Parameter ρ_t allowing to calculate the initial radius of tabu balls
Parameter ρ_{neigh} allowing to calculate the initial size of the hyperrectangular neighborhood

D. *Control parameters calculated*
Initial radius ε_t of tabu balls
Initial radius ε_p of promising balls
Initial size of the hyperrectangular neighborhood

---

ate neighbors in the same way. The reason for using a hyperrectangular neighborhood instead of crown 'balls' is the following: it is mathematically much easier to select a point inside a specified hyperrectangular zone than to select a point inside a specified crown ball. Therefore in the first case, we only have to compare the coordinates of the randomly selected points with the bounds that define the hyperrectangular zone at hand.

Next, we will describe the initialization of some parameters and the tuning of the control parameters. In other words, we give the 'definition' of all the parameters of ECTS. The parameters in part A of Table 5.1 are automatically built by using the parameters fixed at the beginning. The parameters in part B of Table 5.1 are valued in the following way:

(1) the search domain of analytical test functions is set as prescribed in the literature, the initial solution x^* is randomly chosen and checked if it is feasible by the fuzzy rough simulation,

(2) the tabu list is initially empty,

(3) to complete the promising list, the algorithm randomly draws a point. This point is accepted as the center of an initial promising ball, if it does not belong to an already generated ball. In this way the algorithm generates N_p sample points which are uniformly dispersed in the whole space solution S,

(4) the initial threshold for the acceptance of a promising area is taken equal to the average of the objective function values over the previous N_p sample points,

(5) the best point found is taken equal to the best point among the previous N_p,

(6) the number η of neighbors of the current solution investigated at each iteration is set to twice the number of variables, if this number is equal or smaller than five, otherwise η is set to 10;

(7) the maximum number of successive iterations without any detection of a new promising area is equal to twice the number of variables,

(8) the maximum number of successive iterations without any improvement of the objective function value is equal to five times the number of variables,

(9) the maximum number of successive reductions of the hyperrectangular neighborhood and of the radius of tabu balls without any improvement of the objective function value is set to twice the number of variables,

(10) the maximum number of iterations is equal to 50 times the number of variables.

There exist two types of control parameters. Some parameters are chosen by the user. Other ones are deduced from the chosen parameters. The fixed parameters are the length of the tabu list (set to 7, which is the usual tuning advocated by Glover), the length of the promising list (set to 10, like in [325]) and the parameters ρ_t, ρ_p and ρ_{neigh} (set to 100, 50, and 5, respectively). The expressions of ε_t and ε_p are δ/ρ_t and δ/ρ_p respectively, and the initial size of the hyperrectangular neighborhood of the current solution (the more external hyperrectangle) is obtained by dividing δ by the factor ρ_{neigh}.

Diversification. At this stage, the process starts with the initial solution, used as the current one. ECTS generates a specified number of neighbors: one point is selected inside each hyperrectangular zone around the current solution. Each neighbor is accepted only if it does not belong to the tabu list. The best of these neighbors becomes the new current solution, even if it is worse than the previous one. A new promising solution is detected and generated according to the procedure described above. This promising solution defines a new promising area if it does not already belong to a promising ball. If a new promising area is accepted, the worst area of the promising list is replaced by the newly accepted promising area. The use of the promising and tabu lists stimulates the search for solutions far from the starting one and the identified promising areas. The diversification process stops after a given number of successive iterations without any detection of a new promising area. Then the algorithm determines the most promising area among those present in the promising list.

Search for the most promising area. In order to determine the most promising area, we proceed in three steps. First, we calculate the average value of the

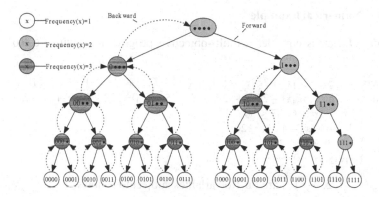

Fig. 5.10 A standard "backtracking"(depth first) branch-and-bound approach

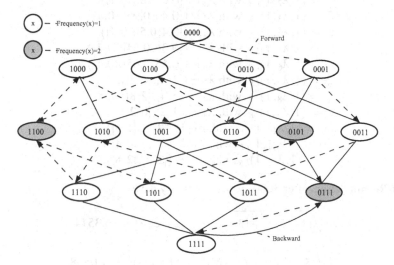

Fig. 5.11 TTS approach

objective function over all the solutions present in the promising list. Secondly, we eliminate all the solutions for which the function value is higher than this average value. Thirdly, we deal with the thus reduced list in the following way. We halve the radius of the tabu balls and the size of the hyperrectangular neighborhood. For each remaining promising solution, we perform the generation of the neighbors and selection of the best. We replace the promising solution by the best neighbor located, yet only if this neighbor is better than that solution. After having scanned the whole promising list, the algorithm removes the least promising solution. This process is reiterated after halving again the above two parameters. It stops when just one promising area remains.

5.3.3.3 Numerical Example

Example 5.4. Let us consider a multi-objective programming with Fu-Ro coefficients.

$$\begin{cases} \max F_1(x,\xi) = 2\xi_1 x_1^2 + 3\xi_2 x_2 - \xi_3 x_3 + \sqrt{(1-\xi_7)^2 + (3-\xi_8)^2 + (2-\xi_9)^2} \\ \max F_2(x,\xi) = 5\xi_4 x_2 - 2\xi_5 x_1 + 2\xi_6 x_3 + \sqrt{(5-\xi_{10})^2 + (2-\xi_{11})^2 + (1-\xi_{12})^2} \\ \text{s.t.} \begin{cases} 5x_1 - 3x_2^2 + 6\sqrt{x_3} \le 50 \\ 4\sqrt{x_1} + 6x_2 - 4.5x_3 \le 20 \\ x_1 + x_2 + x_3 \le 15 \\ x_1, x_2, x_3 \ge 0, \end{cases} \end{cases} \tag{5.34}$$

where $\xi_i (i = 1, 2, \cdots, 12)$ are Fu-Ro variables subject to follows,

$$\begin{aligned}
\xi_1 &= (1, \lambda_1, 1)_{LR}, \text{with } \lambda_1 \vdash ([0.2, 0.5], [0, 1]), \\
\xi_2 &= (1, \lambda_2, 1)_{LR}, \text{with } \lambda_2 \vdash ([0.6, 0.8], [0, 1]), \\
\xi_3 &= (1, \lambda_3, 1)_{LR}, \text{with } \lambda_3 \vdash ([0.45, 0.95], [0, 1]), \\
\xi_4 &= (1, \lambda_4, 1)_{LR}, \text{with } \lambda_4 \vdash ([0.4, 0.5], [0, 1]), \\
\xi_5 &= (1, \lambda_5, 1)_{LR}, \text{with } \lambda_5 \vdash ([0.36, 0.64], [0, 1]), \\
\xi_6 &= (1, \lambda_6, 1)_{LR}, \text{with } \lambda_6 \vdash ([0.55, 0.65], [0, 1]), \\
\xi_7 &= (1, \lambda_7, 1)_{LR}, \text{with } \lambda_7 \vdash ([1, 2], [0, 4]), \\
\xi_8 &= (1, \lambda_8, 1)_{LR}, \text{with } \lambda_8 \vdash ([3, 4], [2, 8]), \\
\xi_9 &= (1, \lambda_9, 1)_{LR}, \text{with } \lambda_9 \vdash ([1, 2], [0, 2]), \\
\xi_{10} &= (1, \lambda_{10}, 1)_{LR}, \text{with } \lambda_{10} \vdash ([2, 3], [1, 5]), \\
\xi_{11} &= (1, \lambda_{11}, 1)_{LR}, \text{with } \lambda_{11} \vdash ([1, 3], [0, 4]), \\
\xi_{12} &= (1, \lambda_{12}, 1)_{LR}, \text{with } \lambda_{12} \vdash ([2, 4], [2, 8]).
\end{aligned}$$

By Fu-Ro simulation, after 3000 cycles, we firstly have

$$E[\sqrt{(1-\xi_7)^2 + (3-\xi_8)^2 + (2-\xi_9)^2}] = 16.3514,$$

$$E[\sqrt{(5-\xi_{10})^2 + (2-\xi_{11})^2 + (1-\xi_{12})^2}] = 7.0568.$$

Next, we apply the tabu search algorithm based on the Fu-Ro simulation to solve the nonlinear programming problem (5.34) with the Fu-Ro parameters.

Step 1. Set the move step $h = 0.5$ and the h neighbor $N(x, h)$ for the present point x is defined as follows,

$$N(x, h) = \left\{ \mathbf{y} \mid \sqrt{(x_1 - y_1)^2 + (x_2 - y_2)^2 + (x_3 - y_3)^2} \le h \right\}.$$

The random move of point x to point $\mathbf{y}$ in its h neighbor along direction s is given by

$$\mathbf{y}_s = x_s + rh,$$

where r is a random number that belongs to $[0,1]$, $s = 1, 2, 3$.

Step 2. Denote $X = \{(x_1, x_2, x_3) | 5x_1 - 3x_2^2 + 6\sqrt{x_3} \leq 50; x_1 + x_2 + x_3 \leq 15; x_i \geq 0, i = 1, 2, 3\}$. Give the step set $H = \{h_1, h_2, \cdots, h_r\}$ and randomly generate a feasible point $x_0 \in X$. One should empty the Tabu list T (the list of inactive steps) at the beginning.

Step 3. For each active neighbor $N(x, h)$ of the present point x, where $h \in H - T$, a feasible random move that satisfies all the constraints in problem (5.34) is to be generated.

Step 4. Construct the single objective function as follows,

$$f(x, \xi) = w_1 \left(2\xi_1 x_1^2 + 3\xi_2 x_2 - \xi_3 x_3 + \sqrt{(1 - \xi_7)^2 + (3 - \xi_8)^2 + (2 - \xi_9)^2} \right)$$
$$+ w_2 \left(5\xi_4 x_2 - 2\xi_5 x_1 + 2\xi_6 x_3 + \sqrt{(5 - \xi_{10})^2 + (2 - \xi_{11})^2 + (1 - \xi_{12})^2} \right),$$

where $w_1 + w_2 = 1$. Compare the $f(x, \xi)$ of the feasible moves with that of the current solution by the fuzzy rough simulation. If an augmenter in new objective function of the feasible moves exists, one should save this feasible move as the updated current one by adding the corresponding step to the Tabu list T and go to the next step; otherwise, go to the next step directly.

Step 5. Stop if the termination criteria are satisfied; other wise, empty T if it is full; then go to Step 3. Here, we set the computation is determined if the better solution doesn't change again.

Table 5.2 Another taxonomy dimension for parallel TS algorithms

w_1	w_2	x_1	x_2	x_3	$f_1(x)$	$f_2(x)$	$f(x)$	Gen
0.1	0.9	0.217	7.861	6.721	219.496	523.976	493.528	1568
0.2	0.8	0.336	7.648	6.712	213.055	511.281	451.638	1470
0.3	0.7	11.356	2.173	1.471	2269.226	-84.484	621.629	1760
0.4	0.6	11.157	2.258	1.573	2194.900	-74.222	833.427	1633
0.5	0.5	11.257	2.151	1.444	2230.700	-84.1427	1073.279	2010
0.6	0.4	11.151	2.244	1.519	2192.769	-75.955	1285.279	1807
0.7	0.3	11.268	2.155	1.532	2234.000	-82.237	1539.189	1834
0.8	0.2	11.205	2.148	1.444	2210.736	-83.245	1752.061	2762
0.9	0.1	11.075	2.288	1.579	2164.917	-71.025	1941.315	1792

5.4 Fu-Ro CCM

Another way to tackle the multi-objective model with Fu-Ro parameters is that we use the chance to measure the uncertainty of fuzzy rough event. In order to compare the degree of occurrence of fuzzy rough events, several kinds of chance measure are introduced.

5.4.1 General Model for Fu-Ro CCM

It has been increasingly recognized that many real-world decision-making problems involve multiple and conflicting objectives which should be considered simultaneously. Fuzzy programming of the multi-objective has been well developed, and as an extension of the fuzzy multi-objective decision-making case, the Fu-Ro multi-objective linear decision-making model is defined as a means for optimizing multiple different objective functions subject to a number of constraints.

Let's introduce the general Fu-Ro CCM to deal with the uncertain model (5.40) as follows.

$$\begin{cases} \max\ [f_1, f_2, \cdots, f_m] \\ \text{s.t.} \begin{cases} Ch\{f_i(x, \xi) \geq f_i\}(\gamma_i) \geq \delta_i,\ i = 1, 2, \cdots, m \\ Ch\{g_r(x, \xi) \leq 0\}(\eta_r) \geq \theta_r,\ r = 1, 2, \cdots, p, \\ x \in X, \end{cases} \end{cases}$$

where Ch is the chance measure of the fuzzy rough events, $\gamma_i, \delta_i, \eta_r, \theta_r$ are the predetermined confidence level, f_i and x_i are the decision variables, $i = 1, 2, \cdots, m$.

According to the definition 5.7 of the primitive chance measure:

$$Ch\{\tilde{\bar{e}}_r^\mathrm{T} x \leq \tilde{\bar{b}}_r\}(\eta_r) \geq \theta_r \Leftrightarrow Appr\{\lambda\,|Pos\{\tilde{\bar{e}}_r(\lambda)^\mathrm{T} x \leq \tilde{\bar{b}}_r(\lambda)\} \geq \theta_r\} \geq \eta_r, \quad (5.35)$$

$$Ch\{\tilde{\bar{c}}_i^\mathrm{T} x \geq f_i\}(\gamma_i) \geq \delta_i \Leftrightarrow Appr\{\lambda\,|Pos\{\tilde{\bar{c}}_i(\lambda)^\mathrm{T} x \geq f_i\} \geq \delta_i\} \geq \gamma_i. \quad (5.36)$$

So we can get the general Fu-Ro CCM,

$$\begin{cases} \max\ [f_1, f_2, \cdots, f_m] \\ \text{s.t.} \begin{cases} Appr\{\lambda\,|Pos\{\tilde{\bar{c}}_i(\lambda)^\mathrm{T} x \geq f_i\} \geq \delta_i\} \geq \gamma_i, & i = 1, 2, \cdots, m \\ Appr\{\lambda\,|Pos\{\tilde{\bar{e}}_r(\lambda)^\mathrm{T} x \leq \tilde{\bar{b}}_r(\lambda)\} \geq \theta_r\} \geq \eta_r, & r = 1, 2, \cdots, p \\ x \geq 0, \end{cases} \end{cases} \quad (5.37)$$

where $\delta_i, \gamma_i, \theta_r, \eta_r \in [0, 1]$ are the predetermined confidence level, $Pos\{\cdot\}$ denotes the possibility of the fuzzy events in $\{\cdot\}$, and $Appr\{\cdot\}$ denotes the approximation degree of the rough events in $\{\cdot\}$.

Definition 5.9. Suppose a feasible solution x^* of the problem (5.43) satisfies

$$Appr\{\lambda\,|Pos\{\tilde{\bar{c}}_i(\lambda)^\mathrm{T} x^* \geq f_i(x^*)\} \geq \delta_i\} \geq \gamma_i, \quad i = 1, 2, \cdots, m,$$

where confidence levels $\delta_i, \gamma_i \in [0, 1]$. x^* is a fuzzy rough efficient solution at $\delta_i - Appr\ \gamma_i - Pos$ levels to the problem (5.40) if and only if there exists no other feasible solution x such that

$$Appr\{\lambda\,|Pos\{\tilde{\bar{c}}_i(\lambda)^\mathrm{T} x \geq f_i(x)\} \geq \delta_i\} \geq \gamma_i, \quad i = 1, 2, \cdots, m,$$

$f_i(x) \geq f_i(x^*)$ for all i and $f_{i_0}(x) \geq f_{i_0}(x^*)$ for at least one $i_0 \in \{1, 2, \cdots, m\}$.

From Definition (5.9), we know that x^* is a fuzzy rough efficient solution at $\delta_i - Appr$ $\gamma_i - Pos$ levels to the problem (5.40) if x^* is a Pareto optimal solution of the problem (5.43).

Remark 5.6. If the fuzzy rough vector $\tilde{\bar{c}}_i$ delegates to rough vector $\bar{c}_i$, then $\bar{c}_i^T x \geq f_i$ is a rough event. For $\lambda \in \Lambda$, $Pos\{\bar{c}_i(\lambda)^T x \geq f_i\} \geq \delta_i$ means $\bar{c}_i(\lambda)^T x \geq f_i$. So,

$$Appr\{\lambda | Pos\{\bar{c}_i(\lambda)^T x \geq f_i\} \geq \delta_i\} \geq \gamma_i$$

is equivalent to $Appr\{\lambda | \bar{c}_i(\lambda)^T x \geq f_i\} \geq \gamma_i, i = 1, 2, \cdots, m$.

If the fuzzy rough vectors $\tilde{\bar{e}}_r$ and $\tilde{\bar{b}}_r$ delegate to rough vectors $\bar{e}_r$ and $\bar{b}_r$ respectively, then the constraint

$$Appr\{\lambda | Pos\{\bar{e}_r(\lambda)^T x \leq \bar{b}_r(\lambda)\} \geq \theta_r\} \geq \eta_r$$

is equivalent to $Appr\{w | \bar{e}_r(\lambda)^T x \leq \bar{b}_r(\lambda)\} \geq \eta_r, r = 1, 2, \cdots, p$. So, the model (5.43) can be rewritten as

$$\begin{cases} \max \ [f_1, f_2, \cdots, f_m] \\ \text{s.t.} \begin{cases} Appr\{\lambda | \bar{c}_i(\lambda)^T x \geq f_i\} \geq \gamma_i, & i = 1, 2, \cdots, m \\ Appr\{\lambda | \bar{e}_r(\lambda)^T x \leq \bar{b}_r(\lambda)\} \geq \eta_r, & r = 1, 2, \cdots, p \\ x \geq 0. \end{cases} \end{cases}$$

Remark 5.7. If the fuzzy rough vector $\tilde{\bar{c}}_i$ delegates to fuzzy vector $\bar{c}_i$, then $Pos\{\bar{c}_i^T x \geq f_i\} \geq \delta_i$ is a crisp event. In order to satisfy $t_i := Appr\{w | Pos\{\bar{c}_i^T x \geq f_i\} \geq \delta_i\} \geq \gamma_i$, the trust t_i should be 1.

So the constraint

$$Appr\{w | Pos\{\bar{c}_i^T x \geq f_i\} \geq \delta_i\} = 1 \geq \gamma_i,$$

is equivalent to $Pos\{\bar{c}_i^T x \geq f_i\} \geq \delta_i, i = 1, 2, \cdots, m$.

And similarly, When the fuzzy rough vectors $\tilde{\bar{e}}_r$ and $\tilde{\bar{b}}_r$ delegate to the fuzzy vector $\bar{e}_r$ and $\bar{b}_r$ respectively, the constraint

$$Appr\{w | Pos\{\bar{e}_r^T x \leq \bar{b}_j\} \geq \theta_r\} \geq \eta_r$$

is equivalent to $Pos\{\bar{e}_r^T x \leq \bar{b}_r\} \geq \theta_r, r = 1, 2, \cdots, p$. So the model (5.43) is equivalent to

$$\begin{cases} \max \ [f_1, f_2, \cdots, f_m] \\ \text{s.t.} \begin{cases} Pos\{\bar{c}_i^T x \geq f_i\} \geq \delta_i, & i = 1, 2, \cdots, m \\ Pos\{\bar{e}_r^T x \leq \bar{b}_r\} \geq \theta_r, & r = 1, 2, \cdots, p \\ x \geq 0. \end{cases} \end{cases}$$

This is coincident to the fuzzy programming introduced in section 2.

Also by the definition 5.7, there are two other kinds of fuzzy rough multi-objective chance-constrained linear decision making models (5.38, 5.38),

$$
\begin{cases}
\max\ [f_1, f_2, \cdots, f_m] \\
\text{s.t.}
\begin{cases}
Appr\{\lambda\,|\,Nec\{\bar{\tilde{c}}_i(\lambda)^\mathrm{T}x \geq f_i\} \geq \delta_i\} \geq \gamma_i, & i = 1,2,\cdots,m \\
Appr\{\lambda\,|\,Nec\{\bar{\tilde{e}}_r(\lambda)^\mathrm{T}x \leq \bar{\tilde{b}}_r(\lambda)\} \geq \theta_r\} \geq \eta_r, & r = 1,2,\cdots,p \\
x \geq 0,
\end{cases}
\end{cases}
\tag{5.38}
$$

$$
\begin{cases}
\max\ [f_1, f_2, \cdots, f_m] \\
\text{s.t.}
\begin{cases}
Appr\{\lambda\,|\,Cr\{\bar{\tilde{c}}_i(\lambda)^\mathrm{T}x \geq f_i\} \geq \delta_i\} \geq \gamma_i, & i = 1,2,\cdots,m \\
Appr\{\lambda\,|\,Cr\{\bar{\tilde{e}}_r(\lambda)^\mathrm{T}x \leq \bar{\tilde{b}}_r(\lambda)\} \geq \theta_r\} \geq \eta_r, & r = 1,2,\cdots,p \\
x \geq 0,
\end{cases}
\end{cases}
\tag{5.39}
$$

where $\delta_i, \gamma_i, \theta_r, \eta_r \in [0,1]$ are the predetermined confidence levels, $Nec\{\cdot\}$ and $Cr\{\cdot\}$ denote the necessary and the credibility of the fuzzy events in $\{\cdot\}$ respectively, and $Appr\{\cdot\}$ denotes the approximations of the rough events in $\{\cdot\}$.

For simpleness, the parameters $\delta, \gamma, \theta, \eta$ can be the same confidence level, i.e. $\delta_i = \delta, \gamma_i = \gamma, \theta_r = \theta, \eta_r = \eta, i = 1,2,\cdots,m, r = 1,2,\cdots,p.$

5.4.2 Linear Fu-Ro CCM and Fuzzy Goal Method

So let's consider the multi-objective linear programming problem with Fu-Ro coefficients:

$$
\begin{cases}
\max\ [\bar{\tilde{c}}_1^\mathrm{T}x, \bar{\tilde{c}}_2^\mathrm{T}x, \cdots, \bar{\tilde{c}}_m^\mathrm{T}x] \\
\text{s.t.}
\begin{cases}
\bar{\tilde{e}}_r^\mathrm{T}x \leq \bar{\tilde{b}}_r, & r = 1,2,\cdots,p \\
x \geq 0.
\end{cases}
\end{cases}
\tag{5.40}
$$

According to the definition 5.7 of the primitive chance measure:

$$
Ch\{\bar{\tilde{e}}_r^\mathrm{T}x \leq \bar{\tilde{b}}_r\}(\eta_r) \geq \theta_r \Leftrightarrow Appr\{\lambda\,|\,Pos\{\bar{\tilde{e}}_r(\lambda)^\mathrm{T}x \leq \bar{\tilde{b}}_r(\lambda)\} \geq \theta_r\} \geq \eta_r, \tag{5.41}
$$

$$
Ch\{\bar{\tilde{c}}_i^\mathrm{T}x \geq f_i\}(\gamma_i) \geq \delta_i \Leftrightarrow Appr\{\lambda\,|\,Pos\{\bar{\tilde{c}}_i(\lambda)^\mathrm{T}x \geq f_i\} \geq \delta_i\} \geq \gamma_i. \tag{5.42}
$$

So we can get linear Fu-Ro CCM,

$$
\begin{cases}
\max\ [f_1, f_2, \cdots, f_m] \\
\text{s.t.}
\begin{cases}
Appr\{\lambda\,|\,Pos\{\bar{\tilde{c}}_i(\lambda)^\mathrm{T}x \geq f_i\} \geq \delta_i\} \geq \gamma_i, & i = 1,2,\cdots,m \\
Appr\{\lambda\,|\,Pos\{\bar{\tilde{e}}_r(\lambda)^\mathrm{T}x \leq \bar{\tilde{b}}_r(\lambda)\} \geq \theta_r\} \geq \eta_r, & r = 1,2,\cdots,p \\
x \geq 0,
\end{cases}
\end{cases}
\tag{5.43}
$$

where $\delta_i, \gamma_i, \theta_r, \eta_r \in [0,1]$ are the predetermined confidence levels, $Pos\{\cdot\}$ denotes the possibility of the fuzzy events in $\{\cdot\}$, and $Appr\{\cdot\}$ denotes the approximation degree of the rough events in $\{\cdot\}$.

5.4.2.1 Crisp Equivalent Model

We introduce the $Appr - Pos$ and $Appr - Nec$ crisp equivalent models for the chance-constrained model, respectively.

Appr-Pos constrained multi-objective linearity model

One way of solving the problem (5.43) is to convert it into its crisp equivalent.

Theorem 5.7. *Assume that $\tilde{\bar{c}}_{ij}$ is a Fu-Ro variable, for any $\lambda \in \Lambda$, the fuzzy variable $\tilde{\bar{c}}_{ij}(\lambda)$ is characterized by the following membership function*

$$\mu_{\tilde{c}_{ij}(\lambda)}(t) = \begin{cases} L\left(\frac{c_{ij}(\lambda)-t}{\alpha_{ij}^c}\right), & t \leq c_{ij}(\lambda), \alpha_{ij}^c > 0 \\ R\left(\frac{t-c_{ij}(\lambda)}{\beta_{ij}^c}\right), & t \geq c_{ij}(\lambda), \beta_{ij}^c > 0 \end{cases} \quad \lambda \in \Lambda, \tag{5.44}$$

where $\alpha_{ij}^c, \beta_{ij}^c$ are positive numbers expressing the left and right spread of $\tilde{c}_{ij}(\lambda)$, reference function $L, R : [0,1] \rightarrow [0,1]$ with $L(1) = R(1) = 0$, and $L(0) = R(0) = 1$ are non-increasing, continuous function. And $(c_{ij}(\lambda))_{n\times 1} = (c_{i1}(\lambda), c_{i2}(\lambda), \cdots, c_{in}(\lambda))^T$ is a rough vector. It follows that $c_i(\lambda)^T x = ([a,b], [c,d])$ (where $c \leq a < b \leq d$) is a rough variable and characterized by the following trust measure function,

$$Appr\{c_i(\lambda)^T x \geq t\} = \begin{cases} 0, & \text{if } d \leq t \\ \frac{d-t}{2(d-c)}, & \text{if } b \leq t \leq d \\ \frac{1}{2}\left(\frac{d-t}{d-c} + \frac{b-t}{b-a}\right), & \text{if } a \leq t < b \\ \frac{1}{2}\left(\frac{d-t}{d-c} + 1\right), & \text{if } c \leq t \leq a \\ 1, & \text{if } t \leq c. \end{cases} \tag{5.45}$$

Then we have $Appr\{\lambda | Pos\{\tilde{\bar{c}}_i(\lambda)^T x \geq f_i\} \geq \delta_i\} \geq \gamma_i$ if and only if

$$\begin{cases} f_i \leq d - 2\gamma_i(d-c) + R^{-1}(\delta_i)\beta_i^{cT}x, & \text{if } b \leq f_i - R^{-1}(\delta_i)\beta_i^{cT}x \leq d \\ f_i \leq \frac{d(b-a)+b(d-c)-2\gamma_i(d-c)(b-a)}{d-c+b-a} + R^{-1}(\delta_i)\beta_i^{cT}x, & \text{if } a \leq f_i - R^{-1}(\delta_i)\beta_i^{cT}x < b \\ f_i \leq d - (d-c)(2\gamma_i - 1) + R^{-1}(\delta_i)\beta_i^{cT}x, & \text{if } c \leq f_i - R^{-1}(\delta_i)\beta_i^{cT}x \leq a \\ f_i \leq c + R^{-1}(\delta_i)\beta_i^{cT}x, & \text{if } f_i - R^{-1}(\delta_i)\beta_i^{cT}x \leq c, \end{cases} \tag{5.46}$$

where $\gamma_i, \delta_i \in [0,1]$ are predetermined confidence levels.

Proof. From the assumption we know that $c_i(\lambda) = (c_{i1}(\lambda), c_{i2}(\lambda), \cdots, c_{in}(\lambda))^T$ and $c_{ij}(\lambda)$ is a rough variable . Let $c_{ij}(\lambda) = ([a_{ij}, b_{ij}], [c_{ij}, d_{ij}])$ and $x = (x_1, x_2, \cdots, x_n)^T$ then

$$x_j c_{ij}(\lambda) = ([x_j a_{ij}, x_j b_{ij}], [x_j c_{ij}, x_j d_{ij}]),$$

$$c_i(\lambda)^T x = \sum_{j=1}^n c_{ij}(\lambda)x_j = \sum_{j=1}^n ([x_j a_{ij}, x_j b_{ij}], [x_j c_{ij}, x_j d_{ij}])$$
$$= ([\sum_{j=1}^n a_{ij}x_j, \sum_{j=1}^n a_{ij}x_j], [\sum_{j=1}^n c_{ij}x_j, \sum_{j=1}^n d_{ij}x_j]).$$

Therefore, $c_i(\lambda)^T x$ is also a rough variable. Now we can assume that

$$a = \sum_{j=1}^n a_{ij}x_j, \ b = \sum_{j=1}^n a_{ij}x_j,$$
$$c = \sum_{j=1}^n c_{ij}x_j, \ d = \sum_{j=1}^n d_{ij}x_j.$$

then $c_i(\lambda)^T x = ([a,b], [c,d])$.

Moreover, we know that $\tilde{c}_{ij}(\lambda)$ is a fuzzy number with the membership function $\mu_{\tilde{c}_{ij}(\lambda)}(t)$ for given $\lambda \in \Lambda$. It follows from the extension principle [22] that the fuzzy number $\widehat{c}_i(\lambda)^T x$ is characterized by the membership function in the following

$$\mu_{\tilde{c}_i(\lambda)^T x}(r) = \begin{cases} L(\frac{c_i(\lambda)^T - r}{\alpha_i^{cT} x}), & r \le c_i(\lambda)^T x \\ R(\frac{r - c_i(\lambda)^T x}{\beta_i^{cT} x}), & r \ge c_i(\lambda)^T x \end{cases} \qquad i = 1,2,\ldots,m.$$

By Lemma 2.2, we have that

$$Pos\{\tilde{c}_i(\lambda)^T x \ge f_i\} \ge \delta_i \Leftrightarrow c_i(\lambda)^T x + R^{-1}(\delta_i)\beta_i^{cT} x \ge f_i, \quad i = 1,2,\ldots,m.$$

For the given confidence level $\delta_i \in [0,1]$, we have

$$Appr\{\lambda | Pos\{\tilde{c}_i(\lambda)^T x \ge f_i\} \ge \delta_i\} \ge \gamma_i$$
$$\Leftrightarrow Appr\{\lambda | c_i(\lambda)^T x \ge f_i - R^{-1}(\delta_i)\beta_i^{cT} x\} \ge \gamma_i$$
$$\Leftrightarrow \gamma_i \le \begin{cases} \frac{d - f_i + R^{-1}(\delta_i)\beta_i^{cT} x}{2(d-c)}, & \text{if } b \le f_i - R^{-1}(\delta_i)\beta_i^{cT} x \le d \\ \frac{1}{2}(\frac{d - f_i + R^{-1}(\delta_i)\beta_i^{cT} x}{d-c} + \frac{b - f_i + R^{-1}(\delta_i)\beta_i^{cT} x}{b-a}), & \text{if } a \le f_i - R^{-1}(\delta_i)\beta_i^{cT} x < b \\ \frac{1}{2}(\frac{d - f_i + R^{-1}(\delta_i)\beta_i^{cT} x}{d-c} + 1), & \text{if } c \le f_i - R^{-1}(\delta_i)\beta_i^{cT} x \le a \\ 1, & \text{if } f_i - R^{-1}(\delta_i)\beta_i^{cT} x \le c \end{cases}$$

$$\Leftrightarrow \begin{cases} f_i \le d - 2\gamma_i(d-c) + R^{-1}(\delta_i)\beta_i^{cT} x, & \text{if } b \le f_i - R^{-1}(\delta_i)\beta_i^{cT} x \le d \\ f_i \le \frac{d(b-a) + b(d-c) - 2\gamma_i(d-c)(b-a)}{d-c+b-a} + R^{-1}(\delta_i)\beta_i^{cT} x, & \text{if } a \le f_i - R^{-1}(\delta_i)\beta_i^{cT} x < b \\ f_i \le d - (d-c)(2\gamma_i - 1) + R^{-1}(\delta_i)\beta_i^{cT} x, & \text{if } c \le f_i - R^{-1}(\delta_i)\beta_i^{cT} x \le a \\ f_i \le c + R^{-1}(\delta_i)\beta_i^{cT} x, & \text{if } f_i - R^{-1}(\delta_i)\beta_i^{cT} x \le c. \end{cases}$$

This completes the proof. $\qquad\qquad\qquad\qquad\qquad\qquad\qquad\qquad\qquad\qquad\square$

Theorem 5.8. *Suppose that $\tilde{e}_{rj}, \tilde{b}_r$ are fuzzy rough variables, for any $\lambda \in \Lambda$, fuzzy variables $\tilde{e}_{rj}(\lambda), \tilde{b}_r(\lambda)$ are characterized by the membership function in the following*

$$\mu_{\tilde{e}_{rj}(\lambda)}(t) = \begin{cases} L(\frac{e_{rj}(\lambda) - t}{\alpha_{rj}^e}), & t \le e_{rj}(\lambda), \alpha_{rj}^e > 0 \\ R(\frac{t - e_{rj}(\lambda)}{\beta_{rj}^m}), & t \ge e_{rj}(\lambda), \beta_{rj}^e > 0 \end{cases} \quad \lambda \in \Lambda \qquad (5.47)$$

and

$$\mu_{\tilde{b}_r(\lambda)}(t) = \begin{cases} L(\frac{b_r(\lambda) - t}{\alpha_r^b}), & t \le b_r(\lambda), \alpha_r^b > 0 \\ R(\frac{t - b_r(\lambda)}{\beta_r^b}), & t \ge b_r(\lambda), \beta_r^b > 0 \end{cases} \quad \lambda \in \Lambda, \qquad (5.48)$$

where $\alpha_{rj}^e, \beta_{rj}^e$ are positive numbers expressing the left and right spread of $\tilde{e}_{rj}(\lambda)$, α_r^b, β_r^b are the left and right spread of $\tilde{b}_r(\lambda)$, and reference functions $L,R :$ $[0,1] \to [0,1]$ with $L(1) = R(1) = 0$, and $L(0) = R(0) = 1$ are non-increasing, continuous functions. And $(e_{rj}(\lambda))_{n\times 1} = (e_{r1}(\lambda), e_{r2}(\lambda), \cdots, e_{rn}(\lambda))^T$ is a rough vector, $e_{rj}(\lambda), b_r(\lambda)$ are rough variables, $r = 1, 2, \cdots, p$, $j = 1, 2, \cdots, n$. By Proposition 5.7, we have $e_r(\lambda)^T x, b_r(\lambda)$ are rough variables, then $e_r(\lambda)^T x - b_r(\lambda) = [(a, b), (c, d)](c \leq a < b \leq d)$ is also a rough variable. We assume that it is characterized by the following trust measure function

$$Appr\{e_r(\lambda)^T x - b_r(\lambda) \leq t\} = \begin{cases} 0, & \text{if } t \leq c \\ \frac{t-c}{2(d-c)}, & \text{if } c \leq t \leq a \\ \frac{1}{2}(\frac{t-c}{d-c} + \frac{t-a}{b-a}), & \text{if } a \leq t < b \\ \frac{1}{2}(\frac{t-c}{d-c} + 1), & \text{if } b \leq t \leq d \\ 1, & \text{if } d \leq t. \end{cases} \quad (5.49)$$

Then, we have that $Appr\{\lambda | Pos\{\tilde{\bar{e}}_r(\lambda)^T x \leq \tilde{\bar{b}}_r(\lambda)\} \geq \theta_r\} \geq \eta_r$ if and only if

$$\begin{cases} W \geq c + 2(d-c)\eta_r, & \text{if } c \leq W \leq a \\ W \geq \frac{2\eta_r(d-c)(b-a)+c(b-a)+a(d-c)}{b-a+d-c}, & \text{if } a \leq W < b \\ W \geq (2\eta_r - 1)(d-c) + c, & \text{if } b \leq W \leq d \\ W \geq d, & \text{if } d \leq W, \end{cases} \quad (5.50)$$

where $W = R^{-1}(\theta_r)\beta_r^b + L^{-1}(\theta_k)\alpha_r^{eT} x$.

Proof. From the assumption, we know

$$Pos\{\tilde{\bar{e}}_r(\lambda)^T x \leq \tilde{\bar{b}}_r(\lambda)\} \geq \theta_r \Leftrightarrow b_r(\lambda) + R^{-1}(\theta_r)\beta_r^b \geq e_r(\lambda)^T x - L^{-1}(\theta_r)\alpha_r^{eT} x.$$

Since $e_r(\lambda)^T x - b_r(\lambda) = [(a, b), (c, d)]$, for given confidence levels $\theta_r, \eta_r \in [0, 1]$, we have that,

$$\begin{aligned} & Appr\{\lambda | Pos\{\tilde{\bar{e}}_r(\lambda)^T x \leq \tilde{\bar{b}}_r(\lambda)\} \geq \theta_r\} \geq \eta_r \\ \Leftrightarrow & Appr\{\lambda | e_r(\lambda)^T x - b_r(\lambda) \leq R^{-1}(\theta_r)\beta_r^b + L^{-1}(\theta_r)\alpha_r^{eT} x\} \geq \eta_r \\ \Leftrightarrow & \eta_r \leq \begin{cases} \frac{W-c}{2(d-c)}, & \text{if } c \leq W \leq a. \\ \frac{1}{2}(\frac{W-c}{d-c} + \frac{W-a}{b-a}), & \text{if } a \leq W < b \\ \frac{1}{2}(\frac{W-c}{d-c} + 1), & \text{if } b \leq W \leq d \\ 1, & \text{if } W \geq d \end{cases} \\ \Leftrightarrow & \begin{cases} W \geq c + 2(d-c)\eta_r, & \text{if } c \leq W \leq a \\ W \geq \frac{2\eta_r(d-c)(b-a)+c(b-a)+a(d-c)}{b-a+d-c}, & \text{if } a \leq W < b \\ W \geq (2\eta_r - 1)(d-c) + c, & \text{if } b \leq W \leq d \\ W \geq d, & \text{if } d \leq W, \end{cases} \end{aligned}$$

where $W = R^{-1}(\theta_r)\beta_r^b + L^{-1}(\theta_r)\alpha_r^{eT} x$.

This completes the proof.   □

From Propositions 5.7 and 5.8, we know that the problem (5.8) is equivalent to the following multi-objective programming problems,

$$\begin{cases} \max \ [f_1, f_2, \cdots, f_m] \\ \text{s.t.} \begin{cases} f_i \leq d - 2\gamma_i(d-c) + R^{-1}(\delta_i)\beta_i^{cT}x, \\ \qquad\qquad \text{if } b \leq f_i - R^{-1}(\delta_i)\beta_i^{cT}x \leq d \\ f_i \leq \frac{d(b-a)+b(d-c)-2\alpha_i(d-c)(b-a)}{d-c+b-a} + R^{-1}(\delta_i)\beta_i^{cT}x, \\ \qquad\qquad \text{if } a \leq f_i - R^{-1}(\delta_i)\beta_i^{cT}x < b \\ f_i \leq d - (d-c)(2\gamma_i - 1) + R^{-1}(\delta_i)\beta_i^{cT}x, \\ \qquad\qquad \text{if } c < f_i - R^{-1}(\delta_i)\beta_i^{cT}x \leq a \\ f_i \leq c + R^{-1}(\delta_i)\beta_i^{cT}x, \\ \qquad\qquad \text{if } f_i - R^{-1}(\delta_i)\beta_i^{cT}x \leq c \\ W \geq c + 2(d-c)\eta_r, \text{ if } c \leq W \leq a \\ W \geq \frac{2\eta_r(d-c)(b-a)+c(b-a)+a(d-c)}{b-a+d-c}, \text{ if } a \leq W < b \\ W \geq (2\eta_r - 1)(d-c) + c, \text{ if } b \leq W \leq d \\ W \geq d, \text{ if } d \leq W \\ x \geq 0, \end{cases} \end{cases} \qquad (5.51)$$

where $W = R^{-1}(\theta_r)\beta_r^b + L^{-1}(\theta_r)\alpha_r^{eT}x$.

Appr-Nec constrained multi-objective linearity model

Similar to the $Appr - Pos$ constrained multi-objective linearity model, we assume that $\tilde{\bar{c}}_{ij}$, $\tilde{\bar{e}}_{rj}$ and $\tilde{\bar{b}}_r$ are fuzzy rough variables, we give the following two theorems to transform the chance-constrained model (5.43) into its crisp model based on $Appr - Nec$ if the decision maker is comparatively pessimistic.

Theorem 5.9. *Assume that $\tilde{\bar{c}}_{ij}$ is a fuzzy rough variable, for any $\lambda \in \Lambda$, the fuzzy variable $\tilde{\bar{c}}_{ij}(\lambda)$ is characterized by the following membership function*

$$\mu_{\tilde{c}_{ij}(\lambda)}(t) = \begin{cases} L\left(\frac{c_{ij}(\lambda)-t}{\alpha_{ij}^c}\right), t \leq c_{ij}(\lambda), \alpha_{ij}^c > 0 \\ R\left(\frac{t-c_{ij}(\lambda)}{\beta_{ij}^c}\right), t \geq c_{ij}(\lambda), \beta_{ij}^c > 0 \end{cases} \quad \lambda \in \Lambda, \qquad (5.52)$$

where $\alpha_{ij}^c, \beta_{ij}^c$ are positive numbers expressing the left and right spread of $\tilde{c}_{ij}(\lambda)$, reference function $L, R : [0,1] \rightarrow [0,1]$ with $L(1) = R(1) = 0$, and $L(0) = R(0) = 1$ are non-increasing, continuous function. And $(c_{ij}(\lambda))_{n\times 1} = (c_{i1}(\lambda), c_{i2}(\lambda), \cdots, c_{in}(\lambda))^T$ is a rough vector. It follows that $c_i(\lambda)^T x = ([a,b],[c,d])$ (where $c \leq a < b \leq d$) is a rough variable and characterized by the following trust measure function,

$$Appr\{c_i(\lambda)^T x \geq t\} = \begin{cases} 0, & \text{if } d \leq t \\ \frac{d-t}{2(d-c)}, & \text{if } b \leq t \leq d \\ \frac{1}{2}(\frac{d-t}{d-c} + \frac{b-t}{b-a}), & \text{if } a \leq t < b \\ \frac{1}{2}(\frac{d-t}{d-c} + 1), & \text{if } c \leq t \leq a \\ 1, & \text{if } t \leq c. \end{cases} \qquad (5.53)$$

Then we have $Appr\{\lambda \,|\, Nec\{\tilde{\tilde{c}}_i(\lambda)^{\mathrm{T}} x \geq f_i\} \geq \delta_i\} \geq \gamma_i$ if and only if

$$
\begin{cases}
f_i \leq d - 2\gamma_i(d-c) - L^{-1}(1-\delta_i)\alpha_i^{cT} x, \\
\qquad\qquad if\ b \leq f_i + L^{-1}(1-\delta_i)\alpha_i^{cT} x \leq d \\
f_i \leq \frac{d(b-a)+b(d-c)-2\gamma_i(d-c)(b-a)}{d-c+b-a} - L^{-1}(1-\delta_i)\alpha_i^{cT} x, \\
\qquad\qquad if\ a \leq f_i + L^{-1}(1-\delta_i)\alpha_i^{cT} x < b \\
f_i \leq d - (d-c)(2\gamma_i - 1) - L^{-1}(1-\delta_i)\alpha_i^{cT} x, \\
\qquad\qquad if\ c \leq f_i + L^{-1}(1-\delta_i)\alpha_i^{cT} x \leq a \\
f_i \leq c - L^{-1}(1-\delta_i)\alpha_i^{cT} x, \\
\qquad\qquad if\ f_i + L^{-1}(1-\delta_i)\alpha_i^{cT} x \leq c,
\end{cases}
\tag{5.54}
$$

where $\gamma_i, \delta_i \in [0,1]$ are predetermined confidence levels.

Proof. From the assumption we know that $c_i(\lambda) = (c_{i1}(\lambda), c_{i2}(\lambda), \cdots, c_{in}(\lambda))^{\mathrm{T}}$ and $c_{ij}(\lambda)$ is a rough variable . Let $c_{ij}(\lambda) = ([a_{ij}, b_{ij}], [c_{ij}, d_{ij}])$ and $x = (x_1, x_2, \cdots, x_n)^{\mathrm{T}}$ then

$$
x_j c_{ij}(\lambda) = ([x_j a_{ij}, x_j b_{ij}], [x_j c_{ij}, x_j d_{ij}]),
$$

$$
\begin{aligned}
c_i(\lambda)^{\mathrm{T}} x &= \textstyle\sum_{j=1}^{n} c_{ij}(\lambda) x_j = \sum_{j=1}^{n} ([x_j a_{ij}, x_j b_{ij}], [x_j c_{ij}, x_j d_{ij}]) \\
&= ([\textstyle\sum_{j=1}^{n} a_{ij} x_j, \sum_{j=1}^{n} a_{ij} x_j], [\sum_{j=1}^{n} c_{ij} x_j, \sum_{j=1}^{n} d_{ij} x_j]).
\end{aligned}
$$

Therefore, $c_i(\lambda)^{\mathrm{T}} x$ is also a rough variable. Now we can assume that

$$
\begin{aligned}
a &= \textstyle\sum_{j=1}^{n} a_{ij} x_j, \ b = \sum_{j=1}^{n} a_{ij} x_j, \\
c &= \textstyle\sum_{j=1}^{n} c_{ij} x_j, \ d = \sum_{j=1}^{n} d_{ij} x_j.
\end{aligned}
$$

then $c_i(\lambda)^{\mathrm{T}} x = ([a,b],[c,d])$.

Moreover, we know that $\tilde{\tilde{c}}_{ij}(\lambda)$ is a fuzzy number with the membership function $\mu_{\tilde{c}_{ij}(\lambda)}(t)$ for given $\lambda \in \Lambda$. It follows from the extension principle that the fuzzy number $\widehat{c}_i(\lambda)^{\mathrm{T}} x$ is characterized by the membership function in the following

$$
\mu_{\tilde{c}_i(\lambda)^{\mathrm{T}} x}(r) =
\begin{cases}
L(\frac{c_i(\lambda)^{\mathrm{T}} - r}{\alpha_i^{cT} x}), & r \leq c_i(\lambda)^{\mathrm{T}} x \\
R(\frac{r - c_i(\lambda)^{\mathrm{T}} x}{\beta_i^{cT} x}), & r \geq c_i(\lambda)^{\mathrm{T}} x
\end{cases}
\qquad i = 1, 2, \ldots, m.
$$

By Lemma 2.2, we have that

$$
Nec\{\tilde{\tilde{c}}_i(\lambda)^{\mathrm{T}} x \geq f_i\} \geq \delta_i \Leftrightarrow c_i(\lambda)^{\mathrm{T}} x - L^{-1}(1-\delta_i)\alpha_i^{cT} x \geq f_i, \quad i = 1, 2, \ldots, m.
$$

For the given confidence level $\delta_i \in [0,1]$, we have

$$Appr\{\lambda | Nec\{\tilde{\tilde{c}}_i(\lambda)^T x \geq f_i\} \geq \delta_i\} \geq \gamma_i$$
$$\Leftrightarrow Appr\{\lambda | c_i(\lambda)^T x \geq f_i + L^{-1}(1-\delta_i)\alpha_i^{cT} x\} \geq \gamma_i$$

$$\Leftrightarrow \gamma_i \leq \begin{cases} \frac{d - f_i - L^{-1}(1-\delta_i)\alpha_i^{cT} x}{2(d-c)}, \\ \qquad\qquad\qquad \text{if } b \leq f_i + L^{-1}(1-\delta_i)\alpha_i^{cT} x \leq d \\ \frac{1}{2}\left(\frac{d - f_i - L^{-1}(1-\delta_i)\alpha_i^{cT} x}{d-c} + \frac{b - f_i - L^{-1}(1-\delta_i)\alpha_i^{cT} x}{b-a}\right), \\ \qquad\qquad\qquad \text{if } a \leq f_i + L^{-1}(1-\delta_i)\alpha_i^{cT} x < b \\ \frac{1}{2}\left(\frac{d - f_i - L^{-1}(1-\delta_i)\alpha_i^{cT} x}{d-c} + 1\right), \\ \qquad\qquad\qquad \text{if } c \leq f_i + L^{-1}(1-\delta_i)\alpha_i^{cT} x \leq a \\ 1, \\ \qquad\qquad\qquad \text{if } f_i + L^{-1}(1-\delta_i)\alpha_i^{cT} x \leq c \end{cases}$$

$$\Leftrightarrow \begin{cases} f_i \leq d - 2\gamma_i(d-c) - L^{-1}(1-\delta_i)\alpha_i^{cT} x, \\ \qquad\qquad \text{if } b \leq f_i + L^{-1}(1-\delta_i)\alpha_i^{cT} x \leq d \\ f_i \leq \frac{d(b-a) + b(d-c) - 2\gamma_i(d-c)(b-a)}{d-c+b-a} - L^{-1}(1-\delta_i)\alpha_i^{cT} x, \\ \qquad\qquad \text{if } a \leq f_i + L^{-1}(1-\delta_i)\alpha_i^{cT} x < b \\ f_i \leq d - (d-c)(2\gamma_i - 1) - L^{-1}(1-\delta_i)\alpha_i^{cT} x, \\ \qquad\qquad \text{if } c \leq f_i + L^{-1}(1-\delta_i)\alpha_i^{cT} x \leq a \\ f_i \leq c - L^{-1}(1-\delta_i)\alpha_i^{cT} x, \\ \qquad\qquad \text{if } f_i + L^{-1}(1-\delta_i)\alpha_i^{cT} x \leq c. \end{cases}$$

This completes the proof.   □

Theorem 5.10. *Suppose that* $\tilde{\tilde{e}}_{rj}, \tilde{\tilde{b}}_r$ *are fuzzy rough variables, for any* $\lambda \in \Lambda$, *fuzzy variables* $\tilde{\tilde{e}}_{rj}(\lambda), \tilde{\tilde{b}}_r(\lambda)$ *are characterized by the membership function in the following*

$$\mu_{\tilde{\tilde{e}}_{rj}(\lambda)}(t) = \begin{cases} L(\frac{e_{rj}(\lambda)-t}{\alpha_{rj}^e}), & t \leq e_{rj}(\lambda), \alpha_{rj}^e > 0 \\ R(\frac{t - e_{rj}(\lambda)}{\beta_{rj}^m}), & t \geq e_{rj}(\lambda), \beta_{rj}^e > 0 \end{cases} \quad \lambda \in \Lambda \qquad (5.55)$$

and

$$\mu_{\tilde{\tilde{b}}_r(\lambda)}(t) = \begin{cases} L(\frac{b_r(\lambda)-t}{\alpha_r^b}), & t \leq b_r(\lambda), \alpha_r^b > 0 \\ R(\frac{t - b_r(\lambda)}{\beta_r^b}), & t \geq b_r(\lambda), \beta_r^b > 0 \end{cases} \quad \lambda \in \Lambda, \qquad (5.56)$$

where $\alpha_{rj}^e, \beta_{rj}^e$ *are positive numbers expressing the left and right spread of* $\tilde{\tilde{e}}_{rj}(\lambda)$, α_r^b, β_r^b *are the left and right spread of* $\tilde{\tilde{b}}_r(\lambda)$, *and reference functions* $L, R : [0,1] \rightarrow [0,1]$ *with* $L(1) = R(1) = 0$, *and* $L(0) = R(0) = 1$ *are non-increasing, continuous functions. And* $(e_{rj}(\lambda))_{n \times 1} = (e_{r1}(\lambda), e_{r2}(\lambda), \cdots, e_{rn}(\lambda))^T$ *is a rough vector,* $e_{rj}(\lambda), b_r(\lambda)$ *are rough variables,* $r = 1, 2, \cdots, p, j = 1, 2, \cdots, n$.

By Theorem 5.7, we have $e_r(\lambda)^T x, b_r(\lambda)$ are rough variables, then $e_r(\lambda)^T x - b_r(\lambda) = [(a,b),(c,d)](c \leq a < b \leq d)$ is also a rough variable. We assume that it is characterized by the following trust measure function

$$Appr\{e_r(\lambda)^T x - b_r(\lambda) \leq t\} = \begin{cases} 0, & \text{if } t \leq c \\ \frac{t-c}{2(d-c)}, & \text{if } c \leq t \leq a \\ \frac{1}{2}(\frac{t-c}{d-c} + \frac{t-a}{b-a}), & \text{if } a \leq t < b \\ \frac{1}{2}(\frac{t-c}{d-c} + 1), & \text{if } b \leq t \leq d \\ 1, & \text{if } d \leq t. \end{cases} \quad (5.57)$$

Then, we have that $Appr\{\lambda | Nec\{\tilde{\tilde{e}}_r(\lambda)^T x \leq \tilde{\tilde{b}}_r(\lambda)\} \geq \theta_r\} \geq \eta_r$ if and only if

$$\begin{cases} W \geq c + 2(d-c)\eta_r, & \text{if } c \leq W \leq a \\ W \geq \frac{2\eta_r(d-c)(b-a)+c(b-a)+a(d-c)}{b-a+d-c}, & \text{if } a \leq W < b \\ W \geq (2\eta_r - 1)(d-c) + c, & \text{if } b \leq W \leq d \\ W \geq d, & \text{if } d \leq W, \end{cases} \quad (5.58)$$

where $W = -R^{-1}(\theta_r)\beta_r^{eT} x - L^{-1}(1-\theta_r)\alpha_r^b$.

Proof. From the assumption, we know

$$Nec\{\tilde{\tilde{e}}_r(\lambda)^T x \leq \tilde{\tilde{b}}_r(\lambda)\} \geq \theta_r \Leftrightarrow b_r(\lambda) - L^{-1}(1-\theta_r)\alpha_r^b \geq e_r(\lambda)^T x + R^{-1}(\theta_r)\beta_r^{eT} x.$$

Since $e_r(\lambda)^T x - b_r(\lambda) = [(a,b),(c,d)]$, for given confidence levels $\theta_r, \eta_r \in [0,1]$, we have that,

$$\begin{aligned} &Appr\{\lambda | Nec\{\tilde{\tilde{e}}_r(\lambda)^T x \leq \tilde{\tilde{b}}_r(\lambda)\} \geq \theta_r\} \geq \eta_r \\ \Leftrightarrow\ &Appr\{\lambda | e_r(\lambda)^T x - b_r(\lambda) \leq -R^{-1}(\theta_r)\beta_r^{eT} x - L^{-1}(1-\theta_r)\alpha_r^b\} \geq \eta_r \\ \Leftrightarrow\ &\eta_r \leq \begin{cases} \frac{W-c}{2(d-c)}, & \text{if } c \leq W \leq a \\ \frac{1}{2}(\frac{W-c}{d-c} + \frac{W-a}{b-a}), & \text{if } a \leq W < b \\ \frac{1}{2}(\frac{W-c}{d-c} + 1), & \text{if } b \leq W \leq d \\ 1, & \text{if } W \geq d \end{cases} \\ \Leftrightarrow\ &\begin{cases} W \geq c + 2(d-c)\eta_r, & \text{if } c \leq W \leq a \\ W \geq \frac{2\eta_r(d-c)(b-a)+c(b-a)+a(d-c)}{b-a+d-c}, & \text{if } a \leq W < b \\ W \geq (2\eta_r - 1)(d-c) + c, & \text{if } b \leq W \leq d \\ W \geq d, & \text{if } d \leq W, \end{cases} \end{aligned}$$

where $W = -R^{-1}(\theta_r)\beta_r^{eT} x - L^{-1}(1-\theta_r)\alpha_r^b$.

This proof is completed. $\square$

From Propositions 5.7 and 5.8, we know that the problem (5.8) is equivalent to the following multi-objective programming problems,

$$
\begin{cases}
\max \ [f_1, f_2, \cdots, f_m] \\
\text{s.t.} \begin{cases}
f_i \leq d - 2\gamma_i(d-c) - L^{-1}(1-\delta_i)\alpha_i^{cT}x, \\
\qquad\qquad \text{if } b \leq f_i + L^{-1}(1-\delta_i)\alpha_i^{cT}x \leq d \\
f_i \leq \frac{d(b-a)+b(d-c)-2\gamma_i(d-c)(b-a)}{d-c+b-a} - L^{-1}(1-\delta_i)\alpha_i^{cT}x, \\
\qquad\qquad \text{if } a \leq f_i + L^{-1}(1-\delta_i)\alpha_i^{cT}x < b \\
f_i \leq d - (d-c)(2\gamma_i - 1) - L^{-1}(1-\delta_i)\alpha_i^{cT}x, \\
\qquad\qquad \text{if } c \leq f_i + L^{-1}(1-\delta_i)\alpha_i^{cT}x \leq a \\
f_i \leq c - L^{-1}(1-\delta_i)\alpha_i^{cT}x, \\
\qquad\qquad \text{if } f_i + L^{-1}(1-\delta_i)\alpha_i^{cT}x \leq c \\
W \geq c + 2(d-c)\eta_r, \text{ if } c \leq W \leq a \\
W \geq \frac{2\eta_r(d-c)(b-a)+c(b-a)+a(d-c)}{b-a+d-c}, \text{ if } a \leq W < b \\
W \geq (2\eta_r - 1)(d-c)+c, \text{ if } b \leq W \leq d \\
W \geq d, \text{ if } d \leq W \\
x \geq 0,
\end{cases}
\end{cases} \tag{5.59}
$$

where $W = -R^{-1}(\theta_r)\beta_r^{eT}x - L^{-1}(1-\theta_r)\alpha_r^b$.

5.4.2.2 Fuzzy Goal Method

In this section, we introduce how to use the fuzzy goal method to solve multi-objective programming problems. As we know, the standard distribution function $\Phi(x)$ is a nonlinear function, so it is difficult to solve using the usual technique. Here we introduce the fuzzy goal method proposed by Sakawa [128] to solve this kind of nonlinear multi-objective programming problems (5.60),

$$
\begin{cases}
\max[H_1(x), H_2(x), \cdots, H_m(x)] \\
\text{s.t. } x \in X.
\end{cases} \tag{5.60}
$$

Assume that decision makers have fixed the membership function $\mu_k(H_k(x))$ and given the goal membership function value $\bar{\mu}_k \ (k = 1, 2, \cdots, m)$. Let's consider the following programming problem,

$$
\begin{cases}
\max \sum_{k=1}^{m} d_k^- \\
\text{s.t.} \begin{cases}
\mu_k(H_k(x)) + d_k^+ - d_k^- = \bar{\mu}_k, k = 1, 2, \cdots, m \\
x \in X \\
d_k^+ d_k^- = 0, d_k^+, d_k^- \geq 0, k = 1, 2, \cdots, m,
\end{cases}
\end{cases} \tag{5.61}
$$

where d_k^+, d_k^- is the positive and negative deviation. Then we have the following result between the optimal solution of the problem (5.61) and the efficient solution of the problem (5.60).

Theorem 5.11. *(Sakawa [128]) (1) If x^* is the optimal solution of the problem (5.61), and $0 < \mu_k(H_k(x^*)) < 1, d_k^+ = 0(k = 1, 2, \cdots, m)$ holds, then x^* is an efficient solution of the problem (5.60).*

(2) If x* is an efficient solution of the problem (5.60), and $0 < \mu_k(H_k(\mathbf{x}^*)) < 1$ ($k = 1, 2, \cdots, m$), then x* is an efficient solution of the problem (5.61) and $d_k^+ = 0$ ($k = 1, 2, \cdots, m$) holds.

5.4.2.3   Numerical Example

Example 5.5. An industry will produce three kinds of products which are seasonal. Because the demand amount is seasonal, the profits are fuzzy rough variables, i.e., the profits are fuzzy variables, but the excepted values of these fuzzy variables are rough variables. When producing every product, the efficiency of the machinery is also a fuzzy rough variable, but the coefficient is different. Each product is no less than 20, and the gross amount is no less than 200 but no more than 250. The other coefficients can be seen in Table 5.3. The problem is how many products to produce in order to get the predetermined levels.

Table 5.3 The resource demand in producing process

product	1	2	3	possible using amount
workman amount	1	1	1	250
storage capacity	1	4	2	600
using efficiency	$c_1\tilde{\bar{\xi}}_4$	$c_2\tilde{\bar{\xi}}_5$	$c_3\tilde{\bar{\xi}}_6$	
profit	$\tilde{\bar{\xi}}_1$	$\tilde{\bar{\xi}}_2$	$\tilde{\bar{\xi}}_3$	

Then we can get the following Appr-pos constrained multi-objective programming problem

$$\max\{f_1, f_2\}$$
$$s.t. \begin{cases} Appr\{\lambda \,|\, Pos\{\tilde{\bar{\xi}}_1 x_1 + \tilde{\bar{\xi}}_2 x_2 + \tilde{\bar{\xi}}_3 x_3 \geq f_1\} \geq \delta_1\} \geq \gamma_1 \\ Appr\{\lambda \,|\, Pos\{c_1\tilde{\bar{\xi}}_4 x_1 + c_2\tilde{\bar{\xi}}_5 x_2 + c_3\tilde{\bar{\xi}}_6 x_3 \geq f_2\} \geq \delta_2\} \geq \gamma_2 \\ x_1 + x_2 + x_3 \leq 250 \\ x_1 + x_2 + x_3 \geq 200 \\ x_1 + 4x_2 + 2x_3 \leq 600 \\ x_1, x_2, x_3 \geq 20, \end{cases}$$

where $c = (c_1, c_2, c_3) = (1.2, 0.8, 1.5)$,

$$\tilde{\bar{\xi}}_1 = (\rho_1, 0.5, 0.5)_{LR}, \text{ with } \rho_1 \vdash ([1,2],[0,3]),$$
$$\tilde{\bar{\xi}}_2 = (\rho_2, 2, 2)_{LR}, \text{ with } \rho_2 \vdash ([2,3],[1,4]),$$
$$\tilde{\bar{\xi}}_3 = (\rho_3, 1, 1)_{LR}, \text{ with } \rho_3 \vdash ([3,4],[2,5]),$$
$$\tilde{\bar{\xi}}_4 = (\rho_4, 1, 1)_{LR}, \text{ with } \rho_4 \vdash ([0,1],[0,3]),$$
$$\tilde{\bar{\xi}}_5 = (\rho_5, 0.5, 0.5)_{LR}, \text{ with } \rho_5 \vdash ([1,2],[0,3]),$$
$$\tilde{\bar{\xi}}_6 = (\rho_6, 0.5, 0.5)_{LR}, \text{ with } \rho_6 \vdash ([2,3],[0,3]).$$

According to the knowledge of fuzzy variable and rough variable, we have that

$$
\begin{aligned}
&\bar{\bar{\xi}}_1 x_1 + \bar{\bar{\xi}}_2 x_2 + \bar{\bar{\xi}}_3 x_3 \\
&= (\rho_1 x_1 + \rho_2 x_2 + \rho_3 x_3, 0.5x_1 + 2x_2 + x_3, 0.5x_1 + 2x_2 + x_3)_{LR}, \\
&c_1 \bar{\bar{\xi}}_4 x_1 + c_2 \bar{\bar{\xi}}_5 x_2 + c_3 \bar{\bar{\xi}}_6 x_3 \\
&= (\rho_4 c_1 x_1 + \rho_5 c_2 x_2 + \rho_6 c_3 x_3, 1.2x_1 + 0.4x_2 + 0.75x_3, \\
&\quad 1.2x_1 + 0.4x_2 + 0.75x_3)_{LR}.
\end{aligned}
\tag{5.62}
$$

and

$$
\begin{aligned}
\rho_1 x_1 &= ([x_1, 2x_1], [0, 3x_1]) \\
\rho_2 x_2 &= ([2x_2, 3x_2], [x_2, 4x_2]) \\
\rho_3 x_3 &= ([3x_3, 4x_3], [2x_3, 5x_3]) \\
\rho_4 c_1 x_1 &= ([0, c_1 x_1], [0, 3c_1 x_1]) \\
\rho_5 c_2 x_2 &= ([c_2 x_2, 2c_2 x_2], [0, 3c_2 x_2]) \\
\rho_6 c_3 x_3 &= ([2c_3 x_3, 3c_3 x_3], [0, 3c_3 x_3]),
\end{aligned}
$$

and

$$
\begin{aligned}
&\rho_1 x_1 + \rho_2 x_2 + \rho_3 x_3 \\
&= ([x_1 + 2x_2 + 3x_3, 2x_1 + 3x_2 + 4x_3], [x_2 + 2x_3, 3x_1 + 4x_2 + 5x_3]) \\
&\rho_4 c_1 x_1 + \rho_5 c_2 x_2 + \rho_6 c_3 x_3 \\
&= ([0.8x_2 + 3x_3, 1.2x_1 + 1.6x_2 + 4.5x_3], [0, 3.6x_1 + 2.4x_2 + 4.5x_3]).
\end{aligned}
$$

Here we consider the case when $b \le f_i - R^{-1}(\delta_i)\beta_i^{cT} x \le d$, and readers can try another three cases through the following method.

According to Propositions 5.7 and 5.8, the problem (5.63) is equivalent to the following multi-objective programming problem,

$$
\begin{cases}
\max\ [f_1, f_2] \\
\text{s.t.}
\begin{cases}
f_1 \le 3x_1 + 4x_2 + 5x_3 - 2\gamma_1(3x_1 + 3x_2 + 3x_3) \\
\qquad + R^{-1}(\delta_1)(0.5x_1 + 2x_2 + x_3) \\
f_2 \le 3.6x_1 + 2.4x_2 + 4.5x_3 - 2\gamma_2(3.6x_1 + 2.4x_2 + 4.5x_3) \\
\qquad + R^{-1}(\delta_2)(1.2x_1 + 0.4x_2 + 0.75x_3) \\
x_1 + x_2 + x_3 \le 250 \\
x_1 + x_2 + x_3 \ge 200 \\
x_1 + 4x_2 + 2x_3 \le 600 \\
x_1, x_2, x_3 \ge 20
\end{cases}
\end{cases}
\tag{5.63}
$$

or equivalently

$$
\begin{cases}
\max\ [H_1(x), H_2(x)] \\
\text{s.t.}
\begin{cases}
x_1 + x_2 + x_3 \le 250 \\
x_1 + x_2 + x_3 \ge 200 \\
x_1 + 4x_2 + 2x_3 \le 600 \\
x_1, x_2, x_3 \ge 20,
\end{cases}
\end{cases}
\tag{5.64}
$$

where

$$H_1(x) := 3x_1 + 4x_2 + 5x_3 - 2\gamma_1(3x_1 + 3x_2 + 3x_3) + R^{-1}(\delta_1)(0.5x_1 + 2x_2 + x_3),$$
$$H_2(x) := 3.6x_1 + 2.4x_2 + 4.5x_3 - 2\gamma_2(3.6x_1 + 2.4x_2 + 4.5x_3)$$
$$+ R^{-1}(\delta_2)(1.2x_1 + 0.4x_2 + 0.75x_3).$$

When $\gamma_1 = \gamma_2 = \delta_1 = \delta_2 = 0.9$, H_i^0 and $H_i^1(i = 1,2)$ are calculated by solving the two single objective model as follows:

$$H_1^1 = -119, (x_1, x_2, x_3) = (20, 20, 160), H_2^0 = -656.9,$$
$$H_2^1 = -455.83, (x_1, x_2, x_3) = (53.33, 126.67, 20), H_1^0 = -283.33.$$

We give the membership functions as follows,

$$\mu_1(H_1(x)) = \frac{H_1(x) - H_1^0}{H_1^1 - H_1^0} = \frac{H_1(x) + 283.33}{164.33},$$
$$\mu_2(H_2(x)) = \frac{H_2(x) - H_2^0}{H_2^1 - H_2^0} = \frac{H_2(x) + 656.9}{200.97}.$$

According to the fuzzy goal method, we construct the fuzzy goal programming model (5.65) as follows,

$$\begin{cases} \max \ d_1^- + d_2^- \\ \text{s.t.} \begin{cases} \frac{H_1(x) + 283.33}{164.33} + d_1^+ + d_1^- = \bar{\mu}_1 \\ \frac{H_2(x) + 656.9}{200.97} + d_2^+ + d_2^- = \bar{\mu}_2 \\ x_1 + x_2 + x_3 \leq 250 \\ x_1 + x_2 + x_3 \geq 200 \\ x_1 + 4x_2 + 2x_3 \leq 600 \\ x_1, x_2, x_3 \geq 20 \\ d_1^+ d_1^- = 0, d_1^+, d_1^- \geq 0 \\ d_2^+ d_2^- = 0, d_2^+, d_2^- \geq 0. \end{cases} \end{cases} \quad (5.65)$$

Set $\bar{\mu}_1 = \bar{\mu}_2 = 0.9$, and we obtain the best solution of model (5.65), this solution is also the efficient solution of model (5.63),

$$(x_1, x_2, x_3) = (210, 20, 20).$$

5.4.3 Non-linear Fu-Ro CCM and Fu-Ro Simulation-Based Parametric TS

For the Fu-Ro CCM, we use the Fu-Ro simulation 2 based parametric TS algorithm to solve.

5.4.3.1 Fu-Ro Simulation 2 for Critical Value

First, we use the Fu-Ro simulation 2 to obtain the critical value which is important in CCM.

Assume that ξ is an n-dimensional fuzzy rough vector defined on the rough space $(\Lambda, \Delta, \mathscr{A}, \pi)$, and $f : \mathbf{R}^n \to \mathbf{R}^m$ is a measurable function. For any real number $\alpha \in (0,1]$, we find the maximal value $\bar{f}$ such that

$$Ch\{f(\xi) \geq \bar{f}\}(\alpha) \geq \beta \qquad (5.66)$$

holds. That is we should compute the maximal value $\bar{f}$ such that

$$Appr\{\lambda \in \Lambda | Cr\{f(\xi(\lambda)) \geq \bar{f}\} \geq \beta\} \geq \alpha. \qquad (5.67)$$

We sample $\underline{\lambda}_1, \underline{\lambda}_2, \cdots, \underline{\lambda}_N$ from Δ and $\overline{\lambda}_1, \overline{\lambda}_2, \cdots, \overline{\lambda}_N$ from Λ according to the measure π. For any number v, let $\underline{N}(v)$ denote the number of $\underline{\lambda}_k$ satisfying $Cr\{f(\xi(\underline{\lambda}_k)) \leq v\} \geq \beta$ for $k = 1, 2, \cdots, N$, and $\overline{N}(v)$ denote the number of $\overline{\lambda}_k$ satisfying

$$Cr\{f(\xi(\overline{\lambda}_k)) \leq v\} \geq \beta, \qquad (5.68)$$

for $k = 1, 2, \cdots, N$, where $Cr\{\cdot\}$ may be estimated by fuzzy simulation. Then we may find the maximal value v such that

$$\frac{\underline{N}(v) + \overline{N}(v)}{2N} \geq \alpha. \qquad (5.69)$$

This value is an estimation of $\bar{f}$. The procedure is as follows:

Step 1. Generate $\underline{\lambda}_1, \underline{\lambda}_2, \cdots, \underline{\lambda}_N$ from Δ according to the measure π.
Step 2. Generate $\overline{\lambda}_1, \overline{\lambda}_2, \cdots, \overline{\lambda}_N$ from Λ according to the measure π.
Step 3. Find the maximal value v such that (5.69) holds.
Step 4. Return v.

Example 5.6. We employ Fu-Ro simulation 2 to find the maximal value $\bar{f}$ such that $Ch\{\xi_1^2 + \xi_2^2 \geq \bar{f}\}(0.9) \geq 0.9$, where xi_1 and ξ_2 are Fu-Ro variables defined as

$$\xi_1 = (\rho_1, \rho_1 + 1, \rho_1 + 2), \text{ with } \rho_1 = ([1,2],[0,3]),$$
$$\xi_2 = (\rho_2, \rho_2 + 1, \rho_2 + 2), \text{ with } \rho_2 = ([2,3],[1,4]).$$

A run of Fu-Ro simulation with 5000 cycles shows that $\bar{f} = 6.39$.

5.4.3.2 Parametric TS

Let's recall the detail of the parametric TS introduced by F. Glover [335]. The solution approach consists of a parametric form of tabu-search utilizing moves based on the approach of parametric branch and bound [336]. Various levels of tabu-search can be used to guide the foregoing processes. We begin by sketching the elements of a basic approach and illustrate its application.

Tabu conditions. At an initial rudimentary level, we attach a tabu restriction to an (R-DN) or (R-UP) response for a particular variable x_j, thereby forbidding the response from being executed, if the opposing response ((R-UP) or (R-DN), respectively) was executed for x_j within the most recent TabuTenure iterations. (That is, we forbid a move in a direction that is contrary to the direction of a move made within the selected span of TabuTenure iterations.) To simplify the discussion, we allow (R-DN) and (R-UP) to refer also to the responses (R-DN^o) and (R-UP^o). The value of TabuTenure varies according to the variable x_j concerned and the history of the search. We represent this value as TabuTenure$_j$(UP) and TabuTenure$_j$(DN) according to whether the tabu condition was launched by an (R-UP) or an (R-DN) response. When such a response is made we use TabuTenure$_j$(UP) or TabuTenure$_j$(DN) and knowledge of the current iteration, which we denote by Iter, to identify the iteration TabuTenure$_j$(UP) or TabuTenure$_j$(DN) that marks the end of x_js tabu tenure. Specifically, when an (R-UP) response occurs, we set

$$TabuEnd_j(DN) = Iter + TabuTenure_j(DN)$$

to forbid the opposing (R-DN) response from being made for the period of TabuTenure$_j$(DN) iterations in the future. Similarly, when an (R-DN) response occurs, we set

$$TabuEnd_j(UP) = Iter + TabuTenure_j(UP)$$

to forbid the opposing (R-UP) response from being made for the period of TabuTenure$_j$ (UP) iterations in the future.

By this means, an (R-DN) response is tabu for x_j as long as the (updated) current iteration satisfies

$$Iter \leq TabuEnd_j(DN)$$

and an (R-UP) response is tabu for x_j as long as the current iteration satisfies

$$Iter \leq TabuEnd_j(UP)$$

Initially, before any responses have been made and before associated tabu conditions have been created, TabuEnd$_j$(UP) and TabuEnd$_j$(DN) are set equal to -1, causing this value to be smaller than every value of Iter and hence assuring that no tabu restrictions will be in effect.

We refer to the values TabuEnd$_j$(UP)-Iter and TabuEnd$_j$(DN)-Iter as *residual tabu tenures*. Hence, a response will be tabu as long as its residual tabu tenure is non-negative. (A negative residual tabu tenure accordingly indicates the response is free from a tabu restriction.) By convention, we refer to the residual tabu tenure of a variable x_j by taking it to be the residual tabu tenure of the response that is selected for this variable. We refer to the variable itself as being tabu when its associated response is tabu. (This reference is unambiguous since each goal infeasible and potentially goal infeasible variable has a single associated response.) Rules for generating the TabuTenure$_j$(DN) and TabuTenure$_j$(UP) values used to determine TabuEnd$_j$(UP) and TabuEnd$_j$(DN) are given in the next section.

In the application of the tabu tenures, a simple form of probabilistic tabu search can be used that replaces $\text{TabuTenure}_j(\text{DN})$ and $\text{TabuTenure}_j(\text{UP})$ in the formulas $\text{TabuEnd}_j(\text{DN})=\text{Iter}$
$+\text{TabuTenure}_j(\text{DN})$ and $\text{TabuEnd}_j(\text{DN})=\text{Iter} +\text{TabuTenure}_j(\text{UP})$ by values that are randomly selected from an interval around the respective tabu tenure values. A fuller use of this type of randomizing effect occurs by making such a replacement each time the inequalities $\text{Iter-TabuEnd}_j(\text{DN})$ and $\text{Iter-TabuEnd}_j(\text{UP})$ are checked.

By design, tabu restrictions are prohibitions against returning to a state previously occupied. We only create these restrictions for states that seek to enforce a goal condition, hence that involve the responses (R-UP) and (R-DN) (understanding these to include reference to the responses (R-UP^o) and (R-DN^o)). Moreover, we only check tabu conditions when at least one variable is goal infeasible. In the case where no explicit goal infeasibility exists, and hence the only responses to consider are those applicable to unrestricted free variables, then no attention is paid to tabu restrictions. The situation where all goal conditions are satisfied (no goal infeasibility exists) may be viewed as meeting the requirements of a special type of aspiration criterion, which overrules all tabu conditions. We now examine the use of criteria that operate when goal infeasibility is present.

Aspiration criteria. As is customary in tabu-search, we allow a tabu response to be released from a tabu restriction if the response satisfies an auxiliary *aspiration criterion* that indicates the response has special merit or novelty (i.e., exhibits a feature not often encountered). A common instance of such a criterion, called *aspiration by objective*, permits the response to be made if it yields a better objective function evaluation than any response previously executed. In the present setting, we find it convenient to additionally consider an *aspiration by resistance*, based on the greatest resistance a particular response has generated in the past.

Specifically, let $\text{Aspire}_j(\text{DN})$ and $\text{Aspire}_j(\text{UP})$ denote the largest goal resistance values $\text{GR}_j(\text{DN})$ and $\text{GR}_j(\text{UP})$ that have occurred for x_j on any iteration, where x_j was selected to execute an (R-DN) or (R-UP) response, respectively. Then we disregard the tabu restriction for an (R-DN) response (identified by $\text{Iter} \leq \text{TabuEnd}_j(\text{DN})$) if

$$\text{GR}_j(\text{DN}) > \text{Aspire}_j(UP)$$

and disregard tabu restriction for an (R-UP) response (identified by $\text{Iter} \leq \text{TabuEnd}_j(\text{UP})$) if

$$\text{GR}_j(\text{UP}) > \text{Aspire}_j(DN)$$

The rationale for these aspiration criteria is that a move can be allowed if its current resistance value, measured by $\text{GR}_j(\text{DN})$ or $\text{GR}_j(\text{UP})$, exceeds the greatest resistance value previously identified for moving in the opposite direction ($\text{Aspire}_j(\text{UP})$ or $\text{Aspire}_j(\text{DN})$, respectively). We initially set $\text{Aspire}_j(\text{UP})$ and $\text{Aspire}_j(\text{DN})$ to a large negative number, so that the first time a variable x_j is evaluated for a potential response (R-UP) or (R-DN), the response will automatically be allowed, and it will continue to be allowed until the opposing response is made, which establishes a resistance to be exceeded.

We call a response admissible if it is either not tabu or else satisfies the aspiration criterion, and call it *inadmissible* otherwise. If the unique available response for a goal infeasible variable is inadmissible, then the variable is not permitted to enter the sets G_P and G_S, even if this makes it impossible for one or both of these sets to attain its targeted size g_P or g_S. The only exception to this rule is that G_P is not permitted to be empty in the case of goal infeasibility. Hence in the extreme case where no variables would enter G_P the typical aspiration by default rule is invoked that allows G_P to contain a variable with a smallest residual tabu tenure. (Probabilistic variations of the aspiration by default rule can also be applied, by assigning larger probabilities to selecting variables with smaller residual tabu tenures.)

As observed earlier, $GR_j = GR_j(DN)$ or $GR_j(UP)$ may be treated as a 2-element vector, with a dominant component for an overt goal infeasibility and a secondary GR_j^o component for potential goal infeasibility. The Aspire$_j$ values are treated in the same way, as 2-element vectors that include a secondary component Aspire$_j^o$ for potential goal infeasibility. Since overt and potential goal infeasibility for a given variable x_j never occur simultaneously, and since overt goal infeasibility is the dominant component, only a single component of the vector is relevant to consider‖the overt component if it exists, and the potential component otherwise.

It is to be emphasized that Aspire$_j$(UP) and Aspire$_j$(DN) do not record the greatest values of GR_j(UP) and GR_j(DN) encountered over the history of the search, but only the greatest values that occurred in the instances where x_j was selected as a variable to be assigned a goal condition, and only in response to overt or potential goal infeasibility (i.e., not in response to integer infeasibility, which occurs only when x_j is an unrestricted fractional variable).

5.4.3.3 Numerical Example

Example 5.7. Let us consider the following problem,

$$
\begin{cases}
\max \; [f_1, f_2] \\
\text{s.t.} \begin{cases}
Appr\{\lambda \,|\, Pos\{\bar{\bar{\xi}}_1 x_1 + \bar{\bar{\xi}}_2 x_2 + \bar{\bar{\xi}}_3 x_3 \geq f_1\} \geq \delta_1\} \geq \gamma_1 \\
Appr\{\lambda \,|\, Pos\{c_1 \bar{\bar{\xi}}_4 x_1 + c_2 \bar{\bar{\xi}}_5 x_2 + c_3 \bar{\bar{\xi}}_6 x_3 \geq f_2\} \geq \delta_2\} \geq \gamma_2 \\
x_1 + x_2 + x_3 \leq 250 \\
x_1 + x_2 + x_3 \geq 200 \\
x_1 + 4x_2 + 2x_3 \leq 600 \\
x_1 \geq 20, x_2 \geq 20, x_3 \geq 20,
\end{cases}
\end{cases}
\tag{5.70}
$$

where $c = (c_1, c_2, c_3) = (1.2, 0.8, 1.5)$,

$$\bar{\xi}_1 = (1, \rho_1, 1)_{LR}, \text{ with } \rho_1 \vdash ([1,2], [0,3]),$$
$$\bar{\xi}_2 = (1, \rho_2, 1)_{LR}, \text{ with } \rho_2 \vdash ([2,3], [1,4]),$$
$$\bar{\xi}_3 = (1, \rho_3, 1)_{LR}, \text{ with } \rho_3 \vdash ([3,4], [2,5]),$$
$$\bar{\xi}_4 = (1, \rho_4, 1)_{LR}, \text{ with } \rho_4 \vdash ([0,1], [0,3]),$$
$$\bar{\xi}_5 = (1, \rho_5, 1)_{LR}, \text{ with } \rho_5 \vdash ([1,2], [0,3]),$$
$$\bar{\xi}_6 = (1, \rho_6, 1)_{LR}, \text{ with } \rho_6 \vdash ([2,3], [0,3]),$$

and $\rho_i (i = 1, 2, \cdots, 6)$ are rough variables. We set $\delta_i = \gamma_i = 0.9$, then $\Phi^{-1}(1 - \delta_i) = -1.28, i = 1, 2$.

Next, we apply the parallel tabu search algorithm based on the Fu-Ro simulation to solve the nonlinear programming problem (5.70) with the fuzzy rough parameters.

Step 1. Set the move step $h = 0.5$ and the h neighbor $N(x,h)$ for the present point x is defined as follows,

$$N(x,h) = \left\{ y \mid \sqrt{(x_1 - y_1)^2 + (x_2 - y_2)^2 + (x_3 - y_3)^2} \leq h \right\}.$$

The random move of point x to point y in its h neighbor along direction s is given by

$$y_s = x_s + rh,$$

where r is a random number that belongs to $[0,1]$, $s = 1, 2, 3$.

Step 2. Give the step set $H = \{h_1, h_2, \cdots, h_r\}$ and randomly generate a feasible point x_0 checked by the fuzzy rough simulation. One should empty the Tabu list T (the list of inactive steps) at the beginning.

Step 3. For each active neighbor $N(x,h)$ of the present point x, where $h \in H - T$, a feasible random move that satisfies all the constraints in problem (5.70) is to be generated.

Step 4. Construct the single objective function as follows,

$$f(x) = w_1 f_1 + w_2 f_2$$

where $w_1 + w_2 = 1$ and $w_i (i = 1, 2)$ is predetermined by the decision maker. Compare the $f(x)$ of the feasible moves with that of the current solution by the fuzzy rough simulation. If an augmenter in new objective function of the feasible moves exists, one should save this feasible move as the updated current one by adding the corresponding step to the Tabu list T and go to the next step; otherwise, go to the next step directly.

Step 5. Stop if the termination criteria are satisfied; other wise, empty T if it is full; then go to Step 3. Here, we set the computation is determined if the better solution doesn't change again.

We apply compute the programming problem (5.70) by the parallel tabu search algorithm. The table 5.4 shows the results.

Table 5.4 The result computed by parallel TS algorithm at different weights

w_1	w_2	x_1	x_2	x_3	H	Gen
0.1	0.9	90.68	25.19	84.13	-2304.55	270
0.2	0.8	90.25	25.08	84.66	-2287.08	240
0.3	0.7	89.82	24.99	85.19	-2269.57	256
0.4	0.6	89.39	24.89	85.72	-2252.01	269
0.5	0.5	88.99	24.79	86.25	-2234.40	294
0.6	0.4	88.53	24.70	86.78	-2216.74	291
0.7	0.3	88.10	24.60	87.30	-2199.03	268
0.8	0.2	87.67	24.50	87.83	-2181.29	281
0.9	0.1	87.24	24.40	88.36	-2163.48	276

5.5 Fu-Ro DCM

This section provides Fu-Ro DCM in which the underling philosophy is based on selecting the decision with the maximum chance to meet the event.

5.5.1 General Model for Fu-Ro DCM

A generally uncertain dependent chance model has the following form,

$$
\begin{cases}
\max \ [Ch\{f_i(x,\xi) \le f_i\}(\alpha_i), i = 1, 2, \cdots, m] \\
\text{s.t.} \begin{cases} g_r(x,\xi) \le 0, r = 1, 2, \cdots, p \\ x \in X, \end{cases}
\end{cases}
\tag{5.71}
$$

where x is an n-dimensional decision vector, ξ is a fuzzy rough vector, the event ξ is characterized by $h_k(x,\xi) \le 0, k = 1, 2, \ldots q$, and the fuzzy rough environment is described by the fuzzy rough constraints $g_r(x,\xi) \le 0, r = 1, 2, \ldots p$. Here, the constraints are all certain. For uncertain constraints, we can deal with them by the technique of chance-constrained programming.

When the fuzzy rough variable degenerates to the single uncertain variable, we obtain the following results.

Remark 5.8. If the fuzzy rough variable ξ degenerates to a fuzzy variable, for any given α_i,

$$
Ch\{f_i(x,\xi) \le f_i\}(\alpha_i) = Cr\{f_i(x,\xi) \le f_i\}(\alpha_i), i = 1, 2, \cdots, m.
$$

Thus, the problem (5.71) is equivalent to

$$\begin{cases} \max \ [Cr\{f_i(x,\xi) \le f_i\}(\alpha_i), i = 1,2,\cdots,m] \\ \text{s.t.} \begin{cases} g_r(x,\xi) \le 0, r = 1,2,\cdots,p \\ x \in X, \end{cases} \end{cases} \quad (5.72)$$

where ξ is a fuzzy variable, and this model is a standard fuzzy DCM.

Remark 5.9. If the fuzzy rough variable ξ degenerates to a rough variable, for any given α_i. This means

$$Ch\{f_i(x,\xi) \le f_i\}(\alpha_i) = Appr\{f_i(x,\xi) \le f_i\}(\alpha_i), i = 1,2,\cdots,m.$$

Thus, the problem (5.71) is converted into

$$\begin{cases} \max \ [Appr\{f_i(x,\xi) \le f_i\}(\alpha_i), i = 1,2,\cdots,m] \\ \text{s.t.} \begin{cases} g_r(x,\xi) \le 0, r = 1,2,\cdots,p \\ x \in X, \end{cases} \end{cases} \quad (5.73)$$

where ξ is a rough variable, and this model is a standard rough DCM.

If there are multiple events in the fuzzy rough environment, a typical formulation of Fu-Ro DCM is given as follows,

$$\begin{cases} \max \ \begin{bmatrix} Ch\{h_{1k}(x,\xi) \le 0, \ k = 1,2,\cdots,q_1\} \\ Ch\{h_{2k}(x,\xi) \le 0, \ k = 1,2,\cdots,q_2\} \\ \cdots \\ Ch\{h_{mk}(x,\xi) \le 0, \ k = 1,2,\cdots,q_m\} \end{bmatrix} \\ \text{s.t.} \begin{cases} g_r(x,\xi) \le 0, \ r = 1,2,\cdots,p \\ x \in X, \end{cases} \end{cases} \quad (5.74)$$

where $h_{ik}(x,\xi) \le 0$, $k = 1,2,\cdots,q_i$ represent events ε_i for $i = 1,2,\cdots,m$, respectively.

Fuzzy rough dependent-chance goal programming is employed to formulate fuzzy rough decision systems according to the priority structure and target levels set by the decision-maker,

$$\begin{cases} \max \ \sum_{j=1}^{l} P_j \sum_{i=1}^{m} (u_{ij}d_i^+ \vee 0 + v_{ij}d_i^- \vee 0) \\ \text{s.t.} \begin{cases} Ch\{h_{1k}(x,\xi) \le 0, \ k = 1,2,\cdots,q_i\} - b_i = d_i^+, i = 1,2,\cdots,m \\ b_i - Ch\{h_{1k}(x,\xi) \le 0, \ k = 1,2,\cdots,q_i\} = d_i^-, i = 1,2,\cdots,m \\ g_r(x,\xi) \le 0, r = 1,2,\cdots,p \\ x \in X, \end{cases} \end{cases} \quad (5.75)$$

where P_j is the preemptive priority factor which expresses the relative importance of various goals, $P_j \gg P_{j+1}$, for all j, u_{ij} is the weighting factor corresponding to positive deviation for goal i with priority j assigned, v_{ij} is the weighting factor corresponding to negative deviation for goal i with priority j assigned, $d_i^+ \vee 0$ is the positive deviation from the target of goal i, $d_i^- \vee 0$ is the negative deviation from the target of goal i, g_j is a function in system constraints, b_i is the target value according to goal i, l is the number of priorities, m is the number of goal constraints, and p is the number of system constraints.

5.5.2 Linear Fu-Ro DCM and ε-Constraint Method

Let's still consider the linear model with Fu-Ro coefficients as follows,

$$\begin{cases} \max \ [\tilde{\bar{c}}_1^T x, \tilde{\bar{c}}_2^T x, \cdots, \tilde{\bar{c}}_m^T x] \\ \text{s.t.} \begin{cases} e_r^T x \leq b_r, \quad r = 1, 2, \cdots, p \\ x \in X, \end{cases} \end{cases} \qquad (5.76)$$

where $\tilde{\bar{c}}_i$ is Fu-Ro vector, $i = 1, 2, \cdots, m$.

5.5.2.1 Crisp Equivalent Model

Appr − *Pos* constrained multi-objective linearity model

Because there are Fu-Ro variables in the model (5.76), so the model doesn't have mathematical meaning. We can use its DCM on *Appr* − *Pos* to deal with it as follows

$$\begin{cases} \max \ [\delta_1, \delta_2, \cdots, \delta_m] \\ \text{s.t.} \begin{cases} Appr\{\lambda | Pos\{\tilde{\bar{c}}_i^T(\lambda)x \leq f_i\} \geq \delta_i\} \geq \gamma_i, \quad i = 1, 2, \cdots, m \\ e_r^T x \leq b_r, \quad r = 1, 2, \cdots, p \\ x \in X, \end{cases} \end{cases} \qquad (5.77)$$

where $\xi = (\xi_1, \xi_2, \cdots, \xi_n)^T$ is a fuzzy rough vector, γ_i is the given confidence level and f_i is the predetermined value.

We can use the following theorem to obtain the equivalent form of the crisp dependent-chance model (5.77).

Theorem 5.12. *Assume that $\tilde{\bar{c}}_{ij}$ is a Fu-Ro variable, for any $\lambda \in \Lambda$, the fuzzy variable $\tilde{\bar{c}}_{ij}(\lambda)$ is characterized by the following membership function*

$$\mu_{\tilde{\bar{c}}_{ij}(\lambda)}(t) = \begin{cases} L\left(\frac{c_{ij}(\lambda)-t}{\alpha_{ij}^c}\right), \ t \leq c_{ij}(\lambda), \alpha_{ij}^c > 0 \\ R\left(\frac{t-c_{ij}(\lambda)}{\beta_{ij}^c}\right), \ t \geq c_{ij}(\lambda), \beta_{ij}^c > 0 \end{cases} \quad \lambda \in \Lambda \qquad (5.78)$$

where $\alpha_{ij}^c, \beta_{ij}^c$ are positive numbers expressing the left and right spread of $\tilde{\bar{c}}_{ij}(\lambda)$, reference function $L, R : [0,1] \to [0,1]$ with $L(1) = R(1) = 0$, and $L(0) = R(0) = 1$ are non-increasing, continuous function. And $(c_{ij}(\lambda))_{n \times 1} = (c_{i1}(\lambda), c_{i2}(\lambda), \cdots, c_{in}(\lambda))^T$ is a rough vector. It follows that $c_i(\lambda)^T x = ([a,b],[c,d])$

(where $c \leq a < b \leq d$) is a rough variable and characterized by the following trust measure function,

$$Appr\{c_i(\lambda)^T x \geq t\} = \begin{cases} 0, & \text{if } d \leq t \\ \frac{d-t}{2(d-c)}, & \text{if } b \leq t \leq d \\ \frac{1}{2}(\frac{d-t}{d-c} + \frac{b-t}{b-a}), & \text{if } a \leq t < b \\ \frac{1}{2}(\frac{d-t}{d-c} + 1), & \text{if } c \leq t \leq a \\ 1, & \text{if } t \leq c. \end{cases} \qquad (5.79)$$

Then we have $Appr\{\lambda \,|\, Pos\{\bar{\bar{c}}_i(\lambda)^T x \geq f_i\} \geq \delta_i\} \geq \gamma_i$ if and only if

$$\begin{cases} R^{-1}(\delta_i) \geq \frac{f_i-d+2(d-c)\gamma_i}{\beta_i^{cT}x}, & \text{if } b \leq f_i - R^{-1}(\delta_i)\beta_i^{cT}x \leq d \\ R^{-1}(\delta_i) \geq \frac{(d-c+b-a)f_i-d(b-a)-b(d-c)+2(d-c)(b-a)\gamma_i}{\beta_i^{cT}x}, & \text{if } a \leq f_i - R^{-1}(\delta_i)\beta_i^{cT}x < b \\ R^{-1}(\delta_i) \geq \frac{f_i-d+2(d-c)(2\gamma_i-1)}{\beta_i^{cT}x}, & \text{if } c \leq f_i - R^{-1}(\delta_i)\beta_i^{cT}x \leq a \\ R^{-1}(\delta_i) \geq \frac{f_i-c}{\beta_i^{cT}x}, & \text{if } f_i - R^{-1}(\delta_i)\beta_i^{cT}x \leq c, \end{cases}$$
$$\qquad (5.80)$$

where $\gamma_i, \delta_i \in [0,1]$ are predetermined confidence levels.

And accordingly we can get the crisp equivalent models in every cases.

Proof. By theorem 5.7, we have

$$\begin{cases} f_i \leq d - 2\gamma_i(d-c) + R^{-1}(\delta_i)\beta_i^{cT}x, & \text{if } b \leq f_i - R^{-1}(\delta_i)\beta_i^{cT}x \leq d \\ f_i \leq \frac{d(b-a)+b(d-c)-2\gamma_i(d-c)(b-a)}{d-c+b-a} + R^{-1}(\delta_i)\beta_i^{cT}x, & \text{if } a \leq f_i - R^{-1}(\delta_i)\beta_i^{cT}x < b \\ f_i \leq d - (d-c)(2\gamma_i-1) + R^{-1}(\delta_i)\beta_i^{cT}x, & \text{if } c \leq f_i - R^{-1}(\delta_i)\beta_i^{cT}x \leq a \\ f_i \leq c + R^{-1}(\delta_i)\beta_i^{cT}x, & \text{if } f_i - R^{-1}(\delta_i)\beta_i^{cT}x \leq c. \end{cases}$$

Because γ_i is a given confidence level between 0 and 1, this is no optimal solution for $L \geq d$. We can discuss the following four cases.

Case 1: $b \leq f_i - R^{-1}(\delta_i)\beta_i^{cT}x \leq d$.

From the assumption we know that $(d-c) > 0$, so we have

$$f_i \leq d - 2\gamma_i(d-c) + R^{-1}(\delta_i)\beta_i^{cT}x$$
$$\Leftrightarrow R^{-1}(\delta_i) \geq \frac{f_i-d+2(d-c)\gamma_i}{\beta_i^{cT}x}.$$

From the assumption, the reference function $R(\cdot)$ is non-increasing continuous function, so $\max \delta_i$ is equivalent to $\min R^{-1}(\delta_i)$, so the problem (5.77) can be transformed into

$$\begin{cases} \max \; [\frac{f_i-d+2(d-c)\gamma_i}{\beta_i^{cT}x}] \\ \text{s.t.} \begin{cases} e_r^T x \leq b_r, & r = 1,2,\cdots,p \\ x \in X \\ x \geq 0. \end{cases} \end{cases}$$

Case 2: $a \le f_i - R^{-1}(\delta_i)\beta_i^{cT}x < b$. We have

$$f_i \le \frac{d(b-a)+b(d-c)-2\gamma_i(d-c)(b-a)}{d-c+b-a} + R^{-1}(\delta_i)\beta_i^{cT}x$$
$$\Leftrightarrow R^{-1}(\delta_i) \ge \frac{(d-c+b-a)f_i-d(b-a)-b(d-c)+2(d-c)(b-a)\gamma_i}{\beta_i^{cT}x},$$

so the problem (5.77) can be transformed into

$$\begin{cases} \max \; [\frac{(d-c+b-a)f_i-d(b-a)-b(d-c)+2(d-c)(b-a)\gamma_i}{\beta_i^{cT}x}] \\ \text{s.t.} \begin{cases} e_r^T x \le b_r, \quad r=1,2,\cdots,p \\ x \in X \\ x \ge 0. \end{cases} \end{cases}$$

Case 3: $c \le f_i - R^{-1}(\delta_i)\beta_i^{cT}x \le a$. We have

$$f_i \le d - (d-c)(2\gamma_i - 1) + R^{-1}(\delta_i)\beta_i^{cT}x$$
$$\Leftrightarrow R^{-1}(\delta_i) \ge \frac{f_i-d+2(d-c)(2\gamma_i-1)}{\beta_i^{cT}x},$$

so the problem (5.77) can be transformed into

$$\begin{cases} \max \; [\frac{f_i-d+2(d-c)(2\gamma_i-1)}{\beta_i^{cT}x}] \\ \text{s.t.} \begin{cases} e_r^T x \le b_r, \quad r=1,2,\cdots,p \\ x \in X \\ x \ge 0. \end{cases} \end{cases}$$

Case 4: $f_i - R^{-1}(\delta_i)\beta_i^{cT}x \le c$. We have

$$f_i \le c + R^{-1}(\delta_i)\beta_i^{cT}x \Leftrightarrow R^{-1}(\delta_i) \ge \frac{f_i-c}{\beta_i^{cT}x},$$

so the problem (5.77) can be transformed into

$$\begin{cases} \max \; [\frac{f_i-c}{\beta_i^{cT}x}] \\ \text{s.t.} \begin{cases} e_r^T x \le b_r, \quad r=1,2,\cdots,p \\ x \in X \\ x \ge 0. \end{cases} \end{cases}$$

This completes the proof. □

Appr-Nec constrained multi-objective linearity model

Also we can use the DCM based on $Appr - Nec$ to deal with model (5.76) as follows

$$\begin{cases} \max\ [\delta_1,\delta_2,\cdots,\delta_m] \\ \text{s.t.}\ \begin{cases} Appr\{\lambda\,|Nec\{\tilde{\bar{c}}_i^T(\lambda)x\le f_i\}\ge \delta_i\}\ge \gamma_i, & i=1,2,\cdots,m \\ e_r^T x\le b_r, & r=1,2,\cdots,p \\ x\in X, \end{cases} \end{cases} \quad (5.81)$$

where $\xi=(\xi_1,\xi_2,\cdots,\xi_n)^T$ is a fuzzy rough vector, γ_i is the given confidence level and f_l is the predetermined value.

We can use the following theorem to obtain the equivalent form of the crisp DCM (5.81).

Theorem 5.13. *Assume that $\tilde{\bar{c}}_{ij}$ is a fuzzy rough variable, for any $\lambda\in\Lambda$, the fuzzy variable $\tilde{\bar{c}}_{ij}(\lambda)$ is characterized by the following membership function*

$$\mu_{\tilde{\bar{c}}_{ij}(\lambda)}(t)=\begin{cases} L\left(\dfrac{c_{ij}(\lambda)-t}{\alpha_{ij}^c}\right), & t\le c_{ij}(\lambda), \alpha_{ij}^c>0 \\ R\left(\dfrac{t-c_{ij}(\lambda)}{\beta_{ij}^c}\right), & t\ge c_{ij}(\lambda), \beta_{ij}^c>0 \end{cases} \quad \lambda\in\Lambda, \quad (5.82)$$

where $\alpha_{ij}^c,\beta_{ij}^c$ are positive numbers expressing the left and right spread of $\tilde{\bar{c}}_{ij}(\lambda)$, reference function $L,R:[0,1]\to[0,1]$ with $L(1)=R(1)=0$, and $L(0)=R(0)=1$ are non-increasing, continuous function. And $(c_{ij}(\lambda))_{n\times1}=(c_{i1}(\lambda),c_{i2}(\lambda),\cdots,c_{in}(\lambda))^T$ is a rough vector. It follows that $c_i(\lambda)^T x=([a,b],[c,d])$ (where $c\le a<b\le d$) is a rough variable and characterized by the following trust measure function,

$$Appr\{c_i(\lambda)^T x\ge t\}=\begin{cases} 0, & \text{if } d\le t \\ \dfrac{d-t}{2(d-c)}, & \text{if } b\le t\le d \\ \dfrac{1}{2}(\dfrac{d-t}{d-c}+\dfrac{b-t}{b-a}), & \text{if } a\le t<b \\ \dfrac{1}{2}(\dfrac{d-t}{d-c}+1), & \text{if } c\le t\le a \\ 1, & \text{if } t\le c. \end{cases} \quad (5.83)$$

Then we have $Appr\{\lambda\,|Nec\{\tilde{\bar{c}}_i(\lambda)^T x\ge f_i\}\ge \delta_i\}\ge \gamma_i$ if and only if

$$\begin{cases} L^{-1}(1-\delta_i)\le \dfrac{-f_i+d-2(d-c)\gamma_i}{\alpha_i^{cT}x}, \\ \qquad\qquad\qquad \text{if } b\le f_i+L^{-1}(1-\delta_i)\alpha_i^{cT}x\le d \\ L^{-1}(1-\delta_i)\le \dfrac{-(d-c+b-a)f_i+d(b-a)-b(d-c)-2(d-c)(b-a)\gamma_i}{\alpha_i^{cT}x}, \\ \qquad\qquad\qquad \text{if } a\le f_i+L^{-1}(1-\delta_i)\alpha_i^{cT}x<b \\ L^{-1}(1-\delta_i)\le \dfrac{-f_i+d-2(d-c)(2\gamma_i-1)}{\alpha_i^{cT}x}, \\ \qquad\qquad\qquad \text{if } c\le f_i+L^{-1}(1-\delta_i)\alpha_i^{cT}x\le a \\ L^{-1}(1-\delta_i)\ge \dfrac{-f_i+c}{\alpha_i^{cT}x}, \\ \qquad\qquad\qquad \text{if } f_i+L^{-1}(1-\delta_i)\alpha_i^{cT}x\le c, \end{cases}$$
$$(5.84)$$

where $\gamma_i,\delta_i\in[0,1]$ are predetermined confidence levels.

And accordingly we can get the crisp equivalent models in every cases.

Proof. By theorem 5.9, we have

$$
\begin{cases}
f_i \leq d - 2\gamma_i(d-c) - L^{-1}(1-\delta_i)\alpha_i^{cT}x, \\
\qquad\qquad \text{if } b \leq f_i + L^{-1}(1-\delta_i)\alpha_i^{cT}x \leq d \\
f_i \leq \frac{d(b-a)+b(d-c)-2\gamma_i(d-c)(b-a)}{d-c+b-a} - L^{-1}(1-\delta_i)\alpha_i^{cT}x, \\
\qquad\qquad \text{if } a \leq f_i + L^{-1}(1-\delta_i)\alpha_i^{cT}x < b \\
f_i \leq d - (d-c)(2\gamma_i-1) - L^{-1}(1-\delta_i)\alpha_i^{cT}x, \\
\qquad\qquad \text{if } c \leq f_i + L^{-1}(1-\delta_i)\alpha_i^{cT}x \leq a \\
f_i \leq c - L^{-1}(1-\delta_i)\alpha_i^{cT}x, \\
\qquad\qquad \text{if } f_i + L^{-1}(1-\delta_i)\alpha_i^{cT}x \leq c.
\end{cases}
$$

Because γ_i is a given confidence level between 0 and 1, this is no optimal solution for $L \geq d$. We can discuss the following four cases.

Case 1: $b \leq f_i + L^{-1}(1-\delta_i)\alpha_i^{cT}x \leq d$.

From the assumption we know that $\alpha_i^{cT}x > 0$, so we have

$$
f_i \leq d - 2\gamma_i(d-c) - L^{-1}(1-\delta_i)\alpha_i^{cT}x
$$
$$
\Leftrightarrow L^{-1}(1-\delta_i) \leq \frac{-f_i+d-2(d-c)\gamma_i}{\alpha_i^{cT}x}.
$$

From the assumption, the reference function $L(\cdot)$ is non-increasing continuous function, so max δ_i is equivalent to max $L^{-1}(1-\delta_i)$, so the problem (5.77) can be transformed into

$$
\begin{cases}
\max \left\{ \frac{f_i-d+2(d-c)\gamma_i}{\beta_i^{cT}x} \right\} \\
\text{s.t.} \begin{cases} e_r^T x \leq b_r, \quad r = 1,2,\cdots,p \\ x \in X \\ x \geq 0. \end{cases}
\end{cases}
$$

Case 2: $a \leq f_i + L^{-1}(1-\delta_i)\alpha_i^{cT}x < b$. We have

$$
f_i \leq \frac{d(b-a)+b(d-c)-2\gamma_i(d-c)(b-a)}{d-c+b-a} - L^{-1}(1-\delta_i)\alpha_i^{cT}x
$$
$$
\Leftrightarrow L^{-1}(1-\delta_i) \leq \frac{-(d-c+b-a)f_i+d(b-a)-b(d-c)-2(d-c)(b-a)\gamma_i}{\alpha_i^{cT}x},
$$

so the problem (5.81) can be transformed into

$$
\begin{cases}
\max \left[\frac{-(d-c+b-a)f_i+d(b-a)-b(d-c)-2(d-c)(b-a)\gamma_i}{\alpha_i^{cT}x} \right] \\
\text{s.t.} \begin{cases} e_r^T x \leq b_r, \quad r = 1,2,\cdots,p \\ x \in X \\ x \geq 0. \end{cases}
\end{cases}
$$

Case 3: $c \leq f_i + L^{-1}(1-\delta_i)\alpha_i^{cT}x \leq a$. We have

$$
f_i \leq d - (d-c)(2\gamma_i-1) - L^{-1}(1-\delta_i)\alpha_i^{cT}x
$$
$$
\Leftrightarrow L^{-1}(1-\delta_i) \leq \frac{-f_i+d-2(d-c)(2\gamma_i-1)}{\alpha_i^{cT}x},
$$

so the problem (5.77) can be transformed into

$$
\begin{cases}
\max \left[\dfrac{-f_i + d - 2(d-c)(2\gamma_i - 1)}{\alpha_i^{cT} x} \right] \\
\text{s.t.} \begin{cases} e_r^T x \le b_r, & r = 1, 2, \cdots, p \\ x \in X \\ x \ge 0. \end{cases}
\end{cases}
$$

Case 4: $f_i + L^{-1}(1 - \delta_i)\alpha_i^{cT} x \le c$. We have

$$
f_i \le c - L^{-1}(1 - \delta_i)\alpha_i^{cT} x \Leftrightarrow L^{-1}(1 - \delta_i) \ge \frac{-f_i + c}{\alpha_i^{cT} x},
$$

so the problem (5.77) can be transformed into

$$
\begin{cases}
\max \left[\dfrac{-f_i + c}{\alpha_i^{cT} x} \right] \\
\text{s.t.} \begin{cases} e_r^T x \le b_r, & r = 1, 2, \cdots, p \\ x \in X \\ x \ge 0. \end{cases}
\end{cases}
$$

This completes the proof.   □

5.5.2.2 ε-Constraint Method

ε-constraint method was proposed by Haimes[5, 6] in 1971. The idea of this method is that we choose a main referenced objective f_{i0}, put the other objective functions into the constraints.

Let's consider the following multi-objective model:

$$
\begin{cases} \min \ [f_i(x), i = 1, 2, \cdots, m] \\ \text{s.t.} \ x \in X. \end{cases} \tag{5.85}
$$

So we use the ε-constraint method, we can get the single objective model (5.86):

$$
\begin{cases} \min \ f_{i0}(x) \\ \text{s.t.} \begin{cases} f_i(x) \le \varepsilon_i, \ i = 1, 2, \cdots, m, i \ne i_0 \\ c \in X, \end{cases} \end{cases} \tag{5.86}
$$

where the parameter ε_i is predetermined by the decision maker, it denote the threshold value that the decision maker will accept, we denote the feasible domain of model (5.86) as X_1.

Theorem 5.14. If $\bar{x}$ is the optimal solution of model (5.86), then $\bar{x}$ is a weak efficient solution of model (5.85).

Proof. Let $\bar{x}$ be the optimal solution of model (5.86), but it is not a weak efficient solution of model (5.85), then there exists $x' \in X$, such that for $\forall i \in \{1,2,\cdots,m\}$, $f_i(x') < f_i(\bar{x})$ holds. Since $\bar{x} \in X_1$, $f_i(\bar{x}) \le \varepsilon_i$ $(i = 1,2,\cdots,m, i \neq i_0)$, So we have

$$f_i(x') < f_i(\bar{x}) \le \varepsilon_i, \ i = 1,2,\cdots,m, i \neq i_0. \tag{5.87}$$

We can obtain from (5.87) that $x' \in X_1$, and $f_{i_0}(x') < f_{i_0}(\bar{x})$. This conflicts with that $\bar{x}$ is the optimal solution. $\qquad\square$

Theorem 5.15. *Let $\bar{x}$ be a efficient solution of model (5.85), then there exists a parameter $\varepsilon_i(i = 1,2,\cdots,m, i \neq i_0)$, such that $\bar{x}$ is the optimal solution of model (5.86).*

Proof. Take $\varepsilon_i = f_i(\bar{x})$ $(i = 1,2,\cdots,m, i \neq i_0)$, by the definition of efficient solution, $\bar{x}$ is a optimal solution of model (5.86). $\qquad\square$

So the advantage of the ε-constraint method is that:
(1). Every efficient solution of model (5.85) can be get by properly choosing parameter $\varepsilon_i(i = 1,2,\cdots,m, i \neq i_0)$.
(2). The i_0th objective are mainly guaranteed, and the other objectives are considered meanwhile.

It is worth for us noticing that the parameter ε_i is important, we should carefully choose it. If the value of every ε_i is too small, then it is possible that the model (5.86) will have no solutions; otherwise, is the value of ε_i is too large, then besides the main objective, the other objective will lose more with higher possibility. Commonly, we can offer the decision maker $f_i^0 = \min_{x \in X} f_i(x)$ $(i = 1,2,\cdots,m)$ and the objective value $(f_1(x), f_2(x), \cdots, f_m(x))^T$ of a certain feasible solution x. And then the decision maker can decide ε_i. For more details, the readers can refer to Chankong [4].

5.5.2.3 Numerical Example

Example 5.8. Let's still consider the Example 5.5. This time we use fuzzy rough dependent chance constrained model to deal with.

$$
\begin{cases}
\max & \left[Ch\{f_1(x, \tilde{\bar{\xi}}) \ge \bar{f}_1\}(\gamma_1), Ch\{f_2(x, \tilde{\bar{\xi}}) \ge \bar{f}_2\}(\gamma_2) \right] \\
s.t. & \begin{cases} x_1 + x_2 + x_3 \le 250 \\ x_1 + x_2 + x_3 \ge 200 \\ x_1 + 4x_2 + 2x_3 \le 600 \\ x_1 \ge 20, x_2 \ge 20, x_3 \ge 20, \end{cases}
\end{cases}
$$

$$f_1 = \tilde{\bar{\xi}}_1 x_1 + \tilde{\bar{\xi}}_2 x_2 + \tilde{\bar{\xi}}_3 x_3 \text{ and } f_2 = c_1 \tilde{\bar{\xi}}_4 x_1 + c_2 \tilde{\bar{\xi}}_5 x_2 + c_3 \tilde{\bar{\xi}}_6 x_3,$$

where $\tilde{\bar{\xi}}$ are Fu-Ro vectors, $Ch\{\cdot\}$ is the chance of the Fu-Ro event, $\gamma_i, i = 1,2$ is the predetermined confidence level, $\bar{f}_i, i = 1,2$ are the ideal levels of every objectives.

Model (5.88) can be written as the following equivalent form (5.88) by introducing $\delta_i, i = 1,2$.

$$\max [\delta_1, \delta_2]$$

$$s.t. \begin{cases} Ch\{\tilde{\bar{\xi}}_1 x_1 + \tilde{\bar{\xi}}_2 x_2 + \tilde{\bar{\xi}}_3 x_3 \geq \bar{f}_1\}(\gamma_1) \geq \delta_1 \\ Ch\{c_1\tilde{\bar{\xi}}_4 x_1 + c_2\tilde{\bar{\xi}}_5 x_2 + c_3\tilde{\bar{\xi}}_6 x_3 \geq \bar{f}_2\}(\gamma_2) \geq \delta_2 \\ x_1 + x_2 + x_3 \leq 250 \\ x_1 + x_2 + x_3 \geq 200 \\ x_1 + 4x_2 + 2x_3 \leq 600 \\ x_1 \geq 20, x_2 \geq 20, x_3 \geq 20. \end{cases}$$

When we use the *Appr − pos* chance measure, model (5.88) and (5.88) can also be written as (5.88).

$$\begin{cases} \max [\delta_1, \delta_2] \\ s.t. \begin{cases} App\{\lambda | Pos\{\tilde{\bar{\xi}}_1 x_1 + \tilde{\bar{\xi}}_2 x_2 + \tilde{\bar{\xi}}_3 x_3 \geq \bar{f}_1\} \geq \delta_1\} \geq \gamma_1 \\ Appr\{\lambda | Pos\{c_1\tilde{\bar{\xi}}_4 x_1 + c_2\tilde{\bar{\xi}}_5 x_2 + c_3\tilde{\bar{\xi}}_6 x_3 \geq \bar{f}_2\} \geq \delta_2\} \geq \gamma_2 \\ x_1 + x_2 + x_3 \leq 250 \\ x_1 + x_2 + x_3 \geq 200 \\ x_1 + 4x_2 + 2x_3 \leq 600 \\ x_1 \geq 20, x_2 \geq 20, x_3 \geq 20. \end{cases} \end{cases}$$

The following is the relevant data:

$$c = (c_1, c_2, c_3) = (1.2, 0.8, 1.5),$$
$$\tilde{\bar{\xi}}_1 = (\rho_1, 0.5, 0.5)_{LR}, \text{ with } \rho_1 \vdash ([1,2],[0,3]),$$
$$\tilde{\bar{\xi}}_2 = (\rho_2, 2, 2)_{LR}, \text{ with } \rho_2 \vdash ([2,3],[1,4]),$$
$$\tilde{\bar{\xi}}_3 = (\rho_3, 1, 1)_{LR}, \text{ with } \rho_3 \vdash ([3,4],[2,5]),$$
$$\tilde{\bar{\xi}}_4 = (\rho_4, 1, 1)_{LR}, \text{ with } \rho_4 \vdash ([0,1],[0,3]),$$
$$\tilde{\bar{\xi}}_5 = (\rho_5, 0.5, 0.5)_{LR}, \text{ with } \rho_5 \vdash ([1,2],[0,3]),$$
$$\tilde{\bar{\xi}}_6 = (\rho_6, 0.5, 0.5)_{LR}, \text{ with } \rho_6 \vdash ([2,3],[0,3]).$$

According to the knowledge of fuzzy variable and rough variable, we have that

$$\begin{aligned} & \tilde{\bar{\xi}}_1 x_1 + \tilde{\bar{\xi}}_2 x_2 + \tilde{\bar{\xi}}_3 x_3 \\ &= (\rho_1 x_1 + \rho_2 x_2 + \rho_3 x_3, 0.5x_1 + 2x_2 + x_3, 0.5x_1 + 2x_2 + x_3)_{LR}, \\ & c_1\tilde{\bar{\xi}}_4 x_1 + c_2\tilde{\bar{\xi}}_5 x_2 + c_3\tilde{\bar{\xi}}_6 x_3 \\ &= (\rho_4 c_1 x_1 + \rho_5 c_2 x_2 + \rho_6 c_3 x_3, 1.2x_1 + 0.4x_2 + 0.75x_3, \\ & \quad 1.2x_1 + 0.4x_2 + 0.75x_3)_{LR}. \end{aligned} \tag{5.88}$$

and

$$\rho_1 x_1 = ([x_1, 2x_1], [0, 3x_1]),$$
$$\rho_2 x_2 = ([2x_2, 3x_2], [x_2, 4x_2]),$$
$$\rho_3 x_3 = ([3x_3, 4x_3], [2x_3, 5x_3]),$$
$$\rho_4 c_1 x_1 = ([0, c_1 x_1], [0, 3c_1 x_1]),$$
$$\rho_5 c_2 x_2 = ([c_2 x_2, 2c_2 x_2], [0, 3c_2 x_2]),$$
$$\rho_6 c_3 x_3 = ([2c_3 x_3, 3c_3 x_3], [0, 3c_3 x_3]),$$

and

$$\rho_1 x_1 + \rho_2 x_2 + \rho_3 x_3$$
$$= ([x_1 + 2x_2 + 3x_3, 2x_1 + 3x_2 + 4x_3], [x_2 + 2x_3, 3x_1 + 4x_2 + 5x_3])$$
$$\rho_4 c_1 x_1 + \rho_5 c_2 x_2 + \rho_6 c_3 x_3$$
$$= ([0.8x_2 + 3x_3, 1.2x_1 + 1.6x_2 + 4.5x_3], [0, 3.6x_1 + 2.4x_2 + 4.5x_3]).$$

Here we consider the case when $b \leq f_i - R^{-1}(\delta_i)\beta_i^{cT} x \leq d$, and readers can try another three cases through the following method.

According to Propositions 5.7 and 5.8, the problem (5.63) is equivalent to the following multi-objective programming problem,

$$
\begin{cases}
\max \ [\delta_1, \delta_2] \\
\text{s.t.} \begin{cases}
R^{-1}(\delta_1) \leq \frac{-\bar{f}_1 + 3x_1 + 4x_2 + 5x_3 - 2\gamma_1(3x_1 + 3x_2 + 3x_3)}{0.5x_1 + 2x_2 + x_3} \\
R^{-1}(\delta_2) \leq \frac{-\bar{f}_2 + 3.6x_1 + 2.4x_2 + 4.5x_3 - 2\gamma_2(3.6x_1 + 2.4x_2 + 4.5x_3)}{1.2x_1 + 0.4x_2 + 0.75x_3} \\
x_1 + x_2 + x_3 \leq 250 \\
x_1 + x_2 + x_3 \geq 200 \\
x_1 + 4x_2 + 2x_3 \leq 600 \\
x_1, x_2, x_3 \geq 0.
\end{cases}
\end{cases}
\tag{5.89}
$$

Since reference function $R(\cdot)$ is non-increasing continuous function, so $\max \delta_i$ is equal to $\min R^{-1}(\delta_i)$, and it is equal to $\min -R^{-1}(\delta_i)$ or model (5.90),

$$
\begin{cases}
\max \ [F_1(x), F_2(x)] \\
\text{s.t.} \begin{cases}
x_1 + x_2 + x_3 \leq 250 \\
x_1 + x_2 + x_3 \geq 200 \\
x_1 + 4x_2 + 2x_3 \leq 600 \\
x_1, x_2, x_3 \geq 0,
\end{cases}
\end{cases}
\tag{5.90}
$$

where

$$F_1(x) := \frac{-\bar{f}_1 + 3x_1 + 4x_2 + 5x_3 - 2\gamma_1(3x_1 + 3x_2 + 3x_3)}{0.5x_1 + 2x_2 + x_3},$$
$$F_2(x) := \frac{-\bar{f}_2 + 3.6x_1 + 2.4x_2 + 4.5x_3 - 2\gamma_2(3.6x_1 + 2.4x_2 + 4.5x_3)}{1.2x_1 + 0.4x_2 + 0.75x_3}.$$

H_i^0 and $H_i^1 (i = 1, 2)$ are calculated as follows:

$$H_1^1 = -599.91, \quad H_1^0 = -850.83, \quad H_2^1 = -751.47, \quad H_2^0 = -1126.83.$$

Then we can use the ε-constraint method to solve it. We suppose that the first objective F_1 is the main objective, and we set $\varepsilon_2 = 4, \bar{f}_1 = -800, \bar{f}_2 = -1000$, and $\gamma_1 = \gamma_2 = 0.9$. We can get the following model (5.91),

$$
\begin{cases}
\max\ F_1(x) \\
\text{s.t.}
\begin{cases}
F_2(x) \geq \varepsilon_2 \\
x_1 + x_2 + x_3 \leq 250 \\
x_1 + x_2 + x_3 \geq 200 \\
x_1 + 4x_2 + 2x_3 \leq 600 \\
x_1, x_2, x_3 \geq 0.
\end{cases}
\end{cases}
\tag{5.91}
$$

After calculating the model (5.91), we obtain the follow efficient solution

$$(x_1, x_2, x_3) = (26.09, 113.04, 60.87).$$

5.5.3 Non-linear Fu-Ro DCM and Fu-Ro Simulation Based Parallel TS

For the Fu-Ro DCM, we adopted the Fu-Ro simulation 3 based parallel TS to solve.

5.5.3.1 Fu-Ro Simulation 3 for Chance

First we introduce the simulation for α-chance of Fu-Ro variables which is very important in Fu-Ro DCM.

Suppose that ξ is an n-dimensional fuzzy rough vector defined on the rough space $(\Lambda, \Delta, \mathscr{A}, \pi)$, and $f : \mathbf{R}^n \to \mathbf{R}^m$ is a measurable function. For any real number $\alpha \in (0,1]$, we design a fuzzy rough simulation to compute the α-chance $Ch\{f(\xi) \leq 0\}(\alpha)$. That is, we should find the supremum $\bar{\beta}$ such that

$$Appr\{\lambda \in \Lambda \,|\, Cr\{f(\xi(\lambda)) \leq 0\} \geq \bar{\beta}\} \geq \alpha. \tag{5.92}$$

We sample $\underline{\lambda}_1, \underline{\lambda}_2, \cdots, \underline{\lambda}_N$ from $\triangle$ and $\overline{\lambda}_1, \overline{\lambda}_2, \cdots, \overline{\lambda}_N$ from Λ according to the measure π. For any number v, let $\underline{N}(v)$ denote the number of $\underline{\lambda}_k$ satisfying $Cr\{f(\xi(\underline{\lambda}_k)) \leq 0\} \geq v$ for $k = 1, 2, \cdots, N$, and $\overline{N}(v)$ denote the number of $\overline{\lambda}_k$ satisfying

$$Cr\{f(\xi(\overline{\lambda}_k)) \leq 0\} \geq v, \tag{5.93}$$

for $k = 1, 2, \cdots, N$, where $Cr\{\cdot\}$ may be estimated by fuzzy simulation. Then we may find the maximal value v such that

$$\frac{\underline{N}(v) + \overline{N}(v)}{2N} \geq \alpha. \tag{5.94}$$

This value is an estimation of $\bar{\beta}$.

The procedure is as follows:

Step 1. Generate $\underline{\lambda}_1, \underline{\lambda}_2, \cdots, \underline{\lambda}_N$ from $\triangle$ according to the measure π.

Step 2. Generate $\overline{\lambda}_1, \overline{\lambda}_2, \cdots, \overline{\lambda}_N$ from Λ according to the measure π.

Step 3. Find the maximal value v such that (5.94) holds.

Step 4. Return v.

Example 5.9. Suppose the Fu-Ro variables ξ_1 and ξ_2 are defined as follows:

$$\xi_1 = (\rho_1, \rho_1 + 1, \rho_1 + 2), \text{ with } \rho_1 = ([1,2],[0,3]),$$
$$\xi_1 = (\rho_2, \rho_2 + 1, \rho_2 + 2), \text{ with } \rho_1 = ([2,3],[1,4]).$$

After a run of Fu-Ro simulation 3 with 5000 cycles, we get that

$$Ch\{\xi_1 + \xi_2\}(0.9) = 0.72.$$

5.5.3.2 Parallel TS

We introduce the parallel tabu search algorithm to solve the multi-objective problem.

TS is an efficient tool to solve the multi-objective problems. However, as the problem size gets larger, TS has some drawbacks:

(a) TS needs to compute the objective function for solution candidates in the neighborhood around a solution at each iteration. The calculation is very time consuming in large-scale problems. The large size problem often gives a large neighborhood even though the neighborhood is defined as a set of solution candidates with the Hamming distance equal to 1.

(b) The complicated non-linear optimal problem has many local minima in large scale problems. That implies that one-point search does not give satisfactory solutions due to the huge search space. Complicated optimal problems require solution diversity.

In this section, the decomposition of the neighborhood accommodates drawback. The neighborhood is decomposed into several sub-neighborhoods. A processor may be assigned to each sub-neighborhood so that the best solution candidate is selected independently in each sub-neighborhood. After selecting the best solution in each sub-neighborhood, the best solution is eventually selected from the best solutions in the sub-neighborhood. Also, the multiple Tabu lengths is proposed to deal with the multi-objective problem with fuzzy rough parameters. TS itself has only one Tabu length. Moreover, it is important to find out better solutions from different directions rather than from only one direction for a longer period. Namely it is effective to make the solution search process more diverse.

Many classifications of parallel TS algorithms have been proposed [326, 327]. They are based on many criteria: number of initial solutions, identical or different parameter settings, control and communications strategies. We have identified two main categories (Figure 5.12).

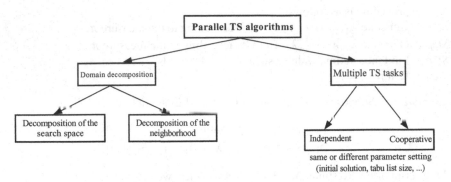

Fig. 5.12 Hierarchical classification of parallel TS strategies.

Domain decomposition: Parallelism in this class of algorithms relies exclusively on:

(1) **The decomposition of the search space:** the main problem is decomposed into a number of smaller subproblems, each subproblem being solved by a different TS algorithm [328].

(2)**The decomposition of the neighborhood:** the search for the best neighbor at each iteration is performed in parallel, and each task evaluates a different subset of the partitioned neighborhood [329, 330].

A high degree of synchronisation is required to implement this class of algorithms.

Multiple tabu search tasks: This class of algorithms consists in executing multiple TS algorithms in parallel. The di.erent TS tasks start with the same or di.erent parameter values (initial solution, tabu list size, maximum number of iterations, etc.). Tabu tasks may be independent (without communication)[331, 332] or cooperative. A cooperative algorithm has been proposed in [327], where each task performs a given number of iterations, then broadcasts the best solution. The best of all solutions becomes the initial solution for the next phase.

Parallelizing the exploration of the search space or the neighborhood is problem-dependent. This assumption is strong and is met only for few problems. The second class of algorithms is less restrictive and then more general. A parallel algorithm that combines the two approaches (two-level parallel organization) has been proposed in [333].

We can extend this classification by introducing a new taxonomy dimension: the way scheduling of tasks over processors is done. Parallel TS algorithms fall into three categories depending on whether the number and/or the location of work (tasks, data) depend or not on the load state of the parallel machine (Table 5.5):

Non-adaptive: This category represents parallel TS in which both the number of tasks of the application and the location of work (tasks or data) are generated at compile time (static scheduling). The allocation of processors to tasks (or data)

Table 5.5 Another taxonomy dimension for parallel TS algorithms

| | Tasks or Data | |
	Number	Location
Non-adaptive	Static	Static
Semi-adaptive	Static	Dynamic
Adaptive	Dynamic	Dynamic

remains unchanged during the execution of the application regardless of the current state of the parallel machine. Most of the proposed algorithms belong to this class.

An example of such an approach is presented in [334]. The neighborhood is partitioned in equal size partitions depending on the number of workers, which is equal to the number of processors of the parallel machine. In [330], the number of tasks generated depends on the size of the problem and is equal to n^2, where n is the problem size.

When there are noticeable load or power differences between processors, the search time of the non-adaptive approach presented is derived by the maximum execution time over all processors (highly loaded processor or the least powerful processor). A significant number of tasks are often idle waiting for other tasks to complete their work.

Semi-adaptive: To improve the performance of the parallel non adaptive TS algorithms, dynamic load balancing must be introduced [333, 334]. This class represents applications for which the number of tasks is fixed at compile-time, but the locations of work (tasks, data) are determined and/or changed at run-time (as seen in Table 5.5). Load balancing requirements are met in [334] by a dynamic redistribution of work between processors. During the search, each time a task finishes its work, it proceeds to a work-demand. Dynamic load balancing through partition of the neighborhood is done by migrating data.

However, the parallelism degree in this class of algorithms is not related to load variation in the parallel system: when the number of tasks exceeds the number of idle nodes, multiple tasks are assigned to the same node. Moreover, when there are more idle nodes than tasks, some of them will not be used.

Adaptive: A parallel adaptive program refers to a parallel computation with a dynamically changing set of tasks. Tasks may be created or killed function of the load state of the parallel machine. Di.erent types of load state dessimination schemes may be used [337]. A task is created automatically when a processor becomes idle. When a processor becomes busy, the task is killed. Next, let's introduce the design about the parallel adaptive TS introduced by Talbi [338].

The programming style used is the master/workers paradigm. The master task generates work to be processed by the workers. Each worker task receives a work from the master, computes a result and sends it back to the master. The master/ workers paradigm works well in adaptive dynamic environments because:

(1) when a new node becomes available, a worker task can be started there,

(2) when a node becomes busy, the master task gets back the pending work which was being computed on this node, to be computed on the next available node.

The master implements a central memory through which passes all communication, and that captures the global knowledge acquired during the search. The number of workers created initially by the master is equal to the number of idle nodes in the parallel platform. Each worker implements a sequential TS task. The initial solution is generated randomly and the tabu list is empty. The parallel adaptive TS algorithm reacts to two events (Figure 5.13):

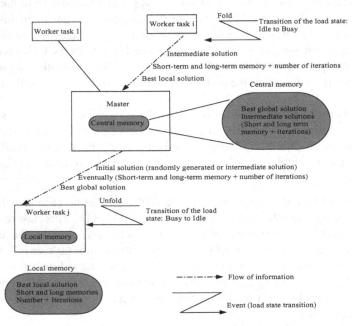

Fig. 5.13 Architecture of the parallel adaptive TS.

Appransition of the load state of a node from idle to busy: If a node hosting a worker becomes loaded, the master folds up the application by withdrawing the worker. The concerned worker puts back all pending work to the master and dies. The pending work is composed of the current solution, the best local solution found, the short-term memory, the long-term memory and the number of iterations done without improving the best solution. The master updates the best global solution if it's worst than the best local solution received.

Appransition of the load state of a node from busy to idle: When a node becomes available, the master unfolds the application by starting a new worker on it. Before starting a sequential TS, the worker task gets the values of the different parameters from the master: the best global solution and an initial solution which may be an intermediate solution found by a folded TS task, which constitutes a "good" initial solution. In this case, the worker receives also the state of the short-term memory, the long-term memory and the number of iterations done without improving the best solution.

The local memory of each TS task which defines the pending work is composed of (Figure 5.13): the best solution found by the task, the number of iterations applied, the intermediate solution and the adaptive memory of the search (short-term and long-term memories). The central memory in the master is then composed of (Figure 5.13): the best global solution found by all TS tasks, the dierent intermediate solutions with the associated number of iterations and adaptive memory.

5.5.3.3 Numerical Example

Example 5.10. Consider the following problem,

$$
\begin{cases}
\max f_1(x) = Ch\{\bar{\bar{\xi}}_1 x_1 + \bar{\bar{\xi}}_2 x_2 + \bar{\bar{\xi}}_3 x_3 \geq f_1\}(\alpha) \\
\max f_2(x) = Ch\{\sqrt{c_1 \bar{\bar{\xi}}_4 x_1 + c_2 \bar{\bar{\xi}}_5 x_2 + c_3 \bar{\bar{\xi}}_6 x_3} \geq f_2\}(\beta) \\
\text{s.t.} \begin{cases}
x_1 + x_2 + x_3 \leq 250 \\
x_1 + x_2 + x_3 \geq 200 \\
x_1 + 4x_2 + 2x_3 \leq 600 \\
x_1 \geq 20, x_2 \geq 20, x_3 \geq 20,
\end{cases}
\end{cases}
$$

where $c = (c_1, c_2, c_3) = (1.2, 0.8, 1.5)$,

$$
\begin{aligned}
\bar{\bar{\xi}}_1 &= (\rho_1, 1, 1), \text{ with } \rho_1 \vdash ([1,2], [0,3]), \\
\bar{\bar{\xi}}_2 &= (\rho_2, 1, 1), \text{ with } \rho_2 \vdash ([2,3], [1,4]), \\
\bar{\bar{\xi}}_3 &= (\rho_3, 1, 1), \text{ with } \rho_3 \vdash ([3,4], [2,5]), \\
\bar{\bar{\xi}}_4 &= (\rho_4, 1, 1), \text{ with } \rho_4 \vdash ([0,1], [0,3]), \\
\bar{\bar{\xi}}_5 &= (\rho_5, 1, 1), \text{ with } \rho_5 \vdash ([1,2], [0,3]), \\
\bar{\bar{\xi}}_6 &= (\rho_6, 1, 1), \text{ with } \rho_6 \vdash ([2,3], [0,3]),
\end{aligned}
$$

and $\bar{\bar{\xi}}_i (i = 1, 2, \cdots, 6)$ are Fu-Ro variables. We set $\alpha = \beta = 0.9$, $f_1 = 1500$, and $f_2 = 1300$.

Next, we apply the tabu search algorithm based on the fuzzy rough simulation to solve the nonlinear programming problem (5.10) with the fuzzy rough parameters.

Step 1. Set the move step $h = 0.5$ and the h neighbor $N(x, h)$ for the present point x is defined as follows,

$$
N(x, h) = \left\{ y \mid \sqrt{(x_1 - y_1)^2 + (x_2 - y_2)^2 + (x_3 - y_3)^2} \leq h \right\}.
$$

The random move of point x to point y in its h neighbor along direction s is given by

$$
y_s = x_s + rh,
$$

where r is a random number that belongs to $[0,1]$, $s = 1, 2, 3$.

Step 2. Give the step set $H = \{h_1, h_2, \cdots, h_r\}$ and randomly generate a feasible point $x_0 \in X$. One should empty the Tabu list T (the list of inactive steps) at the beginning.

Step 3. For each active neighbor $N(x, h)$ of the present point x, where $h \in H - T$, a feasible random move that satisfies all the constraints in problem (5.10) is to be generated.

Step 4. Construct the single objective function as follows,

$$f(x, \xi) = w_1 Ch\{\bar{\bar{\xi}}_1 x_1 + \bar{\bar{\xi}}_2 x_2 + \bar{\bar{\xi}}_3 x_3 \geq f_1\}(\alpha)$$
$$+ w_2 Ch\{c_1 \bar{\bar{\xi}}_4 x_1 + c_2 \bar{\bar{\xi}}_5 x_2 + c_3 \bar{\bar{\xi}}_6 x_3 \geq f_2\}(\beta),$$

where $w_1 + w_2 = 1$. Compare the $f(x, \xi)$ of the feasible moves with that of the current solution by the fuzzy rough simulation. If an augmenter in new objective function of the feasible moves exists, one should save this feasible move as the updated current one by adding the corresponding step to the Tabu list T and go to the next step; otherwise, go to the next step directly.

Step 5. Stop if the termination criteria are satisfied; other wise, empty T if it is full; then go to Step 3. Here, we set the computation is determined if the better solution doesn't change again.

Table 5.6 The result computed by parametric TS algorithm

ω_1	ω_2	ω_3	ω_4	ω_5	x_1	x_2	x_3	x_4	x_5
0.40	0.15	0.15	0.15	0.15	50.48	59.14	80.17	50.12	60.00
0.15	0.40	0.15	0.15	0.15	50.47	59.15	80.17	50.12	60.00
0.15	0.15	0.40	0.15	0.15	50.47	59.14	80.18	50.12	60.00
0.15	0.15	0.15	0.40	0.15	50.47	59.14	80.17	50.13	60.00
0.15	0.15	0.15	0.15	0.40	50.48	59.14	80.17	50.12	60.00

5.6 Application to Integrated Logistics Network Design Problem

Here we consider the problem proposed in section 5.1, and we consider the demand and the amount of the collected recycling packages as triangular fuzzy variables $(\xi - l, \xi, \xi + r)$ from the view point of credibility theory, in which the value of ξ is a rough variable $([a, b], [c, d])$, and l, r are the left spread and the right spread of the triangular fuzzy variable. Therefore, a logistics network design problem with fuzzy rough parameters appears. In this case, a fuzzy rough variable can be used to deal with this kind of combined uncertainty of fuzziness and roughness. Building the model and solving the problem of logistics network design in a fuzzy rough environment is a new area of research interest.

5.6.1 Modelling for Integrated Reuse Logistics Network under Fuzzy Rough Environment

In the following text of this section, we present the details of modelling for the reuse integrated logistics network.

5.6.1.1 Notation

The symbols of the proposed model are defined as follows:

(1) Indices:

i: the location of producers ($i = 1, 2, \cdots, I$),

j: the location of distributors ($i = 1, 2, \cdots, I$),

k: the location of collectors/wholesalers ($i = 1, 2, \cdots, I$),

t: the alternative place of recyclers ($i = 1, 2, \cdots, I$).

(2) Variables:

x_{ij}^{PD}: the quantity of products from producer i to distributor j,

x_{jk}^{DC}: the quantity of products from distributor j to wholesaler k,

x_{kt}^{CR}: the quantity of packages from collector k to recycler t,

x_{kj}^{CD}: the quantity of packages from collector k to distributor j,

x_{ti}^{RP}: the quantity of packages from recycler t to producer i,

x_{ji}^{DP}: the quantity of packages from distributor j to producer i,

x_i: the quantity of new packages bought by producer i,

y_t^R: 0-1 variable, whether the alternative recycler t will be chosen or not, 0 denotes we don't choose, 1 denotes we choose it,

y_j^D: 0-1 variable, whether the distributor j will be expanded or not, 0 denotes we don't expand, 1 denotes we expand it,

y_{jk}^{DC}: 0-1 variable, whether the distributor j will send products to wholesaler k, 0 denotes will not send, 1 denotes will send.

(3) Fu-Ro parameters:

$\tilde{\bar{D}}_k$ denote the demand of wholesaler K,

$\tilde{\bar{R}}_k$ denote the quantity of the recycling packages collected by collector k,

$\tilde{\bar{T}}_k^{Lim}$: the time limit of wholesaler k.

(4) Certain parameters:

C_{ab}: the unit transport cost from a to b (a and b could denote producer, distributor, recycling center or wholesaler),

T_{ab}: the transport time from a to b,

V_t^R: the variable cost of recycler t processing unit package,

V_j^D: the variable cost of expanded distributor j processing unit package,

T_t^R: the time of recycler t processing unit package,

T_j^D: the time of expanded distributor j processing unit package. F_t^R: the fixed cost of building a recycler t,

F_j^D: the fixed cost of expanding a distributor j,

Q_t^R: the capacity of recycler t processing packages,

Q_j^D: the capacity of expanded distributor j processing packages,

α_t^R: the discard proportion after recycler t processing packages,
α_j^D: the discard proportion after expanded distributor j processing packages,
N^R: the ceiling number of recyclers,
N^D: the ceiling number of expanded distributors,
U_k^l: the unit default cost when the demand of wholesaler k are not met,
U_k^e: the processing cost when the supply to wholesaler k are excessive,
P_i: the variable cost of producer i buying unit package,
U_k^D: the disposal cost when the time limits are not satisfied.

5.6.1.2 Modelling

We built the following mathematical model according to the conceptual model.

The first objective is minimizing total costs. After analysis, we conclude that there are six parts which should be included in this objective, as follows: The first part of the objective is the total transportation cost,

$$[\sum_{i\in I}\sum_{j\in J}x_{ij}^{PD}\bar{\bar{C}}_{ij} + \sum_{j\in J}\sum_{k\in K}x_{jk}^{DC}\bar{\bar{C}}_{jk} + \sum_{k\in K}\sum_{t\in T}x_{kt}^{CR}\bar{\bar{C}}_{kt}$$
$$+ \sum_{t\in T}\sum_{i\in I}x_{ti}^{RP}\bar{\bar{C}}_{ti} + \sum_{k\in K}\sum_{j\in J}x_{kj}^{CD}\bar{\bar{C}}_{kj} + \sum_{j\in J}\sum_{i\in I}x_{ji}^{DP}\bar{\bar{C}}_{ji}]_1.$$

The second part is the total fixed cost of building recycling centers and expanding the distribution centers,

$$[\sum_{t\in T}y_t^R F_t^R + \sum_{j\in J}y_j^D F_j^D]_2.$$

The third part is the total variable cost of processing packages,

$$[\sum_{k\in K}\sum_{t\in T}x_{kt}^{CR}V_t^R + \sum_{k\in K}\sum_{j\in J}x_{kj}^{CD}V_j^D]_3.$$

The fourth part is the cost of buying new packages,

$$[\sum_{j\in J}x_i P_i]_4.$$

The fifth part is the cost when there exists an imbalance between supply and demand, when the supply is less than the demand, there will occur default costs, or when the supply is more than the demand, the redundant products will be processed at a cost,

$$[\sum_{k\in K}U_k^l\max(\bar{\bar{D}}_k - \sum_{j\in J}x_{jk}^{DC},0) + \sum_{k\in K}U_k^e\max(\sum_{j\in J}x_{jk}^{DC} - \bar{\bar{D}}_k,0)]_5.$$

The sixth part is the cost of disposing of the un-useable packages,

$$[\sum_{k\in K}\sum_{t\in T}x_{kt}^{CR}\alpha_t^R U^P + \sum_{k\in K}\sum_{j\in J}x_{kj}^{CD}\alpha_j^D U^P]_6.$$

However, minimizing total costs is not the only objective of a logistics company. Shortening the time taken in the distribution and recycling is also required. Hence the second objective is to minimize total time.

$$TotalTime = \sum_{t \in T} \sum_{k \in K} \bar{\bar{T}}_{kt} y_t^R + \sum_{t \in T} (T_t^R \sum_{k \in K} x_{kt}^{CR}) + \sum_{i \in I} \sum_{t \in T} \bar{\bar{T}}_{ti} y_t^R$$
$$\sum_{j \in J} \sum_{k \in K} \bar{\bar{T}}_{kj} y_j^D + \sum_{j \in J} (T_j^D \sum_{k \in K} x_{kj}^{CD}) + \sum_{i \in I} \sum_{j \in J} \bar{\bar{T}}_{ji} y_j^D.$$

Now we can obtain the objectives function as shown in (5.95):

$$\min C =$$
$$\sum_{i \in I} \sum_{j \in J} x_{ij}^{PD} \bar{\bar{C}}_{ij} + \sum_{j \in J} \sum_{k \in K} x_{jk}^{DC} \bar{\bar{C}}_{jk} + \sum_{k \in K} \sum_{t \in T} x_{kt}^{CR} \bar{\bar{C}}_{kt} + \sum_{t \in T} \sum_{i \in I} x_{ti}^{RP} \bar{\bar{C}}_{ti}$$
$$+ \sum_{k \in K} \sum_{j \in J} x_{kj}^{CD} \bar{\bar{C}}_{kj} + \sum_{j \in J} \sum_{i \in I} x_{ji}^{DP} \bar{\bar{C}}_{ji} + \sum_{t \in T} y_t^R F_t^R + \sum_{j \in J} y_j^D F_j^D + \sum_{k \in K} \sum_{t \in T} x_{kt}^{CR} V_t^R$$
$$+ \sum_{k \in K} \sum_{j \in J} x_{kj}^{CD} V_j^D + \sum_{j \in J} x_i P_i + \sum_{k \in K} U_k^I \max(\bar{\bar{D}}_k - \sum_{j \in J} x_{jk}^{DC}, 0)$$
$$+ \sum_{k \in K} U_k^e \max(\sum_{j \in J} x_{jk}^{DC} - \bar{\bar{D}}_k, 0) + \sum_{k \in K} \sum_{t \in T} x_{kt}^{CR} \alpha_t^R U^P + \sum_{k \in K} \sum_{j \in J} x_{kj}^{CD} \alpha_j^D U^P,$$

$$\min T =$$
$$\sum_{t \in T} \sum_{k \in K} \bar{\bar{T}}_{kt} y_t^R + \sum_{t \in T} (T_t^R \sum_{k \in K} x_{kt}^{CR}) + \sum_{i \in I} \sum_{t \in T} \bar{\bar{T}}_{ti} y_t^R$$
$$\sum_{j \in J} \sum_{k \in K} \bar{\bar{T}}_{kj} y_j^D + \sum_{j \in J} (T_j^D \sum_{k \in K} x_{kj}^{CD}) + \sum_{i \in I} \sum_{j \in J} \bar{\bar{T}}_{ji} y_j^D.$$

(5.95)

These two objectives are subject to the following constraints.

(1) Balance constraints:

For every node in Figure 2, the inflow and the outflow must be balanced, such that the total recycling quantity of the recycling centers and expanded distribution centers should be less than or equal to the quantity of used package collected by the collectors. Also the quantity discarded from the recycling center (expanded distribution center) to the disposal place should be less than or equal to the quantity discarded of the recycling center (expanded distribution center), and the total quantity of bottles including recycled bottles and new bottles should be used to produce new products. For one distribution center, the inflow should be equal to the outflow, so we have the following (5.96-5.100) constraints,

$$\sum_{t \in T} x_{kt}^{CR} + \sum_{j \in J} x_{kj}^{CD} \leq \bar{\bar{R}}_k, k \in K,$$ (5.96)

$$\sum_{i \in I} x_{ti}^{RP} \leq (1 - \alpha_t^R) \sum_{k \in K} x_{kt}^{CR}, t \in T,$$ (5.97)

$$\sum_{i \in I} x_{ji}^{DP} \leq (1 - \alpha_j^D) \sum_{k \in K} x_{kj}^{CD}, j \in J,$$ (5.98)

$$\sum_{t \in T} x_{ti}^{RP} + \sum_{j \in J} x_{ji}^{DP} + x_i = \sum_{j \in J} x_{ij}^{PD}, i \in I,$$ (5.99)

$$\sum_{j\in J} x_{ij}^{PD} = \sum_{k\in K} x_{jk}^{DC}, j \in J. \tag{5.100}$$

(2) Capacity constraints:

There are some limits on capacity of the recycling centers and the expanded distribution centers, so we have constraint (5.101) and (5.102),

$$\sum_{k\in K} x_{kt}^{CR} \le y_t^R Q_t^R, t \in T, \tag{5.101}$$

$$\sum_{k\in K} x_{kj}^{CD} \le y_j^D Q_j^D, j \in J. \tag{5.102}$$

(3) Number constraints:

Before setting up a network, because of capital or other reasons, the decision maker will give the numbers of recycling centers and expanded distribution centers, so we have the following (5.103) and (5.104) constraints,

$$\sum_{t\in T} y_t^R \le N^R, \tag{5.103}$$

$$\sum_{j\in J} y_j^D \le N^D. \tag{5.104}$$

(4) Time constraints:

For every wholesaler, the total transport time is required to be under a time limit, so we have constraint (5.105),

$$\sum_{j\in J} \bar{\tilde{T}}_{jk} y_{jk}^{DC} \le T_k^{Lim}, k \in K. \tag{5.105}$$

(5) Logical constraints:

In order to describe some non-negative variables and 0-1 variables in the model, we present constraint (5.106) and (5.107),

$$x_{ij}^{PD}, x_{jk}^{DC}, x_{kt}^{CR}, x_{kj}^{CD}, x_{ti}^{RP}, x_{ji}^{DP}, x_i \ge 0, i \in I, j \in J, k \in K, t \in T, \tag{5.106}$$

$$y_t^R, y_j^D, y_{jk}^{DC} = \{0,1\}, j \in J, k \in K, t \in T. \tag{5.107}$$

5.6.2 Uncertain Linear Multi-objective Model

It's obvious that the above model is non-linear, because the fifth and the seventh part exist in the first objective function. In order to simplify it, we changed it to an uncertain linear multi-objective model by adding the constraints (5.108)-(5.110).

$$e_k^- = \bar{\tilde{D}}_k - \sum_{j\in J} x_{jk}^{DC}, k \in K, \tag{5.108}$$

$$e_k^+ = \sum_{j \in J} x_{jk}^{DC} - \bar{\bar{D}}_k, k \in K, \tag{5.109}$$

$$e_k^-, e_k^+ \geq 0. \tag{5.110}$$

We proposed the Fu-Ro linear multi-objective model for integrated logistics as follows:

$$
\begin{aligned}
\min C = & \sum_{i \in I} \sum_{j \in J} x_{ij}^{PD} \bar{\bar{C}}_{ij} + \sum_{j \in J} \sum_{k \in K} x_{jk}^{DC} \bar{\bar{C}}_{jk} + \sum_{k \in K} \sum_{t \in T} x_{kt}^{CR} \bar{\bar{C}}_{kt} + \sum_{t \in T} \sum_{i \in I} x_{ti}^{RP} \bar{\bar{C}}_{ti} + \sum_{k \in K} \sum_{j \in J} x_{kj}^{CD} \bar{\bar{C}}_{kj} \\
& + \sum_{j \in J} \sum_{i \in I} x_{ji}^{DP} \bar{\bar{C}}_{ji} + \sum_{t \in T} y_t^R F_t^R + \sum_{j \in J} y_j^D F_j^D + \sum_{k \in K} \sum_{t \in T} x_{kt}^{CR} V_t^R + \sum_{k \in K} \sum_{j \in J} x_{kj}^{CD} V_j^D + \sum_{j \in J} x_i P_i \\
& + \sum_{k \in K} U_k^l e_k^- + \sum_{k \in K} U_k^e e_k^+ + \sum_{k \in K} \sum_{t \in T} x_{kt}^{CR} \alpha_t^R U^P + \sum_{k \in K} \sum_{j \in J} x_{kj}^{CD} \alpha_j^D U^P
\end{aligned}
$$

$$
\begin{aligned}
\min T = & \sum_{t \in T} \sum_{k \in K} \bar{\bar{T}}_{kt} y_t^R + \sum_{t \in T} (T_t^R \sum_{k \in K} x_{kt}^{CR}) + \sum_{i \in I} \sum_{t \in T} \bar{\bar{T}}_{ti} y_t^R + \sum_{j \in J} \sum_{k \in K} \bar{\bar{T}}_{kj} y_j^D + \sum_{j \in J} (T_j^D \sum_{k \in K} x_{kj}^{CD}) \\
& + \sum_{i \in I} \sum_{j \in J} \bar{\bar{T}}_{ji} y_j^D
\end{aligned}
$$

$$
s.t. \begin{cases}
\sum_{t \in T} x_{kt}^{CR} + \sum_{j \in J} x_{kj}^{CD} \leq \bar{\bar{R}}_k, k \in K \\
\sum_{i \in I} x_{ti}^{RP} \leq (1 - \alpha_t^R) \sum_{k \in K} x_{kt}^{CR}, t \in T \\
\sum_{i \in I} x_{ji}^{DP} \leq (1 - \alpha_j^D) \sum_{k \in K} x_{kj}^{CD}, j \in J \\
\sum_{t \in T} x_{ti}^{RP} + \sum_{j \in J} x_{ji}^{DP} + x_i = \sum_{j \in J} x_{ij}^{PD}, i \in I \\
\sum_{j \in J} x_{ij}^{PD} = \sum_{k \in K} x_{jk}^{DC}, j \in J \\
\sum_{k \in K} x_{kt}^{CR} \leq y_t^R Q_t^R, t \in T \\
\sum_{k \in K} x_{kj}^{CD} \leq y_j^D Q_j^D, j \in J \\
\sum_{t \in T} y_t^R \leq N^R \\
\sum_{j \in J} y_j^D \leq N^D \\
\sum_{j \in J} \bar{\bar{T}}_{jk} y_{jk}^{DC} \leq T_k^{Lim}, k \in K \\
e_k^- = \bar{\bar{D}}_k - \sum_{j \in J} x_{jk}^{DC}, k \in K \\
e_k^+ = \sum_{j \in J} x_{jk}^{DC} - \bar{\bar{D}}_k, k \in K \\
x_{ij}^{PD}, x_{jk}^{DC}, x_{kt}^{CR}, x_{kj}^{CD}, x_{ti}^{RP}, x_{ji}^{DP}, x_i, e_k^-, e_k^+ \geq 0, i \in I, j \in J, k \in K, t \in T \\
y_t^R, y_j^D, y_{jk}^{DC} = \{0,1\}, j \in J, k \in K, t \in T.
\end{cases}
$$

$$\tag{5.111}$$

The model we proposed is actually a Fu-Ro two-objective linear model, and both of the two objectives are needed for optimization. These two objectives are uncomparable, and there exists inconsistency between them. When we want to reduce the transportation time, but have a large number of recycling centers, we could reduce the number of recycling centers, so it will reduce the cost of building these centers, but the transportation time will inevitably rise.

Since the model (5.111) is including Fu-Ro variables, we need to use the Fu-Ro expected value operator to handle the objective functions and Fu-Ro chance-constrained operator to deal with the constraints.

5.6.3 Application to Beer Company

The beer company Lan Ma was set up in the year 2000 and is located in Xi'an in China's Shanxi province and it has developed successfully for the years of its operation. This enterprise has 2 production plants in Xian Yang and 3 distribution centers, each with the responsibility for a section of Shanxi - Guan Zhong, Shan Bei and Shan Nan. There are 5 main wholesalers and they are located in Wei Nan, Shang Luo, Han Zhong, An Kang and Yan An.

This company wants to establish integrated logistics through building up recycling centers or expanding the existing distribution centers, and integrate the forward logistics and reverse logistics to a loop logistics network which has the abilities of production, distribution, recycle and reuse. So we used this model to help the company to program an integrated logistics network.

At present, according to the survey results, there are four options which could be used to establish new recycling centers, and all three existing distribution centers could be expanded. The alternative locations are Zhou Zhi, Pu Cheng, Zha Shui and Hua Xian. The largest processing capacities of these 4 places are 20000, 23000, 15000, and 27000. The fixed construction costs are 12.5, 16.5, 10 and 19.5(*10000RMB). We suppose the discard proportions are all 0.2 and they want to build 3 recycling centers at the most. We also could expand the 3 distribution centers to process the recycled packages, the expanding costs are 6.6, 5.4 and 7(*10000RMB), their capacities are 11000, 9000, and 12000, the discard proportions are all 0.2. The company has requested that we expand 2 at the most. The price of a new bottle is 0.7(RMB). The other data are as follows.

Table 5.7 Amount of recycling and demand

Wholesaler	Recycling amount	Demand
Wein	$(\xi_1, 100, 100)_{LR}$, $\xi_1 \vdash([8000,10000],[8500,9500])$	$(\xi_6, 100, 100)_{LR}$, $\xi_6 \vdash([10000,12000],[10500,11500])$
Shangl	$(\xi_2, 50, 50)_{LR}$, $\xi_2 \vdash([6000,7000],[6250,6750])$	$(\xi_7, 100, 100)_{LR}$, $\xi_7 \vdash([7000,8000],[7250,7750])$
Hanzh	$(\xi_3, 100, 100)_{LR}$, $\xi_3 \vdash([12000,14000],[12500,13500])$	$(\xi_8, 100, 100)_{LR}$, $\xi_8 \vdash([14000,16000],[14500,15500])$
Ank	$(\xi_4, 50, 50)_{LR}$, $\xi_4 \vdash([10000,11000],[10250,10750])$	$(\xi_9, 100, 100)_{LR}$, $\xi_9 \vdash([11000,13000],[11500,12500])$
Yan an	$(\xi_5, 100, 100)_{LR}$, $\xi_5 \vdash([16000,18000],[16500,17500])$	$(\xi_{10}, 100, 100)_{LR}$, $\xi_{10} \vdash([18000,20000],[18500,19500])$

Transport cost, and time are triangular fuzzy numbers with the left and the right spread 0.02 and 0.1, the middle value of the triangular fuzzy variable are rough variables which shown in following Table. 5.9

We introduced the above data of the company into the proposed model, and got the integrated logistics network model for this Lan Ma beer company. After solving it, we can provide some advice to help the leader make strategic decisions about constructing the integrated logistics network system.

Table 5.8 Default processing cost and time limit (h) of every wholesaler

Wholesaler	Wein	Shangl	Hanzh	Ank	Yan an
Cost of short supply	1.2	1	1.1	1	1.5
Cost of excessive supply	1.8	1.9	1.6	1.3	1.5
Time limit	4	4	3.5	4	3.5
Default cost	3000	3500	4000	3000	4500

Table 5.9 The expected value of transport cost, and time (h) from collectors to recyclers

	Wein	Shangl	Hanzh	Ank	Yan an
Zhouzh	0.1	0.12	0.1	0.05	0.12
	2.2	3.8	4.2	2.7	2.9
Puch	0.13	0.15	0.06	0.11	.08
	2.9	3.0	4.0	4.5	3.3
Zhash	0.11	0.15	0.08	0.13	0.2
	3.5	4.5	2.5	5.0	2.9
Huax	0.12	0.1	0.19	0.1	0.11
	2.8	3.2	4.5	4.0	3.0

Table 5.10 The expected value of transport cost, and time (h) from collectors to distributors (recycled bottles)

	Wein	Shangl	Hanzh	Ank	Yan an
Guanzh	0.08	0.15	0.06	0.12	0.10
	3.5	4.0	2.5	2.0	4.5
Shanb	0.10	0.08	0.10	0.12	0.08
	2.5	2.0	4.5	3.5	5.0
Shann	0.11	0.10	0.08	0.13	0.11
	4.0	2.5	3.0	3.5	3

Table 5.11 The expected value of transport cost, and time (h) from distributors to wholesalers (products)

	Wein	Shangl	Hanzh	Ank	Yan an
Guanzh	0.23	0.31	0.15	0.17	0.3
	3.5	4.0	2.5	2.0	4.5
Shanb	0.17	0.2	0.15	0.18	0.27
	2.5	2.0	4.5	3.5	5.0
Shann	0.13	0.25	0.22	0.18	0.26
	4.0	2.5	3.0	3.5	3

Table 5.12 Transport cost and time from producers to distributors (recycled bottles)

	Guanzh	Shanb	Shann
Plant 1	0.1	0.08	0.15
	2.5	1.0	2.0
Plant 2	0.15	0.2	0.08
	2.0	2.5	1.5

Table 5.13 Transport cost and time from distributors to producers (products)

	Guanzh	Shanb	Shann
Plant 1	0.3	0.25	0.35
	2.5	1.0	2.0
Plant 2	0.35	0.4	0.2
	2.0	2.5	1.5

Table 5.14 The expected value of transport cost from recyclers to producers

	Zhouzh	Puch	Zhash	Huax
Plant 1	0.17	0.13	0.12	0.15
	1.5	2.0	2.5	1.5
Plant 2	0.1	0.16	0.11	0.08
	2.5	1.0	2.0	2.5

Table 5.15 Processing cost, processing time (s) and disposal cost of recyclers

	Zhouzh	Puch	Zhash	Huax
Processing cost	0.25	0.2	0.23	0.18
Processing time	3.0	2.5	3.5	2.0
Disposal cost	0.2	0.15	0.18	0.13

Table 5.16 The expected value of processing cost, time (s) and disposal cost of distributors

	Guanzh	Shanb	Shann
Processing cost	0.28	0.22	0.25
Processing time	5.0	6.0	4.0
Disposal cost	0.2	0.15	0.18

We use the expected value operator and the chance operator to tackle the fuzzy rough objectives and the fuzzy rough constraint, and used the Fu-Ro simulation-based GA to solve this problem under the predetermined confidence level (0.8, 0.8); the corresponding parameters are 100 genetic generation iteration, the population of every generation is 10, the crossover rate 0.3 and the mutation rate is 0.2.

After a run of a genetic algorithm computer program, we obtained the following satisfactory solution: the optimal value of the objective function is $Z^*=456253$ (RMB), $T^*=2200.3$ (hour) and the value of the corresponding location variables are in Table. 5.17.

Table 5.17 Location decision

Zhouzh	Puch	Zhash	Huax	Guanzh	Shanb	Shann
0	1	1	0	1	1	0

Then we could do some sensitivity analysis: we adjusted the weights of these two objectives, and the solutions of the integrated logistics network problem are shown in Table. 5.18.

It shows that small changes in the weights do not significantly influence the location results, and the result is satisfactory to the decision maker of this company. On all accounts, we offered this strategy for Lan Ma beer company - establish the recycling centers in Pu Cheng and Zha Shui, and expand the Guan Zhong and Shan Bei distribution centers. If we consider a given budget, with regard to the number of recycling centers built and the distribution centers expanded, we make the following observations. First, when the location cost factor increases, i.e., the recycling center

Table 5.18 The results of TS (Appr=0.8, Pos=0.8)

w_C	w_T	C^*	T^*	Zhouzh	Puch	Zhash	Huax	Guanzh	Shanb	Shann
0.7	0.3	455353	2289.5	0	1	1	0	1	1	0
0.6	0.4	456253	2200.3	0	1	1	0	1	1	0
0.5	0.5	456696	2186.3	0	1	1	0	1	1	0

location costs increase relative to other costs, the number of opened recycling centers decreases. Second, when the transportation costs between facilities increase, the number of opened recycling centers also decreases. However, since the total cost and time are often conflicting, the handling of multi-objective programming is dependent on the decision-maker's objective. Generally, the solution to this problem often is a balance of multiple objectives.

Chapter 6
Methodological System for FLMODM

Fuzzy-like multiple objective decision making (FLMODM) considers multiple objective decision making under fuzzy-like environments. In this book, we focus on fuzzy-like uncertainty, and develop a series of multiple objective decision making problems:
- Multiple objective decision making under a fuzzy random environment;
- Multiple objective decision making under a bifuzzy environment;
- Multiple objective decision making under a fuzzy rough environment.

In FLMODM, we use the following fuzzy-like variables to describe the coefficients, and consider the fuzzy-like environments:
- Fu-Ra variable;
- Fu-Fu variable;
- Fu-Ro variable.

For the general MODM models with fuzzy-like variables, the meaning is not clear, we have to adopt some philosophy to deal with them, and the following six kinds of fuzzy-like models are proposed:
- Expected value model (EVM);
- Chance constrained model (CCM);
- Dependent chance model (DCM);
- Expectation model with chance constraints (ECM);
- Chance constrained model with expectation constraints (CEM);
- Dependent chance model with expectation constraints (DEM).

The above fuzzy-like model can not be solved directly. So in some special situations, we can use mathematical tools to transform the FLMODM model into crisp equivalent models. For the above fuzzy-like models, we can design a hybrid algorithm to get approximate solutions.

FLMODM has been applied to optimization problems with fuzzy-like variables, for example, the portfolio selection problem [415, 403], supply chain management problem, inventory problem [343, 3], project selection [184], allocation problem [413], the supplier selection problem [402], scheduling problem [404, 405], transportation problem [406], vehicle routing problem [407], location-allocation [408] and so on.

J. Xu and X. Zhou: Fuzzy-Like Multiple Objective Decision Making, STUDFUZZ 263, pp. 375–395.
springerlink.com

6.1　Motivation of Researching FLMODM

Why should we research the FLMODM? Let us recall the two foundations: multiple objective decision making(MODM) and fuzzy set theory, see Figure 6.1.

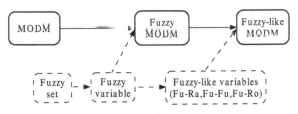

Fig. 6.1 Development of FLMODM

MODM is to find an optimal decision while considering the constraints of limited resources. The research of the certain MODM can be traced back in the 18th century. Franklin introduced how to coordinate the multiple objective problems in 1772. Then Cournot proposed the multiple objective model from an economic point of view in 1836. Pareto [423] first introduced multiple objective decision making models from the mathematical point of view in 1896. Later, Arrow [424] proposed the concept of efficient points in 1953. MODM has gradually been developed since this time.

Generally speaking, there are five elements in a MODM problem:
(1) Decision variable: $x = (x_1, x_2, \cdots, x_n)^{\mathrm{T}}$.
(2) Objective function: $f(x) = (f_1(x), f_2(x), \cdots, f_m(x))$, $m \geq 2$.
(3) Feasible solution set: $X = \{x \in R^n | g_i(x) \leq 0, i = 1, 2, \cdots, p, h_r(x) = 0, r = 1, 2, \cdots, q\}$.
(4) Preference relation: In the image set $f(X) = \{f(x) | x \in X\}$, there is a certain binary relation which could reflect the preference of the decision maker.
(5) Definition of the solution. Define the optimal solution of f in X based on the known preference relation.

So a MODM problem can be described as follows:

$$\begin{cases} \min \ f(x) = [f_1(x), f_2(x), \cdots, f_m(x)] \\ \text{s.t.} \begin{cases} g_i(x) \leq 0, \ i = 1, 2, \cdots, p \\ h_r(x) = 0, r = 1, 2, \cdots, q. \end{cases} \end{cases}$$

The fuzzy set was introduced by L. A. Zadeh [9] in 1965, which is a class of objects with a continuum of grades of membership ranging from 0 to 1. Fuzzy set theory has been applied to operations research, management science, control theory, artificial intelligent expert system and so on. In 1970, Bellman and Zadeh proposed the basic model for fuzzy decision making based on the MODM [416]. A general fuzzy MODM model can be written as follows:

$$\begin{cases} \min\ f(x,\tilde{a}) = [f_1(x,\tilde{a}_1), f_2(x,\tilde{a}_2), \cdots, f_m(x,\tilde{a}_m)] \\ \text{s.t.} \begin{cases} g_i(x,\tilde{b}_i) \leq 0,\ i = 1,2,\cdots,p \\ h_r(x,\tilde{c}_r) = 0, r = 1,2,\cdots,q, \end{cases} \end{cases}$$

where $\tilde{a}_k(k = 1,2,\cdots,m)$, $\tilde{b}_i(i = 1,2,\cdots,p)$, $\tilde{c}_r(r = 1,2,\cdots,q)$ are the vectors of fuzzy coefficients. And it should be noted that "$\leq$" denote "basically less than or equal to", "$=$" denote "basically equal to", and "min" denote "minimize the value of the objective functions as much as possible".

Actually, in order to make a satisfactory decision in practice, an important problem is to determine the type and accuracy of information. If complete information is required in the decision making process, it will mean the expenditure of some extra time and money. If incomplete information is used to make a decision quickly, then it is possible to take nonoptimal action. In fact, we cannot have complete accuracy in both information and decision because the total cost is the sum of the cost spent for running the target system and the cost spent for getting decision information. Since we have to balance the advantage of making better decisions against the disadvantages of getting more accurate information, incomplete information will almost surely be used in the real-life decision process, and uncertain programming is an important tool in dealing with the decision making with imperfect information. Among all of the uncertain programming, the fuzzy programming approach [194, 410, 412] is useful and efficient in handling a programming problem with uncertainty. While classical and random/stochastic programming approaches may require a lot of cost to obtain the exact coefficient value or distribution, the fuzzy programming approach does not [411]. From this aspect, fuzzy programming is more advantageous when the coefficients cannot be specified, but vaguely estimated by human experiences.

In 1978, H. Kwakernaak [215] combined randomness with fuzziness and initialized the concept of the Fu-Ra variable. If the value of a fuzzy variable ξ is a random variable, then ξ is called Fu-Ra variable. In 1971, Zadeh [11] proposed the level-2 fuzzy set which is the basis of the Fu-Fu variable, and Gottwald, Dubois [21, 32] developed it. If the value of a fuzzy variable ξ is also a fuzzy variable, then ξ is called the Fu-Fu variable. Similarly, if the value of fuzzy variable ξ is a rough variable, then ξ is called the Fu-Ro variable. In realistic problems, the information may be described as the Fu-Ra variable, Fu-Fu variable or Fu-Ro variable. For example, we already know that some information is the random variable, fuzzy variable and rough variable, and when we want to integrate the experiences and the knowledge of human beings, it's a good way to add some tolerance interval to the former random variable, fuzzy variable or rough variable, and thus Fu-Ra variable, Fu-Fu variable or Fu-Ro variable will exist. So how to deal with the MODM with those fuzzy-like coefficients? It is very necessary and important for us to research fuzzy-like multiple objective decision making.

6.2 Physics-Based Model System

Fuzzy-like multiple objective decision making deals with the multiple objective decision making problems under fuzzy-like environments. In other words, when some parameters or the coefficients for a multiple objective decision making problem are some fuzzy-like variables, then this multiple objective decision making problem is called a fuzzy like multiple objective decision making problem, which includes problems under Fu-Ra, Fu-Fu, and Fu-Ro environments. In this book, we use three typical problems-selection problems, the purchasing problem and logistic problem to illustrate the Fu-Ra, Fu-Fu and Fu-Ro multiple objective decision making, respectively. Among all kinds of typical problems, we chose the portfolio selection problem, raw material purchasing problem, and integrated logistic problem to clarify corresponding fuzzy-like multiple objective decision making in detail, see Figure 6.2.

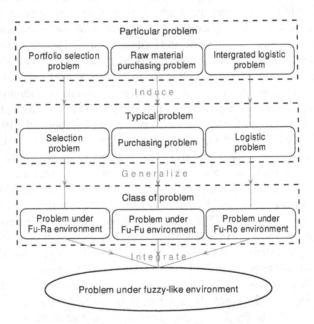

Fig. 6.2 Problem under fuzzy-like environment

In the portfolio selection problem, We know the basic assumption behind Markowitz's mean variance model is that the situation of the stock market in the future can be correctly reflected by securities data from the past, that is, the mean and covariance of a portfolio of securities in the future are similar to the past ones. However, there are so many uncertain factors that this assumption cannot be guaranteed for real stock markets. Since stock experts possess enough information and experience about the stock market, it is a good method to let them provide their rough estimation about the future returns of securities, and the certain mean value could extend to a fuzzy number. In this case, the return of securities are fuzzy random variables. When we use fuzzy random variables to describe the future return of the securities, we can use the historical data and advice about the historical returns of the experts. And it is rational for people to consider that the future return of every security is a fuzzy variable which is around a value with left and right spreads, but here the middle value is usually not a certain number, but a random variable, so the future return is a fuzzy random variable. So it is reasonable for people to believe that the fuzzy random portfolio selection problem is more realistic and proper.

In the raw material purchasing problem, the cost, scrap ratio, tardy-delivery fraction and the demand are some coefficients in the future. According to the anticipative data, we can get the initial data with the expression of a fuzzy variable for the above information, then people can estimate the situation and make a forecast. The former fuzzy variable can be deemed as middle value of a new fuzzy variable, that is, the initial fuzzy variable is extended to Fu-Fu variable. In this situation, in order to deal with the raw material purchasing problem using these data, we have to solve a Fu-Fu multiple objective decision making model.

In the integrated logistics problem, both the forward logistics system and the reverse logistics system are considered. We focus on the integrated logistics problem in a very special beer company. People usually drink more beer in summer and autumn, and less beer in winter and spring, that is, the demand for beer is seasonal. When we forecast the demand in a period, we can use the fuzzy variable to estimate, for example, we give a middle value μ, two spread α and β. Further more, the middle value μ is usually not a certain number, because when we design the network of the network of a reuse integrated logistics network, the period we consider will definitely cover the whole season, so it is appropriate to use a rough variable to describe the middle value μ. So until now, in this situation, we can use Fu-Ro variables to describe the demand of the beer. Because the amount of used packages is relevant to the consumtion of the product, so it is natural to consider the amount of used packages as a fuzzy rough variable also, just as that of the demand of the products. Thus the Fu-Ro multiple objective model should be built for the integrated logistic decision making problem under a Fu-Ro environment.

It is noted that the problems introduced in this book are just some example of typical problems, so readers can extend the application into other areas.

6.3 Mathematical Model System

The initial fuzzy-like multiple objective decision making model is as follows:

$$\begin{cases} \max \ [f_1(x,\xi), f_2(x,\xi), \cdots, f_m(x,\xi)] \\ \text{s.t.} \begin{cases} g_r(x,\xi) \le 0, \ r = 1, 2, \cdots, p \\ x \in X \end{cases} \end{cases}$$

or

$$\begin{cases} \min \ [f_1(x,\xi), f_2(x,\xi), \cdots, f_m(x,\xi)] \\ \text{s.t.} \begin{cases} g_r(x,\xi) \le 0, \ r = 1, 2, \cdots, p \\ x \in X, \end{cases} \end{cases}$$

where ξ are fuzzy-like variables, that is, the Fu-Ra variables, Fu-Fu variables and Fu-Ro variables.

It is necessary for us to know that the above models are conceptual models rather than mathematical models, because we cannot maximize an uncertain quantity. There does not exist a natural order in an uncertain world. Since fuzzy-like variables exist, the above models are ambiguous. The meaning of maximizing/minimizing $f_1(x,\xi), f_2(x,\xi), \cdots, f_m(x,\xi)$ is unclear, and the constraints $g_r(x,\xi) \leq 0, r = 1, 2, \cdots, p$ do not define a deterministic feasible set. So we need to adopt some philosophies to deal with and make the above model solvable. Philosophies 1-5 are used to deal with decision making models under a fuzzy-like environment.

First, let us consider the objective functions

$$\max \ [f_1(x,\xi), f_2(x,\xi), \cdots, f_m(x,\xi)].$$

where ξ is the fuzzy-like variables.

There are three types of philosophy to handle the objectives.

Philosophy 1: Making the decision by optimizing the expected value of the objectives. That is, maximizing the expected values of the objective functions for the Max problem, or minimizing the expected values of the objective functions for the Min problem.

$$\max \ E[f_1(x,\xi), f_2(x,\xi), \cdots, f_m(x,\xi)].$$

or

$$\min \ E[f_1(x,\xi), f_2(x,\xi), \cdots, f_m(x,\xi)].$$

Philosophy 2: Making the decision which provides the best optimal objective values with a given confidence level. That is, maximizing the referenced objective values $\bar{f}_i$ subjects to $f_i(x,\xi) \geq \bar{f}_i$ with a confidence level α_i, or minimizing the referenced objective values $\bar{f}_i$ subjects to $f_i(x,\xi) \leq \bar{f}_i$ with a confidence level α_i.

$$\max \ [\bar{f}_1, \bar{f}_2, \cdots, \bar{f}_m]$$
$$\text{s.t.} \ Ch\{f_i(x,\xi) \geq \bar{f}_i\} \geq \alpha_i, \ i = 1, 2, \cdots, m.$$

or

$$\min \ [\bar{f}_1, \bar{f}_2, \cdots, \bar{f}_m]$$
$$\text{s.t.} \ Ch\{f_i(x,\xi) \leq \bar{f}_i\} \geq \alpha_i, \ i = 1, 2, \cdots, m.$$

where α_i should be predetermined, $\bar{f}_1, \bar{f}_2, \cdots, \bar{f}_m$ are called critical values.

Philosophy 3: Making the decision by maximizing the chance of the events. That is, maximizing the chance of the events $f_i(x,\xi) \geq \bar{f}_i$ or $f_i(x,\xi) \leq \bar{f}_i$.

$$\max \ \begin{bmatrix} Ch\{f_1(x,\xi) \geq \bar{f}_1\}, \\ Ch\{f_2(x,\xi) \geq \bar{f}_2\}, \\ \cdots \\ Ch\{f_m(x,\xi) \geq \bar{f}_m\}, \end{bmatrix}$$

or

$$\max \begin{bmatrix} Ch\{f_1(x,\xi) \le \bar{f}_1\}, \\ Ch\{f_2(x,\xi) \le \bar{f}_2\}, \\ \cdots \\ Ch\{f_m(x,\xi) \le \bar{f}_m\}, \end{bmatrix}$$

where $\bar{f}_i$ should be predetermined.

Then, let us consider the constraints

$$\text{s.t.} \begin{cases} g_r(x,\xi) \le 0, \ r = 1,2,\cdots,p \\ x \in X, \end{cases}$$

where ξ is the fuzzy-like variables.

There are two types of philosophy to handle the constraints.

Philosophy 4: Making the optimal decision subjects to the expected constraints. That is,

$$E[g_r(x,\xi) \le 0], \ r = 1,2,\cdots,p.$$

Philosophy 5: Making the optimal decision subjects to the chance constraints.

$$Ch\{g_r(x,\xi) \le 0\} \ge \beta_r, \ r = 1,2,\cdots,p.$$

where β_r is predetermined.

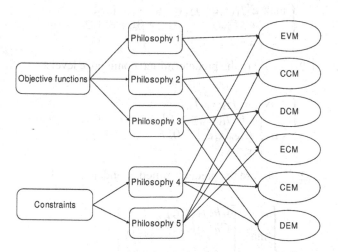

Fig. 6.3 Fuzzy-like model system

By combining the 3 philosophies for the objective functions and 2 philosophies for the constraints, we can get six types of models which can deal with the initial fuzzy-like multiple objective decision making models: EVM, CCM, DCM, ECM, CEM and DEM, see Figure 6.3.

EVM

$$\begin{cases} \max\ E[f_1(x,\xi), f_2(x,\xi), \cdots, f_m(x,\xi)] \\ \text{s.t.}\ \begin{cases} E[g_r(x,\xi)] \leq 0,\ r = 1,2,\cdots,p \\ x \in X. \end{cases} \end{cases}$$

CCM

$$\begin{cases} \max\ [\bar{f}_1, \bar{f}_2, \cdots, \bar{f}_m] \\ \text{s.t.}\ \begin{cases} Ch\{f_i(x,\xi) > \bar{f}_i\} \geq \alpha_i,\ i = 1,2,\cdots,m \\ Ch\{g_r(x,\xi) \leq 0\} \geq \beta_r,\ r = 1,2,\cdots,p \\ x \in X, \end{cases} \end{cases}$$

where $\alpha_i(i = 1,2,\cdots,m), \beta_r(r = 1,2,\cdots,p)$ are the predetermined confidence levels.

DCM

$$\begin{cases} \max\ \begin{bmatrix} Ch\{f_1(x,\xi) \geq \bar{f}_1\}, \\ Ch\{f_2(x,\xi) \geq \bar{f}_2\}, \\ \cdots \\ Ch\{f_n(x,\xi) \geq \bar{f}_m\}, \end{bmatrix} \\ \text{s.t.}\ \begin{cases} Ch\{g_r(x,\xi) \leq 0\} \geq \bar{\beta}_r,\ r = 1,2,\cdots,p \\ x \in X, \end{cases} \end{cases}$$

where $\bar{f}_i(i = 1,2,\cdots,m), \beta_r(r = 1,2,\cdots,p)$ are the predetermined referenced objective values and confidence levels.

ECM

$$\begin{cases} \max\ E[f_1(x,\xi), f_2(x,\xi), \cdots, f_m(x,\xi)] \\ \text{s.t.}\ \begin{cases} Ch\{g_r(x,\xi) \leq 0\} \geq \beta_r,\ r = 1,2,\cdots,p \\ x \in X, \end{cases} \end{cases}$$

where $\beta_r(r = 1,2,\cdots,p)$ are the predetermined confidence levels.

CEM

$$\begin{cases} \max\ [\bar{f}_1, \bar{f}_2, \cdots, \bar{f}_m] \\ \text{s.t.}\ \begin{cases} Ch\{f_i(x,\xi) \geq \bar{f}_i\} \geq \alpha_i,\ i = 1,2,\cdots,m \\ E[g_r(x,\xi)] \leq 0,\ r = 1,2,\cdots,p. \\ x \in X, \end{cases} \end{cases}$$

where $\alpha_i(i = 1,2,\cdots,m)$ are the predetermined confidence levels.

DEM

$$\begin{cases} \max\ \begin{bmatrix} Ch\{f_1(x,\xi) \geq \bar{f}_1\}, \\ Ch\{f_2(x,\xi) \geq \bar{f}_2\}, \\ \cdots \\ Ch\{f_n(x,\xi) \geq \bar{f}_m\}, \end{bmatrix} \\ \text{s.t.}\ \begin{cases} E[g_r(x,\xi)] \leq 0,\ r = 1,2,\cdots,p. \\ x \in X, \end{cases} \end{cases}$$

where $\bar{f}_i(i = 1,2,\cdots,m)$ are the predetermined referenced objective values and confidence levels.

In this book, we mainly discuss the first three models, and the techniques are all incorporated when we deal with EVM, CCM and DCM. The rest of the models, ECM, CEM and DEM, can be handled in the same way. The reader can use the model when they use different philosophies, and it is possible to use every model.

6.4   Model Analysis System

For linear fuzzy-like multiple objective decision making models,

$$
\begin{cases}
\max & \left[\sum\limits_{j=1}^{n} \tilde{c}_{1j}x_j, \cdots, \sum\limits_{j=1}^{m} \tilde{c}_{mj}x_j \right] \\
\text{s.t.} & \begin{cases} \tilde{a}_{rj}x_j \leq \tilde{b}_r, \ r = 1,2,\cdots,p \\ x \in X, \end{cases}
\end{cases}
$$

where $\tilde{c}_{ij}, \tilde{a}_{rj}, \tilde{b}_r, (i = 1,2,\cdots,m; r = 1,2,\cdots,p)$ are special fuzzy-like coefficients, that is, the Fu-Ra variables, Fu-Fu variables and Fu-Ro variables. We introduced how to transform the 3 types of objective functions and the 2 types of constraints into their crisp equivalent formulas in detail. In this book, we introduced the equivalent models for EVM, CCM and DCM in detail, and we simplified them as EEVM, ECCM and EDCM. See Figure 6.4.

For the Fu-Ra linear multi-objective models, there are 5 basic theorems for handling the objective functions and the constraints: Theorem 3.3, Theorem 3.4, Theorem 3.6 (Theorem 3.8), Theorem 3.7 (Theorem 3.9) and Theorem 3.10 (Theorem 3.14). And according to these 5 theorems, we can get the crisp equivalent models for Fu-Ra EVM, CCM and DCM.

Fu-Ra EEVM

$$
\begin{cases}
\max & \left[\frac{1}{4}\sum\limits_{t=1}^{4}\sum\limits_{j=1}^{n}\sum \mu_{1jt}x_j, \frac{1}{4}\sum\limits_{t=1}^{4}\sum\limits_{j=1}^{n}\sum \mu_{2jt}x_j, \cdots, \frac{1}{4}\sum\limits_{t=1}^{4}\sum\limits_{j=1}^{n}\sum \mu_{mjt}x_j \right] \\
\text{s.t.} & \begin{cases} \sum\limits_{t=1}^{4}\sum\limits_{j=1}^{n}\mu_{rjt}x_j \leq \sum\limits_{t=1}^{4}\mu_{rt}, r = 1,2,\cdots,p \\ x \in X. \end{cases}
\end{cases}
$$

Fu-Ra ECCM 1 based on Pr-Pos

$$
\begin{cases}
\max & [f_1, f_2, \cdots, f_m] \\
\text{s.t.} & \begin{cases} f_i \leq R^{-1}(\delta_i)\beta_i^{cT}x + d_i^{cT}x + \Phi^{-1}(1-\gamma_i)\sqrt{x^TV_i^cx}, \ i = 1,2,\cdots,m \\ R^{-1}(\theta_r)\beta_r^b + L^{-1}(\theta_r)\alpha_r^{eT}x - (d_r^{eT}x - d_r^b) \\ \qquad\qquad - \Phi^{-1}(\eta_r)\sqrt{x^TV_r^ex + (\sigma_r^b)^2} \geq 0, r = 1,2,\cdots,p \\ x \in X. \end{cases}
\end{cases}
$$

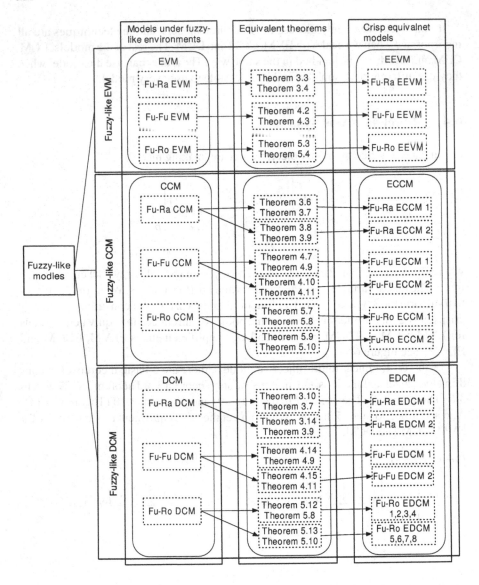

Fig. 6.4 Transformation to crisp equivalent models

Fu-Ra ECCM 2 *based on Pr-Nec*

$$
\begin{cases}
\max\ [f_1, f_2, \cdots, f_m] \\
\text{s.t.}
\begin{cases}
f_i \le d_i^{cT}x - L^{-1}(1-\delta_i)\alpha_i^{cT}x + \Phi^{-1}(1-\gamma_i)\sqrt{x^T V_i^c x},\ i=1,2,\cdots,m \\
\Phi^{-1}(1-\eta_r)\sqrt{x^T V_r^e x + (\sigma_r^b)^2} - L^{-1}(1-\theta_r)\alpha_r^b \\
\qquad - R^{-1}(\theta_r)\beta_r^{eT}x + (d_r^b - d_r^{eT}x) \ge 0, r=1,2,\cdots,p \\
x \in X.
\end{cases}
\end{cases}
$$

Fu-Ra EDCM 1 *based on Pr-Pos*

$$
\begin{cases}
\max \left[\dfrac{\Phi^{-1}(1-\gamma_i)\sqrt{x^{\mathrm{T}}V_i^c x}+d_i^{c\mathrm{T}}x-\bar{f}_i}{\beta_i^{c\mathrm{T}}x}, i=1,2,\cdots,m \right] \\
\text{s.t.} \begin{cases}
R^{-1}(\theta_r)\beta_r^b + L^{-1}(\theta_r)\alpha_r^{e\mathrm{T}}x - (d_r^{e\mathrm{T}}x - d_r^b) \\
\qquad\qquad - \Phi^{-1}(\eta_r)\sqrt{x^{\mathrm{T}}V_r^e x + (\sigma_r^b)^2} \geq 0, \quad r=1,2,\cdots,p \\
x \in X.
\end{cases}
\end{cases}
$$

Fu-Ra EDCM 2 *based on Pr-Nec*

$$
\begin{cases}
\max \left[\dfrac{\Phi^{-1}(1-\gamma_i)\sqrt{x^{\mathrm{T}}V_i^c x}-\bar{f}_i+d_i^{c\mathrm{T}}x}{\alpha_i^{c\mathrm{T}}x}, i=1,2,\cdots,m \right] \\
\text{s.t.} \begin{cases}
\Phi^{-1}(1-\eta_r)\sqrt{x^{\mathrm{T}}V_r^e x + (\sigma_r^b)^2} - L^{-1}(1-\theta_r)\alpha_r^b - R^{-1}(\theta_r)\beta_r^{e\mathrm{T}}x \\
\qquad\qquad + (d_r^b - d_r^{e\mathrm{T}}x) \geq 0, \quad r=1,2,\cdots,p \\
x \in X.
\end{cases}
\end{cases}
$$

For the Fu-Fu linear multi-objective models, there are 5 important theorems for handling the objective functions and the constraints: Theorem 4.2, Theorem 4.3, Theorem 4.7 (Theorem 4.10), Theorem 4.9 (Theorem 4.11) and Theorem 4.14 (Theorem 4.15). And according to these 5 theorems, we can get the crisp equivalent models for Fu-Fu EVM, CCM and DCM.

Fu-Fu EEVM

$$
\begin{cases}
\max \left[\dfrac{1}{8}\sum_{j=1}^{n}\sum_{t=1}^{4}\sum_{k=1}^{2} c_{1jtk}x_j, \dfrac{1}{8}\sum_{j=1}^{n}\sum_{t=1}^{4}\sum_{k=1}^{2} c_{2jtk}x_j, \cdots, \dfrac{1}{8}\sum_{j=1}^{n}\sum_{t=1}^{4}\sum_{k=1}^{2} c_{mjtk}x_j \right] \\
\text{s.t.} \begin{cases}
\sum_{j=1}^{n}\sum_{t=1}^{4}\sum_{k=1}^{2} a_{rjtk}x_j \leq \sum_{t=1}^{4}\sum_{k=1}^{2} b_{rtk}, r=1,2,\cdots,p \\
x \in X.
\end{cases}
\end{cases}
$$

Fu-Fu ECCM 1 *based on Pos-Pos*

$$
\begin{cases}
\max [f_1, f_2, \cdots, f_m] \\
\text{s.t.} \begin{cases}
f_i \leq c_i^{\mathrm{T}}x + R^{-1}(\delta_i)\beta_{i1}^{c\mathrm{T}}x + R^{-1}(\gamma_i)\beta_{i2}^{c\mathrm{T}}x, \quad i=1,2,\cdots,m \\
R^{-1}(\theta_r)\beta_{r1}^b + L^{-1}(\theta_r)\alpha_{r1}^{e\mathrm{T}}x - e_r^{\mathrm{T}}x + b_r \\
\qquad\qquad + L^{-1}(\eta_r)(\alpha_{r2}^{e\mathrm{T}}x + \beta_{r2}^b) \geq 0, r=1,2,\cdots,p \\
x \in X.
\end{cases}
\end{cases}
$$

Fu-Fu ECCM 2 *based on Nec-Nec*

$$
\begin{cases}
\max \{f_1, f_2, \cdots, f_m\} \\
\text{s.t.} \begin{cases}
f_i \leq c_i^{\mathrm{T}}x + L^{-1}(1-\delta_i)\alpha_{i1}^{c\mathrm{T}}x + L^{-1}(1-\gamma_i)\alpha_{i2}^{c\mathrm{T}}x, \quad i=1,2,\cdots,m \\
b_r - e_r^{\mathrm{T}}x - L^{-1}(1-\eta_r)(\alpha_{r2}^b + \beta_{r2}^{e\,\mathrm{T}}x) - L^{-1}(1-\theta_r)\alpha_{1r}^b \\
\qquad\qquad - R^{-1}(\theta_r)\beta_{r1}^{e\,\mathrm{T}}x \geq 0, r=1,2,\cdots,p \\
x \in X.
\end{cases}
\end{cases}
$$

Fu-Fu EDCM 1 *based on Pos-Pos*

$$
\begin{cases}
\max \left[\dfrac{f_i - c_i^{\mathrm{T}}x - R^{-1}(\gamma_i)\beta_{i2}^{c\mathrm{T}}x}{\beta_{i1}^{c\mathrm{T}}x}, \ i = 1,2,\cdots,m \right] \\
\text{s.t.} \begin{cases} R^{-1}(\theta_r)\beta_{r1}^{b} + L^{-1}(\theta_r)\alpha_{r1}^{e\mathrm{T}}x - e_r^{\mathrm{T}}x + b_r + L^{-1}(\eta_r)(\alpha^{e_{r2}^{\mathrm{T}}}x + \beta_{r2}^{b}) \geq 0 \\ \qquad\qquad\qquad r = 1,2,\cdots,p \\ x \in X. \end{cases}
\end{cases}
$$

Fu-Fu EDCM 2 *based on Nec-Nec*

$$
\begin{cases}
\max \left[\dfrac{c_i^{\mathrm{T}}x - L^{-1}(1-\gamma_i)\alpha_{i2}^{c\mathrm{T}}x - f_i}{\alpha_{i1}^{c\mathrm{T}}x}, \ i = 1,2,\cdots,m \right] \\
\text{s.t.} \begin{cases} b_r - e_r^{\mathrm{T}}x - L^{-1}(1-\eta_r)(\alpha_{r2}^{b} + \beta^{e_{r2}^{\mathrm{T}}}x) - L^{-1}(1-\theta_r)\alpha_{1r}^{b} - R^{-1}(\theta_r)\beta_{1r}^{e\mathrm{T}}x \geq 0, \\ \qquad\qquad\qquad r = 1,2,\cdots,p \\ x \in X. \end{cases}
\end{cases}
$$

For the Fu-Ro linear multi-objective models, there are 5 important theorems for handling the objective functions and the constraints: Theorem5.3, Theorem5.4, Theorem5.7 (Theorem5.9), Theorem5.8 (Theorem5.10) and Theorem5.12 (Theorem 5.13). And according to these 5 theorems, we can get the crisp equivalent models for Fu-Ro EVM, CCM and DCM.

Fu-Ro EEVM

$$
\begin{cases}
\max \left[\dfrac{1}{16}\sum\limits_{j=1}^{n}\sum\limits_{t=1}^{4}\sum\limits_{k=1}^{4} c_{1jtk}x_j, \ \dfrac{1}{16}\sum\limits_{j=1}^{n}\sum\limits_{t=1}^{4}\sum\limits_{k=1}^{4} c_{2jtk}x_j, \cdots, \dfrac{1}{16}\sum\limits_{j=1}^{n}\sum\limits_{t=1}^{4}\sum\limits_{k=1}^{4} c_{mjtk}x_j \right] \\
\text{s.t.} \begin{cases} \sum\limits_{j=1}^{n}\sum\limits_{t=1}^{4}\sum\limits_{k=1}^{4} a_{rjtk}x_j \leq \sum\limits_{t=1}^{4}\sum\limits_{k=1}^{4} b_{rtk}, r = 1,2,\cdots,p \\ x \in X. \end{cases}
\end{cases}
$$

Fu-Ro ECCM 1 *based on Appr-Pos*

$$
\begin{cases}
\max [f_1, f_2, \cdots, f_m] \\
\text{s.t.} \begin{cases}
f_i \leq d - 2\gamma_i(d-c) + R^{-1}(\delta_i)\beta_i^{cT}x, \\
\qquad\qquad \text{if } b \leq f_i - R^{-1}(\delta_i)\beta_i^{cT}x \leq d \\
f_i \leq \dfrac{d(b-a)+b(d-c)-2\alpha_i(d-c)(b-a)}{d-c+b-a} + R^{-1}(\delta_i)\beta_i^{cT}x, \\
\qquad\qquad \text{if } a \leq f_i - R^{-1}(\delta_i)\beta_i^{cT}x < b \\
f_i \leq d - (d-c)(2\gamma_i - 1) + R^{-1}(\delta_i)\beta_i^{cT}x, \\
\qquad\qquad \text{if } c \leq f_i - R^{-1}(\delta_i)\beta_i^{cT}x \leq a \\
f_i \leq c + R^{-1}(\delta_i)\beta_i^{cT}x, \\
\qquad\qquad \text{if } f_i - R^{-1}(\delta_i)\beta_i^{cT}x \leq c \\
W \geq c + 2(d-c)\eta_r, \text{ if } c \leq W \leq a \\
W \geq \dfrac{2\eta_r(d-c)(b-a)+c(b-a)+a(d-c)}{b-a+d-c}, \text{ if } a \leq W < b \\
W \geq (2\eta_r - 1)(d-c) + c, \text{ if } b \leq W \leq d \\
W \geq d, \text{ if } d \leq W \\
x \in X.
\end{cases}
\end{cases}
$$

Fu-Ro ECCM 2 based on Appr-Nec

$$
\begin{cases}
\max \ [f_1, f_2, \cdots, f_m] \\
\text{s.t.} \begin{cases}
f_i \le d - 2\gamma_i(d-c) - L^{-1}(1-\delta_i)\alpha_i^{cT}x, \\
\qquad\qquad \text{if } b \le f_i + L^{-1}(1-\delta_i)\alpha_i^{cT}x \le d \\
f_i \le \dfrac{d(b-a)+b(d-c)-2\gamma_i(d-c)(b-a)}{d-c+b-a} - L^{-1}(1-\delta_i)\alpha_i^{cT}x, \\
\qquad\qquad \text{if } a \le f_i + L^{-1}(1-\delta_i)\alpha_i^{cT}x < b \\
f_i \le d - (d-c)(2\gamma_i - 1) - L^{-1}(1-\delta_i)\alpha_i^{cT}x, \\
\qquad\qquad \text{if } c \le f_i + L^{-1}(1-\delta_i)\alpha_i^{cT}x \le a \\
f_i \le c - L^{-1}(1-\delta_i)\alpha_i^{cT}x, \\
\qquad\qquad \text{if } f_i + L^{-1}(1-\delta_i)\alpha_i^{cT}x \le c \\
W' \ge c + 2(d-c)\eta_r \text{ if } c \le W' \le a \\
W' \ge \dfrac{2\eta_r(d-c)(b-a)+c(b-a)+a(d-c)}{b-a+d-c} \text{ if } a \le W' < b \\
W' \ge (2\eta_r - 1)(d-c) + c \text{ if } b \le W' \le d \\
W' \ge d \text{ if } d \le W' \\
x \in X,
\end{cases}
\end{cases}
$$

where $W = R^{-1}(\theta_r)\beta_r^b + L^{-1}(\theta_r)\alpha_r^{eT}x$, and $W' = -R^{-1}(\theta_r)\beta_r^{eT}x - L^{-1}(1-\theta_r)\alpha_r^b$.

Fu-Ro EDCM 1 based on Appr-Pos

$$
\begin{cases}
\max \ \left[\dfrac{f_i - d + 2(d-c)\gamma_i}{\beta_i^{cT}x}, \ i = 1, 2, \cdots, m\right] \\
\text{s.t.} \begin{cases}
e_r^T x \le b_r, \quad r = 1, 2, \cdots, p \\
b \le f_i - R^{-1}(\delta_i)\beta_i^{cT}x \le d \\
x \in X \\
x \ge 0.
\end{cases}
\end{cases}
$$

Fu-Ro EDCM 2 based on Appr-Pos

$$
\begin{cases}
\max \ \left[\dfrac{(d-c+b-a)f_i - d(b-a) - b(d-c) + 2(d-c)(b-a)\gamma_i}{\beta_i^{cT}x}, \ i = 1, 2, \cdots, m\right] \\
\text{s.t.} \begin{cases}
e_r^T x \le b_r, \quad r = 1, 2, \cdots, p \\
a \le f_i - R^{-1}(\delta_i)\beta_i^{cT}x < b \\
x \in X.
\end{cases}
\end{cases}
$$

Fu-Ro EDCM 3 based on Appr-Pos

$$
\begin{cases}
\max \ \left[\dfrac{f_i - d + 2(d-c)(2\gamma_i - 1)}{\beta_i^{cT}x}, \ i = 1, 2, \cdots, m\right] \\
\text{s.t.} \begin{cases}
e_r^T x \le b_r, \quad r = 1, 2, \cdots, p \\
c \le f_i - R^{-1}(\delta_i)\beta_i^{cT}x \le a \\
x \in X.
\end{cases}
\end{cases}
$$

Fu-Ro EDCM 4 *based on Appr-Pos*

$$
\begin{cases}
\max \left[\frac{f_i - c}{\beta_i^{cT} x}, \ i = 1, 2, \cdots, m \right] \\
\text{s.t.} \begin{cases}
e_r^T x \le b_r, \quad r = 1, 2, \cdots, p \\
f_i - R^{-1}(\delta_i)\beta_i^{cT} x \le c \\
x \in X.
\end{cases}
\end{cases}
$$

Fu-Ro EDCM 5 *based on Appr-Nec*

$$
\begin{cases}
\max \left[\frac{f_i - d + 2(d-c)\gamma_i}{\beta_i^{cT} x}, \ i = 1, 2, \cdots, m \right] \\
\text{s.t.} \begin{cases}
e_r^T x \le b_r, \quad r = 1, 2, \cdots, p \\
b \le f_i + L^{-1}(1 - \delta_i)\alpha_i^{cT} x \le d \\
x \in X.
\end{cases}
\end{cases}
$$

Fu-Ro EDCM 6 *based on Appr-Nec*

$$
\begin{cases}
\max \left[\frac{-(d-c+b-a)f_i + d(b-a) - b(d-c) - 2(d-c)(b-a)\gamma_i}{\alpha_i^{cT} x}, \ i = 1, 2, \cdots, m \right] \\
\text{s.t.} \begin{cases}
e_r^T x \le b_r, \quad r = 1, 2, \cdots, p \\
a \le f_i + L^{-1}(1 - \delta_i)\alpha_i^{cT} x < b \\
x \in X.
\end{cases}
\end{cases}
$$

Fu-Ro EDCM 7 *based on Appr-Nec*

$$
\begin{cases}
\max \left[\frac{-f_i + d - 2(d-c)(2\gamma_i - 1)}{\alpha_i^{cT} x}, \ i = 1, 2, \cdots, m \right] \\
\text{s.t.} \begin{cases}
e_r^T x \le b_r, \quad r = 1, 2, \cdots, p \\
c \le f_i + L^{-1}(1 - \delta_i)\alpha_i^{cT} x \le a \\
x \in X.
\end{cases}
\end{cases}
$$

Fu-Ro EDCM 8 *based on Appr-Nec*

$$
\begin{cases}
\max \left[\frac{-f_i + c}{\alpha_i^{cT} x}, \ i = 1, 2, \cdots, m \right] \\
\text{s.t.} \begin{cases}
e_r^T x \le b_r, \quad r = 1, 2, \cdots, p \\
f_i + L^{-1}(1 - \delta_i)\alpha_i^{cT} x \le c \\
x \in X.
\end{cases}
\end{cases}
$$

6.5 Algorithm System

After we get the crisp equivalent models, we can employ the basic solution methods to get the solution. There are 12 solution methods introduced indetail in the book, which include:

- two-stage method,
- goal programming method,

- ideal point method,
- fuzzy satisfied method,
- surrogate worth trade-off method,
- satisfying trade-off method,
- step method,
- lexicographic method,
- weight sum method,
- minimax point method,
- fuzzy goal method,
- ε-constraint method.

The above 12 solution methods are the most popular methods for multiple objective decision making. The decision maker can choose different methods when they have different requests or are under different conditions.

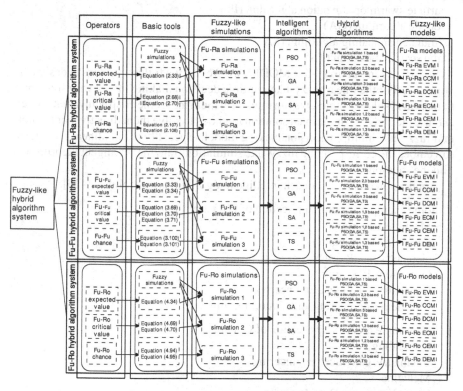

Fig. 6.5 Fuzzy-like hybrid algorithm system

For the nonlinear fuzzy-like multiple objective decision making models, it is very difficult to transform the 3 types of objective functions and 2 types of constraints into their crisp equivalences. So we propose several fuzzy-like simulations to simulate the objective functions and constraints. There are three kinds of simulations for each kind of fuzzy-like uncertainty. Based on the fuzzy simulations and Equations (3.33), (3.68), (3.70), (3.107), (3.108), we could propose Fu-Ra simulation 1 for expected value, Fu-Ra simulation 2 for critical value and Fu-Ra simulation 3 for chance. Based on fuzzy simulations and Equations (4.33), (4.34), (4.69), (4.70), (4.71), (4.100), (4.101), we can propose Fu-Fu simulation 1 for expected value, Fu-Fu simulation 2 for critical value, and Fu-Fu simulation 3 for chance. Based on fuzzy simulations and Equations (5.33), (5.68), (5.69), (5.93), (5.94), we can propose Fu-Ro simulation 1 for expected value, Fu-Ro simulation 2 for critical value, and Fu-Ro simulation 3 for chance. By combining fuzzy-like simulations and intelligent algorithms, we can create some hybrid algorithms. Then for the six kinds of models: EVM, CCM, DCM, ECM, CEM, DEM, we can obtain several kinds of fuzzy-like hybrid algorithms, see Figure 6.5.

These fuzzy-like simulations can be embedded into 4 types of basic intelligent algorithms, which includes
- particle swarm optimization algorithm (PSO)
- genetic algorithm (GA)
- simulated annealing algorithm (SA)
- tabu search algorithm (TS).

So for the general fuzzy-like MODM, we present the following ideas to design the algorithm. For the linear fuzzy-like multiple decision making model with some particular fuzzy-like variables, we can transform them into some crisp equivalent models and use the above 12 traditional solution methods to solve them directly. For the normal fuzzy-like multiple decision making model, especially the nonlinear model, we can embed the corresponding fuzzy-like simulations into the intelligent algorithm to find the solutions.

Application domains for each intelligent algorithm are as follows [418].

Some example areas of application of PSO are:
· Machine Learning
· Function Optimization
· Geometry and Physics
· Operations Research
· Chemistry, Chemical Engineering
· Electrical Engineering and Circuit Design.

Some example areas of application of GA are:
· Scheduling
· Chemistry, Chemical Engineering
· Medicine
· Data Mining and Data Analysis
· Geometry and Physics
· Economics and Finance
· Networking and Communication

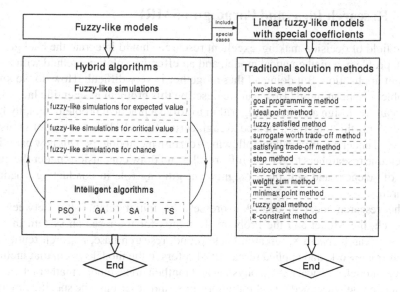

Fig. 6.6 Algorithm design

- · Electrical Engineering and Circuit Design
- · Image Processing
- · Combinatorial Optimization.

Some example areas of application of Simulated Annealing are:

- · Combinatorial Optimization
- · Function Optimization
- · Chemistry, Chemical Engineering
- · Image Processing
- · Economics and Finance
- · Electrical Engineering and Circuit Design
- · Machine Learning
- · Geometry and Physics
- · Networking and Communication.

Some example areas of application of TS are:

- · Combinatorial Optimization
- · Machine Learning
- · Biochemistry
- · Operations Research
- · Networking and Communication .

Although we used these 4 algorithms in the book, there are some other excellent intelligent algorithms, like the ant colony optimization algorithm (ACO), artificial neural network (ANN), immune algorithms (IA) and so on. We expect more advanced intelligent algorithms, and we are willing to use them if it is appropriate in our future research.

6.6   Research Ideas and Paradigm: 5MRP

In the field of decision making, excellent research should integrate the background
of the problem, a mathematical model, and an effective solution method with a significant application. But doing all these together is very difficult. How do we know
a problem is meaningful? How can we describe this problem in scientific language?
How can we design an efficient algorithm to solve a practical problem? Finally how
can we apply this integrated method to engineering fields? All these questions must be
answered under a new paradigm following a certain methodology. This new paradigm
will enable researchers to get scientific results and draw conclusions under the guidance of science, and will play a significant guiding role in conducting scientific
research.

The research ideal of 5RMP expresses the initial relationship between the
Research, the Model and the Problem. R stands for the research system that includes research specifics, research background, research base, research reality, re-
search framework, and applied research; M refers to the model system that includes
concept model, physical model, physical and mathematical model, mathematical and
physical model, designed model for algorithms, and describing the specific model. P
represents a problem system that includes a particular problem, a class of problems,
abstract problems, problem restoration, problem solution, and problem settlement.

Let us summarize the research ideas and the framework of our research work, see
Figure 6.7.

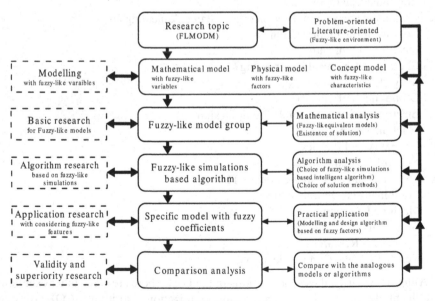

Fig. 6.7 Ideal route

Then we use the following Figure 6.8-6.10 to describe the relationship among between the problem system, the model system and the research system. Figure 6.8 emphasizes the problem system, and presents the train of thought in dealing with the problem.

Figure 6.9 emphasizes the model system, and present a series of models which are used in dealing with the corresponding problems.

Figure 6.10 emphasizes the research system, and presents the technological process when we confront a problem and conduct research work.

Let us propose the steps for 5MRP. When we start research, we usually study a particular problem, which has research value and can be described as a concept model. This is the introduction to the research; After studying the particular problem or a

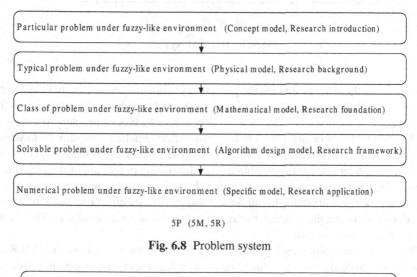

Particular problem under fuzzy-like environment (Concept model, Research introduction)

Typical problem under fuzzy-like environment (Physical model, Research background)

Class of problem under fuzzy-like environment (Mathematical model, Research foundation)

Solvable problem under fuzzy-like environment (Algorithm design model, Research framework)

Numerical problem under fuzzy-like environment (Specific model, Research application)

5P (5M, 5R)

Fig. 6.8 Problem system

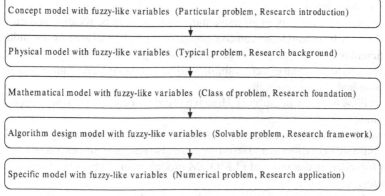

Concept model with fuzzy-like variables (Particular problem, Research introduction)

Physical model with fuzzy-like variables (Typical problem, Research background)

Mathematical model with fuzzy-like variables (Class of problem, Research foundation)

Algorithm design model with fuzzy-like variables (Solvable problem, Research framework)

Specific model with fuzzy-like variables (Numerical problem, Research application)

5M (5P, 5R)

Fig. 6.9 Model system

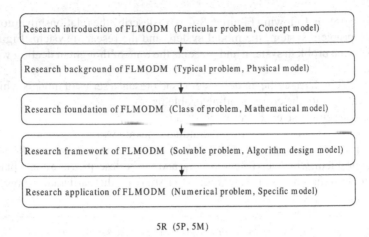

5R (5P, 5M)

Fig. 6.10 Research system

problem with the same essence as the particular problem, then we can obtain a typical problem which has universality and can be abstracted to a physical model. This is the background to the research; Then we can generalize the typical problem to a class of problems which can be abstracted to a common mathematical problem, from which we can propose a mathematical model. This is the foundation of the research; Then we can design an algorithm and obtain a model for the procedure of the algorithm. This is the framework of the research; Finally, we can apply the above models to a practical problem and establish a numerical model for a specific problem, and employ the algorithm to get the solution to illustrate the efficiency and validity. This is the application of the research.

We employ the following Figure 6.11 to illustrate how we can use the 5MPR research ideal to do our research: fuzzy random, bifuzzy and fuzzy rough multiple objective decision making.

In conclusion, 5RMP is an effective paradigm that can be widely used in various fields of scientific research and can contribute to research in all areas in a standardized and efficient manner. In the area of decision making, 5RMP is well reflected because of its rigorous logical and effective applicability, and it plays a dominant guiding role in the practical side of research.

FLMODM is a growing subject. Here we provide some further research problems in this area.

For mathematical properties, we should consider sensitivity analysis, dual theorems, optimality conditions, and so on. We can also research the crisp equivalent conditions for other special types of FLMODM models in the light of their mathematical properties.

From the viewpoint of solution methods, we can design more effective and powerful algorithms. We have integrated fuzzy-like simulations, genetic algorithms, simulated annealing, particle swarm optimization and tabu search to produce a series of hybrid intelligent algorithms to derive the solution to a FLMODM problem.

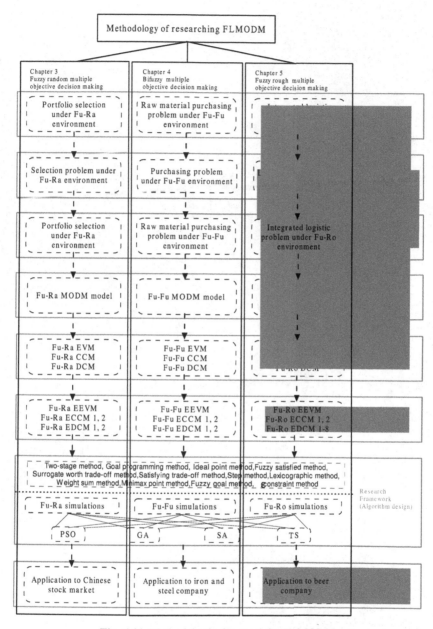

Fig. 6.11 Methodological system for FLMODM

In the perspective of applications, FLMODM can be applied to any decision making problems with fuzzy-like factors, for example, finance, supply chain management, manufacturing system, engineering management and so on.

We expect more development in every aspect of FLMODM.

References

1. Li, J., Xu, J., Gen, M.: A class of multiobjective linear programming model with fuzzy random coefficients. Mathematical and Computer Modelling 44(11-12), 1097–1113 (2006)
2. Li, J., Xu, J.: A novel portfolio selection model in a hybrid uncertain environment. Omega 37(2), 439–449 (2009)
3. Xu, J., Liu, Y.: Multi-objective decision making model under fuzzy random environment and its application to inventory problems. Information Sciences 178(14), 2899–2914 (2008)
4. Chankong, V., Haimes, Y.Y.: Multuobjective decision making: theory and methodology. North-Holland Publ., Amesterdam (1983)
5. Wismer, D.A., Haimes, Y.Y., Lason, L.S.: On bicriterion formulation of the intergrated systems identification and system optimization. IEEE transactions on systems, Man, and Cybernetics SMC-1, 296–297 (1971)
6. Haimes, Y.Y., Hall, W.A.: Multiobjective in water resources systens analysis: The surrogate worth trade-off (SWT) method. Water Resource Reaserch 10, 615–624 (1974)
7. Dombi, J.: A Fuzzy, Heuristic, Interactive approach to the Optimal Network Problem, pp. 253–275 (1983)
8. Smets, P., Magrez, P.: The measure of the degree of truth and of the grade of membership. Fuzzy sets and systems 25, 67–72 (1988)
9. Zadeh, L.A.: Fuzzy sets. Information and control 8, 338–353 (1965)
10. Zadeh, L.A.: Posibility measure of fuzzy events. Journal of Mathematical Analysis and Applications 8, 421–427 (1965)
11. Zadeh, L.A.: Quantitative fuzzy semantics. Information Sciences 3(2), 177–200 (1971)
12. Zadeh, L.A.: A fuzzy-set-theoretic interpretation of linguistic hedges. Journal of Cybernetics 2(3), 4–34 (1965)
13. Zadeh, L.A.: Calculus of Fuzzy restrictions. In: Zadeh, L.A., Fu, K.S., Tanaka, K., Shimura, M. (eds.) Fuzzy Sets and Their Applications to Cognitive and Decision Processes, p. 1-39. Academic Press, New York (1975)
14. Giles, R.: The concept of grade of membership. Fuzzy sets and systems 25, 297–323 (1988)
15. Bezdek, J.C., Hathaway, R.J.: Clustering with relational c-means partitions from pairwise distance data. Mathematical modelling 9, 435–439 (1987)

16. Pedrycz, W.: Fuzzy models and relational equations. Mathematical modelling 9, 427–434 (1987)

17. Bandemer, H.: From fuzzy data to functional relationships. Mathematical modelling 9, 419–426 (1987)

18. Rapoport, A., Wallsten, T.S., Cox, J.A.: Direct and indirect scaling of membership functions of probability phrase. Mathematical modelling 9, 397–417 (1987)

19. Zadeh, L.A.: The concept of a linguistic variable and its application to approximate reasoning, Part 1. Information Sci. 8, 199–249 (1975); Part II, Information Sci. 8, 301–357 (1975); Part 1, Information Sci. 9, 43–80 (1975)

20. Zadeh, L.A.: Similarity relations and fuzzy orderings. Information Sci. 3, 177–200 (1971)

21. Gottwald, S.: Set theory for fuzzy sets of higher level. Fuzzy Sets and Systems 2(2), 125–151 (1979)

22. Zadeh, L.A.: Fuzzy sets as a basis for a theory of possibility. Fuzzy Sets and Systems 1, 3–28 (1978)

23. Dubios, D., Prade, H.: Fuzzy Sets and Systerms: Theory and Application. Academic Press, New York (1980)

24. Bellman, R.E., Zadeh, L.A.: Local and fuzzy logics, ERL Memo M-584. University of California, Berkeley (1976); Epstein, D. (ed.): Modern Uses of Multiple-Valued Logics. D. Reidel, Dordrecht (1977)

25. Lai, Y.J., Hwang, C.L.: Interactive fuzzy linearing programming. Fuzzy sets and systerms 45, 169–183 (1992)

26. Verdegay, J.L.: Fuzzy mathematical programming, pp. 231–256 (1982)

27. Verdegay, J.L.: S dual approach to solve the fuzzy linear programming problem. Fuzzy sets and systerms 14, 131–141 (1984)

28. Tanaka, H., Okuda, T., Asai, K.: On fuzzy mathematical programming. Journal of Cybernetics 3, 37–46 (1974)

29. Orlovski, S.A.: On programming with fuzzy constraints set. Kebernetes 6, 197–202 (1977)

30. Dubois, D., Prade, H.: The mean value of a fuzzy number. Fuzzy Sets and Systems 24, 279–300 (1987)

31. Dubois, D., Prade, H.: Possibility Theory: An Approach to Computerized Processing of Uncertainty. Plenum, New York (1988)

32. Dubois, D., Prade, H.: The three semantics of fuzzy sets. Fuzzy Sets and Systems 90(2), 141–150 (1997)

33. Campos, L., Verdegay, J.L.: Linear programming problems and ranking of fuzzy numbers. Fuzzy Sets and Systems 32, 1–11 (1989)

34. González, A.: A study of the ranking function approach through mean values. Fuzzy Sets and Systems 35, 29–41 (1990)

35. Yager, R.R.: A procedure for ordering fuzzy subsets of the unit interval. Information Sciences 24, 143–161 (1981)

36. Yager, R.R.: On the evaluation of uncertain courses of action. Fuzzy Optimization and Decision Making 1, 13–41 (2002)

37. Lu, M.: On crisp equivalents and solutions of fuzzy programming with different chance measures, Technical Report (2001)

38. DeFinetti, B.: Probability Theory. Wiley, New York (1974)

39. Fine, T.: Theories ol Probability. Academic Press, New York (1973)

40. Dempster, A.: Upper and lower probabilities induced by multi-valued mapping. Ann. Math. Statist. 38, 325–339 (1967)
41. Sharer, G.: A Mathematical Theory or Evidence. Princeton University Press, Princeton (1976)
42. Shorflifle, E.H.: A model ofinexact reasoning in medicine. Math. Biosciences 23, 351–379 (1975)
43. Duda, R.O., Hart, P.F., Nilsson, N.J.: Subjective Bayesian methods for rule-based inference systems, Stanford Research Institute Tech. Note 124, Stanford, CA (1976)
44. Zimmermann, H.-J.: Using fuzzy sets in operational research. European Journal of Operational Research 13, 244–260 (1983)
45. Zimmermann, H.-J.: Fuzzy programming and linear programming with several objective functions. Fuzzy sets and systerms 1, 45–55 (1978)
46. Markowitz, H.: Portfolio selection. Journal of Finance 7, 77–91 (1952)
47. Markowitz, H.: Portfolio Selection: Efficient Diversification of Investments. Wiley, New York (1959)
48. Dubios, D., Prade, H.: Possibility theory. Plenum Press, New York (1988)
49. Tanaka, H., Guo, P., Turksen, I.B.: Portfolio selection based on fuzzy probabilities and possibility distributions. Fuzzy Sets and Systems 111, 387–397 (2000)
50. Inuiguchi, M., Tanino, T.: Portfolio selection under independent possibilistic information. Fuzzy Sets and Systems 115, 83–92 (2000)
51. Parra, M.A., Terol, A.B., Uría, M.V.R.: A Fuzzy Goal Programming Approach to Portfolio Selection. European Journal of Operational Research 133, 287–297 (2001)
52. Zhang, W.G., Nie, Z.K.: On admissible efficient portfolio selection problem. Applied Mathematics and Computation 159, 357–371 (2004)
53. Ramaswamy, S.: Portfolio selection using fuzzy decision theory, Working Paper of Bank for International Settlements, No. 59 (1998)
54. Fang, Y., Lai, K.K., Wang, S.Y.: Portfolio rebalancing model with transaction costs based on fuzzy decision theory. European Journal of Operational Research 175, 879–893 (2006)
55. Huang, X.X.: Fuzzy chance-constrained portfolio selection. Applied Mathematics and Computation 177, 500–507 (2006)
56. Perez Gladish, B., Jones, D.F., Tamiz, M., Bilbao Terol, A.: An interactive three-stage model for mutual funds portfolio selection. Omega 35, 75–88 (2007)
57. Sakawa, M., Yano, H.: An interactive fuzzy satisfying method for generalized multiobjective linear programming problems with fuzzy parameters. Fuzzy Sets and Systems 35, 125–142 (1990)
58. Wang, J., Hwang, W.L.: A fuzzy set approach for R&D portfolio selection using a real options valuation model. Omega 35, 247–257 (2007)
59. Wang, S.Y., Xia, Y.S.: Portfolio Selection and Asset Pricing. Springer, Berlin (2002)
60. Pareto, V.: Manuale di Economica Polittica. Societa Editrice Libraia, Milan, Italy (1906)
61. Xu, J., Liu, Q., Wang, R.: A class of multi-objective supply chain networks optimal model under random fuzzy enviroment and its application to the industry of Chinese liquor. Information Sciences 178, 2022–2043 (2008)
62. Holland, H.: Adaption in Natural and Artifical Systems. University of Michigan, Ann Arbor (1975)
63. Ishibuchi, H., Murata, T.: A multiobjective genetic local search algorithm and its application to flowshop scheduling. IEEE Transactions on Systems, Man and Cybernetics 28(3), 392–403 (1969)

64. Goldberg, D., Richardson, J.: Genetic algorithms with sharing for multi-model function optimization. In: Grefenstette [267], pp. 41–49
65. Ishibuchi, H., Murata, T.: A multi-objective genetic local search algorithm and its application to flowshop shceduling. IEEE transactions on System, Man and Cybernetics 28(3), 392–403 (1998)
66. Cheng, R., Gen, M.: An adaptive superplane approach for multiple objective optimizatio problems, Techinique report, Ashikaga Institute of Technology (1998)
67. Cheng, R., Gen, M.: An evolution program for the resource constraintd project scheduling problem. Computer itegrated manufacturing 11(3), 274–287 (1998)
68. Gen, M., Cheng, R.: A survey of penalty techniques in genetic algorithms. In: Fogel [195], pp. 804–809
69. Cheng, R., Gen, M.: Evolation program for resource constrained project scheduling problem. In: Fogel [197], pp. 736–741
70. Gen, M., Cheng, R.: Gennetic Algorithms and Engineering Optimization. Wiley, New York (2000)
71. Dubois, D., Prade, H.: Operations on fuzzy numbers. International Journal of System Sciences 9, 613–626 (1978)
72. Liu, B.: Dependent-chance programming: A class of stochastic programming. Computers & Mathematics with Applications 34(12), 89–104 (1997)
73. Liu, B.: Fuzzy random chance-constraint programming. IEEE transactions on fuzzy systems 9(5), 713–720 (2001)
74. Li, P., Liu, B.: Entropy of credibility distributions for fuzzy variables. IEEE transactions on Fuzzy systerms 133, 7–21 (2008)
75. Liu, B.: Random fuzzy dependent-chance programming and its hybrid intelligent algorithm. Information Sciences 141, 259–271 (2002)
76. Xu, J., Liu, Y.: Multi-objective decision making model under fuzzy random environment and its application to inventory problems. Information sciences (2008)
77. Lu, M.: On crisp equivalents and solutions of fuzzy programming with different chance measures, Technical report (2001)
78. Lu, M.: Some mathematical properties of fuzzy random programming, Technical report (2001)
79. Wei, Q., Yan, H.: Generalized Optimization Theory and models. Science publishing company, Beijing (2003)
80. Luhandjula, M.K.: Optimisation under hybrid uncertainty. Fuzzy Sets and Systems 146, 187–203 (2004)
81. Stock, J.R.: Reverse logistics. Council of Logistics Management (1992)
82. Stock, J.R.: Development and Implementation of Reverse Logistics Programs. Council of Logistics Management (1998)
83. Spengler, T., Puchert, H., Penkuhn, T., Rentz, O.: Environmental integrated introduction and recycling management. European Journal of Operation Research 2, 308–326 (1997)
84. Barros, A.I., Dekekr, R., Scholten, V.: A two-level network for recycling sand: a case study. European Journal of Operation Research 110, 199–215 (1998)
85. Marin, A., Pelegrin, B.: The return plant location problem: modeling and resolution. European Journal of Operational research 104, 375–392 (1998)
86. Krikkle, H.R., Van Harten, A., Schuur, P.C.: Business case one: reverse logistics redesign for copiers, OR Spectrum, 3, pp. 381–409 (1999)
87. Jayaraman, V., Patterson, R.A., Rolland, E.: The design of reverse distribution networks: Models and solution procedures. European Journal of Operational Research 150, 128–149 (2003)

88. Liste, O.: A generic stochastic model for supply-and-return network design. Computers & Operations Research 34, 417–442 (2007)

89. Salema, M.I.G., Barbosa-Povoa, A.P., Novais, A.Q.: An optimization model for the design of a capacitated multi-product reverse logistics network with uncertainty. European Journal of Operational Research 179, 1063–1077 (2007)

90. Felischmann, M., Bloemhof-Ruwaard, J.M., Dekker, R., van der Laan, E., van Nunen, J.A.E.E., Van Wassenhove, L.N.: Quantitative models for reverse logistics: A review. European Journal of Operational Research 103, 1–17 (1997)

91. Ko, H.J., Evans, G.W.: A genetic algorithm-based heuristic for the dynamic integrated forward/reverse logistics network for 3PLs. Computers & Operations Research 34, 346–366 (2007)

92. Xu, J., Li, J.: Multiple Objective Decision Making Theory and Methods. Tsinghua University Press, Beijing (2005) (in Chinese)

93. Xu, J., Yao, L.: A class of multiobjective linear programming model with random rough coeffients. Mathematical and Computer Moddelling 49, 189–206 (2009)

94. Chen, M.L., Lu, Y.Z.: A novel elitist multiobjective optimization algorithm: Multiobjective extremal optimization. European Journal of Operational Research 188, 637–651 (2008)

95. Kim, K., Gen, M., Kim, M.: Adaptive genetic algorithms for multi-recource constrained project scheduling problem with multiple modes. International Journal of Innovative Computing, Information and Control 2(1), 41–49 (2006)

96. Mendesa, J., Goncalvesb, J., Resendec, M.: A random key based genetic algorithm for the resource constrained project scheduling problem. Computers & Operations Research 36, 92–109 (2007) (available online July 25, 2007)

97. Ke, H., Liu, B.: Project scheduling problem with mixed uncertainty of randomness and fuzziness. European Journal of Operational Research 183, 135–147 (2007)

98. Holland, J.: Adaptation in Natural and Artificial. University of Michigan Press (1975)

99. Kumar, A., Pathak, R., et al.: A genetic algorithm for distributed system topology design. Computers and Industrial Engineering 28(659-670) (1995)

100. Blazewicz, J., Lenstra, J., Rinnooy, K.: Scheduiling subject to resource constrains: Classification & complexity. Discrete Applied Mathematics 5, 11–24 (1983)

101. Jozefowska, J., Mika, M., et al.: Solving the discrete-continuous project scheduling problem via its discretization. Mathematical Methods of Operations Research 52, 489–499 (2000)

102. Yun, Y., Gen, M.: Advanced scheduling problem using constrained programming techniques in scm environment. Computer & Industrial Engineering 43, 213–229 (2002)

103. Michalewicz, Z.: Genetic Algorithm + Data Structure = Evolution Programs, 3rd edn. Springer, New York (1996)

104. Mak, K., Wong, Y., Wang, X.: An adaptive genetic algorithm for manufacturing cell formation. International Journal of Manufacturing Technology 16, 491–497 (2000)

105. Dubois, D., Prade, H.: Operations of fuzzy numbers. Internat Journal Systems Science 96, 613–626 (1978)

106. Lushu, S., Li, K., Nair, P.K.: Fuzzy models for single-period inventory problem. Fuzzy Sets and Systems 132, 273–289 (2002)

107. Zimmermann, H.J.: Fuzzy linear programming with several objective functions. Fuzzy Sets and System 1, 46–55 (1978)

108. Chen, E.J., Lee, L.H.: A multi-objective selection procedure of determining a Pareto set. Computers & OperationsResearch 36, 1872–1879 (2009)

109. Youness, Y.E.: European Journal of Operational Research. 81, 440–443 (1995)

110. Hu, Y.: Practical Multiobjective Optimization. Shanghai Science and Technology Press, Shanghai (1990) (in Chinese)

111. Lin, C., Dong, J.: Methods and Theort for Multiobjective Optimization. Jilin Education Press, Changchun (1992) (in Chinese)

112. Nakayama, H., Sawaragi, Y., Tanino, T.: Theory of Multiobjective Optimization. Academic Press, Inc., London (1985)

113. Luc, D.T.: On the domination property in vector optimization. Optim. Theory Appl. 43, 327–330 (1984)

114. Ijiri, Y.: Management Goals and Accounting for Control. Rand McNally, Chicago (1965)

115. Szidarovszky, F., Gershon, M.E., Duckstein, L.: Techniques for Multiobjective Decision Making in Systems Management. Elsevier, Amsterdam (1986)

116. Werners, B.: Interactive multiple objective programming subject to flexible constraints. European Journal of Operational Research 31, 342–349 (1987)

117. Leberling, H.: On finding compromise solution in multicriteria problems using the fuzzy min-operator. Fuzzy Sets and Systems 6, 105–118 (1981)

118. Zimmermann, H.J.: Fuzzy programming and linear programming with several objective functions. Fuzzy Sets and Systems 1, 45–55 (1978)

119. Wierzbicki, A.: Multiple Creteria Decision Making Theory and Application. Springer, Berlin (1980)

120. Choo, E.V., Atkins, D.R.: An interactive algorithm for multicriteria programming. Computers and Operations Research 7, 81–88 (1980)

121. Shimizu, K.: Theory of Multiobjective and Conflict. Kyoritsu Syuppan (1982) (in Japanese)

122. Krickeberg, K.: Probability theory. Addison-Wesley, Reading (1965)

123. Billingsley, P.: Probability and measure. Wiley, New York (1979)

124. Shiryaev, A.N.: Probability, 2nd edn. Springer, Heidelberg (1995)

125. Teargny, J., Benayoun, R., Montgolfier, J., Larichev, O.: Linear programming with multiobjective functions: Step Method (STEM). Math. Programming 1(2), 366–375 (1971)

126. Lee, S.M.: Goal Programming for Decision Analysis. Auerbach Publishers, Philadelphia (1972)

127. Kendall, K.E., Lee, S.M.: Formulating blood rotation policies with multiple objectives. Management Sciences 26(11), 1145–1157 (1980)

128. Sakawa, M.: Interactive fuzzy goal programming for multiobjective nonlinear programming problems and its applications to water quality management. Control and Cybernetics 13, 217–228 (1984)

129. Sakawa, M., Yumine, T., Yano, Y.: An interactive fuzzy satisfying method for multiobjective linear programming problems and its applications. IEEE Transaction on Systems, Man, and Cybernetics 17, 654–661 (1987)

130. Sakawa, M., Yano, Y.: Interactive fuzzy decision making for multiobjective nonlinear programming using augmented minimax problems. Fuzzy Sets and Systems 20, 31–43 (1986)

131. Sakawa, M., Seo, F.: Multiple Criteria Decision Analysis in Regional Planning-Concepts, Methods and Applications. D. Reidel, Dordrecht (1988)

132. Bazaraa, M.S., Shetty, C.M.: Nonlinear Programming: Theory and Algorithms. Wiley, New York (1979)

133. Li, J., Xu, J.P., Gen, M.: A class of fuzzy random multiobjective programming problem. Mathematical and Computer Modelling 44, 1097–1113 (2006)

134. Liu, B.: Uncertainty Theory: An Introduction to its Axiomatic Foundations. Springer, Berlin (2004)

135. Yoshida, Y.: The valuation of European options in uncertain environment. European Journal of Operational Research 145, 221–229 (2003)

136. Wu, H.C.: Pricing European options based on the fuzzy pattern of Black-Scholes formula. Computers & Operations Research 31, 1069–1081 (2004)

137. Zmeskal, Z.: Application of the fuzzy-stochastic methodology to appraising the firm value as a european call option. European Journal of Operational Research 135, 303–310 (2001)

138. Zmeskal, Z.: Value at risk methodology under soft conditions approach (fuzzy-stochastic approach). European Journal of Operational Research 161, 337–347 (2005)

139. Puri, M.L., Ralescu, D.A.: Fuzzy random variables. J. Math. Annal. Appl. 114, 409–422 (1986)

140. Luhandjula, M.K., Gupta, M.M.: On fuzzy stochastic optimization. Fuzzy Sets and Systems 81, 41–55 (1996)

141. Feng, Y.H., Hu, L.J., Shu, H.S.: The variance and convariance of fuzzy random variables and their applications. Fuzzy Sets and Systems 120, 487–497 (2001)

142. Korner, R.: On the variance of fuzzy random variable. Fuzzy Sets and System 92, 83–93 (1997)

143. de Campos, L.M., González, A.: A subjective approach for ranking fuzzy numbers. Fuzzy Sets and Systems 29, 145–153 (1989)

144. López-Díaz, M., Gil, M.A.: The λ-average value and the fuzzy expectation of a fuzzy random variable. Fuzzy Sets and Systems 99, 347–352 (1998)

145. Liu, B.: Uncertainty Theory: An Introduction to its Axiomatic Foundations. Springer, Berlin (2004)

146. Keown, A.J., Martin, J.D.: A chance constrained goal programming model for working capital management. The Engineering Economists 22, 153–174 (1977)

147. Charnes, A., Cooper, W.W.: Chance-constrained programming. Management Science 6, 73–79 (1959)

148. Xu, J., Li, J.: A novel portfolio selection model in hybrid uncertain environment, Omega (2008)

149. Ammar, E., Khalifa, H.A.: Fuzzy portfolio optimization a quadratic programming approach. Chaos, Solitons and Fractals 18, 1045–1054 (2003)

150. Antczak, T.: A modified objective function method for solving nonlinear multiobjective fractional programming problems. J. Math. Anal. Appl. 322, 971–989 (2006)

151. Ben-Israel, A., Robers, P.D.: A Decomposition method for interval linear programming. Operations Research 21, 1154–1157 (1973)

152. Bitran, G.R.: Linear multiple objective problems with interval coefficient. Management Science 26, 694–706 (1980)

153. Black, F., Scholes, M.: The pricing of options and corporate liabilities. Journal of Political Economy 81, 637–654 (1973)

154. Buckley, J.J.: Stochastic versus possibilistic programming. Fuzzy Sets and Systems 34, 173–177 (1990)

155. Carlsson, C., Fuller, R.: On possibilistic mean value and variance of fuzzy numbers. Fuzzy Sets and Systems 122, 315–326 (2001)

156. Chakraborty, M., Gupta, S.: Fuzzy mathematical programming for multi-objective linear fractional programming problem. Fuzzy Sets and Systems 125, 335–342 (2002)

157. Chanas, S., Kuchta, D.: Multiple objectiveprogramming in optimization of the interval objective function—a generalized approach. European Journal of Operational Research 94, 594–598 (1996)

158. Chang, T.J., Meade, N., Beasley, J.B., Sharaiha, Y.: Heuristic for cardinality constrained portfolio optimization. Computers and Operations Research 27, 1271–1302 (2000)

159. Chankong, V., Haimes, Y.Y.: Multiobjective Decision Making: Theory and Methodology. North-Holland, New York (1983)

160. Charnes, A., Cooper, W.W.: Chance-constrained programming. Management Science 6(1), 73–79 (1959)

161. Charnes, A., Cooper, W.W.: Management Models and Application of Linear Programming. Wiley, New York (1961)

162. Charnes, A., Granot, F., Philips, F.: An algorithm for solving interval linear programming probelms. Operations Research 25, 688–695 (1977)

163. Chen, M.S., Yao, J.S., Lu, H.F.: A fuzzy stochastic single-period model for cash management. European Journal of Operational Research 170, 72–90 (2006)

164. Chen, Y.J., Liu, Y.K.: Portfolio selection in fuzzy environment. In: Proceeedings of the Fourth International Conference on Machine Learning and Cybernetics, Guangzhou, pp. 2694–2699 (2005)

165. Chiodi, L., Mansini, R., Speranza, M.G.: Semi-absolute deviation rule for mutual funds portfolio selection. Annals of Operations Research 124, 245–265 (2003)

166. Colubi, A., Domonguez-Menchero, J.S., Lopez-Diaz, M., Ralescu, D.A.: On the formalization of fuzzy random variables. Information Sciences 133, 3–6 (2001)

167. Costa, J.P.: Computing non-dominated solutions in MOLFP. European Journal of Operational Research 181(3), 1464–1475 (2007)

168. Deb, K.: Multi-Objective Optimization Using Evolutionary Algorithms. John Wiley, New York (2001)

169. Doerner, K., Gutjahr, W.J., Hartl, R.F., Strauss, C., Stummer, C.: Pareto ant colony optimization: a metaheuristic approach to multiobjective portfolio Selection. Annals of Operations Research 131, 79–99 (2004)

170. Dubois, D., Prade, H.: The mean value of a fuzzy number. Fuzzy Sets and Systems 24, 279–300 (1987)

171. Dubois, D., Prade, H.: Possibility Theory. Plenum Press, New York (1988)

172. Elton, E.J., Gruber, M.J.: The multi-period consumption investment problem and single period analysis. Oxford Economics Papers 9, 289–301 (1974)

173. Elton, E.J., Gruber, M.J.: On the optimality of some multiperiod portfolio selection criteria. Journal of Business 7, 231–243 (1974)

174. Fonseca, C., Fleming, P.: An overview of evolutionary algorithms in multiobjective optimization. Evolutionary Computation 3(1), 1–16 (1995)

175. Fuller, R., Majlender, P.: On weighted possibilistic mean and variance of fuzzy numbers. Fuzzy Sets and Systems 136, 363–374 (2003)

176. Gao, J., Liu, B.: New primitive chance measures of fuzzy random event. International Journal of Fuzzy Systems 3(4), 527–531 (2001)

177. Gen, M., Cheng, R.: Genetic Algorithms & Engineering Optimization. John Wiley & Sons, New York (2000)

178. Gil, M.A., Lopez-Diaz, M., Ralescu, D.A.: Overview on the development of fuzzy random variables. Fuzzy Sets and Systems 157, 2546–2557 (2006)

179. Giove, S., Funari, S., Nardelli, C.: An interval portfolio selection problem based on regret function. European Journal of Operational Research 170, 253–264 (2006)

180. Gladish, B.P., Parra, M.A., Terol, A.B., Uria, M.V.R.: Solving a multiobjective possibilistic problem through compromise programming. European Journal of Operational Research 164, 748–759 (2005)

181. Hamza, F., Janssen, J.: The mean-semivariances approach to realistic portfolio optimization subject to transaction costs. Applied Stochastic Models and Data Analysis 14, 275–283 (1998)

182. Horn, J.: Multicriterion Decision Making. Oxford University Press, New York (1997)

183. Huang, J.J., Tzeng, G.H., Ong, C.S.: A novel algorithm for uncertain portfolio selection. Applied Mathematics and Computation 173, 350–359 (2006)

184. Huang, X.: A new perspective for optimal portfolio selection with random fuzzy returns. Information Sciences 177, 5404–5414 (2007)

185. Alefeld, G., Herzberger, J.: Introduction to Interval Computations. Academic Press, New York (1983)

186. Hansen, E.: Global Optimization Using Interval Analysis. Marcel Dekker, New York (1992)

187. Halmos, P.R.: Measure theory. Van Nostrand, Princeton (1950) (republished by Spinger in 1974)

188. Ichihashi, H., Inuiguchi, M., Kume, Y.: Modality constrained programming models: a unified approach to fuzzy mathematical programming problems in the setting of possibilisty theory. Information Sciences 67, 93–126 (1993)

189. Ida, M.: Portfolio selelction problem with interval coefficients. Applied Mathematics Letters 16, 709–713 (2003)

190. Ida, M.: Solutions for the portfolio selection problem with interval and fuzzy coefficients. Reliable Computing 10, 389–400 (2004)

191. Dubois, D., Prade, H.: Operations of fuzzy numbers. Internat. J. Systems Sci. 96, 613–626 (1978)

192. Ignizio, J.P.: Linear Programming in Single & Multi-Objective Systems. Prentice-Hall, Englewood Cliffs (1982)

193. Inuiguchi, M., Kume, Y.: Goal programming problems with interval coefficients and target intervals. European Journal of Operational Reseach 52, 345–360 (1991)

194. Inuiguchi, M., Ramik, J.: Possibilistic linear programming: A brief review of fuzzy mathematical programming and a comparison with stochastic programming in portfolio selection problem. Fuzzy Sets and Systems 111, 3–28 (2000)

195. Inguiguchi, M., Ramik, J., Tanino, T., Vlach, M.: Satisficing solutions and duality in interval and fuzzy linear programming. Fuzzy Sets and Systems 135, 151–177 (2003)

196. Inuiguchi, M., Sakawa, M.: Minimax regret solution to linear programming problems with an interval objective function. European Journal of Operational Research 86, 526–536 (1995)

197. Ishibuchi, H., Tanaka, H.: Formulation and analysis of linear programming problems with interval coefficients. J. Jpn. Ind. Manage. Assoc. 40, 320–329 (1989)

198. Ishibuchi, H., Tanaka, H.: Multiobjective programming in optimization of the interval objective function. European Journal of Operational Research 48, 219–225 (1990)

199. Itoh, T., Ishii, H.: One machine scheduling problem with fuzzy random due-dates. Fuzzy Optimization and Decision Making 4, 71–78 (2005)

200. Chih, H.H.: Optimization of fuzzy production inventory models. Information Sciences 146, 29–40 (2002)

201. Roy, T.K., Maiti, M.: Multi-objective inventory models of deterioration items with some constraints in a fuzzy envirinment. Computers and Operations Researchs, 1085–1095 (1998)

202. Luhandjula, M.K., Gupta, M.M.: On fuzzy stochastic optimization. Fuzzy Sets and Systems 81, 41–55 (1996)

203. Korner, R.: On the variance of fuzzy random variable. Fuzzy Sets and System 92, 83–93 (1997)

204. Katagiri, H., Ishii, H.: Fuzzy portfolio selection problem. In: Proceedings IEEE SMC 1999 Conference, vol. 3, pp. 973–978 (1999)

205. Katagiri, H., Ishii, H.: Linear programming problem with fuzzy random constraint. Mathematica Japonica 52, 123–129 (2000)

206. Katagiri, H., Sakawa, M., Ishii, H.: Fuzzy random bottleneck spanning tree problems using possibility and necessary measures. European Journal of Operational Research 152, 88–95 (2004)

207. Katagiri, H., Sakawa, M., Kato, K., Nishizaki, I.: A fuzzy random multiobjective 0-1 programming based on the expectation optimization model using possibility and necessary measures. Mathematical and Computer Modeling 40, 411–421 (2004)

208. Kataoka, S.: A stochastic programming model. Econometrica 31, 181–196 (1963)

209. Kato, K., Katagiri, H., Sakawa, M., Ohsaki, S.: An interactive fuzzy satisficing method based on the fractile optimization model using possibility and necessity measures for a fuzzy random multiobjective linear programming problem. Electronics and Communications in Japan 88(5), 20–28 (2005)

210. Kaufmann, A.: Introduction to the Theory of Fuzzy Subsets, vol. I. Academci Press, New York (1975)

211. Kita, H., Tamaki, H., Kobayashi, S.: Multiobjective optimization by genetic algorithms: A review. In: Fogel, D. (ed.) Proceedings of the IEEE International Conference on Evolutionary Computation, pp. 517–522. IEEE Press, Piscataway (1996)

212. Konno, K., Yamazika, H.: Mean absolute deviation portfolio optimization model and its application to Tokyo stock market. Mangement Science 37, 519–531 (1991)

213. Korner, R.: On the variance of fuzzy random variables. Fuzzy Sets and Systems 92, 83–93 (1997)

214. Kruse, R., Meyer, K.D.: Statistics with Vague Data. Reidel Publishing Company, Dordrecht (1987)

215. Kwakernaak, H.: Fuzzy random variables-definitions and theorems. Information Science 15, 1–29 (1978)

216. Lai, K.K., Wang, S.Y., Xu, J.P., Fang, Y.: A class of linear interval programming problems and its applications to portfolio selection. IEEE Transaction on Fuzzy Systems 10, 698–704 (2002)

217. Li, R.J., Lee, E.S.: Ranking fuzzy numbers-A comparison. In: Proceedings of North American Fuzzy Information Processing Society Workshop, West Lafayette, IL, pp. 169–204 (1987)

218. Liu, B.: Fuzzy random chance-constraint programming. IEEE Transactions on Fuzzy Systems 9(5), 713–720 (2001)

219. Liu, B.: Fuzzy random dependent-chance programming. IEEE Transactions on Fuzzy Systems 9(5), 721–726 (2001)

220. Liu, B.: Theory and Practice of Uncertain Programming. Physica Verlag, Heidelberg (2002)

221. Liu, B., Iwamura, K.: Chance constrained programming with fuzzy parameters. Fuzzy Sets and Systems 94, 227–237 (1998)

222. Liu, B., Iwamura, K.: A note on chance constrained programming with fuzzy coefficients. Fuzzy Sets and Systems 100, 229–233 (1998)

223. Liu, W.A., Zhang, W.G., Wang, Y.L.: On admissible efficient portfolio selection: Models and algorithms. Applied Mathematics and Computation 176, 208–218 (2006)

224. Liu, Y.K.: Convergent results about the use of fuzzy simulation in fuzzy optimization problems. IEEE Transactions on Fuzzy Systems 14(2), 295–304 (2006)

225. Liu, Y.K., Liu, B.: A class of fuzzy random optimization: Expected value models. Information Sciences 155, 89–102 (2003)

226. Liu, Y.K., Liu, B.: Fuzzy random variables: A scalar expected value operator. Fuzzy Optimization and Decision Making 2, 143–160 (2003)

227. Liu, Y.K., Liu, B.: On minimum-risk problems in fuzzy random decision systems. Computers & Operations Research 32, 257–283 (2005)

228. Lopez-Diaz, M., Angeles Gil, M.: Constructive definitions of fuzzy random variables. Stochastics & probability Letters 36, 135–143 (1997)

229. Luhandjula, M.K.: Linear programming under randomness and fuzziness. Fuzzy Sets and Systems 10, 45–55 (1983)

230. Luhandjula, M.K.: Fuzziness and randomness in an optimization framework. Fuzzy Sets and Systems 77, 291–297 (1996)

231. Luhandjula, M.K., Gupta, M.M.: On fuzzy stochastic optimization. Fuzzy Sets and Systems 81, 47–55 (1996)

232. Luhandjula, M.K.: Optimisation under hybrid uncertainty. Fuzzy Sets and Systems 146, 187–203 (2005)

233. Luhandjula, M.K.: Fuzzy stochastic linear programming: Survey and future research directions. European Journal of Operational Research 74, 1353–1367 (2006)

234. Markowitz, H.: Portfolio selection. Journal of Finance 7, 77–91 (1952)

235. Matlab. Optimization Toolbox for Use with Matlab, Version 7.0.1 (R14). Mathworks, Inc., Natick, Massachusetts (2004)

236. Mausser, H.E., Laguna, M.: A heuristic to minimax absolute regret for linear programs with interval objective function coefficients. European Journal of Operational Research 117, 157–174 (1999)

237. Merton, R.C.: Lifetime portfolio selection under uncertainty: The continuous time case. Review of Economics and Statistics 51, 247–257 (1969)

238. Merton, R.C.: The theory of rational option pricing. Bell Journal of Economics and Management Science 4, 141–183 (1973)

239. Metev, B.: Use of reference points for solving MONLP problems. European Journal of Operational Research 80, 193–203 (1995)

240. Metev, B., Gueorguieva, D.: A simple method for obtaining weakly efficient points in multiobjective linear fractional programming problems. European Journal of Operational Research 126, 386–390 (2000)

241. Michalewicz, Z.: Genetic Algorithms + Data Structures = Evolution Programs. Springer, New York (1994)

242. Mohan, C., Nguyen, H.T.: An interactive satisfying method for solving mixed fuzzy-stochastic programming problems. Fuzzy Sets and Systems 117, 67–79 (2001)

243. Moore, R.E.: Method and Applications of Interval Analysis. SIAM, Philadelphia (1979)

244. Nahmias, S.: Fuzzy variables. Fuzzy Sets and Systems 1, 97–110 (1978)

245. Negoita, C.V., Ralescu, D.: Simulation, Knowledge-Based Computing, and Fuzzy Statistics. Van Nostrand Reinhold, New York (1987)

246. Nguyen, V.H.: Fuzzy stochastic goal programming problems. European Journal of Operational Research 176, 77–86 (2007)

247. Ogryczak, W.: Multiple criteria linear programming model for portfolio selection. Annals of Operations Research 97, 143–162 (2000)

248. Perold, A.F.: Large-scale portfolio optimization. Management Science 31(10), 1143–1159 (1984)

249. Pratap, A., Deb, K., Agrawal, S., Meyarivan, T.: A fast and elitist multi-objective genetic algorithm: NSGA-II. IEEE Transactions on Evolutionary Computation 6(2), 182–197 (2002)

250. Qi, Y., Hirschberger, M., Steuer, R.E.: Tri-criterion quadratic-linear programming. Technical report, Working paper, Department of Banking and Finance, University of Georgia, Athens (2006)

251. Qi, Y., Steuer, R.E., Hirschberger, M.: Suitable-portfolio investors, nondominated frontier sensitivity, and the effect of multiple objectives on standard portfolio selection. Annals of Operations Research 152(1), 297–317 (2007)

252. Qiao, Z., Wang, G.: On solutions and distributions problems of the linear programming with fuzzy random variable coefficients. Fuzzy Sets and Systems 58, 155–170 (1993)

253. Qiao, Z., Zhang, Y., Wang, G.: On fuzzy random linear programming. Fuzzy Sets and Systems 65, 31–49 (1994)

254. Ross, S.A.: The arbitrage theory of capital asset pricing. Journal of Economic Theory 13, 341–360 (1976)

255. Roubens, M., Teghem, J.: Comparisons of methodologies for fuzzy and stochastic multiobjective programming problems. Fuzzy Sets and Systems 42, 119–132 (1991)

256. Roy, A.D.: Safety-first and the holding of assets. Econometrics 20, 431–449 (1952)

257. Sakawa, K.: Fuzzy Sets and Interactive Multiobjective Optimization. Plenum Press, New York (1993)

258. Sawaragi, Y., Nakayama, H., Tanino, T.: Theory of Multiobjective Optimization. Academic Press, New York (1985)

259. Schaerf, A.: Local search techniques for constrained portfolio selection problems. Computational Economics 20, 177–190 (2002)

260. Schrage, L.: LINGO User's Guide. Lindo Publishing, Chicago (2004)

261. Sengupta, A., Pal, T.K.: On comparing interval numbers. European Journal of Operational Research 127, 28–43 (2000)

262. Sengupta, A., Pal, T.K., Chakraborty, D.: Interpretation of inequality constraints involving interval coefficients and a solution to interval linear programming. Fuzzy Sets and Systems 119, 129–138 (2001)

263. Sharpe, W.: Capital asset prices: a theory of market equilibrium under conditions of risk. Journal of Finance 19, 425–442 (1964)

264. Shih, C.J., Wangsawidjaja, R.A.S.: Mixed fuzzy-probabilistic programming approach for multiobjective engineering optimization with random variables. Computers and Structures 59(2), 283–290 (1996)

265. Shing, C., Nagasawa, H.: Interactive decision system in stochastic multiobjective portfolio selection. Int. J. Production Economics, 60–61, 187–193 (1999)

266. Shukla, P.K., Deb, K.: On finding multiple pareto-optimal solutions using classical and evolutionary generating methods. European Journal of Operational Research 181(3), 1630–1652 (2007)

267. Siddharha, S.S.: A dual ascent method for the portfolio selection problem with multiple constraints and linked proposals. European Journal of Operational Research 108, 196–207 (1998)

268. Slowinski, R., Teghem, J.: Stochastic versus Fuzzy Approaches to Multiobjective Mathematical Programming under Uncertainty. Kluwer Academic Publishers, Dordrecht (1990)

269. Stancu-Minasian, I.M.: Stochastic Programming with Mutiple Objective Functions. Rediel, Dordrecht (1984)

270. Stancu-Minasiana, I.M., Pop, B.: On a fuzzy set approach to solving multiple objective linear fractional programming problem. Fuzzy Sets and Systems 134, 397–405 (2003)

271. Steuer, R.E.: Mulitple objective linear programming with interval criterion weights. Management Science 23, 305–316 (1976)

272. Steuer, R.E., Qi, Y., Hirschberger, M.: Developments in Multi-Attribute Portfolio Selection. Working Paper, Department of Banking and Finance. University of Georgia, Athens (2006)

273. Streichert, F., Ulmer, H., Zell, A.: Evolutionary algorithms and the cardinality constrained portfolio selection problem. In: Ahr, D., Fahrion, R., Oswald, M., Reinelt, G. (eds.) Operations Research Proceedings 2003, Selected Papers of the International Conference on Operations Research (OR 2003), pp. 253–260. Springer, Berlin (2003)

274. Sturm, J.: Using SeDuMi 1.02, a MATLAB Toolbox for Optimization Over Symmetric Cones. Optimization Methods and Software 11, 625–653 (1999)
275. Tanaka, H., Okuda, T., Asai, K.: On fuzzy mathematical programming. Journal of Cybernetics 3, 37–46 (1974)
276. Tanaka, H., Guo, P., Turksen, I.B.: Portfolio selection based on fuzzy probabilities and possibility distributions. Fuzzy Sets and Systems 111, 387–397 (2000)
277. Telser, L.G.: Safety first and hedging. Review of Economic Studies 23, 1–16 (1955)
278. Tiryaki, F., Ahlatcioglu, M.: Fuzzy stock selection using a new fuzzy ranking and weighting algorithm. Applied Mathematics and Computation 170, 144–157 (2005)
279. Ulmer, H., Streichert, F., Zell, A.: Evaluating a hybrid encoding and three crossover operations on the constrained portfolio selection problem. In: Congress of Evolutionary Computation (CEC 2004), Portland, Oregon, USA, pp. 932–939. IEEE Press, Los Alamitos (2004)
280. Vandenberghe, L., Boyd, S.: Semidefinite Programming. SIAM Review 38, 49–95 (1996)
281. Wang, G., Qiao, Z.: Linear programming with fuzzy random variable coefficients. Fuzzy Sets and Systems 57, 3295–3311 (1993)
282. Wang, S.Y., Zhu, S.S.: On Fuzzy Portfolio Selection Problem. Fuzzy Optimization and Decision Making 1, 361–377 (2002)
283. Wierzbicki, A.: A mathematical basis for satisficing decision making. In: Morse, J.N. (ed.) Organizations: Multiple Agents with Multiple Criteria, Proceedings, pp. 465–485. Springer, Berlin (1981)
284. Wierzbicki, A.: On the completeness and constructiveness of parametric characterization to vector optimization problems. OR Spektrum 8, 73–87 (1986)
285. Williams, J.O.: Maximizing the probability of achieving investment goals. Journal of Portfolio Management 24, 77–81 (1997)
286. Yazenin, A.V.: Fuzzy and stochastic programming. Fuzzy Sets and Systems 22, 171–188 (1987)
287. Yoshimoto, A.: The mean-variance approach to portfolio optimization subject to transaction costs. Journal of Operations Research Society of Japan 39, 99–117 (1996)
288. Zadeh, L.A.: Fuzzy sets. Inform. and Control 8, 338–353 (1965)
289. Zadeh, L.A.: The concept of a linguistic variable and its application to approximate reasoning. Information Science 8, 199–251 (1975)
290. Zadeh, L.A.: Fuzzy sets as a basis for a theory of possibility. Fuzzy Sets and Systems 1, 3–28 (1978)
291. Nahmias, S.: FFuzzy variables. Fuzzy Sets and Systems 1, 97–110 (1979)
292. Zeleny, M.: Compromise programming in multiple criteria decision making. In: Cochrane, J.L., Zeleny, M. (eds.), University of South Carolina Press, Columbia (1973)
293. Kumar, A., Pathak, R., et al.: A genetic algorithm for distributed system topology design. Computers and Industrial Engineering 28, 659–670 (1995)
294. Mendesa, J., Goncalvesb, J., Resendec, M.: A random key based genetic algorithm for the resource constrained project scheduling problem. Computers & Operations Research 36, 92–109 (2009)
295. Ke, H., Liu, B.: Project scheduling problem with mixed uncertainty of randomness and fuzziness. European Journal of Operational Research 183, 135–147 (2007)
296. Blazewicz, J., Lenstra, J., Rinnooy, K.: Scheduiling subject to resource constrains: Classification & complexity. Discrete Applied Mathematics 5, 11–24 (1983)
297. Jozefowska, J., Mika, M., et al.: Solving the discrete-continuous project scheduling problem via its discretization. Mathematical Methods of Operations Research 52, 489–499 (2000)

298. Kim, K., Gen, M., Kim, M.: Adaptive genetic algorithms for multi-recource constrained project scheduling problem with multiple modes. International Journal of Innovative Computing, Information and Control 2(1), 41–49 (2006)

299. Yun, Y., Gen, M.: Advanced scheduling problem using constrained programming techniques in scm environment. Computer & Industrial Engineering 43, 213–229 (2002)

300. Jain, R.: Decision making in the presence of fuzzy variables. IEEE Trans. On Systens, Man and Cybernetics 6, 698–703 (1976)

301. Mizumoto, M., Tanaka, K.. Algebraic properties of fuzzy numbers. In: IEEE International Conference on Cybernatics and Society, pp. 559–563 (1976)

302. Mizumoto, M., Tanaka, K.: Algebraic properties of fuzzy numbers. In: Gupta, M.M. (ed.) Advances in fuzzy set theory and applications, pp. 153–164. North-Holland, Amsterdam (1979)

303. Baas, S.M., Kwakernaak, H.: Rating and ranking of multiple aspect alternative using fuzzy sets. Automatica 13, 47–58 (1977)

304. Dubois, D., Prade, H.: Operations on fuzzy numbers. International Journal of System Science 9, 613–626 (1978)

305. Dubois, D., Prade, H.: Operations on fuzzy numbers. Fuzzy Sets and System: Theory and Applications. Academic Press, New York (1980)

306. Dubois, D., Prade, H.: Possibility theorey: An apporach to computerized processing of uncertainty. Plenum, New York (1988)

307. Dijkman, J., Van Haeringen, H., Delange, S.J.: Fuzzy numbers. Journal of Mathematical Analysis and Applications 92(2), 302–341 (1983)

308. Gupta, M.M.: Fuzzy information in decision making. In: International Conference on Advances in Information Sciences and Thechnology, Golden Jubilee Conference at the Indian Conference at the Indian Statistical Institute, Calcutta, Indian (1982)

309. Kaufmann, A., Gupta, M.M.: Introduction to Fuzzy Arithmetic. Van Nostrand, New York (1985)

310. Laarhoven, P.J.M., Pedrycz, W.: A fuzzy extension of Satty's priority theory. Fuzzy sets and Systems 11(3), 229–241 (1983)

311. Buckley, J.J.: The multiple-judge, multiple-criteria ranking problem: A fuzzy-set approach. Fuzzy sets and Systems 13(1), 139–147 (1984)

312. Buckley, J.J.: Generalized and extended fuzzys sets with applications. Fuzzy sets and Systems 25(2), 159–174 (1988)

313. Bonissone, P.P.: A fuzzy set based linguistic approach: Theory and applications. In: Proceedings of the 1980 Winter Simulation Conference, Orlando, Florida, pp. 99–111 (1980)

314. Bonissone, P.P.: A fuzzy set based linguistic approach: Theory and applications. In: Gupta, M.M., Sanchez, E. (eds.) Approximate Reasoning in Decision Making, pp. 329–339. North-Holland, Amsterdam (1982)

315. Kennedy, J., Eberhart, R.C.: Swarm Intelligence. Morgan Kaufmann, San Francisco

316. Kennedy, J., Eberhart, R.C.: Particle swarm optimization. In: Proc. IEEE Int'l Conf. on neural networks (1995)

317. Jacqueline, M., Richard, C.: Application of particle swarm to multiobjective optimization. Department of Computer Science and Software Engineering, Auburn University (1999)

318. Clerc, M., Kennedy, J.: The particle swarmłexplosion, stability, and convergence in a multidimensional complex space. IEEE Transactions on Evolutionary Computation 6, 58–73 (2002)

319. Parsopoulos, K., Vrahatis, M.: Particle swarm optimization method in multiobjective problems. In: Proceedings of the 2002 ACM Symposium on Applied Computing (SAC 2002), pp. 603–607. ACM Press, New York (2002)

320. Fieldsend, J.E., Singh, S.: A multiobjective algorithm based upon particle swarm optimisation, an efficient data structure and turbulence. In: Proceedings of the 2002 UK Workshop on Computational Intelligence, Birmingham, UK, September 2002, pp. 37–44 (2002)

321. Li, X.: A Non-dominated Sorting Particle Swarm Optimizer for Multiobjective Optimization. In: Cantú-Paz, E., Foster, J.A., Deb, K., Davis, L., Roy, R., O'Reilly, U.-M., Beyer, H.-G., Kendall, G., Wilson, S.W., Harman, M., Wegener, J., Dasgupta, D., Potter, M.A., Schultz, A., Dowsland, K.A., Jonoska, N., Miller, J., Standish, R.K. (eds.) GECCO 2003. LNCS, vol. 2723, pp. 37–48. Springer, Heidelberg (2003)

322. Reyes Sierra, M., Coello Coello, C.: Multi-objective particle swarm optimizers: A survey of the state-of-the-art. International Journal of Computational Intelligence Research 2(3), 287–308 (2006)

323. Chelouah, R., Siarry, P.: Tabu Search applied to global optimization. European Journal of Operational Research 123, 256–270 (2000)

324. Siarry, P., Berthiau, G.: Fitting of tabu search to optimize functions of continuous variables. Interational Journal for Numerical Methods in Engineering 40, 2449–2457 (1997)

325. Cvijovic, D., Klinowski, J.: Taboo search: An approach to the multiple minima problem. Science 667, 664–666 (1995)

326. Voss, S.: Tabu search: Applications and prospects, Technical report Technische Hochschule Darmstadt, Germany (1992)

327. Crainic, T.D., Toulouse, M., Gendreau, M.: Towards a taxonomy of parallel tabu search algorithms, Technical Report CRT-993, Centre de Recherche sur les Transports, Université de Montreal (1993)

328. Taillard, E.: Parallel iterative search methods for vehicle routing problem. Networks 23, 661–673 (1993)

329. Taillard, E.: Robust taboo search for the quadratic assignment problem. Parallel Computing 17, 443–455 (1991)

330. Chakrapani, J., Skorin-Kapov, J.: Massively parallel tabu search for the quadratic assignment problem. Annals of Operations Research 41, 327–341 (1993)

331. Malek, M., Guruswamy, M., Pandya, M., Owens, H.: Serial and parallel simulated annealing and tabu search algorithms for the traveling salesman problem. Annals of Operations Research 21, 59–84 (1989)

332. Rego, C., Roucairol, C.: A parallel tabu search algorithm using ejection chains for the vehicle routing problem. In: Proc. of the Metaheuristics Int. Conf., Breckenridge, pp. 253–295 (1995)

333. Badeau, P., Gendreau, M., Guertin, F., Potvin, J.-Y., Taillard, E.: A parallel tabu search heuristic for the vehicle routing problem with time windows, RR CRT-95-84, Centre de Recherche sur les Transports, Université de Montréal (1995)

334. Porto, S.C.S., Ribeiro, C.: Parallel tabu search message-passing synchronous strategies for task scheduling under precedence constraints. Journal of heuristics 1(2), 207–223 (1996)

335. Glover, F.: Parametric tabu-search for mixed integer programs. Computers & Operations Research 33, 2449–2494 (2006)

336. Glover, F.: A template for scatter search and path relinking. Mathematical Programming 8, 161–188 (1998)

337. Casavant, T.L., Kuhl, J.G.: A taxonomy of scheduling in general-purpose distributed computing systems. IEEE Transactions on Software Engineering 14(2), 141–154 (1988)

338. Talbi, E.G., Hafidi, Z., Geib, J.-M.: A parallel adaptive tabu search approach. Parallel Computing 24, 2003–2019 (1998)

339. Glover, F.: Parametric tabu-search for mixed integer programs. Computers & Operations Research 33, 2449–2494 (2006)

340. Pawlak, Z.: Rough sets. International Journal of Information and Computer Sciences 11(5), 341–356 (1982)

341. Pawlak, Z.: Rough sets and fuzzy sets. Fuzzy Sets and Systems 17, 99–102 (1985)

342. Pawlak, Z., Slowinski, R.: Rough set approach to multi-attribute decision analysis. European Journal of Operational Research 72, 443–459 (1994)

343. Xu, J., Zhao, L.: A multi-objective decision making model with fuzzy rough coefficients and its application to the inventory problem. Information Sciences 180(5), 679–696 (2010)

344. Siarry, P., Berthiau, G.: Fitting of tabu search to optimize functions of continuous variables. Interational Journal for Numerical Methods in Engineering 40, 2449–2457 (1997)

345. Reck, R.F., Long, B.G.: Purchasing: A competitive weapon. Journal of Purchasing and Materials Management 24, 2–8 (1988)

346. Browning, J.M., Zabriskie, N.B., Huellmantel, A.B.: Strategy purchasing planning. Journal ofPurchasin g and Materials Management 19, 19–24 (1983)

347. Anthony, T.F., Buffa, F.P.: Strategy purchasing scheduling. Journal of Purchasing and Materials Management 13, 27–31 (1977)

348. Bender, P.S., Brown, R.W., Isaac, H., Shapiro, J.F.: Improving purchasing productivity at IBM with a normative decision support system. Interfaces 15, 106–115 (1985)

349. Roy, R.N., Guin, K.K.: A proposed model ofJIT purchasing in an integrated steel plant. International Journal of Production Economics 59, 179–187 (1999)

350. Virolainen, V.M.: A survey ofprocuremen t strategy development in industrial companies. International Journal of Production Economics, 56–57, 677–688 (1998)

351. Gunasekaran, V.: Just-in-time purchasing: An investigation for research and applications. International Journal of Production Economics 59, 77–84 (1999)

352. Jahnukainen, V., Lahti, M.: Efficient purchasing in make-to-order supply chains. International Journal of Production Economics 59, 103–111 (1999)

353. Weber, C.A., Current, J.R., Benton, W.C.: Vendor selection criteria and methods. European Journal ofOperational Research 50, 1–17 (1991)

354. Weber, C.A., Current, J.R.: A multi-objective approach to vendor selection. European Journal of Operation al Research 68, 173–184 (1993)

355. Pan, A.C.: Allocation oforder quantity among suppliers. Journal of Purchasing and Materials Management 25, 36–39 (1989)

356. Verma, R., Pullman, M.E.: An analysis ofthe supplier selection process. Omega 26(6), 739–750 (1998)

357. Yahya, S., Kingsman, B.: Vendor rating for an entrepreneur development programme: A case study using the analytic hierarchy process method. Journal ofthe Operational Research Society 50, 916–930 (1999)

358. Metropolis, N., Osenbluth, A.R., Rosenbluth, M., Teller, A., Teller, E.: Equation of state calculations by fast computing machines. Journal of Chemical Physics 21, 1087–1092 (1953)

359. Černý, V.: A thermodynamical approach to the travelling salesman problem: An efficient simulation algorithm, Preprint, Inst. Phys. and Biophys. Comenius University, Bratislava (1982)

360. Kirpatrick, S., Gelatt Jr., C.D., Vecchi, M.P.: Optimization by simulated annealing. Science 220, 671–680 (1983)

361. Geman, S., Geman, D.: Stochastic relaxation, Gibbs distributions, and the Bayesian restoration of images. IEEE Proceedings Pattern Analysis and Machine Intelligence PAMI 6, 721–741 (1984)

362. Ingber, L.: Very fast simulated re-annealing. Math. Comput. Model 12, 967–973 (1989)

363. Ingber, L.: Simulated annealing: practice versus theory. Mathematical and Computer Modelling 18(11), 29–57 (1993)

364. Ingber, L.: Adaptive simulated annealing (ASA): lessons learned. J. Control and Cybernetics 25(1), 33–54 (1996)

365. Chen, S., Luk, B.L., Liu, Y.: Application of adaptive simulated annealing to blind channel identification with HOC fitting. Electronics Letters 34(3), 234–235 (1998)

366. Rosen, B.E.: Rotationally parameterized very fast simulated reannealing. IEEE Trans. Neural Networks (1997) (submitted)

367. Teargny, T., Renayoun, R., de Montgolfier, J., Larichev, O.: Linear programming with multiobjective functions step method (STEM). Math. Programming 1(2), 366–375 (1971)

368. Sawaragi, Y., Nakayama, H., Tanino, T.: Theory of Multiobjective optimization. Academic Press, New York (1985)

369. Shimizu, K.: Theory of Multiobjective and Conflict, Kyoritsu Syuppan (1982) (in Japanese)

370. Nijkamp, P., Spronk, J.: Interactive multiple goal programming: An evaluation and some results. In: Fandel, G., Gal, T. (eds.) Multiple Criteria Decision Making: Theory and Applications, pp. 278–293. Springer, Berlin (1980)

371. Spronk, J., Telgen, J.: An ellipsoidal interactive multiple goal programming method. In: Multiple Criteria Decision Making Conference at University of Delware, Newwark, August 10-15 (1980)

372. Choo, E.V., Atkins, D.R.: An interactive algorithm for multiobjective programming. Computers and Operations Reasearch 7, 81–88 (1980)

373. Dubois, D., Prade, H.: Rough fuzzy sets and fuzzy rough sets. International Journal of Information & Int. J. of General Systems 17, 191–200 (1990)

374. Dubois, D., Prade, H.: Intelligent Decision Support, Handbook of Applications and Advances of the Rough Sets Theory, pp. 203–233. Kluwer, Dordrecht (1992)

375. Krusinska, E., Slowinski, R., Stefanowski, J.: Discriminant versus rough set approach to vague data analysis. Applied Stochastic Models and Data Analysis 8, 43–56 (1992)

376. Pawlak, Z.: Rough probability. Bull. Polish Acad. Scis., Technical Sci. 33, 9–10 (1985)

377. Pawlak, Z.: Rough Sets. Theoretical Aspects of Reasoning about Data. St. Kluwer, Dordrecht (1985)

378. Polkowski, L., Skowron, A.: Rough mereology. In: Raś, Z.W., Zemankova, M. (eds.) ISMIS 1994. LNCS (LNAI), vol. 869, Springer, Heidelberg (1994)

379. Skowron, A., Grzymala-Busse, W.J.: From the rough set theory to the evidence theory. In: Fedrizzi, M., Kacprzyk, J., Yager, R.R. (eds.) Advances in the Dempster-Shafer Theory of Evidence, pp. 193–236. Wiley, New York (1994)

380. Slowinski, R.: Rough set processing of fuzzy information. In: Lin, T.Y., Wildberger, A. (eds.) Soft Computing: Rough Sets, Fuzzy Logic, Neural Networks, Uncertainty Management, Knowledge Discovery, Simulation Councils, San Diego, CA, pp. 142–145 (1995)

381. Slowinski, R., Vanderpooten, D.: A generalized definition of rough approximations based on similarity. IEEE Transactions on Knowledge and Data Engineering 12(2), 331–336 (2000)

382. Greco, S., Matrazzo, B., Slowinski, R.: Rough sets theory for multicriteria decision analysis. European Journal of Operational Research 129, 1–47 (2001)

383. Xu, J., Yao, L.: A class of expected value multi-objective programming problems with random rough coefficients. Mathematical and Computer Modelling 50, 141–158 (2009)

384. Xu, J., Yao, L.: Random rough variable and random rough programming. European Journal of Operational Research (under review)

385. Glover, F.: Future paths for integer programming and links to artificial intelligence. Comp. Oper. Res. 13, 533–549 (1986)

386. Hansen, P.: The steepest ascent mildest descent heuristic for combinatorial programming. In: Congress on Numerical Methods in Combinatorial Optimization, Capri, Italy (1986)

387. Szidarovszky, F., Gershon, M.E., Dukstein, L.: Techniques for Multiobjective Decision Making in Systems Management. Elsevier, Amsterdam (1986)

388. Goicoeche, A.G., Hansen, D.R., Dukstein, L.: Multiobjective Decision Analysis with Engineering and Business Applications. Wiley, New York (1982)

389. Das, I.: A preference ordering among various Pareto optimal alternatives. Structural and Multidisciplinary Optimization 18(1), 30–35 (1999)

390. di Pierro, F., Khu, S.T., Savi, D.A.: An investigation on preference order ranking scheme for multiobjective evolutionary optimization. IEEE Transactions on Evolutionary Computation 11(1), 17–45 (2007)

391. Tripathi, P.K., Bandyopadhyay, S., Pal, S.K.: Multi-objective particle swarm optimization with time variant inertia and acceleration coefficients. Information Sciences 177(22), 5033–5049 (2007)

392. Li, R.J.: Multiple Objective Decision Making in a Fuzzy Enviroment. Ph.D. thesis, Department of Industrial Engineering, Kansas State University (1990)

393. Zimmermann, H.J.: Decision Making and Expert System. Kluwer Academic, Norwell (1987)

394. Ijiri, Y.: Management Goals and Accounting for Control. Road McNally, Chicago (1965)

395. Kendall, K.E., Lee, S.M.: Formulating blood rotations policies with multiple objectives. Management Sciences 26, 1145–1157 (1980)

396. Ignizio, J.P.: Goal Programming and Extensions. D. C. Health, Massachusetts (1976)

397. Nakayama, H.: Proposal of satisfying trade-off method for multiobjective programming. Trans. Soc. Inst. Contorl Eng. 20, 29–53 (1984) (in Japanese)

398. Xu, G.: Basic Handbook of Operation. Science Press, Peking (1999) (in Chinese)

399. Haimes, Y.Y., Hall, W.A.: Multiobjective in water resources system analysis: The surrogate worth trade-off (SWT) method. Warter Research 10, 615–624 (1974)

400. Chankong, V., Haimes, Y.Y.: Multiobjective Decision Making: Theory and Methodology. North-Holland, Amsterdam (1983)

401. Iwamura, K., Horiike, M.: Lambda credibility. In: Proceedings of the Fifth International Conference on Information and Management Sciences, Series of Information and Management Sciences, vol. 5, pp. 508–513 (2006)

402. Zhou, X., Xu, J.: A class of multi-objective linear programming with bifuzzy coefficients and its application to supplier selection and order allocation, Sichuan University Research Report (2010)

403. Zhou, X., Xu, J., Steven, L.: A Class of Chance Constrained Multi-objective Portfolio Selection Model under Fuzzy Random Environment, Sichuan University Research Report (2010)

404. Xu, J., Zhang, Z.: A Fuzzy Random Resource-constrained Scheduling Model with Multiple Projects and Its Application to a Working Procedure in a Large-Scale Water Conservancy and Hydropower Construction Project. Journal of Scheduling, doi:10.1007/s10951-010-0173-1

405. Gang, J., Xu, J.: The resource-constraint project scheduling with multi-mode under fuzzy random environment in the drainage engineering of LT hydropower. International Journal of Logistics and Transport Special Issue on Joint Seminar on Uncertainty Decision-Making and Engineering Network

406. Xu, J., Gan, L.: Retrofitting transportation network of large-scale construction project under seismic risk using a fuzzy random multi-objective bilevel programming approach. Sichuan University research report (2010)

407. Xu, J., Yang, Y.: A class of multiobjective vehicle routing problem with time windows in a fuzzy random environment and its application to petrol station replenishment problem. Sichuan University research report (2008)

408. Liu, Q., Xu, J.: A study on facility location-allocation problem in mixed random and fuzzy environment. Journal of Intelligent Manufacturing, doi:10.1007/s10845-009-0297-3

409. Huang, X.: Optimal project selection with random fuzzy parameters. International Journal of Production Economics 106(2), 513–522 (2007)

410. Rommelfanger, H.: Fuzzy linear programming and applications. European J. Oprl. Res. 92(3), 512–527 (1996)

411. Rommelfanger, H.: The advantages of fuzzy optimization models in practical use. Fuzzy Optim. Decision Making 3(4), 295–309 (2004)

412. Slowiński, R., Teghem, J. (eds.): Stochastic versus Fuzzy Approaches to Multiobjective Mathematical Programming under Uncertainty. Kluwer Academic Publishers, Dordrecht (1990)

413. Xu, J., Zhou, X.: A class of fuzzy expectation multi-objective model with chance constraints based on rough approximation and it's application in allocation problem, Information Sciences (2010) (accepted)

414. Zhou, X., Xu, J.: A class of integrated logistics network model under random fuzzy environment and its application to Chinese beer company. International Journal of Uncertainty, Fuzziness and Knowledge-Based Systems 17(6), 807–831 (2009)

415. Xu, J., Zhou, X., DashWu, D.: Portfolio selection using λ mean and hybrid entropy. Annals of operations research, doi:10.1007/s10479-009-0550-3

416. Bellman, R.E., Zadeh, L.A.: Decision making in a fuzzy environment. Management Science 17, 141–164 (1970)

417. Slowinski, et al.: Rough sets approach to analysis of data from peritoneal lavage in acute pancreatitis. Medical Informatics 13, 143–159 (1988)

418. Weise, T.: Global Optimization Algorithms C Theory and Application, http://www.it-weise.de/

419. Mendel, J.M.: Computing with words when words can mean different things to different people. Presented at Internat. ICSC Congress on Computational Intelligence: Methods & Applications, 3rd Annual Symp. on Fuzzy Logic and Applications, Rochester, New York, June 22-25 (1999)

420. Mendel, J.M., John, R.I.B.: Type-2 Fuzzy Sets Made Simple. IEEE Transactions on Fuzzy Systems 10(2), 117–127 (2002)

421. John, R.I.: Type 2 fuzzy sets: an appraisal of theory and applications. Internat. J. Uncertainty, Fuzziness Knowledge-Based Systems 6(6), 563–576 (1998)

422. Parsons, S.: Current approaches to handling imperfect information in data and knowledge bases. IEEE Transactions on Knowledge and Data Engineering 8(3), 353–372 (1996)

423. Pareto, V.: Cours d'économie politique professé à l'université de Lausanne. 3 volumes (1896)
424. Arrow, K.J., Barankin, E.M., Blackwell, D.: Admissible points of convex sets (1953)
425. Lai, Y.-J., Hwang, C.-L.: Fuzzy multiple objective decision making: methods and applications. Springer, Heidelberg (1994)
426. Weile, D.S., Michielssen, E., Goldberg, D.E.: Genetic algorithm design of Pareto optimal broadband microwave absorbers. IEEE Transactions on Electromagnetic Compatibility 38(3), 518–525 (1996)
427. Horn, J., Nafpliotis, N., Goldberg, D.E.: A niched Pareto genetic algorithm for multiobjective optimization. In: Proceedings of the First IEEE Conference on Evolutionary Computation, vol. 1, pp. 82–87 (1994)

Appendix A
Procedures

A.1 Procedures for Example 2.6

There are 1 main procedure and 6 sub-procedures.

main

```
tic global popsize exetime N pos v pbest gbest t
    c_i c_p c_g w1 w2 gen v_max best_fitness
    best_in_history pos_max pos_min

gen=1000;
popsize=10;
N=2;
pos=zeros(popsize,N);
v=zeros(popsize,N);
pbest=zeros(popsize,N+2);
gbest=zeros(1,N);
c_i=1;
c_p=2;
c_g=2;
w1=0.7;
w2=0.3;
v_max=0.5;
pos_max=[30 17];
pos_min=[0 0];
initialization;
for exetime=1:gen
  fitness;
  update;
end
toc
```

initialization

```
i=0;
k=1;
while (k<=popsize+1)
  while (i<1)
  x1=random('uniform',0,30);
  x2=random('uniform',0,17);
  t=constraint_check(x1,x2);
  if(t==1)
    i=i+1;
  end
  end
  pos(k,:)=[x1 x2];
  v(k,1:N)=random('uniform',1,5);
  pbest(k,1:N)=pos(k,1:N);
  pbest(k,N+1:N+2)=inf;
  k=k+1;
end
gbest(1,1:N)=pos(1,1:N);
```

fitness

```
global exetime
for i=1:popsize
  pbest(i,4)=w1*Obj1(pos(i,1),pos(i,2))
      +w2*Obj2(pos(i,1),pos(i,2));
  if (pbest(i,3)<pbest(i,4))
    pbest(i,3)=pbest(i,4);
    pbest(i,1:N)=pos(i,1:N);
  end
end

if (best_fitness<max(pbest(:,3)))
  best_fitness=max(pbest(:,3));
  for j=1:N
    gbest(1,j)=pbest(find(pbest(:,3)==max(pbest
    (:,3))),j);
  end
end
```

Obj1

```
function [Obj1]=Obj1(x1,x2)
N=10;
Obj1=0;
F1=[];
for i=1:N
  T=0;
  %~~~~~~~~~~initial value~~~~~~~~~~
   r1=unifrnd(1.8,2.2);
   if 1.8<r1&&r1<2
     mu1=(r1-1.8)/0.2;
   end
   if 2<=r1&&r1<2.2
     mu1=(2.2-r1)/0.2;
   end

   r2=unifrnd(1,2);
   if 1<=r2&&r2<1.5
     mu2=(r2-1)/0.5;
   end
   if 1.5<=r2&&r2<=2
     mu2=(2-r2)/0.5;
   end
   F1(i)=r1^2*x1+r2^2*x2;
   mu=[mu1,mu2];
   MU(i)=min(mu);
end

MIN=F1(1);
MAX=F1(1);
for j=2:N
  if MIN>F1(i)
    MIN=F1(i);
  end
  if MAX<F1(i)
    MAX=F1(i);
  end
end

for k=1:N
  r=unifrnd(MIN,MAX);
  b1=0;
  b2=0;
  if r>=0
```

```
    for i=1:N
      if F1(i)>=r&&b1<=MU(i)
        b1=MU(i);
      end
      if F1(i)<r&&b2<=MU(i)
        b2=MU(i);
      end
    end
    Obj1=Obj1+(b1+1-b2)/2;
  else
    for i=1:N
      if F1(i)<=r&&b1<=MU(i)
        b1=MU(i);
      end
      if F1(i)>r&&b2<=MU(i)
        b2=MU(i);
      end
    end
    Obj1=Obj1-(b1+1-b2)/2;
  end
end

if MIN<=0
  a=0;
else
  a=MIN;
end if MAX<=0
  b=0;
else
  b=MAX;
end
Obj1=Obj1*(MAX-MIN)/N+a+b;
```

Obj2

```
function [Obj2]=Obj2(x1,x2)
N=10;
Obj2=0;
F2=[];
for i=1:N
  T=0;
  r1=unifrnd(1.8,2.2);
  if 1.8<r1&&r1<2
    mu1=(r1-1.8)/0.2;
```

```
      end
    if 2<=r1&&r1<2.2
      mu1=(2.2-r1)/0.2;
    end

    r2=unifrnd(1,2);
    if 1<=r2&&r2<1.5
      mu2=2*(r2-1)/0.5;
    end
    if 1.5<=r2&&r2<=2
      mu2=(2-r2)/0.5;
    end
    F2(i)=1.2*r1^2*x1-0.5*r2^2*x2;
    mu=[mu1,mu2];
    MU(i)=min(mu);
  end

MIN=F2(1);
MAX=F2(1);
for j=2:N
  if MIN>F2(i)
    MIN=F2(i);
  end
  if MAX<F2(i)
    MAX=F2(i);
  end
end

for k=1:N
  r=unifrnd(MIN,MAX);
  b1=0;
  b2=0;
  if r>=0
    for i=1:N
      if F2(i)>=r&&b1<=MU(i)
        b1=MU(i);
      end
      if F2(i)<r&&b2<=MU(i)
        b2=MU(i);
      end
    end
    Obj2=Obj2+(b1+1-b2)/2;
  else
    for i=1:N
      if F2(i)<=r&&b1<=MU(i)
```

```
      b1=MU(i);
    end
    if F2(i)>r&&b2<=MU(i)
      b2=MU(i);
    end
  end
  Obj2=Obj2-(b1+1 b2)/2;
  end
end

if MIN<=0
  a=0;
else
  a=MIN;
end if MAX<=0
  b=0;
else
  b=MAX;
end
Obj2=Obj2*(MAX-MIN)/N+a+b;
```

constraint_check

```
function [t]=constraint_check(x1,x2)
t=1;
if ((x1<0)&(x2<0))
  t=0;
end
if (x1+x2>=30)
  t=0;
end
if (3*x1-2*x2<=8)
  t=0;
end
```

update

```
c_i=0.4+0.5*(gen-exetime)/gen;
for i=1:popsize
  for j=1:N
    v(i,j)=c_i*rand()*v(i,j)+c_p*rand()
    *(pbest(i,j)-pos(i,j))+c_g*rand()*
    (gbest(1,j)-pos(i,j));
  end
end
```

```
for i=1:popsize
  for j=1:N
    pos(i,j)=pos(i,j)+v(i,j);
    if pos(i,j)>pos_max(1,j)
      pos(i,j)=pos_max(1,j);
    elseif pos(i,j)<pos_min(1,j)
      pos(i,j)=pos_min(1,j);
    else
    end
  end
end
```

A.2   Procedures for Fu-Ra Portfolio Selection Problem

The procedures for fuzzy random portfolio selection problem are programmed in MATLAB language.

Creating the optimistic, pessimistic and neutral efficient frontier function lambda efficient

```
% FuzzyReturn1 Denote the left endpoint of fuzzy
expected value of fuzzy random return
% FuzzyReturn2 Denote the right endpoint of fuzzy
expected value of fuzzy random return
% parapess, paraoptim and paraneutral denote the
optimistic, pessimistic and neutral attitude,

FuzzyReturn1=load('FuzzyReturn1.txt');
FuzzyReturn2=load('FuzzyReturn2.txt');
ExpCovariance=load('ExpCovariance.txt');
parapess=0;
paraoptim=1;
paraneutral=0.5;

ExpReturnwgt(:,1)=parapess*FuzzyReturn2
        +(1-parapess)*FuzzyReturn1;
ExpReturnwgt(:,2)=paraoptim*FuzzyReturn2
        +(1-paraoptim)*FuzzyReturn1;
ExpReturnwgt(:,3)=paraneutral*FuzzyReturn2
        +(1-paraneutral)*FuzzyReturn1;
```

```
for numplot=1:3
  ExpReturn=ExpReturnwgt(:,numplot);
  NASSETS=length(ExpReturn);
  W0=ones(NASSETS,1)/NASSETS;
  Aeq=ones(1,NASSETS);
  Beq=1;
  LB=zeros(NASSETS,1);
  UB=ones(NASSETS,1);
  Aineq=ones(1,NASSETS);
  Bineq=1000;
  options=optimset('display','off','largescale',
     'off');
  [MaxReturnWeights, Fval, ErrorFlag]=linprog(
   -ExpReturn, Aineq, Bineq, Aeq, Beq, LB, UB,
   W0, options);
  if ErrorFlag~=1
    error('No portfolios satisfy all the
       input constraints');
  end
  MaxReturn=transpose(MaxReturnWeights)*ExpReturn;
% Find the minimum variance return.
  F=zeros(NASSETS,1);
  [MinVarWeights,Fval,ErrorFlag]=quadprog
    (ExpCovariance, F, Aineq, Bineq, Aeq, Beq, LB,
    UB, W0, options);
  if ErrorFlag~=1
    error('A solution was not feasible for the
      minimum variance portfolio.');
  end
  MinVarReturn=transpose(MinVarWeights)*ExpReturn;
  NumFrontPoints=50;
  MinVarStd=sqrt(transpose(MinVarWeights)
     *ExpCovariance *MinVarWeights);
  PortReturn=linspace(MinVarReturn,MaxReturn,
     NumFrontPoints);
  PortfOptResults=zeros(NumFrontPoints, 2+NASSETS);
  PortfOptResults(1,:)=[MinVarReturn MinVarStd
          transpose(MinVarWeights(:))];
%PortfOptResults(1, :) include the minimal return,
the minimal standard deviation,
and the coefficients for every stock
  StartPoint=2;
  EndPoint=NumFrontPoints;
  FrontPointConstraint=-ExpReturn';
  Aeq=[FrontPointConstraint; Aeq];
```

```
%Add a new equality constraint
  Beq=[0; Beq];
%Add a new equality constraint
  W0=MaxReturnWeights;
% Set the options:
 options=optimset(options,'largescale','off');
 for Point=StartPoint:EndPoint
  Beq(1)=-PortReturn(Point);
  [Weights,Fval,ErrorFlag]=quadprog(
  ExpCovariance,F,Aineq,Bineq,Aeq,
  Beq,LB,UB,W0,options);
  if ErrorFlag~=1
    PortfOptResults(Point,:)=[Beq(2)
    nan*ones(1,NASSETS+1)];
  else
  Return=dot(Weights,ExpReturn);
  Std=sqrt(Weights'*ExpCovariance*Weights)
  PortfOptResults(Point,:)=[Return Std Weights(:)'];
  end
 end
 PortReturn=PortfOptResults(:,1);
 PortRisk=PortfOptResults(:,2);
 PortWts=PortfOptResults(:,3:size(PortfOptResults,2));
 hold on
 if nargout==0
  if numplot==1
    outcolor='-+k';
  elseif numplot==2
    outcolor='-k';
  else
  outcolor='--k';
  end
  plot(PortRisk,PortReturn,outcolor);
 end
 title('\lambda Efficient Frontier','Color','k');
 xlabel('Risk(Standard Deviation)');
 ylabel('\lambda Average Value of Expected Return');
 grid on;
 hold off
end
```

Creating λ mean-variance efficient frontier

It is similar to the above program, we just skip it over.

Creating histogram of historical exchange rate

```
function PSbarfigure
x=0:0.2:4;
data=load('data.txt');
y=zeros(1,21);
for j=1:719
 for k=1:21
 if (data(j)>0.2*(k-1))&(data(j)<=0.2*k)
   y(k)=y(k)+1;
 end
 end
end bar(x,y)
```

Genetic algorithm for Fu-Ra multi-objective portfolio selection model

(1) Checking constraints

```
function [feas_chromosome_row]=check_constraint
  (chromosome_row)
Knum=20;
nvars=30;
low=0.01;
roundlot=0.0001;
epsilon=0.0000000001;
[temp_chromosome_row,cindex]=sort(chromosome_row,
 'descend');
chromosome_row(cindex(Knum+1:nvars))=zeros(1,
  nvars-Knum);
chromosome_row=chromosome_row/sum(chromosome_row);
for ii=1:Knum
  tt=Knum+1-ii;
   if (chromosome_row(cindex(tt))<low)
     chromosome_row(cindex(tt))=0;
     chromosome_row=chromosome_row/sum
  (chromosome_row);
   end
end

bb=mod(chromosome_row,roundlot);
chromosome_row=chromosome_row-mod(chromosome_row,
  roundlot);
dd=sum(bb);
[out,idx]=sort(bb,'descend');
nn=1;
```

```
while ((dd>epsilon)&(nn<nvars))
  chromosome_row(idx(nn))=chromosome_row(idx(nn))
    +floor((dd+roundlot/10)/roundlot)*roundlot;
  dd=dd-floor((dd+roundlot/10)/roundlot)*roundlot;
  nn=nn+1;

end
feas_chromosome_row=chromosome_row;
```

(2) Creating initialized solution

```
function parents=create_conMOPS() P
OP_SIZE=100;
nvars=30;
limit=0.2;
range=zeros(2,nvars);
range(2,:)=ones(1,nvars)*limit;
lowerBound=range(1,:);
span=range(2,:)-lowerBound;
parents=repmat(lowerBound,POP_SIZE,1)
    +repmat(span,POP_SIZE,1).*rand(POP_SIZE,nvars);
for m=1:POP_SIZE
 parents(m,:)=check_constraint_conMOPS(parents(m,:));
end
```

(3) Computing the fitness

```
function [fitness]=FitnessFcn(pareto_x,obj_x)
weight=[0.4 0.4 0.2];
fitness=sqrt(weight(1)^2*abs(pareto_x(1)-obj_x(1))^2
+weight(2)^2*abs(pareto_x(2)-obj_x(2))^2+weight(3)^2
*abs(pareto_x(3)-obj_x(3))^2);
```

(4) Crossover

```
function parents=crossover_permutation_PS(parents)
POP_SIZE=100;
P_CROSSOVER=0.3;
fixn=fix(POP_SIZE/2);

for j=1:fixn
 r1=rand;
 if (r1<P_CROSSOVER)
   k=fix(rand*POP_SIZE)+1;
   kk=fix(rand*POP_SIZE)+1;
   r2=rand;
   child1=r2*parents(k,:)+(1-r2)*parents(kk,:);
```

```
    child2=(1-r2)*parents(k,:)+r2*parents(kk,:);
    parents(k,:)=check_constraint_conMOPS(child1);
    parents(kk,:)=check_constraint_conMOPS(child2);
  else
  continue;
  end
end
```

(5) Mutation

```
function parents=mutate_permutation_PS(parents)
POP_SIZE=100;
P_MUTATION=0.2;
nvars=30;
INFTY=0.001;
for j=1:POP_SIZE
  rmut=rand;
  if (rmut<P_MUTATION)
    mutx=parents(j,:);
    direction=unifrnd(-1,1,1,nvars);
    infty=rand*INFTY;
    muty=mutx+infty*direction;
    zz=1;
    for zz=1:nvars
      if (muty(zz)<0)
        muty(zz)=0;
      end
    end
    parents(j,:)=check_constraint_conMOPS(muty);
  else
  continue;
  end
end
```

(6) Evaluation

```
function [optim_parents,optim_objective,expect,regr]=
      evaluatation_conMOPS(parents)
nvars=30;
optim_objective=1000;
POP_SIZE=100;
centReturn=load('centReturn.txt');
leftReturn=load('leftReturn.txt');
rightReturn=load('rightReturn.txt');
lambda=ones(1,nvars).*1;
covmatrix=load('covmatrix.txt');
centliquidity1=load('centliquidity1.txt');
```

```
centliquidity2=load('centliquidity2.txt');
leftwidth=load('leftwidth.txt');
rightwidth=load('rightwidth.txt');
for m=1:POP_SIZE
objresu(m,2)=-parents(m,:)*covmatrix
  *transpose(parents(m,:));
obj_mean=0;
obj_liqu=0;
for t=1:nvars
 F1=@(x)(((1-x).*centReturn(t)+x.*rightReturn(t))
  *lambda(t)+((1-x).*centReturn(t)'
  +x.*leftReturn(t))*(1-lambda(t)))*parents(m,t);
 obj_mean=obj_mean+quad(F1,0,1);
 F3=@(x)x.*((centliquidity1(t)-(1-x)*leftwidth(t))+
  (centliquidity2(t)+(1-x)
    .*rightwidth(t)))*parents(m,t);
 obj_liqu=obj_liqu+quad(F3,0,1);
end
objresu(m,1)=obj_mean;
objresu(m,3)=obj_liqu;
end
[gg,idxgg]=sort(objresu);
paretoff=gg(POP_SIZE,:);

for r=1:POP_SIZE
 regr(r)=FitnessFcn_conMOPS(paretoff,objresu(r,:));
end [regr,idx]=sort(regr);
parents=parents(idx,:);
objresu=objresu(idx,:);
if (regr(1)<optim_objective)
  optim_parents=[parents(1,:) objresu(1,:)];
  optim_objective=regr(1);
end
maxregr=max(regr);
minregr=min(regr);
for tt=1:POP_SIZE
rr=rand;
 eval(tt)=(maxregr-regr(tt)+rr)/(maxregr-minregr+rr);
end
expect(1)=eval(1);
for j=2:POP_SIZE
  expect(j)=expect(j-1)+eval(j);
end
```

(7) Selection

```
function parents=selection_conMOPS(expect,parents)
POP_SIZE=100;

for j=1:POP_SIZE
  r=rand*expect(POP_SIZE);
  for uu=1:POP_SIZE
    if (r<=expect(uu))
      temp(j,:)=parents(uu,:);
      break;
    end
  end
end
parents=temp;
```

(8). Main file

```
function conMOPS_1
GEN=200;
nvars=30;
chromosome=create_conMOPS();
figure;
for i=1:GEN
 [bestchromosome, bestobjective, expectation, regret]
   =evaluatation_conMOPS(chromosome);
chromosome=selection_conMOPS(expectation,chromosome);
chromosome=crossover_permutation_conMOPS(chromosome);
chromosome=mutate_permutation_conMOPS(chromosome);
disp('Generation=');
disp(i);
disp('f(x)=');
disp(bestobjective);
disp('x=');
disp(bestchromosome);
disp('sum(x)=');
disp(sum(bestchromosome(1:nvars)));
hold on
set(gca,'xlim',[0,GEN]);
xlabel('Generation','interp','none');
ylabel('Fitness value','interp','none');
plotbest=plot(i,bestobjective,'+r');
plotmean=plot(i,mean(regret),'+black');
end
hold off;
```

Program for Fu-Ra CCM of portfolio selection problem

```
function possprobCCP
nvars=5;
GEN=20;
POP_SIZE=10;
P_MUTATION=0.2;
P_CROSSOVER=0.3;
INFTY=0.005;
objnum=2;
probabnum=4;
possnum=4;
LARGENUMBER=-100000000;
precision=0.000001;
constrposs=0.90;
checkposs=0.75;
constrprob=0.9;
range=[20 35 45 19 19;300 100 80 80
25];
mu=[113 241 87 56 92; 1.2*628 0.5*143 1.3*476
   0.8*324 0.9*539];
stddelta=[1 4 1 2 1;1 2 2 2 2];
spread=[3 8 3 7 5; 1.2*10 0.5*7
1.3*12 0.8*5 0.9*8];
weight=[0.9 0.1];
constraintmatrix=[1 1 1 1 1;
 1 1 1 1 1;4 2 1.5 1 2;
 1 4 2 5 3;1 0 0 0 0;
 0 1 0 0 0;0 0 1 0 0;
 0 0 0 1 0;0 0 0 0 1];
Rconst=[350; 300; 1085; 660; 20; 20; 20; 20;
20];
testposs=0.10;
lowerBound=range(1,:);
span=range(2,:)-lowerBound;
chromosome=repmat(lowerBound,POP_SIZE,1)+
     repmat(span,POP_SIZE,1).
     *rand(POP_SIZE,nvars);

for m=1:POP_SIZE
  ww=constraintmatrix*transpose(chromosome(m,:))
    -Rconst;
while ((chromosome(m,1)<=0)|(chromosome(m,2)<=0)|
    (chromosome(m,3)<=0)|(chromosome(m,4)<=0)|
    (chromosome(m,5)<=0)|(ww(1)>0)|(ww(2)<0)|
```

```
        (ww(3)>0) | (ww(4)>0) | (ww(5)<0) | (ww(6)<0) |
        (ww(7)<0) | (ww(8)<0) | (ww(9)<0) )
    chromosome(m,:)=repmat(lowerBound,1,1)
          +repmat(span,1,1).*rand(1,nvars);
    ww=constraintmatrix*transpose(chromosome(m,:))
      -Rconst;
  end
end

for r=1:POP_SIZE
  progoalf=zeros(probabnum,objnum);
  for i=1:probabnum
    outx=normrnd(mu,stddelta,objnum,nvars);
    goalf=LARGENUMBER.*ones(1,objnum);
    for k=1:possnum
      fcrisp=(outx-(1-constrposs)*spread)
        +2*(1-constrposs).*spread
        .*rand(objnum,nvars);
      objresu=chromosome(r,:)*transpose(fcrisp);
      for ss=1:objnum
        if (objresu(1,ss)>=goalf(1,ss));
          goalf(1,ss)=objresu(1,ss);
        end
      end
    end
    progoalf(i,:)=goalf;
  end
  [gg,idxgg]=sort(progoalf);
  ix=fix(probabnum.*constrprob);
  for w=1:objnum
    keepff(r,w)=gg(ix,w);
  end
end

[pare,iid]=sort(keepff);
paretoff=pare(POP_SIZE,:);
for r=1:POP_SIZE
  regr(r)=weight(1)*abs(paretoff(1)-keepff(r,1))
      +weight(2)*abs(paretoff(2)-keepff(r,2));
end
[regr,idx]=sort(regr);
chromosome=chromosome(idx,:);
keepff=keepff(idx,:);
bestchromosome=[chromosome(1,:)
keepff(1,:)];
```

```
bestobjective=regr(1);
constraintvalues=constraintmatrix*
  transpose(bestchromosome(1,1:nvars))-Rconst;
maxregr=max(regr);
minregr=min(regr);
for j=1:POP_SIZE
 rr=rand;
 eval(j)=(maxregr-regr(j)+rr)/(maxregr-minregr+rr);
end
expectation(1)=eval(1);
for j=2:POP_SIZE
  expectation(j)=expectation(j-1)+eval(j);
end
figure;
for i=1:GEN
 for j=1:POP_SIZE
  r=rand*expectation(POP_SIZE);
  for k=1:POP_SIZE
    if (r<=expectation(k))
      temp(j,:)=chromosome(k,:);
      break;
    end
  end
 end
end
chromosome=temp;
fixn=fix(POP_SIZE/2);
for j=1:fixn
 r1=rand;
 if (r1>P_CROSSOVER)
 k=fix(rand*POP_SIZE)+1;
 kk=fix(rand*POP_SIZE)+1;
 r2=rand;
 temp1=zeros(1,nvars);
 temp2=zeros(1,nvars);
 temp1=r2*chromosome(k,:)+(1-r2)*chromosome(kk,:);
 temp2=(1-r2)*chromosome(k,:)+r2*chromosome(kk,:);
 ww=constraintmatrix*transpose(temp1(1,:))-Rconst;
 if ((temp1(1)>0)&(temp1(2)>0)&(temp1(3)>0)&
     (temp1(4)>0)&(temp1(5)>0)&(ww(1)<=0)&
     (ww(2)>=0)&(ww(3)<=0)&(ww(4)<=0)&
     (ww(5)>=0)&(ww(6)>=0)&(ww(7)>=0)&
     (ww(8)>=0)&(ww(9)>=0))
   chromosome(k,:)=temp1;
 end
 ww=constraintmatrix*transpose(temp2(1,:))-Rconst;
```

```
    if ((temp2(1)>0)&(temp2(2)>0)&(temp2(3)>0)&
       (temp2(4)>0)&(temp2(5)>0)&(ww(1)<=0)&
       (ww(2)>=0)&(ww(3)<=0)&(ww(4)<=0)&
       (ww(5)>=0)&(ww(6)>=0)&(ww(7)>=0)
       &(ww(8)>=0)&(ww(9)>=0))
      chromosome(kk,:)=temp2;
    end
    else
      continue;
    end
  end
  for j=1:POP_SIZE
    r1=rand;
    if (r1>P_MUTATION)
      mutx=chromosome(j,:);
      direction=unifrnd(-1,1,1,nvars);
      infty=rand*INFTY;
      muty=mutx+infty*direction;
      ww=constraintmatrix*transpose(muty(1,:))
        -Rconst;
      if ((muty(1)>0)&(muty(2)>0)&(muty(3)>0)&
         (muty(4)>0)&(muty(5)>0)&(ww(1)<=0)&
         (ww(2)>=0)&(ww(3)<=0)&(ww(4)<=0)&
         w(5)>=0)&(ww(6)>=0)&(ww(7)>=0)
         &(ww(8)>=0)&(ww(9)>=0))
        chromosome(j,:)=muty;
      end
      else
        continue;
      end
  end
  for r=1:POP_SIZE
   progoalf=zeros(probabnum,objnum);
   for q=1:probabnum
     outx=normrnd(mu,stddelta,objnum,nvars);
     goalf=LARGENUMBER.*ones(1,objnum);
     for k=1:possnum
       fcrisp=(outx-(1-constrposs)*spread)
          +2*(1-constrposs)
          .*spread.*rand(objnum,nvars);
       objresu=chromosome(r,:)*transpose(fcrisp);
       for ss=1:objnum
         if (objresu(1,ss)>=goalf(1,ss))
         goalf(1,ss)=objresu(1,ss);
         end
```

```
      end
     end
     progoalf(q,:)=goalf;
   end
    [gg,idxgg]=sort(progoalf);
    ix=fix(probabnum*constrprob);
    for w=1:objnum
      keepff(r,w)=gg(ix,w);
    end
  end
  [pare,iid]=sort(keepff);
  paretoff=pare(POP_SIZE,:);

  for r=1:POP_SIZE
    regr(r)=weight(1)*abs(paretoff(1)-keepff(r,1))
        +weight(2)*abs(paretoff(2)-keepff(r,2));
  end
  [regr,idx]=sort(regr);
  chromosome=chromosome(idx,:);
  keepff=keepff(idx,:);
  if (regr(1)<bestobjective)
  bestchromosome=[chromosome(1,:) keepff(1,:)];
   bestobjective=regr(1);
  end
  constraintvalues=constraintmatrix
        *transpose(bestchromosome(1,1:nvars))
        -Rconst;
  maxregr=max(regr);
  minregr=min(regr);
 for j=1:POP_SIZE
 rr=rand;
 eval(j)=(maxregr-regr(j)+rr)/(maxregr-minregr+rr);
 end
  expectation(1)=eval(1);
  for j=2:POP_SIZE
    expectation(j)=expectation(j-1)+eval(j);
  end

M=2000;
MM=2000;
number=zeros(1,objnum);
for kk=1:M
 test=normrnd(mu,stddelta,objnum,nvars);
 tempmembership=zeros(MM,objnum);
 for qq=1:MM
```

```
realization=(test-(1-testposs)*spread)
  +2*(1-testposs).*spread
  .*rand(objnum,nvars);
realizeff=realization*transpose
  (bestchromosome(1,1:nvars));
if (realizeff(1)>=bestchromosome(1,nvars+1))
  for bb=1:nvars
  if (realization(1,bb)>test(1,bb))
    membershipvalue(1,bb)=(test(1,bb)
      +spread(1,bb)-realization(1,bb))
      /spread(1,bb);
  else
    membershipvalue(1,bb)=(realization(1,bb)
      -(test(1,bb)-spread(1,bb)))
      /spread(1,bb);
  end
  end
  tempmembership(qq,1)=min(membershipvalue(1,:));
end
if (realizeff(2)>=bestchromosome(1,nvars+2))
  for bb=1:nvars
    if (realization(2,bb)>test(2,bb))
      membershipvalue(2,bb)=(test(2,bb)
        +spread(2,bb)-realization(2,bb))
        /spread(2,bb);
    else
      membershipvalue(2,bb)=(realization(2,bb)
        -(test(2,bb)-spread(2,bb)))
        /spread(2,bb);
    end
  end
  tempmembership(qq,2)=min(membershipvalue(2,:));
end
testmembership=max(tempmembership);
if (testmembership(1)>=checkposs)
  number(1)=number(1)+1;
end
if (testmembership(2)>=checkposs)
  number(2)=number(2)+1;
end
outmembership(kk,:)=testmembership;
end
probability(1)=number(1)/M;
probability(2)=number(2)/M;
```

```
disp('Generation=');
disp(i);
disp('averagemembership=');
disp(mean(outmembership));
disp('probability=');
disp(probability);
disp('f(x)=');
disp(bestobjective);
disp('x=');
disp(bestchromosome);
disp('ww=');
disp(constraintvalues);
hold on
set(gca,'xlim',[0,GEN]);
xlabel('Generation','interp','none');
ylabel('Fitness value','interp','none');
title(['Best: ',' Mean: '],'interp','none')
plotbest=plot(i,bestobjective,'+red');
plotmean=plot(i,mean(regr),'+black');
end
LegnD=legend('Best fitness','Mean fitness',4);
set(LegnD,'FontSize',8);
hold off;
```

Program for Fu-Ra DCM

It's similar to the program for Fu-Ra CCM, so we skip over.

Index